# Personalized Cancer Therapy

Mrinal K. Ghosh • Malini Basu
Editors

# Personalized Cancer Therapy

## Advances in Precision Medicine for Targeted Cancer Treatment

Springer

*Editors*
Mrinal K. Ghosh
Department of Cancer Biology
and Inflammatory Disorder Division
CSIR-Indian Institute of Chemical
Biology (IICB)
Kolkata, West Bengal, India

Malini Basu
Department of Microbiology, Dhruba
Chand Halder (DCH) College,
University of Calcutta,
Kolkata, West Bengal, India

ISBN 978-981-95-6844-4 ISBN 978-981-95-6845-1 (eBook)
https://doi.org/10.1007/978-981-95-6845-1



This Springer imprint is published by the registered company Springer Nature Singapore Pte Ltd.
The registered company address is: 152 Beach Road, #21-01/04 Gateway East, Singapore 189721, Singapore

# Preface

This proposed book aims to serve as a definitive and comprehensive resource on the dynamic and rapidly advancing field of precision medicine. It is meticulously structured into six detailed sections to provide readers with a thorough understanding of both foundational concepts and cutting-edge developments. The first section introduces the basic principles and concepts of precision medicine, setting the stage by highlighting the significant shift from traditional, one-size-fits-all medical approaches to highly personalized strategies that consider individual variability in genes, environment, and lifestyle. The second section delves deeply into the core scientific pillars of precision medicine: genomics, epigenomics, and pharmacogenomics. It examines key genomic alterations that drive disease, the role of epigenetic changes in cancer, and how pharmacogenomics tailors treatments to individual genetic profiles. This section also covers the integration of these disciplines into precision cancer medicine, including innovative approaches to clinical trial design that enhance the efficacy and relevance of treatment regimens. In the third section, the book provides a comprehensive analysis of cancer-specific molecular profiling, diagnosis, and treatment. Each chapter is dedicated to a major cancer type—such as gliomas, head and neck cancers, breast cancer, non-small cell lung cancer, prostate cancer, colon cancer, pancreatic cancer, and other miscellaneous cancers—offering in-depth insights into the unique molecular characteristics, diagnostic techniques, and therapeutic strategies pertinent to each malignancy. The fourth section addresses the multifaceted challenges and limitations of precision medicine, including biomarker identification, and targeted delivery for cancer therapy and real-time monitoring using nanotherapeutics. The fifth section explores the technical hurdles in implementing advanced technologies and clinical challenges associated with integrating personalized approaches into routine practice, social issues, ethical considerations, and benefits that arise from these personalized strategies, emphasizing the importance of addressing these aspects to maximize the impact of precision medicine on public health. The final section looks forward, examining the novel innovations and future perspectives in the field. It discusses emerging technologies and their potential to revolutionize personalized medicine, the application of bioinformatics and computational biology in interpreting complex data, and the role of artificial intelligence (AI) and machine learning (ML) in advancing personalized treatment strategies. The book concludes with an exploration of potential breakthroughs and future research directions that promise to further enhance patient care and outcomes.

Kolkata, West Bengal, India

Mrinal K. Ghosh
Malini Basu

# Acknowledgments

The Editors gratefully acknowledge Sunny Kumar (MS Pharm.) and Srija Roy (MSc) doctoral students of CSIR-IICB, for their continuous support in organizing the chapters of this book. We also extend our sincere thanks to CSIR-Indian Institute of Chemical Biology (CSIR-IICB), the Academy of Scientific and Innovative Research (AcSIR), and Dhruba Chand Halder (DCH) College (University of Calcutta). Additionally, we acknowledge Springer for valuable support and collaboration during development of this interesting book for global readers.

# Contents

# About the Editors

**Mrinal K. Ghosh** is a chief scientist, In-Charge of CSIR-IICB (TRUE), Salt Lake Campus and head (2019–2025) of the Cancer Biology and Inflammatory Disorder Division at CSIR- Indian Institute of Chemical Biology (IICB), Kolkata, India; Chairman of management council at CSIR-IICB (TRUE); and a Professor at the Academy of Scientific and Innovative Research (AcSIR), India. He completed his PhD from the University of Calcutta and Post-doctoral research (1997–2005) from Lerner Research Institute, Cleveland Clinic Foundation, Ohio, USA. His research interests include signal transduction in cancer focusing on the regulation of various oncogenes and tumor suppressor proteins at transcriptional and post-translational levels, and developing nanotechnology-based targeted drug delivery systems to the targets (identified molecular vulnerabilities) in brain, breast, and colon cancers. He has written several papers in peer-reviewed international and national journals, authored books and book chapters, delivered numerous lectures, and is a reviewer of various national and international journals and research project proposals of World Cancer Research, Ireland Research Board, ICMR, DBT, and DST. He is a lifetime member of the Indian Society of Translational Research (ISTR), the Chemical Biology Society (CBS), the Indian Association of Cancer Research (IACR), the Society of Biological Chemistry (SBC), and the Indian Science Congress Association (ISCA), India.

**Malini Basu** is an Assistant Professor and Head, Department of Microbiology, Dhruba Chand Halder College, University of Calcutta, Kolkata, India. She completed her PhD from the University of Calcutta and Post-doctoral research from Case Western Reserve University (CWRU), USA, in collaboration with Rainbow Babies and Children Hospital, Ohio, USA. Her research interests include heavy metals, microbial bioremediation and bioindicators, unique virulence, pathogenicity, and host immune response of *Helicobacter pylori* in gastric cancer. She has written several papers in peer-reviewed national and international journals and some book chapters. She is a lifetime member of the Microbiologists Society India (MBSI), the Association of Microbiologists of India (AMI), and ISNA. She is also the state vice-president of the Microbiologists Society of India.

# Introduction to Precision Medicine

# Basic Principles and Concepts of Precision Medicine in Cancer

Sunny Kumar and Malini Basu

**Abstract**

The emergence of precision medicine marks a transformative shift in cancer care, moving away from conventional, generalized therapeutic strategies toward a more individualized approach that considers the unique genetic, molecular, and environmental context of each patient. This chapter provides an in-depth exploration of the foundational principles and concepts of precision oncology. It begins with a discussion of the historical evolution of cancer treatment and the limitations of traditional models, setting the stage for the rise of personalized medicine. Core components such as genomic profiling, biomarker discovery, molecular diagnostics, and the role of predictive and prognostic indicators are examined in detail. Additionally, this chapter highlights the critical role of technologies such as NGS, liquid biopsies, and artificial intelligence in driving forward the field. Case examples of successful targeted therapies (e.g., EGFR inhibitors, ALK inhibitors, PARP inhibitors) are presented to illustrate clinical applications. The chapter also addresses the major challenges facing precision medicine, including tumor heterogeneity, drug resistance, cost, data interpretation, and ethical concerns and discusses future directions aimed at improving accessibility, refining predictive models, and integrating multi-omics data for holistic patient care. By synthesizing current knowledge and emerging trends, this chapter serves as a foundational resource for understanding how precision medicine is reshaping the diagnosis, treatment, and prevention of cancer.

S. Kumar
Cancer Biology and Inflammatory Disorder Division, Council of Scientific and Industrial Research-Indian Institute of Chemical Biology (CSIR-IICB), TRUE Campus, Kolkata, India

M. Basu (✉)
Department of Microbiology, Dhruba Chand Halder (DCH) College, University of Calcutta, Kolkata, West Bengal, India

 
M. K. Ghosh, M. Basu (eds.), *Personalized Cancer Therapy*,
https://doi.org/10.1007/978-981-95-6845-1_1

**Keywords**

Precision medicine · Cancer · Principles · Molecular profiling · Biomarkers

## 1 Introduction

Cancer, historically perceived as a single disease entity, is now understood to be a diverse group of disorders driven by complex and dynamic alterations at the genetic, epigenetic, transcriptomic, proteomic, and metabolic levels. Traditional oncology has relied heavily on population-based treatment protocols, which offer standardized regimens based on tumor type, anatomical location, and clinical staging. However, these approaches often overlook the molecular heterogeneity that exists even among tumors of the same histological type, resulting in suboptimal therapeutic outcomes for many patients [1].

The advent of precision medicine—sometimes referred to as personalized or individualized medicine—has catalyzed a major paradigm shift in cancer care. At its core, precision medicine aims to deliver the right treatment to the right patient at the right time by leveraging molecular and genetic information to guide decision-making. This approach encompasses several interconnected elements: molecular diagnostics, the use of predictive and prognostic biomarkers, targeted therapies, risk stratification, and the integration of patient-specific data into treatment planning. Technological breakthroughs, particularly in genomics and bioinformatics, have made it feasible to characterize tumors at an unprecedented depth. Next-generation sequencing (NGS), high-throughput proteomics, and single-cell analysis now allow for comprehensive profiling of cancer genomes, enabling clinicians and researchers to identify actionable mutations and tailor therapies accordingly. Liquid biopsies and circulating tumor DNA (ctDNA) further enhance the ability to monitor tumor evolution in real time, paving the way for adaptive treatment strategies. This chapter will delve into the fundamental concepts that form the basis of precision oncology. It will examine how genomic and molecular profiling informs therapeutic decisions, how predictive biomarkers are used to match patients with targeted agents, and how large-scale data integration is transforming clinical workflows. By understanding these principles, clinicians and researchers can better navigate the complexities of modern cancer care and contribute to the ongoing advancement of precision medicine [2] (Fig. 1).

## 2 Historical Background

The concept of precision medicine in cancer treatment has evolved over several decades, driven by key discoveries in molecular biology and technological advancements. From the identification of cancer-associated genes to the

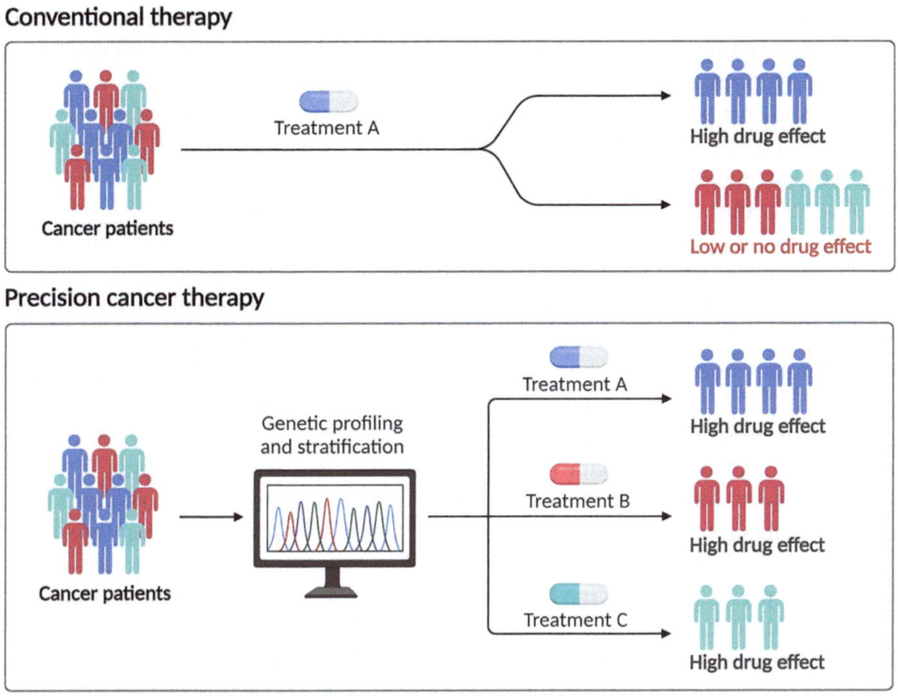


**Fig. 1** Concept of conventional and precision medicine-based therapy systems

development of high-throughput sequencing and biomarker-driven clinical trials, precision medicine has transformed the landscape of cancer treatment, offering new avenues for more personalized and effective therapies. One of the earliest and most significant milestones in the development of precision medicine was the identification of the Philadelphia chromosome in chronic myeloid leukemia (CML) in 1960. This genetic abnormality, which results from a translocation between chromosomes 9 and 22, creates the BCR-ABL fusion gene, leading to the uncontrolled proliferation of white blood cells. The discovery of the Philadelphia chromosome was groundbreaking because it identified a specific molecular target for therapeutic intervention. This paved the way for the development of imatinib, a targeted therapy that specifically inhibits the BCR-ABL protein. Imatinib's success in treating CML marked a paradigm shift, demonstrating that targeting specific genetic mutations in cancer could lead to more effective and less toxic therapies compared to traditional chemotherapy. The next major leap in precision medicine came with the advent of high-throughput sequencing technologies in the early 2000s. Technologies like NGS revolutionized our ability to characterize cancer genomes comprehensively. These innovations enabled researchers to sequence entire cancer genomes, identifying not only driver mutations but also the genetic variations that contribute to tumor growth and metastasis. High-throughput sequencing made it possible to perform detailed molecular profiling of tumors, which is the cornerstone of modern precision oncology. This capability has allowed for the identification of various molecular targets, such as mutations in EGFR, HER2, and BRAF, all of which are now targeted by specific therapies in various cancers [2–5].

As molecular profiling technologies advanced, the integration of biomarker-driven clinical trials became a pivotal step in the evolution of precision medicine. The Cancer Genome Atlas (TCGA), launched in 2006, was a landmark initiative that systematically characterized the molecular landscape of various cancers, contributing to a better understanding of the genetic underpinnings of cancer and providing a valuable resource for future clinical applications. Following TCGA, the precision medicine initiative (PMI), launched in 2015, further advanced the concept of personalized medicine by aiming to gather large-scale genomic, environmental, and lifestyle data to develop targeted treatment strategies. The PMI focused on the idea that cancer treatment could be optimized by considering the genetic makeup of both the patient and their tumor, thereby improving therapeutic efficacy and minimizing side effects. These key milestones in the historical development of precision medicine have set the stage for an era of highly personalized cancer treatment, where therapies are tailored to the specific molecular characteristics of an individual's cancer [6–8]. As the field continues to evolve, the integration of molecular diagnostics, advanced sequencing technologies, and biomarker-driven treatment strategies will continue to reshape the future of oncology.

## 3 Principles of Precision Medicine in Cancer

The principles of precision medicine in cancer revolve around the concept of tailoring medical care to the individual characteristics of each patient and their tumor. This approach emphasizes the integration of genomic, transcriptomic, proteomic, and epigenetic data to identify molecular alterations that drive tumor growth and response to therapy. By leveraging this information, clinicians can select targeted therapies, predict treatment outcomes, minimize toxicity, and monitor disease progression with greater accuracy. This section outlines the core principles that form the foundation of precision oncology, including molecular profiling, biomarker-guided therapy, patient stratification, and dynamic treatment adaptation [9] (Fig. 2) (Box 1).

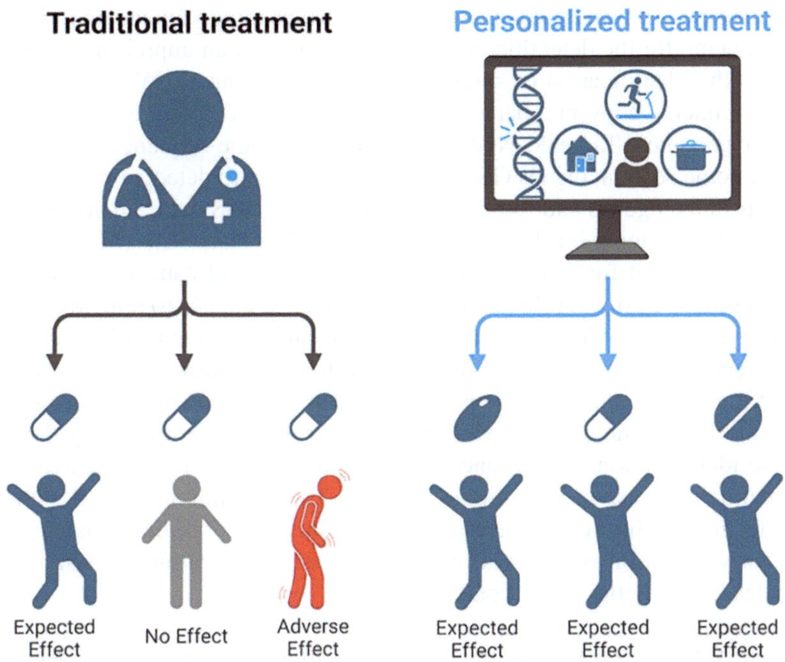


**Fig. 2** Therapeutic impact of traditional and personalized medicine systems

**Box 1: Principles of Precision Medicine in Cancer**

A. Molecular profiling
   Next-generation sequencing (NGS)
   WES & WGS
   Transcriptomics
   Epigenetic profiling
   Proteomics and metabolomics
B. Targeted therapies
C. Immunotherapy
D. Pharmacogenomics

## 3.1 Molecular Profiling

Understanding the genetic and molecular alterations in tumors is critical for precision medicine. Molecular profiling involves a comprehensive analysis of the genetic, transcriptomic, epigenetic, and proteomic features of cancer cells. By characterizing these molecular features, researchers and clinicians can identify oncogenic drivers, predict treatment responses, and develop targeted therapies.

**NGS** It is a high-throughput technology that enables rapid sequencing of DNA and RNA, allowing for the detection of genetic alterations at an unprecedented scale. It facilitates the identification of single nucleotide variations (SNVs), insertions and deletions (indels), copy number variations (CNVs), and structural rearrangements in cancer genomes. Targeted NGS panels, whole-exome sequencing (WES), and whole-genome sequencing (WGS) are commonly used to detect clinically relevant mutations in oncogenes and tumor suppressor genes. NGS data can also be used to infer tumor mutational burden (TMB), microsatellite instability (MSI), and other biomarkers relevant for immunotherapy. In the context of cancer precision medicine, NGS plays a crucial role in tailoring treatment strategies to individual patients by identifying actionable genetic alterations that can be targeted with specific therapies. By enabling comprehensive tumor profiling, NGS allows for the selection of targeted therapies such as EGFR inhibitors, ALK inhibitors, and PARP inhibitors, improving therapeutic efficacy while minimizing toxicity. Additionally, NGS supports the identification of resistance mechanisms, guiding the adaptation of treatment regimens in response to tumor evolution. The integration of NGS with liquid biopsy techniques further enhances real-time monitoring of tumor dynamics, facilitating early detection of minimal residual disease and treatment resistance, ultimately leading to better patient outcomes [10].

**WES and WGS** WES focuses on the coding regions (exons) of genes, which constitute approximately 1–2% of the genome but harbor the majority of known disease-related mutations. It provides a comprehensive view of both coding and non-coding regions, enabling the discovery of regulatory mutations and structural variations that may influence cancer progression. These techniques help identify novel oncogenic mutations, potential resistance mechanisms, and germline mutations associated with hereditary cancer syndromes. In precision medicine, WES and WGS allow for a deeper understanding of the genetic landscape of individual tumors, facilitating the identification of rare and novel mutations that may not be captured by targeted sequencing approaches. By uncovering alterations in non-coding regions, WGS provides insights into gene regulation, enhancer activity, and non-coding RNA involvement in cancer progression. These sequencing approaches are instrumental in uncovering tumor heterogeneity, tracking clonal evolution, and informing treatment decisions by revealing potential targets for personalized therapies. Furthermore, WES and WGS support pharmacogenomic studies, helping to predict patient responses to specific drugs based on genetic variations, thus optimizing therapeutic strategies and minimizing adverse effects. The integration of WES and WGS with other omics data, such as transcriptomics and proteomics, enhances the ability to develop more effective and tailored therapeutic interventions for cancer patients [11].

**Transcriptomics** RNA-seq assesses gene expression levels, alternative splicing events, and fusion transcripts that may contribute to oncogenesis. This approach helps identify differentially expressed genes in tumors compared to normal tissues, providing insights into dysregulated pathways and tumor subtypes. Single-cell

RNA sequencing is a powerful technique that enables the characterization of tumor heterogeneity, immune cell infiltration, and the tumor microenvironment at a single-cell resolution. RNA-seq can also detect non-coding RNAs, including miRNAs and lncRNAs, which play crucial roles in cancer development and the regulation of gene expression. By analyzing transcriptomic data, researchers can identify potential therapeutic targets, such as genes that are overexpressed in specific cancer subtypes and may be susceptible to targeted inhibition. RNA-seq is also used to monitor dynamic changes in gene expression in response to treatments, allowing for the identification of biomarkers that predict drug response or resistance. Integration of transcriptomic data with genomic and epigenetic analyses enhances the understanding of gene regulation networks in cancer, providing a more comprehensive approach to precision medicine [12–14].

**Epigenetic Profiling** Epigenetic modifications, such as DNA methylation and histone modifications, play a crucial role in regulating gene expression without altering the DNA sequence. DNA methylation profiling can identify hypermethylated or hypomethylated regions associated with gene silencing or activation in cancer. Chromatin immunoprecipitation sequencing (ChIP-seq) and assay for transposase-accessible chromatin using sequencing (ATAC-seq) provide insights into histone modifications and chromatin accessibility, revealing regulatory elements that drive cancer progression. Epigenetic profiling helps in classifying tumor subtypes and identifying epigenetic biomarkers that can serve as diagnostic and prognostic indicators. Histone modification analysis allows for the identification of epigenetic changes that contribute to oncogenic transformation and therapeutic resistance. Targeting epigenetic modifications using drugs such as DNA methyltransferase inhibitors (e.g., azacitidine) and histone deacetylase inhibitors (e.g., vorinostat) represents a promising therapeutic approach in precision oncology. The integration of epigenomic data with other molecular profiling approaches enhances the understanding of gene regulatory networks, providing a more comprehensive landscape of tumor biology and potential therapeutic targets [15–17].

**Proteomics and Metabolomics** Proteomics involves the large-scale study of protein expression, post-translational modifications, and protein-protein interactions in cancer cells. Mass spectrometry-based proteomics enables the identification of protein biomarkers that can aid in diagnosis, prognosis, and treatment selection. Proteomic analyses also help in understanding the functional consequences of genetic mutations and post-translational modifications, such as phosphorylation, ubiquitination, and glycosylation, which can influence cancer progression and therapeutic resistance. Metabolomics focuses on metabolic alterations in cancer cells, providing insights into tumor metabolism, metabolic dependencies, and potential metabolic vulnerabilities that can be targeted therapeutically. By profiling metabolites, metabolomics helps identify dysregulated metabolic pathways, such as altered glycolysis (Warburg effect), glutaminolysis, and lipid metabolism, which are characteristic of cancer cells. Metabolomic studies also play a key role in understanding the tumor microenvironment, including interactions between cancer cells and stromal or immune cells that influence tumor progression. Integration of proteomic and metabolomic data with genomic, transcriptomic, and epigenomic information offers a holistic understanding of tumor biology, facilitating the identification of novel drug targets and biomarkers. Advancements in proteomics and metabolomics are contributing to the development of personalized therapeutic strategies, including the use of proteomic-based companion diagnostics to guide targeted therapies and metabolomics-based approaches for predicting drug efficacy and resistance [18–21].

### 3.2 Targeted Therapies

Precision medicine has revolutionized cancer treatment by enabling the development of targeted therapies that specifically inhibit key molecular pathways driving tumorigenesis. These therapies are designed to block the activity of cancer-related proteins or signaling pathways, offering more effective treatment with fewer side effects compared to traditional chemotherapy. One prominent example is the use of EGFR inhibitors, such as erlotinib, in NSCLC. Mutations in the EGFR gene lead to uncontrolled cell growth, and EGFR inhibitors specifically target this mutated protein, improving patient outcomes. Similarly, in breast cancer, HER2-targeted therapies such as trastuzumab have dramatically improved survival for patients with HER2-positive tumors. These therapies block the HER2 receptor, a driver of cancer growth. BRAF inhibitors like vemurafenib have shown efficacy in melanoma patients with the BRAF V600E mutation, targeting the aberrantly activated BRAF protein to inhibit tumor cell proliferation. Targeted therapies offer significant advancements in cancer treatment by selectively targeting cancer-driving mutations, reducing off-target effects and enhancing treatment precision [22–24].

### 3.3 Immunotherapy

Immunotherapy represents a transformative approach to cancer treatment, leveraging the body's immune system to detect and destroy cancer cells. One of the most significant classes of immunotherapies are immune checkpoint inhibitors, such as pembrolizumab and nivolumab. These drugs block checkpoint proteins like PD-1 or PD-L1, which tumors use to evade immune detection. By inhibiting these checkpoints, immune checkpoint inhibitors reinvigorate T cells, allowing the immune system to recognize and attack cancer cells more effectively. The success of immune checkpoint inhibitors has been particularly notable in cancers such as NSCLC, melanoma, and bladder cancer. The development of predictive biomarkers has further refined patient selection for immunotherapy. PD-L1 expression, the level of PD-L1 found on tumor cells or immune cells, serves as a key biomarker for identifying patients most likely to benefit from checkpoint inhibitors. Additionally, TMB measures the number of mutations in a tumor's genome, has been shown to correlate with a higher likelihood of responding to immune therapies. The integration of these biomarkers into clinical practice has significantly enhanced the ability to select patients who are most likely to respond to immunotherapy, optimizing therapeutic outcomes [5, 25, 26].

### 3.4 Pharmacogenomics

Pharmacogenomics is the study of how genetic variations influence an individual's response to drugs, aiming to tailor drug prescriptions based on genetic makeup to maximize efficacy and minimize adverse effects. This field is especially important in cancer treatment, where drugs with narrow therapeutic windows, such as chemotherapy agents, can have significant side effects. By analyzing genetic variants in key enzymes or receptors, pharmacogenomics helps identify patients who may be at risk of severe drug reactions or suboptimal responses. For example, TPMT (thiopurine methyltransferase) genotyping can predict how a patient will metabolize thiopurine drugs used in chemotherapy. Patients with low TPMT activity are at higher risk of severe toxicity, and adjusting the dose accordingly can improve safety. Similarly, DPYD (dihydropyrimidine dehydrogenase) genotyping is crucial for patients receiving 5-fluorouracil, as variations in the DPYD gene can lead to severe toxicity. By incorporating pharmacogenomic data into clinical practice, healthcare providers can optimize drug dosages, reduce adverse drug reactions, and ultimately improve patient outcomes in cancer therapy.

## 4 Clinical Applications of Molecular Profiling

Precision medicine represents a revolutionary shift in the treatment of cancer, enabling healthcare providers to tailor therapies based on the genetic and molecular characteristics of an individual's cancer. Molecular profiling, as an integral component of precision medicine, involves the systematic analysis of a tumor's genetic and molecular alterations, identifying specific mutations, gene expressions, and other biomarkers that influence disease progression and response to treatment. These molecular alterations can be detected through technologies like NGS, microarrays, and proteomics, which have made it possible to understand cancer at a molecular level. This approach has not only improved early detection and diagnosis but has also significantly impacted the development of targeted therapies, contributing to personalized treatment strategies that increase the chances of favorable outcomes for cancer patients [27–29].

One of the primary clinical applications of molecular profiling in cancer treatment is the identification of actionable mutations that can guide therapy selection. For instance, in lung cancer, the detection of mutations in the EGFR gene has led to the use of targeted therapies such as TKIs, which specifically inhibit the growth of cancer cells harboring these mutations. Similarly, in breast cancer, molecular profiling can detect mutations in the HER2 gene, which is overexpressed in some tumors and can be targeted by drugs such as trastuzumab (Herceptin). These drugs have been shown to improve survival outcomes by selectively targeting cancer cells while minimizing damage to healthy cells. The ability to identify such mutations and selectively target them allows for the implementation of treatments that are far more effective and less toxic than traditional chemotherapy, which indiscriminately targets both cancerous and normal cells. Another significant clinical application of molecular profiling is the ability to predict and monitor treatment resistance. Tumors can evolve and acquire resistance to initially effective therapies, and molecular profiling can help identify the mechanisms behind this resistance. For example, in colorectal cancer, the BRAF V600E mutation is often associated with resistance to EGFR inhibitors. In cases where patients initially respond to EGFR inhibitors but later experience relapse, testing for the BRAF V600E mutation can help guide the use of alternative therapies, such as BRAF inhibitors or combination therapies. Similarly, in melanoma, molecular profiling has uncovered mutations in the BRAF gene, and the use of BRAF inhibitors has revolutionized the treatment of patients with metastatic melanoma. However, resistance to these inhibitors often develops due to secondary mutations in the tumor's genome, which can be detected through molecular profiling, allowing clinicians to adjust the treatment regimen accordingly [27, 29].

Molecular profiling is also critical in identifying tumor heterogeneity, a phenomenon where different areas of a single tumor or metastases exhibit different genetic and molecular characteristics. Tumor heterogeneity can complicate treatment, as some areas of the tumor may be resistant to certain therapies while others may remain sensitive. By performing molecular profiling on multiple tumor samples or monitoring the tumor over time, clinicians can gain a more comprehensive understanding of the tumor's evolving molecular landscape. This approach can help in the development of more effective combination therapies and inform decisions on the optimal sequencing of treatments to overcome the challenges posed by heterogeneity. In addition to improving treatment outcomes, molecular profiling plays a crucial role in early detection and diagnosis of cancer. Many cancers, such as pancreatic cancer, are often diagnosed at advanced stages due to the lack of early symptoms. However, advances in liquid biopsy technologies, which involve analyzing circulating tumor DNA (ctDNA), have enabled the detection of molecular alterations in a patient's blood or other body fluids. Liquid biopsy offers a non-invasive alternative to traditional tissue biopsy and can be used to detect genetic mutations, epigenetic changes, and other molecular markers that are indicative of the presence of cancer. This technology has the potential to identify cancers at earlier stages when they are more treatable, providing a valuable tool for screening high-risk populations or for monitoring disease progression in patients undergoing treatment. Molecular profiling is also instrumental in guiding decisions about the use of immunotherapy, a treatment approach that has shown promising results in a variety of cancers, including melanoma, NSCLC, and bladder cancer. Immunotherapies, such as immune checkpoint inhibitors (e.g., pembrolizumab, nivolumab), work by enhancing the body's immune system to recognize and attack cancer cells. However, not all patients respond to immunotherapy, and molecular profiling can help identify biomarkers that predict which patients are most likely to benefit. For example, tumors with high microsatellite instability (MSI-H) or deficient mismatch repair (dMMR) are more likely to respond to immune checkpoint inhibitors. Similarly, the expression of PD-L1 on tumor cells can serve as a biomarker to predict response to PD-1/PD-L1 inhibitors. By identifying these biomarkers through molecular profiling, clinicians can personalize immunotherapy treatments, improving the likelihood of a successful outcome [4, 24, 27, 28].

In addition to identifying predictive biomarkers, molecular profiling also aids in understanding the underlying mechanisms of cancer progression. For example, the identification of specific gene fusions or translocations, such as the BCR-ABL fusion in CML or the EML4-ALK fusion in non-small cell lung cancer, can offer insights into the biological processes driving cancer. Targeted therapies that inhibit these specific gene fusions have demonstrated remarkable efficacy in treating patients with these alterations, as seen with the use of tyrosine kinase inhibitors like imatinib in CML and crizotinib in ALK-positive NSCLC. Molecular profiling, therefore, helps to uncover the precise molecular underpinnings of individual cancers, leading to the development of therapies that directly target the causes of the disease. Furthermore, molecular profiling has applications in cancer recurrence monitoring and post-treatment surveillance. Cancer recurrence remains a significant challenge in oncology, as many patients experience relapse even after seemingly successful treatments. By continuously monitoring the molecular profile of a patient's tumor through liquid biopsies or other diagnostic tools, clinicians can detect the emergence of minimal residual disease (MRD) or the presence of subclonal populations of cancer cells that may be resistant to therapy. Early detection of recurrence through molecular profiling enables timely interventions, such as adjusting treatment regimens or administering additional therapies, which can help improve patient outcomes and extend survival. While molecular profiling has undoubtedly transformed cancer treatment, several challenges remain. The complexity of interpreting large amounts of genetic and molecular data, the cost of testing, and the variability in treatment response remain hurdles that must be addressed. Furthermore, the development of new therapies based on molecular profiling requires rigorous clinical trials to demonstrate their safety and efficacy. Nonetheless, the progress made thus far highlights the immense potential of molecular profiling in advancing precision medicine and improving cancer care [2]. In conclusion, molecular profiling has become a cornerstone of precision medicine in cancer treatment, offering personalized, targeted therapies that are more effective and less toxic than traditional treatments. Its clinical applications span a wide range of areas, including treatment selection, monitoring of resistance, early detection, and prediction of therapeutic response. As technology continues to evolve and the understanding of cancer biology deepens, molecular profiling will undoubtedly play an even more prominent role in reshaping the future of cancer treatment.

## 5 Biomarkers in Precision Oncology

Biomarkers play a pivotal role in precision medicine by aiding in cancer diagnosis, prognosis, and treatment selection. These molecular indicators provide essential information about the biological state of tumors and the potential benefit from specific therapies. Biomarkers are classified into several categories based on their clinical utility (Box 2):

**Box 2: Biomarkers in Precision Oncology**

A. Predictive biomarkers
B. Prognostic biomarkers
C. Companion diagnostics
D. Emerging and exploratory biomarkers
E. Biomarker validation and clinical implementation

### 5.1 Predictive Biomarkers

Predictive biomarkers play a crucial role in the era of precision medicine, enabling the selection of patients most likely to benefit from specific therapeutic interventions. These biomarkers are particularly valuable in the context of targeted therapies and immunotherapies, where they guide personalized treatment strategies to improve clinical outcomes. One well-established predictive biomarker is the presence of ALK (anaplastic lymphoma kinase) rearrangements in NSCLC. Patients harboring ALK gene fusions demonstrate significant clinical benefit from ALK inhibitors such as crizotinib, ceritinib, and alectinib. The PROFILE 1014 clinical trial, among others, has shown that these targeted agents significantly improve progression-free survival in ALK-positive NSCLC compared to standard chemotherapy. Similarly, activating mutations in the EGFR gene serve as predictive markers for response to EGFR-TKIs, including gefitinib and erlotinib. These agents selectively inhibit mutated EGFR, leading to tumor regression and improved survival in patients with EGFR-mutant NSCLC. In melanoma, the BRAF V600E mutation is another well-characterized predictive biomarker. This mutation results in constitutive activation of the MAPK pathway and is associated with a robust response to BRAF inhibitors such as vemurafenib and dabrafenib, especially when combined with MEK inhibitors, which help overcome resistance mechanisms and prolong therapeutic response. In the field of immuno-oncology, biomarkers such as PD-L1 expression and TMB have emerged as valuable predictors of response to immune checkpoint inhibitors. High levels of PD-L1 expression on tumor cells or immune cells may indicate an increased likelihood of response to therapies targeting the PD-1/PD-L1 axis, such as pembrolizumab or nivolumab. Additionally, tumors with a high TMB tend to produce more neoantigens, enhancing the visibility of cancer cells to the immune system and thus improving the efficacy of immune checkpoint blockade across a variety of cancer types [30–32]. Collectively, these predictive biomarkers are instrumental in personalizing treatment approaches and optimizing therapeutic efficacy while minimizing unnecessary toxicity.

## 5.2 Prognostic Biomarkers

Prognostic biomarkers offer critical insights into the natural course of a disease, independent of any specific treatment, and are instrumental in risk stratification, patient counseling, and decision-making regarding monitoring and therapeutic intensity. These biomarkers help clinicians identify patients at higher risk of disease progression or recurrence, thereby enabling more personalized and effective care planning.

One prominent prognostic biomarker is the presence of KRAS mutations in colorectal cancer. These mutations are associated with a more aggressive disease phenotype and generally poorer overall survival. Moreover, KRAS-mutated tumors are resistant to EGFR-targeted monoclonal antibody therapies such as cetuximab and panitumumab, although this resistance is a predictive consideration, the mutation's primary significance lies in its association with unfavorable prognosis. Another widely recognized prognostic marker is the TP53 gene mutation. TP53, often referred to as the "guardian of the genome," plays a key role in regulating cell cycle arrest, DNA repair, and apoptosis. Mutations in this gene are found in a broad range of cancers and typically correlate with higher-grade tumors, treatment resistance, and reduced survival, reflecting the aggressive nature of TP53-mutant malignancies. In the realm of breast cancer, multigene assays such as Oncotype DX and MammaPrint have transformed prognostic assessment. These genomic tools evaluate the expression of multiple genes associated with tumor behavior to estimate the risk of recurrence. Oncotype DX provides a recurrence score that helps predict the likelihood of distant recurrence in estrogen receptor-positive, HER2-negative breast cancer and assists in determining the potential benefit of adjuvant chemotherapy. Similarly, MammaPrint classifies patients into low- or high-risk categories for recurrence, supporting decisions about the need for additional systemic therapy. These assays exemplify how molecular profiling can refine prognostication and optimize treatment strategies in clinical practice [33].

## 5.3 Companion Diagnostics

Companion diagnostics are specialized in vitro diagnostic tools that provide critical information necessary for the safe and effective use of a corresponding therapeutic product. These diagnostics are often mandated by regulatory bodies such as the US-FDA to ensure that targeted therapies are prescribed only to patients most likely to benefit from them. By identifying specific molecular characteristics of a patient's tumor, companion diagnostics enable personalized treatment approaches, reduce exposure to ineffective drugs, and minimize the risk of adverse effects [34].

A well-known example of a companion diagnostic is HER2 testing in breast and gastric cancers. Overexpression or amplification of the HER2 gene is a prerequisite for the use of HER2-targeted therapies such as trastuzumab. Only patients with HER2-positive tumors, as confirmed by IHC or FISH, are eligible to receive these therapies, which have been shown to significantly improve outcomes in this subgroup. Another important example is BRCA1/2 mutation testing for the use of PARP inhibitors. Germline or somatic mutations in these genes impair DNA repair mechanisms, rendering tumors more susceptible to synthetic lethality induced by PARP inhibitors like olaparib. Testing for BRCA mutations is essential in identifying patients with breast, ovarian, and prostate cancers who are most likely to respond to these agents. PD-L1 expression testing is also a widely used companion diagnostic, particularly in the context of immunotherapy. The level of PD-L1 expression on tumor cells or tumor-infiltrating immune cells can help predict response to immune checkpoint inhibitors such as pembrolizumab and nivolumab. This biomarker is used to stratify patients with cancers like NSCLC, bladder cancer, and melanoma, guiding the selection of those most likely to benefit from immunotherapy. Overall, companion diagnostics are pivotal in precision oncology, aligning the right treatment with the right patient at the right time [5, 25].

## 5.4 Emerging and Exploratory Biomarkers

Emerging and exploratory biomarkers represent the forefront of precision oncology, driven by rapid advancements in molecular and genomic technologies. These biomarkers hold the potential to revolutionize cancer diagnosis, monitoring, and treatment personalization, offering new avenues for early detection, dynamic disease tracking, and immunotherapeutic targeting. One of the most promising developments is the use of ctDNA, which consists of small fragments of tumor-derived DNA found in the bloodstream. ctDNA analysis allows for non-invasive, real-time monitoring of tumor dynamics, providing insights into tumor burden, detection of MRD, and identification of resistance mutations as they arise. Liquid biopsies based on ctDNA are increasingly being integrated into clinical practice, offering a less invasive alternative to tissue biopsies and enabling longitudinal tracking of disease progression or response to therapy. miRNAs are another category of emerging biomarkers gaining attention for their diagnostic and prognostic potential. These small, non-coding RNA molecules regulate gene expression post-transcriptionally and exhibit distinct expression profiles in different cancers. Aberrant miRNA expression can be detected in both tumor tissues and various body fluids, such as blood and urine, making them attractive candidates for non-invasive cancer detection, prognosis, and treatment response monitoring. Additionally, neoantigens represent a highly personalized class of biomarkers arising from tumor-specific somatic mutations. These novel peptides are not present in normal tissues and can be recognized by the immune system as foreign, making them ideal targets for immunotherapeutic approaches such as personalized cancer vaccines and adoptive T-cell therapies. By harnessing neoantigen profiles unique to an individual's tumor, researchers aim to develop customized treatments that elicit robust and specific anti-tumor immune responses [33]. Collectively, these emerging biomarkers are poised to expand the capabilities of precision oncology, moving beyond traditional molecular markers to more dynamic and individualized approaches to cancer management.

### 5.5 Biomarker Validation and Clinical Implementation

The integration of biomarkers into clinical practice requires a rigorous, multi-phase validation process to ensure their reliability and clinical relevance. Analytical validation assesses the technical performance of the assay, confirming accuracy, precision, sensitivity, and reproducibility. This ensures the biomarker can be consistently and reliably measured. Clinical validation establishes a clear association between the biomarker and a specific clinical outcome, typically through well-designed studies across diverse populations. Finally, clinical utility demonstrates that the biomarker provides actionable insights that improve patient management and outcomes. Global initiatives support this process. The U.S. NCI's Biomarker, Imaging, and Quality of Life Studies Funding Program (BIQSFP) facilitates biomarker integration into clinical trials, while the European Innovative Medicines Initiative (IMI) advances development and regulatory approval through collaborative research [6, 9]. Together, these efforts ensure that validated biomarkers are both scientifically robust and clinically impactful, driving forward the goals of precision medicine.

## 6 Challenges

Despite remarkable progress in the field of precision medicine, several persistent challenges continue to hinder its full clinical implementation. These obstacles span biological, technological, logistical, and socioeconomic domains and must be addressed to ensure that the benefits of precision oncology are accessible, effective, and equitable (Box 3).

**Box 3: Major Challenges in Precision Medicine**

A. Tumor's heterogeneity
B. Resistance mechanisms
C. Data integration and interpretation
D. Cost and accessibility

### 6.1 Tumor Heterogeneity

One of the most significant biological barriers to precision medicine is intratumoral heterogeneity, where multiple genetically distinct subclones coexist within a single tumor. These subpopulations may harbor different mutations, exhibit diverse phenotypes, and respond variably to therapy. This complexity complicates the identification of universal therapeutic targets and may result in partial treatment responses, disease persistence, or relapse following initially effective treatment. Pioneering studies by McGranahan and Swanton (2017) have underscored the evolutionary dynamics of tumors, highlighting how selective pressures—such as therapy—can drive clonal evolution and therapeutic resistance. To better understand and address tumor heterogeneity, advanced techniques like multi-region sequencing and single-cell genomics are being employed, as demonstrated in the landmark study by Gerlinger et al. (2012). These approaches enable the dissection of spatial and temporal variations within tumors and support the development of multi-targeted or adaptive therapies that can anticipate and counteract clonal diversification [25, 33].

## 6.2 Resistance Mechanisms

The emergence of therapy resistance remains a major obstacle in sustaining the benefits of targeted treatments. Tumors often adapt through various mechanisms, including the acquisition of secondary mutations, activation of alternative signaling pathways, or phenotypic shifts such as the epithelial–mesenchymal transition (EMT). For instance, Yu et al. (2013) described how NSCLC patients initially responsive to EGFR inhibitors developed resistance via T790M mutations or MET amplification. These resistance mechanisms necessitate a dynamic and multifaceted treatment approach. Strategies under investigation include combination therapies that target multiple pathways simultaneously, sequential therapy designed to prompt or delay resistance, and real-time molecular monitoring to detect emerging mutations and guide timely adjustments in treatment. Such personalized and adaptive approaches are crucial to improving long-term outcomes in precision oncology [5, 25, 35].

## 6.3 Data Integration and Interpretation

The expansion of multi-omics technologies—encompassing genomics, transcriptomics, proteomics, and metabolomics—has led to the generation of vast and complex datasets. While these data offer rich insights into tumor biology, their integration and interpretation remain challenging. Advanced computational tools and interdisciplinary expertise in bioinformatics, statistics, and clinical oncology are required to extract meaningful and clinically actionable information. A major bottleneck lies in the standardization of bioinformatics pipelines, which is essential to ensure reproducibility and comparability across studies and platforms. Furthermore, the lack of universally accepted frameworks for data sharing and clinical interpretation hinders collaborative efforts and delays the translation of discoveries into practice. Addressing these issues will require the development of robust data ecosystems, regulatory guidelines, and machine learning models capable of sifting through high-dimensional data to prioritize relevant biomarkers and therapeutic targets [36, 37].

### 6.4 Cost and Accessibility

The promise of precision medicine is often limited by issues of cost and accessibility. Comprehensive genomic testing, targeted therapies, and the infrastructure needed for molecular diagnostics are expensive and not uniformly available, particularly in low- and middle-income countries. These financial and logistical barriers contribute to disparities in healthcare access, preventing many patients from benefiting from personalized treatment approaches. Moreover, insurance coverage for genetic testing and targeted therapies can be inconsistent, even in high-income settings. To overcome these challenges, it is essential to invest in the development of affordable point-of-care diagnostics, expand public health initiatives, and implement equitable healthcare policies that ensure broader access to precision medicine. International collaboration and public-private partnerships will be key in scaling innovations and delivering precision oncology globally. Addressing these multifaceted challenges will be critical to unlocking the full potential of precision medicine. Continued research, technological innovation, policy reform, and global cooperation are all necessary to overcome current limitations and deliver personalized cancer care to all patients, regardless of geography or socioeconomic status [6, 38, 39].

## 7 Future Directions

The future of precision medicine in cancer is centered on innovation and integration. Artificial intelligence (AI) and machine learning technologies can analyze complex datasets to uncover novel biomarkers, predict treatment outcomes, and optimize clinical decision-making. AI-driven algorithms are being developed to assist in diagnostics, drug discovery, and patient stratification. Multi-Omics Integration of genomics, epigenomics, transcriptomics, proteomics, and metabolomics offers a holistic understanding of tumor biology. This integrative approach facilitates the identification of novel therapeutic targets and resistance mechanisms, paving the way for more precise interventions. Liquid Biopsies are non-invasive techniques for detecting ctDNA, RNA, and exosomes are gaining traction. Liquid biopsies allow for real-time monitoring of disease progression, minimal residual disease, and emerging resistance, enabling dynamic treatment adjustments. Further, the advancement in the neoantigen identification, cancer vaccines, and engineered T-cell therapies (e.g., CAR-T cells) are enhancing the specificity and efficacy of immunotherapeutic approaches. Global collaborative research initiatives that foster data sharing, harmonized trial designs, and cross-institutional collaboration are crucial for validating precision medicine strategies and ensuring reproducibility [33, 37]. Overall, these advancements promise to refine the precision medicine paradigm, delivering more effective, personalized, and accessible cancer care.

## 8 Conclusion

Precision medicine has rapidly progressed from a theoretical ideal to a practical and increasingly essential component of modern oncology. Through the integration of molecular profiling, data-driven analytics, and patient-centered approaches, it offers the promise of more effective, less toxic, and highly individualized cancer care. The successes of targeted therapies, such as tyrosine kinase inhibitors in lung cancer or PARP inhibitors in BRCA-mutated ovarian cancer, underscore the clinical potential of matching treatment strategies with molecular drivers of disease. Despite its promise, precision medicine faces several significant hurdles. Tumor heterogeneity—both inter- and intra-patient—remains a major barrier to consistent treatment response. The emergence of resistance mechanisms often limits the long-term efficacy of targeted therapies. Additionally, the accessibility of molecular diagnostics, the interpretation of complex genomic data, and the integration of findings into routine clinical practice pose ongoing challenges. Socioeconomic and geographical disparities can further limit the reach of precision medicine, particularly in low-resource settings. Moreover, ethical considerations regarding data privacy, incidental findings, and equitable access must be carefully navigated as genomic technologies become more widely used. The future of precision oncology lies in addressing these challenges through collaborative, multidisciplinary efforts that span basic science, clinical research, data science, and public health. Looking forward, the continued advancement of multi-omics technologies, artificial intelligence, and systems biology will deepen our understanding of cancer biology and enable more comprehensive and dynamic models of treatment. As precision medicine evolves, it is poised to not only improve survival outcomes but also enhance the quality of life for cancer patients by minimizing unnecessary treatments and optimizing therapeutic efficacy. Ultimately, precision medicine is not a replacement for existing cancer treatments but an evolution of them—one that embraces complexity, promotes personalization, and strives for equity in cancer care. By adhering to its principles and continuing to innovate, the medical community can move closer to the goal of truly individualized oncology.

## References

1. Iyer K, Saini S, Bhadra S, Kulavi S, Bandyopadhyay J (2023) Precision medicine advancements in glioblastoma: a systematic review. Biomedicine 13(2):1
2. Naithani N, Atal AT, Tilak T, Vasudevan B, Misra P, Sinha S (2021) Precision medicine: uses and challenges. Med J Armed Forces India 77(3):258–265
3. Xu X, Li L, Luo L et al (2021) Opportunities and challenges of glioma organoids. Cell Commun Signal 19(1):102. https://doi.org/10.1186/s12964-021-00777-0
4. Kumar S, Basu M, Ghosh MK (2022) Chaperone-assisted E3 ligase CHIP: a double agent in cancer. Genes Dis 9(6):1521–1555
5. Kumar S, Chatterjee M, Ghosh P, Ganguly KK, Basu M, Ghosh MK (2023) Targeting PD-1/PD-L1 in cancer immunotherapy: an effective strategy for treatment of triple-negative breast cancer (TNBC) patients. Genes Dis 10(4):1318–1350

6. Collins FS, Varmus H (2015) A new initiative on precision medicine. N Engl J Med 372(9):793–795. https://doi.org/10.1056/NEJMp1500523
7. Ashley EA (2015) The precision medicine initiative: a new national effort. JAMA 313(21):2119–2120
8. Porche DJ (2015) Precision medicine initiative. Am J Mens Health 9(3):177–177. https://doi.org/10.1177/1557988315574512
9. Larijani B, Aghaei Meybodi HR, Sarhangi N, Hasanzad M (2022) Principles of precision medicine. In: Hasanzad M (ed) Precision medicine in clinical practice. Springer Nature Singapore, pp 1–11. https://doi.org/10.1007/978-981-19-5082-7_1
10. Dong L, Wang W, Li A et al (2015) Clinical next generation sequencing for precision medicine in cancer. Curr Genomics 16(4):253–263
11. Nakagawa H, Fujita M (2018) Whole genome sequencing analysis for cancer genomics and precision medicine. Cancer Sci 109(3):513–522. https://doi.org/10.1111/cas.13505
12. Khodadadian A, Darzi S, Haghi-Daredeh S et al (2020) Genomics and transcriptomics: the powerful technologies in precision medicine. Int J Gen Med 13:627–640. https://doi.org/10.2147/IJGM.S249970
13. Mubarak G, Zahir FR (2022) Recent major transcriptomics and epitranscriptomics contributions toward personalized and precision medicine. J Pers Med 12(2):199
14. Supplitt S, Karpinski P, Sasiadek M, Laczmanska I (2021) Current achievements and applications of transcriptomics in personalized cancer medicine. Int J Mol Sci 22(3):1422
15. Heyn H, Méndez-González J, Esteller M (2013) Epigenetic profiling joins personalized cancer medicine. Expert Rev Mol Diagn 13(5):473–479. https://doi.org/10.1586/erm.13.36
16. Verma M, Kumar V (2018) Epigenetic drugs for cancer and precision medicine. In: Epigenetics of aging and longevity. Elsevier, pp 439–451. https://www.sciencedirect.com/science/article/pii/B9780128110607000218. Accessed 23 Apr 2025
17. Ghosh MK, Kumar S, Ganguly KK et al (2023) Covid-19 and cancer: insights into their association and influence on genetic and epigenetic landscape. Epigenomics 15(4):227–248. https://doi.org/10.2217/epi-2023-0052
18. Su M, Zhang Z, Zhou L, Han C, Huang C, Nice EC (2021) Proteomics, personalized medicine and cancer. Cancer 13(11):2512
19. Jacob M, Lopata AL, Dasouki M, Abdel Rahman AM (2019) Metabolomics toward personalized medicine. Mass Spectrom Rev 38(3):221–238. https://doi.org/10.1002/mas.21548
20. Puchades-Carrasco L, Pineda-Lucena A (2017) Metabolomics applications in precision medicine: an oncological perspective. Curr Top Med Chem 17(24):2740–2751
21. Wu C, Zheng L (2019) Proteomics promises a new era of precision cancer medicine. Signal Transduct Target Ther 4(1):13
22. Habeeb NWA, Kulasingam V, Diamandis EP et al (2016) The use of targeted therapies for precision medicine in oncology. Clin Chem 62(12):1556–1564
23. Saeed RF, Awan UA, Saeed S, Mumtaz S, Akhtar N, Aslam S (2023) Targeted therapy and personalized medicine. In: Qazi AS, Tariq K (eds) Therapeutic approaches in cancer treatment, Cancer treatment and research, vol 185. Springer International Publishing, pp 177–205. https://doi.org/10.1007/978-3-031-27156-4_10
24. Sarkar S, Kumar S, Saha G, Basu M, Ghosh MK (2024) Glioma nanotherapy: unleashing the synergy of dual-loaded DIM and TMZ. Int J Pharm 665:124697
25. Ghosh MK, Kumar S, Begam S, Ghosh S, Basu M (2024) GBM immunotherapy: exploring molecular and clinical frontiers. Life Sci. Published online 2024:123018
26. Kumar S, Basu M, Ghosh MK (2024) E3 ubiquitin ligases and deubiquitinases in colorectal cancer: emerging molecular insights and therapeutic opportunities. Biochim Biophys Acta BBA-Mol Cell Res. Published online 2024:119827
27. Malone ER, Oliva M, Sabatini PJB, Stockley TL, Siu LL (2020) Molecular profiling for precision cancer therapies. Genome Med 12(1):8. https://doi.org/10.1186/s13073-019-0703-1
28. Zhang Q, Fu Q, Bai X, Liang T (2020) Molecular profiling–based precision medicine in cancer: a review of current evidence and challenges. Front Oncol 10:532403

29. Zardavas D, Piccart-Gebhart M (2015) Clinical trials of precision medicine through molecular profiling: focus on breast cancer. Am Soc Clin Oncol Educ Book 35:e183–e190. https://doi.org/10.14694/EdBook_AM.2015.35.e183
30. Badve S, Kumar GL (2018) Predictive biomarkers in oncology: applications in precision medicine. Springer. https://books.google.com/books?hl=en&lr=&id=Gvx9DwAAQBAJ&oi=fnd&pg=PR7&dq=Predictive+biomarkers+in+precision+medicine+in+cancer&ots=2nZRahhkJL&sig=BeQBUmoNQAwwpmh9_sJIz-agspg. Accessed April 23, 2025
31. Fang B, Mehran RJ, Heymach JV, Swisher SG (2015) Predictive biomarkers in precision medicine and drug development against lung cancer. Chin J Cancer 34(3):26. https://doi.org/10.1186/s40880-015-0028-4
32. Rodrigues-Ferreira S, Nahmias C (2022) Predictive biomarkers for personalized medicine in breast cancer. Cancer Lett 545:215828
33. Koulis C, Yap R, Engel R et al (2020) Personalized medicine—current and emerging predictive and prognostic biomarkers in colorectal cancer. Cancer 12(4):812
34. Jørgensen JT (2015) Companion diagnostics: the key to personalized medicine. Expert Rev Mol Diagn 15(2):153–156. https://doi.org/10.1586/14737159.2015.1002470
35. Lo HW (2010) EGFR-targeted therapy in malignant glioma: novel aspects and mechanisms of drug resistance. Curr Mol Pharmacol 3(1):37–52
36. Martínez-García M, Hernández-Lemus E (2022) Data integration challenges for machine learning in precision medicine. Front Med 8:784455
37. Gligorijević V, Malod-Dognin N, Pržulj N (2016) Integrative methods for analyzing big data in precision medicine. Proteomics 16(5):741–758. https://doi.org/10.1002/pmic.201500396
38. Cooper KE, Abdallah KE, Angove RS, Gallagher KD, Bonham VL (2022) Navigating access to cancer care: identifying barriers to precision cancer medicine. Ethn Dis 32(1):39
39. Doble B (2016) Budget impact and cost-effectiveness: can we afford precision medicine in oncology? Scand J Clin Lab Invest 76(sup245):S6–S11. https://doi.org/10.1080/00365513.2016.1206437

# Shift from Traditional to Personalized Approaches

Srija Roy, Tathagata Jha, and Mrinal K. Ghosh

**Abstract**

Since the nineteenth century, cancer has been treated mainly by surgery, radiation, and chemotherapy. However, limitations of these modalities, like the inability to surgically resect risky anatomical positions, side effects of radiation, and non-specificity of chemotherapy, have hindered clinicians from achieving optimal results. Two patients with the same type and grade of tumor were found to respond differently to the same chemotherapy. The reason behind some patients achieving meaningful results while others had no effect was investigated to be due to the uniqueness of each individual's molecular code, immunological landscape, lifestyle habits, and/or genetic makeup. Since we aim to deliver maximum benefit to all patients, there came the need to precisely evaluate each patient's tumor in terms of genomics, transcriptomics, proteomics, pharmacogenomics, metabolomics, and epigenomics to find out key vulnerabilities that could be targeted to achieve better outcomes compared to the "one size fits all" model of traditional therapy. Currently, researchers, clinicians, and bioinformaticians focus mainly on the genomics of individual tumors and design treatment regimens or clinical trials. So far, many such treatment strategies and clinical trials have achieved improved progression-free survival and overall response rates compared to standard first-line chemotherapy. However, challenges like high-cost sequencing platforms, availability of skilled data analysts, and socioeconomic inequities have slowed down the pace of precision oncology to become the mainstay in clinical settings.

S. Roy · M. K. Ghosh (✉)
Cancer Biology and Inflammatory Disorder Division, Council of Scientific and Industrial Research-Indian Institute of Chemical Biology (CSIR-IICB), Kolkata, West Bengal, India
e-mail: mrinalghosh@iicb.res.in

T. Jha
Department of Medicine, Medical College & Hospital, Kolkata, West Bengal, India



M. K. Ghosh, M. Basu (eds.), *Personalized Cancer Therapy*,
https://doi.org/10.1007/978-981-95-6845-1_2

**Keywords**

Precision oncology · Personalized medicine · Cancer therapy · Tumor sequencing · Cancer multi-omics

## 1 Introduction

For most of modern medical history, anti-cancer therapy has included standardized procedures with three main modalities- surgery, radiation, and chemotherapy. This model of treatment follows a "one size fits all" principle, where tumors are classified based on their anatomical origin and histological features, thus giving clinicians an idea of the extent of progression of the cancer. This system was often ideally suited for early-stage treatment naïve patients, which may be because, along with progression of the disease, additional redundant pathways get activated that may counteract the effectiveness of chemotherapy. The TNM staging system, which will be discussed later, was established in the nineteenth century. This was a landmark for cancer therapy as it helped clinicians to identify the severity of the tumor burden [1]. Gradually, various drugs were developed targeting tumors of particular origin, e.g., hormone blockers for hormone receptor-upregulated or overactive cancers. While this model has led to improvements in survival and quality of life (QoL) for many patients, a portion of the patient pool either do not have any response, or have an even more aggressive relapse, and/or have poor QoL [2].

Scientific investigations have clarified the fact that two patients with the same type and stage of cancer may respond differently to a particular therapy. This is primarily due to the different genetic makeup, immunological features, comorbidity, lifestyle, and inherent uncertainty of tumor biology. This paradigm shift from treating cancers based on their origin to treating them based on peripheral factors that drive them is the foundation of precision oncology. It is driven by genomic sequencing, biomarker discovery, bioinformatics, and targeted drug development to move beyond the trial-and-error approach of traditional cancer care. Instead of broad-spectrum cytotoxic drugs like doxorubicin or cisplatin, clinicians are now opting for targeted therapy, which specifically inhibits the molecular drivers of an individual's cancer [3–5].

This chapter will explore this transformative evolution in cancer care, viz., the scientific foundations of personalized oncology, including the role of genomics, molecular diagnostics, and tumor profiling. It will also review the rising role of biomarker-driven clinical trials and the collective role of clinicians, researchers, and patients being crucial in determining the fruitfulness of the transition from traditional to personalized approaches.

# 2 Traditional Approaches to Cancer Treatment

## 2.1 Tumor Type-Based Classification and Therapeutic Approaches

For most of modern oncology, the diagnosis and treatment of cancer predominantly depend on the anatomical site of origin of the tumor. This method assumes that cancers originating from the same organ, like the lung, breast, or colon, have common properties and therapeutic vulnerabilities. This organ-based classification was designed on the basis of gross pathology, histology, and biostatistics, which led to the development of staging systems during the twentieth century. Since then, histopathological features mainly shaped the identification of the presence of tumors and their severity, i.e., cell shape, organization, and level of differentiation were examined in tumors under the microscope. For example, lung cancers were broadly divided into small cell lung carcinoma (SCLC) and non-small cell lung carcinoma (NSCLC) based on the size of cancer cells, with further subtypes like adenocarcinoma, squamous cell carcinoma, etc., each subtype having varying treatment regimens and prognostic factors.

TNM staging was developed by the French surgical oncologist, Pierre Denoix, at the Gustave Roussy Institute in the 1940s to assess the extent of cancer spread in patients. Precisely, TNM stands for Tumor size, regional lymph Node involvement, and distant Metastasis, respectively. Each component was assigned a number or alphabet to signify the stage of a particular tumor based on its histological, anatomical, and radiological features [6], tabulated in Table 1.

Such classification aimed to aid effective treatment strategies, evaluate treatment results, disseminate information among cancer care centers worldwide, and support cancer research. Such a staging system should be simple yet effectively distinguish between patient groups and be different for each cancer type. Denoix's TNM classification was later adopted by the American Joint Committee on Cancer (AJCC) and the Union for International Cancer Control (UICC), the chronology of which is depicted in Fig. 1.

This system allowed physicians to group patients based on disease severity and choose treatment plans accordingly [7]. For example, Stage I NSCLC was usually treated with surgery, while more advanced stages needed additional systemic therapies like chemotherapy and/or radiation. The choice of treatment regime depended on large randomized clinical trials that usually group patients based on tumor stage and specific criteria like molecular features. Upon achieving significant positive results, the drug and/or regimen gets approved by the Food and Drug Administration (FDA) as standardized management protocols. For example, Stage III colorectal cancers typically receive fluoropyridine-based chemotherapy regimens like FOLFOX [8, 9]. As a result, standardized treatment guidelines, issued by national health organizations like the National Comprehensive Cancer Network (NCCN) and international bodies like the World Health Organization (WHO issuing the "BlueBooks"), were developed for each cancer type based on the organ system involved, histology, and specific immunohistochemical markers of the tumor. This

**Table 1** Pierre Denoix's clinical classification of anatomic tumor extent published by the National Institute of Hygiene, Paris, in 1950

| **Tumor extent** | |
|---|---|
| *Stage* | *Description* |
| Stage I | Tumor is confined to the organ of origin and is relatively small in size |
| Stage II | Tumor remains within the original organ but is either larger in size or shows limited extension beyond the organ |
| Stage III | Tumor shows significant invasion into nearby structures or organs |
| Stage IV | Tumor extensively invades surrounding tissues or multiple adjacent organs |
| **Regional lymph node involvement** | |
| *Code* | *Description* |
| 0 | No clinically detectable lymph node enlargement |
| a | Lymph nodes were surgically removed and confirmed histologically negative (non-cancerous) |
| b | Lymph nodes were surgically removed and confirmed histologically positive (cancerous) |
| c | Lymph nodes appear cancerous based on clinical assessment but not confirmed by pathology |
| d | Lymph node involvement is uncertain or cannot be assessed |
| **Distant metastasis** | |
| *Code* | *Description* |
| 0 | No signs of distant metastasis |
| 1 | Metastasis to distant lymph nodes |
| 2 | Reserved for site-specific or special types of metastases |
| 3 | Metastasis to the lungs |
| 4 | Metastasis to the liver |
| 5 | Metastasis to bone |
| 6 | Metastasis to the skin |
| 7 | Metastasis to the brain |
| 8 | Metastasis to other unspecified organs |
| 9 | Multiple distant metastases affecting various organs |

reduced variability in cancer therapy across institutions. These classification frameworks were used to develop various traditional modalities- surgery, chemotherapy, and radiotherapy (Fig. 2). Surgery remains very crucial for managing localized solid tumors. Technological advancements like robotic-assisted surgery and fluorescence-guided imaging improved precision, but still, challenges like incomplete resection and microscopic residual disease are prevalent. Advanced radiotherapy through techniques like 3D-conformal radiation therapy (3D-CRT), intensity-modulated radiation therapy (IMRT), and stereotactic body radiation therapy (SBRT) offered better local and regional control. However, this was limited by the sensitivity of surrounding normal tissues and tumor-specific radio-resistance [10–13]. Next,

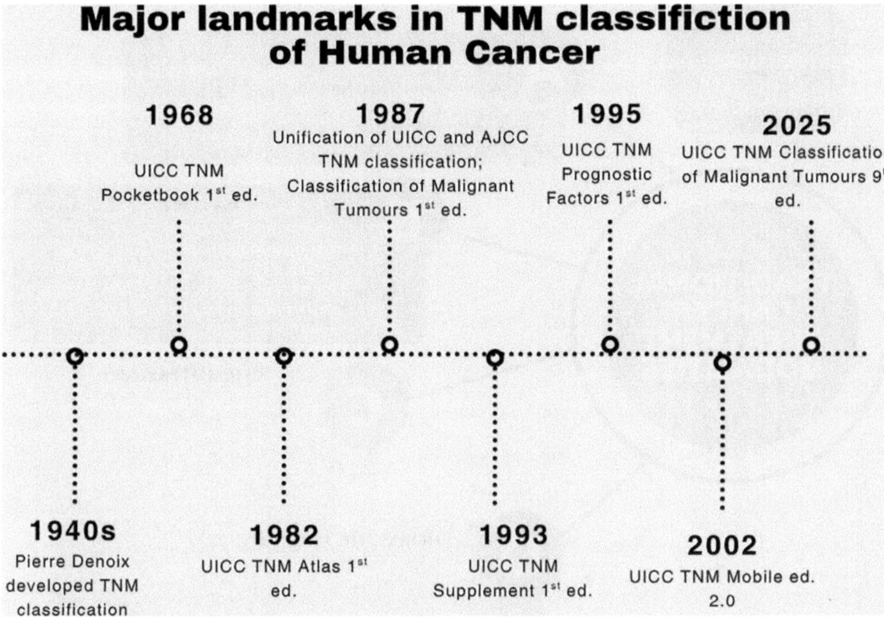


**Fig. 1** Major landmarks of TNM Classification of Human Cancer from its first establishment by Pierre Denoix through adoption and modifications by UICC and AJCC to the current format, which is easily accessible to clinicians and the first line of diagnosis in traditional cancer therapy

chemotherapy originated in the mid-twentieth century, starting with early agents like nitrogen mustards and antifolates. Over time, various regimens combining platinum agents, taxanes, anti-metabolites, and anthracyclines were developed, which led to modest improvements in survival [14, 15]. However, chemotherapy was inherently non-specific, which targets all rapidly dividing cells in the body. This lack of specificity causes common side effects like neutropenia, alopecia, mucositis, and peripheral neuropathy [16–18]. Often, tumors that shrank at first would come back stronger due to resistance to treatments [19]. Hormonal therapies like anti-estrogens, androgen deprivation were also developed for hormone-driven cancers like breast and prostate cancer. These helped control the disease, but resistance and long-term side effects such as osteoporosis and metabolic changes remained concerns [20, 21]. In some cases, a combination of different treatment modalities, like surgery and chemoradiation, led to curative outcomes for selected early-stage cancers [22]. But, for most advanced solid tumors, the treatments remained incurable, and long-term survival gains were modest. For example, even after decades of developments and updates in cytotoxic therapy, late relapses were common, especially in tumors with micrometastatic spread. These patterns highlighted a typical drawback of the traditional model, that is, treatments were applied broadly to patient populations, with limited ability to predict individual outcomes [23].

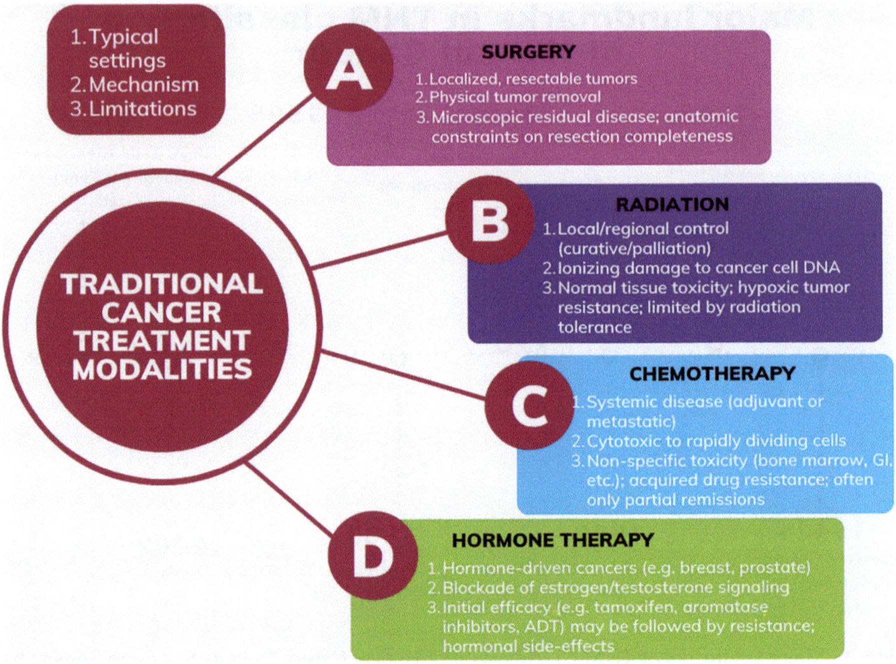


**Fig. 2** The four modalities of the traditional cancer therapy (**a**) surgery, (**b**) radiation, (**c**) chemotherapy, and (**d**) hormonal therapy with their typical settings, mechanism, and limitations/ challenges listed respectively

## 2.2 Limitations of the Traditional Model

Empirical treatment protocols based on tumor types and staging may simplify cancer management in clinical settings, however with recent understanding of the disease has rendered the traditional model of cancer treatment inadequate. Research over the last decade has unraveled that cancer and the tumor microenvironment are a complex challenge to overcome. Such findings were triggered by the fact that chemotherapy led to several side effects in patients like neutropenia, hair loss, mouth sores and nerve damage, thus reducing QoL because such drugs were mostly cell cycle inhibitors that target fast-dividing cells, irrespective of whether it is cancerous or not. Therefore, while these drugs seem to reduce tumor burden initially, frequent recurrence of a severe form of cancer happens because of the selection of more aggressive "drug-adapted" cancer cells, a phenomenon known as chemoresistance, which hinders the long-term clinical success of such drugs. The treatment choices were based on average outcomes from large clinical trials, but not on individual tumor biology. As a result, therapies were given in a trial-and-error fashion, with some patients responding well and others experiencing side effects without much benefit. For example, NSCLC patients traditionally received platinum-based chemotherapy, but only about 20–30% showed meaningful responses and there was minimal reasoning on who would benefit [24].

This model also overlooked molecular diversity within the same tumor type. For example, the traditional model included treatment of breast cancers with doxorubicin, paclitaxel, capecitabine, or tamoxifen. However, today depending on the expression of estrogen receptor (ER), progesterone receptor (PR), and human epidermal growth factor receptor 2 (HER2), breast cancer is subdivided into ER-positive, PR-positive, HR-positive, HR-negative, HER2-positive, and triple-negative breast cancers (TNBC), each having its treatment protocols and prognosis [25, 26]. Similarly, microsatellite instability-high (MSI-H) and microsatellite-stable colorectal cancers were both treated identically until researchers identified that the parameter was crucial for effective treatment since MSI-H tumors responded better to immune checkpoint inhibitors [27, 28]. These limitations hindered the identification of suitable druggable targets. Tumorigenic alterations like neurotrophic tyrosine receptor kinase (NTRK) gene fusions, which occur across multiple tumor types, were only identified and targeted after genomic profiling became more routine. First-generation TRK inhibitors like larotrectinib and entrectinib were among the first to be approved based on molecular features (NTRK fusion) rather than tumor origin [29, 30]. Moreover, traditional approaches struggled with rare tumors and cancers of unknown primary (CUP), where the anatomical origin is unclear. In such cases, standard classification tools failed to provide therapeutic guidance, leading to poor outcomes and treatment uncertainty [31, 32]. Finally, these limitations contributed to non-uniformity in cancer treatment outcomes. Since population-level data, which accounted for approved management protocols, could not account for individual variability in germline genetics, drug metabolism, or comorbid conditions, some patients benefited from standard care while others experienced ineffective interventions. In some cases, the percentage of beneficiaries may be greater; however, that should not mean that individuals with unique molecular features should be treated with any less effort.

In conclusion, while the traditional ways of treating cancer provided the first foundation for cancer therapy, offering structure, standardization, and a basis for clinical decision-making that has saved millions of lives, its organ-based framework, non-specific treatments, and population-guided protocols are no longer sufficient in an era that recognizes the complexity and diversity of cancer. As we will see in the subsequent sections, these limitations have driven the evolution of personalized medicine, which seeks to match therapies not only to where a tumor begins, but to the unique molecular features that define it.

## 3 The Emergence of Personalized Medicine

The emergence of precision medicine or personalized medicine in cancer care marks an important shift from the conventional organ- and staging-based therapy to clinically significant molecular feature-based treatment (Fig. 3). This approach comes with a need for genomic profiling of individual patients to determine targeted

**Fig. 3** Comparison of Traditional and Precision Oncology in terms of their treatment selection, biomarker used, clinical trials, expected outcomes, and examples

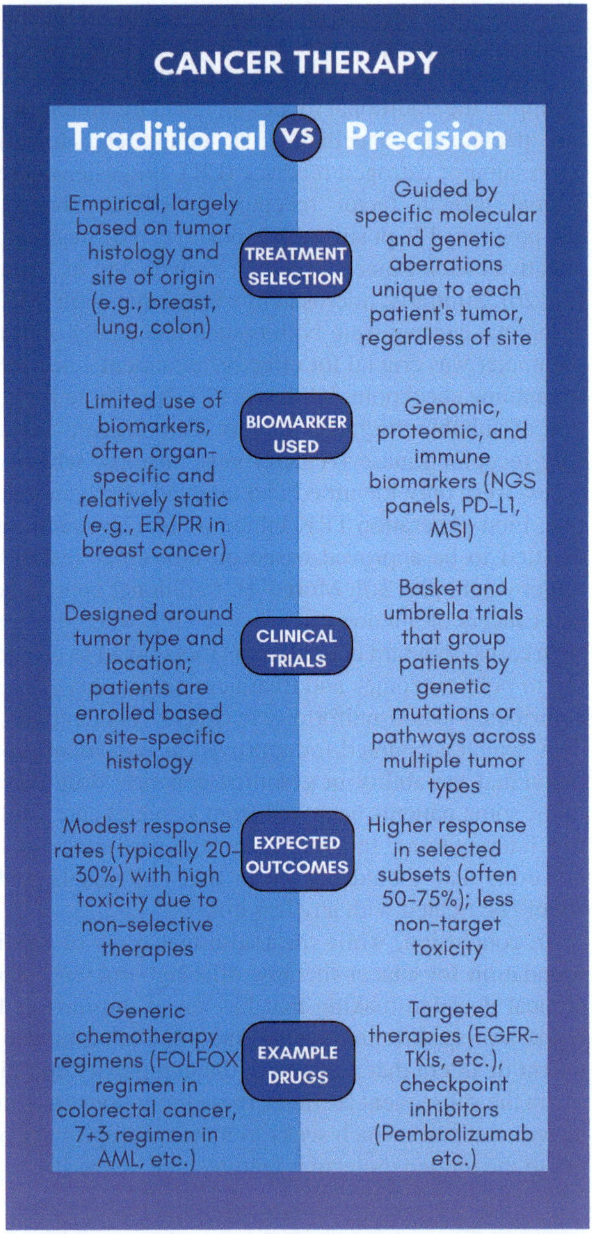


therapy based on a specific immunological profile and, molecular signature of the tumor. Additionally, such a system also assesses the individual drug pharmacogenomics, thus predicting treatment efficacy and adverse effects altogether leading to a better prognosis and patient QoL.

## 3.1 Definition and Scope

Precision oncology refers to tailored treatment regimens using the genomic, proteomic, and immunological profiles, along with, in certain cases, germline genetic variants of a particular [33, 34]. Molecular diagnostic approaches such as next-generation sequencing (NGS), immunohistochemistry (IHC), and digital PCR have helped researchers and clinicians to establish the real-time scenario of tumor genome, proteome, and immunological features and identify targetable mutations, gene fusions, and protein expressions leading to better therapy selection. This model includes several important elements, including targeted therapy (e.g., tyrosine kinase inhibitors (TKIs), monoclonal antibodies, and antibody-drug conjugates, which are directed at specific oncogenic drivers), predictive biomarkers (e.g., PD-L1 expression in immunotherapy and MSI-H status for immune checkpoint blockade), pharmacogenomics (addresses inter-individual differences in drug metabolism), and tumor-agnostic therapy (where drugs are approved based on molecular alterations independent of tumor origin) [29, 35–38].

Essentially, there are three types of precision oncology clinical trials- basket trials, umbrella trials, and platform trials. Basket trials implement drugs that target major genetic alterations across tumor types, e.g., NTRK fusion-bearing cancers, tumor mutational burden-high (TMB-H) cancers, or MSI-H cancers. This evolution has fundamentally changed clinical guidelines, which led to better informed diagnoses, better patient risk stratification, and more rational therapeutic decision-making.

## 3.2 Key Drivers of the Shift

### 3.2.1 Genomic Technology and NGS

Although a single biomarker assay (e.g., ER status) was a step towards precision oncology, ongoing technological advancement has brought forth sequencing platforms like NGS, which can analyze hundreds of cancer-related genes using a single tumor. The strategy behind this landmark technology is sequencing the genome of tumor cells and identifying mutations, copy number alterations, and fusions—all of which are crucial for tumorigenesis and cancer progression. Then, based on such information, clinicians will determine therapeutic vulnerabilities that may comprise a single or multiple drug administration to achieve the best possible outcome for the individual patient. However, such facilities are quite expensive and not prevalent worldwide. Among the first tests that got FDA approval in 2017 were FoundationOne (tests for 324 genes and 2 genomic signatures) and MSK-IMPACT (tests 468 genes), which showed that over 40% of advanced solid tumors have at least one targetable mutation [39]. Apart from sequencing solid tumors, an NGS platform like the FoundationOne Liquid CDx test is used for liquid biopsy samples [40]. Hopefully, with improvements in modern technology, tumor sequencing will become affordable and easily available to patients worldwide.

### 3.2.2 Bioinformatics and Data Interpretation

We heavily depend on bioinformatics tools that interpret sequencing data. These "key drivers" determine the clinical utility of the genomic profiling tests. Bioinformatics tools also help to quantify biomarkers such as TMB, which helps in immunotherapy decision-making. The Genome Analysis Toolkit (GATK) (https://gatk.broadinstitute.org/hc/en-us), developed by the Broad Institute, is one such programming algorithm that has enabled analysts and developers to write NGS tools catering to specific needs in a clinical setting. It has already been used to analyze data from sequencing projects like the 1000 Genome Project and The Cancer Genome Atlas (TCGA) [41]. OncoKB (https://www.oncokb.org/), maintained by Memorial Sloan Kettering Cancer Centre, is an expert-guided precision oncology repository of annotated mutations in human cancers, their implications, and possible targeting options. It helps in streamlining the huge information generated from patient tumors into meaningful patterns based on which that particular tumor could be precisely targeted. To date, 7952 genetic alterations of 941 genes across 143 cancer types have been annotated in OncoKB along with 156 targeted drugs, aimed at supporting oncologists to design therapeutic regimes for each patient [42]. Similarly, other variant databases, including dbSNP, DoCM, ClinVar, COSMIC, OMIM, ICGC, and TCGA/GDC, produce a bulk of data regarding various alterations of the cancer genome, its associated phenotype, prognosis, and actionable targets. For instance, the Catalogue of Somatic Mutations in Cancer (COSMIC) (https://cancer.sanger.ac.uk/), maintained by Wellcome Sanger Institute, contains information regarding mutations of tumor samples as well as cancer cell lines with projects like COSMIC-3D, Cancer Gene Census, Cancer Mutation Census, and Actionability under its radar [43]. Similarly ClinVar (https://www.ncbi.nlm.nih.gov/clinvar/), maintained by the National Institute of Health (NIH), accumulates genetic variation in humans associated with various health conditions including cancer [44]. Clinical Interpretation of Variants in Cancer (CIViC) knowledgebase (https://civicdb.org/), hubs information from all these large-scale databases and attempts to provide an opportunity for precision oncology for all cancer patients. Precisely, the gaps in the databases are addressed by researchers and clinicians while going through a cyclic process of sampling, sequencing, analysis, interpretation, intervention, evaluation, and publication [45].

### 3.2.3 Pharmacogenomics

Germline genetics plays a role in predicting drug metabolism and toxicity. For instance, a study on tamoxifen-treated ER-positive breast cancer in Thai women has found that a mutation in the CYP2D6 gene is associated with a shorter disease-free survival rate. Tamoxifen is a prodrug that is oxidized by cytochrome P450 enzyme CYP2D6, thus producing several metabolites, including endoxifen, which is the most potent suppressor of ER. Mutations of CYP2D6 render patients with breast cancer poorly responding to tamoxifen therapy, as evidenced by low levels of endoxifen in their plasma. Therefore, detection of such mutations and/or any other possibility that might affect CYP2D6 activity should be taken into consideration before treatment [46]. Similarly, irinotecan is used to treat advanced CRC patients.

It is also a prodrug that is hydrolyzed to 7-ethyl-10-hydroxy-camptothecin (SN-38) by carboxylesterases. SN-38 is the active form of irinotecan, which inhibits topoisomerase I and later gets metabolized by uridine diphosphate glucuronosyltransferase 1A1 (UGT1A1). Mutations of UGT1A1, which may be relevant to different ethnic groups, are responsible for SN-38 accumulation and severe toxicity in patients, including neutropenia [47]. Mutational status of thiopurine methyltransferase (TPMT) and NUDT15 determines thiopurine dosing and efficacy in hematologic cancers [48]. Therefore, integrating pharmacogenomics testing into clinical oncology practice helps in dose adjustment and the appropriate management of potential side effects.

### 3.2.4 Tumor-Agnostic Drug Development

The development of tumor-agnostic therapies has been one of the most significant developments in this era. These agents are approved based on the molecular features of the tumor and not its anatomical origin [37]. For example, pembrolizumab is FDA-approved for any solid tumor exhibiting high microsatellite instability (MSI-H) or mismatch repair deficiency, with overall response rates of around 40% across various cancers. Larotrectinib and entrectinib are TRK inhibitors that are effective in tumors having NTRK fusions, achieving response rates of around 75% regardless of histology. These approvals highlight a new model of "mutation-guided therapy", where clinical benefit is linked to shared genomic signatures.

### 3.2.5 Impact on Clinical Trials and Practice

Another "key driver" of precision oncology encompasses innovative trial designs. Basket trials like NCI-MATCH and TAPUR enroll patients based on shared mutations across tumor types, while umbrella trials like ALCHEMIST explore multiple treatments within a single tumor type stratified by molecular subtype. These trials not only improve treatment efficacy but also help in reaching conclusions faster. Accordingly, clinical practice is restructured to include an early testing of molecular patterns in the diagnosis, leading to targeted therapy rather than being considered as a last resort. In conclusion, personalized oncology has fundamentally changed cancer treatment. By shifting focus from anatomical and histological classification to molecular profiling, thus expanding treatment opportunities, improving treatment outcomes, and minimizing time to reach successful endpoints, which is crucial for cancer patients. Nevertheless, challenges remain, including the interpretation of variants of uncertain significance (VUSs), uneven access to genomic testing, and the need for sustainable health system integration. Addressing these hurdles is essential to ensure that the benefits of precision medicine are equitably realized across the global cancer care landscape.

### 3.2.6 Examples of Successful Personalized Therapies

Multiple targeted agents are now central to first-line treatment in genetically defined subgroups. The following examples show how biomarker-based therapy is now integral to routine oncology. EGFR-mutated NSCLC: Osimertinib was found to improve median progression-free survival (10.1 months vs 4.4 months), objective response rate (75% vs 31%), and reduce treatment-related adverse effects (23% vs 47%) compared to platinum-based chemotherapy in patients with EGFR T790M-positive advanced NSCLC [49]. ALK-rearranged NSCLC: Anaplastic lymphoma kinase (ALK) inhibitor alectinib was shown to achieve a better response rate and reduced side effects, although not significant compared to crizotinib. The 12-month event-free survival rate was 68.4% vs 48.7% in alectinib vs. crizotinib-treated NSCLC patients with ALK mutation [50]. HER2-positive breast cancer: Monoclonal antibodies against HER2, like trastuzumab and pertuzumab, and newer agents like Trastuzumab deruxtecan (antibody-drug conjugate) have significantly improved survival in HER2-positive metastatic breast cancer [51]. Finally, BRAF V600-mutated melanoma: Combined BRAF/MEK inhibition using dabrafenib and trametinib yields response rates around 64% vs 51%, and median PFS of 11.4 months vs. 7.3 months in metastatic melanoma with BRAF V600E or V600K mutations compared to dabrafenib treatment alone [52].

## 4 Case Study: NSCLC

The case of NSCLC markedly shows the paradigm shift from traditional to personalized medicine. Previously, advanced NSCLC was treated with platinum-based chemotherapy and often with a second agent. Optional radiation or surgery was used with several limitations. These therapies were not tailored, leading to poor outcomes such as overall survival of around 6–9 months. However, with recent technological and research advancements, routine molecular profiling has discovered several targets. In one study of metastatic NSCLC, 71% of tumors had at least 1 NCCN-recognized driver mutation. Of them, the most common were KRAS (32%), EGFR (20%), and smaller percentages of ALK, MET, ROS1, and BRAF mutations. The NCCN guidelines now recommend testing all these genes along with PD-L1 at the time of diagnosis of the patient [53, 54].

These days, patients who have a targetable mutation receive targeted therapy as a first-line treatment [55]. For example, EGFR-mutated NSCLC is now treated with osimertinib or some other EGFR-TKIs, which increased the progression-free survival of the disease from around 4–5 months (with older TKIs or chemotherapy) to around 10–11 months [49]. Also, ALK inhibitors like alectinib are being given to ALK-rearranged NSCLC patients. This has led to response rates of greater than 80% in front-line trials [50]. ROS1 fusions, BRAF V600E mutations, RET and NTRK fusions each have specific actionable inhibitors that can be used and have high response rates [30, 56]. For example, crizotinib for ROS1, dabrafenib + trametinib for BRAF [57, 58]. IHC for PD-L1 guides therapy when no actionable mutations are found. First-line pembrolizumab (PD-1 inhibitor) is used if the tumor PD-L1 is more than or equal to 50% [59]. The KEYNOTE-024 trial showed that first-line pembrolizumab for PD-L1-high NSCLC results in a response rate of around 45% when compared to platinum-based chemotherapy (around 28%). Further, it had a significantly longer survival rate (hazard ratio 0.60). These days, many patients either receive a targeted TKI or an immune checkpoint inhibitor right up front, instead of chemotherapy, leading to significantly better outcomes for the patients. Some examples of precision medicine in NSCLC are summarized in Table 2. These advances have transformed some NSCLC subgroups from poor prognosis to relatively indolent disease. Still, some challenges include the management of resistance cases, for example, due to new mutations and secondary pathways; and the fact that many patients still have no targetable mutations.

Nonetheless, the journey of NSCLC care emphasizes how genomic stratification enables tailored therapy, which leads to better prognosis and lower side effects than conventional treatments (Fig. 4).

## 5 Implications for Clinicians

The rise of personalized oncology has widespread clinical implications. Practicing doctors must adapt by integrating genomic medicine into routine care. This begins with ordering appropriate molecular tests and understanding/interpreting complex results. The use of molecular tumor boards (MTBs), which are multidisciplinary teams including medical/radiation oncologists, pathologists, geneticists, pharmacists, and bioinformaticians, has become an important strategy. In MTBs, specialists review patients' molecular profiles to better recommend treatment regimens. MTB-recommended targeted therapy significantly improved progression-free survival in 3328 cancer patients compared to non-targeted treatment protocols [60]. Specifically, a case-control study on NSCLC patients found that those who were reviewed by an MTB had better treatment outcomes than those who were not [61]. Thus, clinicians should participate in or refer to MTBs when possible. In this regard, doctors are expected to stay updated with testing guidelines laid out by relevant boards like NCCN or ASCO (American Society of Clinical Oncology), which recommends panels for genomic profiling for many cancers. Timely collection of tumor tissue and/or liquid biopsy samples should be practiced in all institutions. Raised awareness of the importance of early genomic testing is the need of the hour. A recent study among pediatric oncologists found that delayed implementation of MTB recommendations was due to unfamiliarity of clinicians [62]. Thus, healthcare providers must be educated on the recent advancements of the targeted therapy regimen. New training programs and precision oncology fellowships could fill this gap [63]. Collaborative efforts of genetic counselors, research teams, and patient educators might be helpful for patients to understand their situation better. Shared decision-making is crucial to interpret the complexities of VUSs. Clinicians may also refer eligible patients to precision oncology-driven clinical trials and encourage them to participate in large-scale databases and consortia like the American Association of Cancer Research (AACR) Project GENIE, ASCO TAPUR, etc., to increase the chances of documenting rare variants. An important understanding is that the doctors' role has changed from solely treating disease by organ to acting as a care coordinator for an individualized precision treatment plan. Practically, this means adopting workflows that incorporate genomic results early in care.

**Table 2** Precision medicine-oriented treatment of NSCLC focusing on key molecular targets and therapies

| Target/mutation | Drug | Median PFS (months) | HR (PFS) | ORR (%) | HR (CNS progression) | Grade ≥ 3 AEs | Ref. |
|---|---|---|---|---|---|---|---|
| EGFR-mutant NSCLC | Osimertinib vs. Platinum + Pemetrexed | 10.1 vs 4.4 | 0.30 (95% CI: 0.23–0.41) | 71% vs 31% | 0.32 (95% CI: 0.21–0.49) | 23% vs 47% | [44] |
| ALK-rearranged NSCLC | Alectinib vs. Crizotinib | Not reached vs ~11 (Crizotinib) | 0.47 (95% CI: 0.34–0.65) | 82.9% vs 75.5% | 0.16 (95% CI: 0.10–0.28) | 41% vs 50% | [45] |
| ROS1 fusion-positive NSCLC | Crizotinib | 20.0 (overall), 7.0 (TP53-mutant) vs 24.1 (TP53-wt) | – | 70% overall | – | 97% had any AEs | [55] |
| BRAF V600E-mutated cancers | Dabrafenib + Trametinib (first-line) | – | – | 64% (2 CR, 21 PR) | – | 69% | [52] |
| NSCLC, PD-L1 ≥ 50% | Pembrolizumab vs. Chemotherapy | 10.3 vs 6.0 | 0.50 (95% CI: 0.37–0.68) | 44.8% vs 27.8% | – | 26.6% vs 53.3% | [54] |
| RET fusion–positive NSCLC | Selpercatinib (after platinum pre-treatment) | 18.4 months (95% CI: 16.4–24.8) vs. NE (95% CI: 13.8–NE) (Platinum pre-treated vs. previously untreated) | – | 64% (pretreated), 85% (untreated), 91% (CNS) | – | 14% (HTN), 12% (↑ALT), 10% (↑AST), 6% (hyponatremia, lymphopenia) | [56] |

*PFS* progression-free survival, *HR* hazard ratio, *ORR* overall response rate, *CNS* central nervous system, *AE* adverse event, *CI* confidence interval, *CR* complete response, *PR* partial response, *HTN* hypertension, *ALT* alanine transaminase, *AST* aspartate transaminase

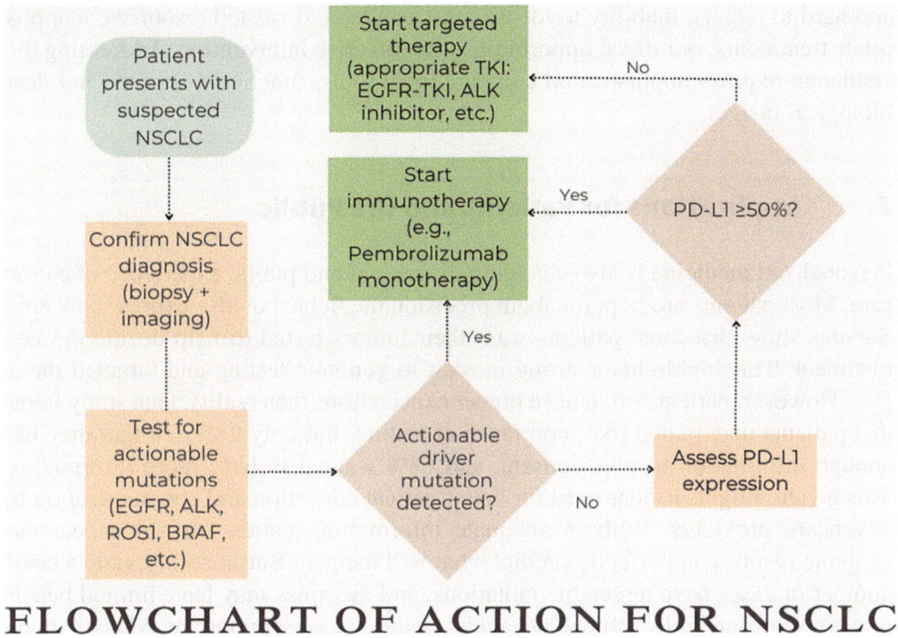


**Fig. 4** A precision oncology goal-oriented flowchart of action to sample, analyze, and treat NSCLC patients

## 6 Implications for Researchers

Researchers are an essential part of the development of personalized medicine. They play crucial roles in identifying novel drug targets, developing diagnostic tests, and designing better clinical trials. Personalized oncology can only be successful with the collective efforts of experts from various fields—doctors, researchers, and technical experts. Oncologists provide clinical insights into patient demographics, treatment response, and disease progression. Genomic researchers analyze a patient's tumor's molecular blueprint. Data analysts and AI developers are also essential for handling and interpreting large-scale patient data, involving thousands of mutations, gene expression levels, and other biomarkers. Additionally, pharmacologists decipher how these genetic alterations will affect drug effectiveness and metabolism. The future of precision medicine and cancer research in general requires real-time collaboration across these diverse fields. The huge genomic data derived from patient tumors may reveal missing links in a research question. This may be further analyzed in research laboratories to establish a pathway crucial for understanding pathways of tumorigenesis or chemoresistance. However, a challenge that researchers might face is to differentiate between driver mutations (which can be easily targeted) from passenger mutations (which are often biologically inert and hard to target). Inability to identify this can lead to wasted resources, inappropriate treatments, or missed opportunities for effective intervention. Addressing this challenge requires sophisticated computational tools, functional studies, and deep biological insight.

## 7 Implications for Patients and the Public

Personalized medicine is also changing the patient and public experience of cancer care. Most patients are hopeful about precision medicine but also express concerns. Surveys show that most patients want their tumors tested to help decide the best treatment. This highlights a strong interest in genomic testing and targeted therapies. However, patients often have higher expectations than reality. One study found that patients understood the importance of testing, but only 28% felt that they had enough information to give consent, and 68% wanted to have more information. This gap highlights a clear need for better patient education and communication by healthcare providers. Without adequate information, patients might expect that genomic results can perfectly predict what will happen. But in reality, only a small number of cases have targetable mutations, and therapies may have limited benefit or adverse events [64]. Ethical and social issues are very important. Many patients show concerns about privacy and equity. According to a systematic review, common worries include misuse of data by insurers and employers, risk of being judged or treated unfairly due to genetic test results, and the possibility that underprivileged groups may lack access to precision treatments. Even if personalized medicine proves effective, justice requires that it be available to all: as one review mentioned, "unlimited access to precision medicine for all people" is an ethical duty once benefits are proven. The public must also understand the limits of such testing: incidental germline findings and uncertain variants can cause anxiety, so clear guidelines for data security and result disclosure are needed [65].

Awareness of precision oncology among the public is growing. Patient advocacy organizations now educate about genomics, and many campaigns focus on "the right drug for the right patient." But media hype can cause unrealistic hopes of a cure. Patient educational initiatives must explain that personalized therapy may improve outcomes but is not capable of curing all cases. Government and professional guidelines are now addressing these issues by asking for informed consent that clearly explains the potential benefits and risks of genomic tests.

## 8 Challenges in the Transition

The transition to precision oncology, despite being promising, comes with economic, technical, and ethical challenges. The main pillar of precision medicine is tumor genetic profiling through comprehensive genomic profiling (CGP), targeted gene panel testing (TGPT), whole-exome sequencing (WES), or RNA-sequencing. A study on metastatic lung adenocarcinoma found that for such genomic profiling, the lifetime cost per quality-adjusted life years (QALY) gained estimates up to $445,545, which does not fall within the willingness-to-pay threshold. In other words, tumor genetic profiling as a first line of treatment does not produce enough gains in QALYs to cut down the expense of these therapies [66]. The price tag of this approach increases further due to the high cost of targeted therapy and immunotherapy compared to traditional chemotherapy [67, 68], placing precision oncology out of reach for most patients. Moreover, many effective drugs may not be included in the national approved list, which makes it inequitable, where access is determined by geography and socioeconomic status of a country, rather than by need.

Despite the economic burden, many medical institutions do have adequate infrastructure and funds to carry out genomic sequencing, which brings the next challenge. NGS generates a bulk of genomic variant data, many of which cannot be interpreted in terms of their significance in cancer or drug response (VUS) [69]. VUSs are not considered to be interpreted as a cause of an observed genetic condition. But if any VUS is in a gene causing a clinically diagnosed disease and there is a strong genotype-phenotype correlation, such VUSs should be considered as clinically significant [70]. However, the sheer uncertainty complicates clinical decision-making, leading to hesitation in choosing targeted treatments or, in some cases, misapplication of therapies.

Precision oncology also encompasses ethical considerations. For instance, genome sequencing can find mutations that are unrelated to cancer but potentially linked to other diseases, which gives rise to a dilemma about the extent of information to be revealed to the patients, as well as who decides what is relevant in the clinical findings. Additionally, since research is ever-evolving, there might be a shift in focus on the clinically relevant mutation. This requires updating the patients about new findings. The European Society of Human Genetics (ESHG) has issued guidelines about re-contacting patients and collectively deciding the progress of treatment [71].

Ensuring the security and privacy of genomic data is another challenge in precision oncology. There is also a possibility of misuse of genetic information by employers, insurers, or third-party data organizations. Therefore, robust data protection policies should be laid out by hospitals and medical institutions. Finally, global implementation of precision oncology faces the inevitable drawback of inequity of availability. Since genetic information-guided research is mainly conducted in first-world countries, this leads to failure of proof-of-concept in the case of treating patients in low-income countries. This can be solved in two ways: the government of low-income countries has to intensively fuel research on their population's genetic pool; and/or the government may provide financial support for the welfare of less privileged patients. Interestingly, as pointed out by Jonsson and Stephansdottir, should the government then demand people to follow a protocolized lifestyle in return [72]?

## 9 Future Directions

The evolution of precision oncology demands the amalgamation of a deep understanding of molecular biology and technological advancements. Recently, there has been a rapid advancement of artificial intelligence (AI) and machine learning (ML)/deep learning (DL) platforms. Since precision oncology deals with a huge repertoire of genomic, transcriptomic, and proteomic data as well as potential drugs, the need for AI and ML has become immense to handle, analyze and interpret the results based on scientifically established principles [73]. In short, AI uses mathematics, computer science, and data analysis to learn, reason, and solve problems like human intelligence does. AI uses various models and algorithms, like clustering algorithms, principal component analysis, etc., to find out subtle patterns and correlations that are beyond human recognition or at least within the time constraint. ML is a broader term encompassing AI and neural networks. AI-based DL models were found to correctly predict oncogenic drivers in NSCLC patients from hematoxylin and eosin (H&E)-stained whole slide images (WSIs) [74]. Additionally, a group developed a DL model to predict the TMB in lung adenocarcinoma (LUAD) patients using H&E WSIs from TCGA [75]. Similarly, another DL model name TR-MAMIL was developed to predict TMB from WSIs in endometrial cancer patients [76]. Therefore, AI can be used to identify patterns in histological images and correlate them to cancer progression based on evaluated data. This ability of AI can be further used to predict not only TMB but also potential therapy and the tendency to develop chemoresistance. However, rigorous validations and ethical considerations must follow after AI-based prediction to avoid possible miscalculations.

Genomics has been the main focus of the current form of precision oncology. However, we predict that a convergence of genomics, transcriptomics, proteomics, epigenomics, and metabolomics would only increase the effective goal toward complete cancer care. This is because genomics is a major but only one piece of the puzzle of the complex disease. Multi-omics approach will give us a complete portrait of the patient's tumor and potentially predict resistance mechanisms.

A rising star on the horizon of cancer therapy is the cancer vaccine. A cancer vaccine can be prophylactic or therapeutic. Vaccines against oncogenic viruses like HPV are popular nowadays; however, they cannot be categorized strictly into precision oncology. Vaccines developed from neoantigens from the patient's tumor are another driver of personalized cancer therapy. Sipuleucel-T is used for metastatic castration-resistant prostate cancer therapy [77]. Bladder cancer is treated by injecting the BCG vaccine to stimulate an immunogenic response to bladder cells. Nadofaragene firadonevec is used in case BCG fails to inhibit tumor progression [78, 79]. Besides, many clinical trials are currently testing cancer vaccines against pancreatic cancer, head and neck cancer, etc. [80, 81].

## 10 Concluding Remarks

The transition from traditional to precision oncology is remodeling our understanding of the way cancer should be treated. However, several obstacles have imperatively slowed down the process. Addressing highly priced genomic sequencing platforms, improving clinical data analysis, resolving ethical conflicts, and ensuring global equity are all important to fully realize the potential of personalized medicine in cancer. Additionally, handling AI to get optimal resourcefulness in terms of prediction of TMB, targeted therapy, and possible chemoresistance pathway activation is also essential to drive this field. Finally, we must collectively put our efforts into including each and every patient, irrespective of their socioeconomic status, ethnic background, and financial status, under the umbrella of personalized cancer therapy (Fig. 5).

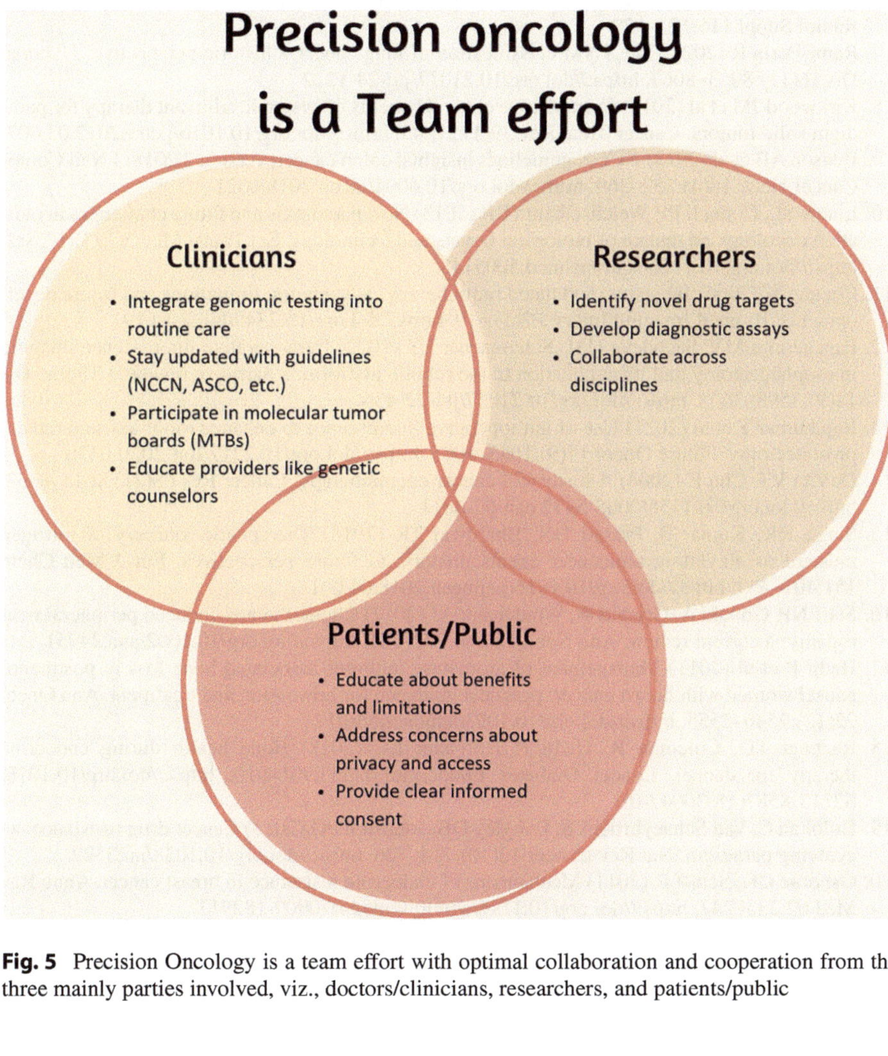


**Fig. 5** Precision Oncology is a team effort with optimal collaboration and cooperation from the three mainly parties involved, viz., doctors/clinicians, researchers, and patients/public

## References


1. Brierley J, National Cancer Institute of Canada Committee on Cancer Staging (2006) The evolving TNM cancer staging system: an essential component of cancer care. CMAJ 174(2):155–156. https://doi.org/10.1503/cmaj.045113
2. Anand U et al (2023) Cancer chemotherapy and beyond: current status, drug candidates, associated risks and progress in targeted therapeutics. Genes Dis 10(4):1367–1401. https://doi.org/10.1016/j.gendis.2022.02.007
3. Kim DH (2025) Personalized medical approach in gastrointestinal surgical oncology: current trends and future perspectives. J Pers Med 15(5):175. https://doi.org/10.3390/jpm15050175
4. Strobl M a R, Gallaher J, Robertson-Tessi M, West J, Anderson ARA (2023) Treatment of evolving cancers will require dynamic decision support. Ann Oncol 34(10):867–884. https://doi.org/10.1016/j.annonc.2023.08.008
5. Garcia-Ortega DY (2024) Comprehensive treatment strategy for improving surgical resection rate of retroperitoneal sarcomas: a histology-specific approach narrative review. Front Oncol 14:1432900. https://doi.org/10.3389/fonc.2024.1432900
6. Denoix PF (1954) Importance of a classification common to the various forms of cancer. Acta Radiol Suppl 116:122–129
7. Rami-Porta R (2024) The TNM classification of lung cancer-a historic perspective. J Thorac Dis 16(11):8053–8067. https://doi.org/10.21037/jtd-24-1212
8. Kirkwood JM et al (2013) Comparative clinical benefits of systemic adjuvant therapy for paradigm solid tumors. Cancer Treat Rev 39(1):27–43. https://doi.org/10.1016/j.ctrv.2012.03.007
9. Benson AB et al (2018) NCCN guidelines insights: colon cancer, version 2.2018. J Natl Compr Cancer Netw 16(4):359–369. https://doi.org/10.6004/jnccn.2018.0021
10. Liauw SL, Connell PP, Weichselbaum RR (2013) New paradigms and future challenges in radiation oncology: an update of biological targets and technology. Sci Transl Med 5(173):173sr2. https://doi.org/10.1126/scitranslmed.3005148
11. Cheung K (2006) Intensity modulated radiotherapy: advantages, limitations and future developments. Biomed Imaging Interv J 2(1):e19. https://doi.org/10.2349/biij.2.1.e19
12. Papageorge MV, Sachdeva UM, Schumacher LY (2022) Intraoperative fluorescence imaging in esophagectomy and its application to the robotic platform: a narrative review. J Thorac Dis 14(9):3598–3605. https://doi.org/10.21037/jtd-22-456
13. Rajakumar T et al (2021) Use of intraoperative fluorescence to enhance robot-assisted radical prostatectomy. Future Oncol 17(9):1083–1095. https://doi.org/10.2217/fon-2020-0370
14. DeVita VT, Chu E (2008) A history of cancer chemotherapy. Cancer Res 68(21):8643–8653. https://doi.org/10.1158/0008-5472.can-07-6611
15. Singh RK, Kumar S, Prasad DN, Bhardwaj TR (2018) Therapeutic journery of nitrogen mustard as alkylating anticancer agents: historic to future perspectives. Eur J Med Chem 151:401–433. https://doi.org/10.1016/j.ejmech.2018.04.001
16. Staff NP, Grisold A, Grisold W, Windebank AJ (2017) Chemotherapy-induced peripheral neuropathy: a current review. Ann Neurol 81(6):772–781. https://doi.org/10.1002/ana.24951
17. Hadji P et al (2011) Management of aromatase inhibitor-associated bone loss in postmenopausal women with breast cancer: practical guidance for prevention and treatment. Ann Oncol 22(12):2546–2555. https://doi.org/10.1093/annonc/mdr017
18. Rachner TD, Coleman R, Hadji P, Hofbauer LC (2018) Bone health during endocrine therapy for cancer. Lancet Diabetes Endocrinol 6(11):901–910. https://doi.org/10.1016/S2213-8587(18)30047-0
19. Holohan C, Van Schaeybroeck S, Longley DB, Johnston PG (2013) Cancer drug resistance: an evolving paradigm. Nat Rev Cancer 13(10):714–726. https://doi.org/10.1038/nrc3599
20. Osborne CK, Schiff R (2011) Mechanisms of endocrine resistance in breast cancer. Annu Rev Med 62:233–247. https://doi.org/10.1146/annurev-med-070909-182917

21. Smith MR et al (2002) Changes in body composition during androgen deprivation therapy for prostate cancer. J Clin Endocrinol Metab 87(2):599–603. https://doi.org/10.1210/jcem.87.2.8299
22. Shapiro J et al (2015) Neoadjuvant chemoradiotherapy plus surgery versus surgery alone for oesophageal or junctional cancer (CROSS): long-term results of a randomised controlled trial. Lancet Oncol 16(9):1090–1098. https://doi.org/10.1016/S1470-2045(15)00040-6
23. Davies C et al (2013) Long-term effects of continuing adjuvant tamoxifen to 10 years versus stopping at 5 years after diagnosis of oestrogen receptor-positive breast cancer: ATLAS, a randomised trial. Lancet 381(9869):805–816. https://doi.org/10.1016/S0140-6736(12)61963-1
24. Soon YY et al (2023) Clinical trial and real-world outcomes of patients with metastatic NSCLC in the post-platinum-based chemotherapy failure setting. JTO Clin Res Rep 4(11):100579. https://doi.org/10.1016/j.jtocrr.2023.100579
25. Tsang JYS, Tse GM (2020) Molecular classification of breast cancer. Adv Anat Pathol 27(1):27–35. https://doi.org/10.1097/PAP.0000000000000232
26. Prat A, Perou CM (2011) Deconstructing the molecular portraits of breast cancer. Mol Oncol 5(1):5–23. https://doi.org/10.1016/j.molonc.2010.11.003
27. Guven DC et al (2024) The efficacy of immune checkpoint inhibitors in microsatellite stable colorectal cancer: a systematic review. Oncologist 29(5):e580–e600. https://doi.org/10.1093/oncolo/oyae013
28. Niu C-G, Zhang J, Rao A-V, Joshi U, Okolo P (2024) Comparative effectiveness of immunotherapy and chemotherapy in patients with metastatic colorectal cancer stratified by microsatellite instability status. World J Clin Oncol 15(4):540–547. https://doi.org/10.5306/wjco.v15.i4.540
29. Gouda MA, Thein KZ, Hong DS (2024) Tissue-agnostic targeting of neurotrophic tyrosine receptor kinase fusions: current approvals and future directions. Cancers (Basel) 16(19):3395. https://doi.org/10.3390/cancers16193395
30. Cocco E, Scaltriti M, Drilon A (2018) NTRK fusion-positive cancers and TRK inhibitor therapy. Nat Rev Clin Oncol 15(12):731–747. https://doi.org/10.1038/s41571-018-0113-0
31. Krämer A et al (2024) Molecularly guided therapy versus chemotherapy after disease control in unfavourable cancer of unknown primary (CUPISCO): an open-label, randomised, phase 2 study. Lancet 404(10452):527–539. https://doi.org/10.1016/S0140-6736(24)00814-6
32. Laprovitera N, Riefolo M, Ambrosini E, Klec C, Pichler M, Ferracin M (2021) Cancer of unknown primary: challenges and progress in clinical management. Cancers (Basel) 13(3):451. https://doi.org/10.3390/cancers13030451
33. Khan A et al (2025) Genomic medicine and personalized treatment: a narrative review. Ann Med Surg (Lond) 87(3):1406–1414. https://doi.org/10.1097/MS9.0000000000002965
34. Ghoreyshi N et al (2025) Next-generation sequencing in cancer diagnosis and treatment: clinical applications and future directions. Discov Oncol 16(1):578. https://doi.org/10.1007/s12672-025-01816-9
35. Restrepo JC, Martínez Guevara D, Pareja López A, Montenegro Palacios JF, Liscano Y (2024) Identification and application of emerging biomarkers in treatment of non-small-cell lung cancer: systematic review. Cancers (Basel) 16(13):2338. https://doi.org/10.3390/cancers16132338
36. Su M, Zhang Z, Zhou L, Han C, Huang C, Nice EC (2021) Proteomics, personalized medicine and cancer. Cancers (Basel) 13(11):2512. https://doi.org/10.3390/cancers13112512
37. Nguyen K, Fama K, Mercado G, Myat Y, Thein K (2024) Histology agnostic drug development: an updated review. Cancers (Basel) 16(21):3642. https://doi.org/10.3390/cancers16213642
38. Fountzilas E, Tsimberidou AM, Vo HH, Kurzrock R (2022) Clinical trial design in the era of precision medicine. Genome Med 14(1):101. https://doi.org/10.1186/s13073-022-01102-1
39. Colomer R, Mondejar R, Romero-Laorden N, Alfranca A, Sanchez-Madrid F, Quintela-Fandino M (2020) When should we order a next generation sequencing test in a patient with cancer? EClinicalMedicine 25:100487. https://doi.org/10.1016/j.eclinm.2020.100487
40. "FoundationOne Liquid CDx test"

41. McKenna A et al (2010) The genome analysis toolkit: a MapReduce framework for analyzing next-generation DNA sequencing data. Genome Res 20(9):1297–1303. https://doi.org/10.1101/gr.107524.110
42. Chakravarty D et al (2017) OncoKB: a precision oncology knowledge base. JCO Precis Oncol 2017:PO.17.00011. https://doi.org/10.1200/PO.17.00011
43. Forbes SA et al (2017) COSMIC: somatic cancer genetics at high-resolution. Nucleic Acids Res 45(D1):D777–D783. https://doi.org/10.1093/nar/gkw1121
44. Landrum MJ et al (2018) ClinVar: improving access to variant interpretations and supporting evidence. Nucleic Acids Res 46(D1):D1062–D1067. https://doi.org/10.1093/nar/gkx1153
45. Griffith M et al (2017) CIViC is a community knowledgebase for expert crowdsourcing the clinical interpretation of variants in cancer. Nat Genet 49(2):170–174. https://doi.org/10.1038/ng.3774
46. Sukasem C et al (2012) CYP2D6 polymorphisms influence the efficacy of adjuvant tamoxifen in Thai breast cancer patients. PGPM:149. https://doi.org/10.2147/PGPM.S32160
47. Xu C, Tang X, Qu Y, Keyoumu S, Zhou N, Tang Y (2016) UGT1A1 gene polymorphism is associated with toxicity and clinical efficacy of irinotecan-based chemotherapy in patients with advanced colorectal cancer. Cancer Chemother Pharmacol 78(1):119–130. https://doi.org/10.1007/s00280-016-3057-z
48. Relling MV et al (2019) Clinical pharmacogenetics implementation consortium guideline for thiopurine dosing based on TPMT and NUDT15 genotypes: 2018 update. Clin Pharmacol Ther 105(5):1095–1105. https://doi.org/10.1002/cpt.1304
49. Mok TS et al (2017) Osimertinib or platinum-pemetrexed in EGFR T790M-positive lung cancer. N Engl J Med 376(7):629–640. https://doi.org/10.1056/NEJMoa1612674
50. Peters S et al (2017) Alectinib versus Crizotinib in untreated ALK-positive non-small-cell lung cancer. N Engl J Med 377(9):829–838. https://doi.org/10.1056/NEJMoa1704795
51. Swain SM et al (2015) Pertuzumab, Trastuzumab, and docetaxel in HER2-positive metastatic breast cancer. N Engl J Med 372(8):724–734. https://doi.org/10.1056/NEJMoa1413513
52. Johnson DB et al (2014) Combined BRAF (Dabrafenib) and MEK inhibition (Trametinib) in patients with BRAFV600-mutant melanoma experiencing progression with single-agent BRAF inhibitor. J Clin Oncol 32(33):3697–3704. https://doi.org/10.1200/JCO.2014.57.3535
53. Olivares-Hernández A et al (2024) Response to platinum-based therapies in second-line after immunotherapy in advanced or metastatic non-small-cell lung cancer PD-L1 ≥50. Transl Lung Cancer Res 13(10):2649–2659. https://doi.org/10.21037/tlcr-24-513
54. Suh JH et al (2016) Comprehensive genomic profiling facilitates implementation of the National Comprehensive Cancer Network Guidelines for lung cancer biomarker testing and identifies patients who may benefit from enrollment in mechanism-driven clinical trials. Oncologist 21(6):684–691. https://doi.org/10.1634/theoncologist.2016-0030
55. Berardi R (2019) KEYNOTE-024: goodbye to chemotherapy? J Thorac Dis 11(Suppl 3):S428–S432. https://doi.org/10.21037/jtd.2018.11.65
56. Drilon A et al (2020) Efficacy of selpercatinib in *RET* fusion–positive non–small-cell lung cancer. N Engl J Med 383(9):813–824. https://doi.org/10.1056/NEJMoa2005653
57. Planchard D et al (2017) Dabrafenib plus trametinib in patients with previously untreated BRAFV600E-mutant metastatic non-small-cell lung cancer: an open-label, phase 2 trial. Lancet Oncol 18(10):1307–1316. https://doi.org/10.1016/S1470-2045(17)30679-4
58. Michels S et al (2019) Safety and efficacy of Crizotinib in patients with advanced or metastatic ROS1-rearranged lung cancer (EUCROSS): a European phase II clinical trial. J Thorac Oncol 14(7):1266–1276. https://doi.org/10.1016/j.jtho.2019.03.020
59. Reck M et al (2016) Pembrolizumab versus chemotherapy for PD-L1-positive non-small-cell lung cancer. N Engl J Med 375(19):1823–1833. https://doi.org/10.1056/NEJMoa1606774
60. Larson KL et al (2021) Clinical outcomes of molecular tumor boards: a systematic review. JCO Precis Oncol 5:PO.20.00495. https://doi.org/10.1200/PO.20.00495
61. Huang B et al (2021) Molecular tumor board review and improved overall survival in non-small-cell lung cancer. JCO Precis Oncol 5:PO.21.00210. https://doi.org/10.1200/PO.21.00210
62. Liu A et al (2023) Molecular tumor boards: the next step towards precision therapy in cancer care. Hematol Rep 15(2):244–255. https://doi.org/10.3390/hematolrep15020025

63. Masucci M, Del Villar Pérez J, Mazzocato P, Ernberg I, Brommels M (2025) Implementing personalized cancer medicine: insights from a qualitative interview study. J Pers Med 15(4):150. https://doi.org/10.3390/jpm15040150
64. Dehar N, Abedin T, Tang P, Bebb G, Cheung WY (2022) A comparison of patients' and physicians' knowledge and expectations regarding precision oncology tests. Curr Oncol 29(12):9916–9927. https://doi.org/10.3390/curroncol29120780
65. Erdmann A, Rehmann-Sutter C, Bozzaro C (2021) Patients' and professionals' views related to ethical issues in precision medicine: a mixed research synthesis. BMC Med Ethics 22(1):116. https://doi.org/10.1186/s12910-021-00682-8
66. Dong OM et al (2022) Cost-effectiveness of tumor genomic profiling to guide first-line targeted therapy selection in patients with metastatic lung adenocarcinoma. Value Health 25(4):582–594. https://doi.org/10.1016/j.jval.2021.09.017
67. Li A et al (2024) Cost-effectiveness of first-line enfortumab vedotin in addition to pembrolizumab for metastatic urothelial carcinoma in the United States. Front Immunol 15:1464092. https://doi.org/10.3389/fimmu.2024.1464092
68. Lei J, Zhang J, You C, Liu M, Li N (2024) First-line treatment with Atezolizumab plus bevacizumab and chemotherapy for US patients with metastatic, persistent, or recurrent cervical cancer: a cost-effectiveness analysis. Value Health 27(11):1528–1534. https://doi.org/10.1016/j.jval.2024.07.013
69. Gunawardena K, Sirisena ND, Anandagoda G, Neththikumara N, Dissanayake VHW (2023) Germline variants of uncertain significance, their frequency, and clinico-pathological features in a cohort of Sri Lankan patients with hereditary breast cancer. BMC Res Notes 16(1):95. https://doi.org/10.1186/s13104-023-06365-4
70. Levkova M, Stoyanova M, Benkova-Petrova M, Georgieva M, Angelova L (2022) Variants of uncertain significance in the era of next-generation sequencing. J Am Assoc Nurse Pract 34(8):1018–1021. https://doi.org/10.1097/JXX.0000000000000745
71. Carrieri D et al (2019) Recontacting patients in clinical genetics services: recommendations of the European Society of Human Genetics. Eur J Hum Genet 27(2):169–182. https://doi.org/10.1038/s41431-018-0285-1
72. Jonsson JJ, Stefansdottir V (2019) Ethical issues in precision medicine. Ann Clin Biochem 56(6):628–629. https://doi.org/10.1177/0004563219870824
73. Hashem H, Sultan I (2025) Revolutionizing precision oncology: the role of artificial intelligence in personalized pediatric cancer care. Front Med (Lausanne) 12:1555893. https://doi.org/10.3389/fmed.2025.1555893
74. Parra-Medina R et al (2025) Deep learning in histopathology images for prediction of oncogenic driver molecular alterations in lung cancer: a systematic review and meta-analysis. Transl Lung Cancer Res 14(5):1756–1769. https://doi.org/10.21037/tlcr-2024-1196
75. Niu Y et al (2022) Predicting tumor mutational burden from lung adenocarcinoma histopathological images using deep learning. Front Oncol 12:927426. https://doi.org/10.3389/fonc.2022.927426
76. Wang C-W et al (2024) Deep learning for endometrial cancer subtyping and predicting tumor mutational burden from histopathological slides. NPJ Precis Oncol 8(1):287. https://doi.org/10.1038/s41698-024-00766-9
77. Kantoff PW et al (2010) Sipuleucel-T immunotherapy for castration-resistant prostate cancer. N Engl J Med 363(5):411–422. https://doi.org/10.1056/NEJMoa1001294
78. Lamm DL, Morales A (2021) A BCG success story: from prevention of tuberculosis to optimal bladder cancer treatment. Vaccine 39(50):7308–7318. https://doi.org/10.1016/j.vaccine.2021.08.026
79. Lee A (2023) Correction to: Nadofaragene Firadenovec: first approval. Drugs 83(10):951. https://doi.org/10.1007/s40265-023-01902-8
80. Rojas LA et al (2023) Personalized RNA neoantigen vaccines stimulate T cells in pancreatic cancer. Nature 618(7963):144–150. https://doi.org/10.1038/s41586-023-06063-y
81. Memorial Sloan Kettering. A phase 1/2 study in patients with HPV16+ recurrent/metastatic head and neck squamous cell carcinoma and other cancers

# Genomics, Epigenomics and Pharmacogenomics

# Genomics and Key Alterations

A. Esmaeilzadeh, F. Ebrahimi, M. H. Makoui, K. Hadiloo,
S. Taremi, K. Esmaeilzadeh, F. Afshari, and P. M. Yeganeh

**Abstract**

Cancer is a complex disease with various components influenced by genomic changes that initiate tumor development and result in unpredictable growth patterns. Advances in high-throughput genomic technologies have transformed our understanding of cancer biology, enabling the identification of key mutations that drive tumor growth amidst the intricate variations within tumors. This chapter highlights the crucial role of genomics in tailoring cancer treatments to individual patients. It also provides a comprehensive overview of the range of somatic and germline alterations, including mutations, copy number variations, and structural rearrangements. The chapter emphasizes the functional significance of driver mutations in oncogenes and tumor suppressor genes, as well as their impact on targeted therapies. Emerging technologies like liquid biopsies, single-cell genomics, CRISPR-based genome editing, and next-generation sequencing are discussed as groundbreaking tools for advancing personalized medicine and cancer treatment. The convergence of multi-omics approaches, integrating genomics with transcriptomics, proteomics, and metabolomics, is also explored. This integration uncovers the intricate complexity of cancer, leading to the identification of more effective strategies and goals for combating tumors. This chapter further addresses the role of the tumor microenvironment

A. Esmaeilzadeh (✉)
Pficell R&D Canadian Institution & Corporation, Profound Future Focused Innovative Cell and Gene Therapy, Toronto, ON, Canada

Cancer Gene Therapy Research Center, Zanjan University of Medical Sciences, Zanjan, Iran

Cancer Immunology Project (CIP), Universal Scientific Education and Research Network (USERN), Tehran, Iran
e-mail: a46reza@zums.ac.ir

F. Ebrahimi
Zanjan Pharmaceutical Biotechnology Research Center, Zanjan University of Medical Sciences, Zanjan, Iran

M. H. Makoui
Student Research Committee, Zanjan University of Medical Sciences, Zanjan, Iran



M. K. Ghosh, M. Basu (eds.), *Personalized Cancer Therapy*,
https://doi.org/10.1007/978-981-95-6845-1_3

and genomic interactions in shaping immune responses and in developing biomarkers for immunotherapy. Despite the breakthroughs achieved in this field, challenges persist, including ethical and practical considerations in genomic testing, tumor heterogeneity, and the development of treatment resistance. Addressing these challenges, the chapter also explores future directions, such as the application of artificial intelligence, CRISPR-based functional genomics, and personalized cancer vaccines. Overall, this chapter underscores the central role of genomics in personalized cancer therapies, emphasizing the vital connection between basic research and clinical practice to improve treatment strategies.

**Keywords**

Cancer · Genomics · Precision medicine

## 1 Introduction

Cancer is a genetic disease caused by the accumulation of mutations in key genes involved in cellular processes such as proliferation, cell division, survival, differentiation, apoptosis, and genome maintenance [1]. Cancer has now become one of the most significant medical challenges and, in addition to high mortality, it has also imposed great psychological and economic pressure on human society. Advancements in genomic technologies have uncovered several molecular alterations associated with cancer. In the past, when cancer classification was mainly based on the type of cell or tissue of origin and histopathological appearance, today, however, the integration of genomic information with clinical characteristics provides a more accurate classification of cancer [2]. This classification system, based on the specific molecular profile of the tumor, has enabled faster diagnosis, the selection of the most optimal treatment method, more accurate prediction of the risk of recurrence or cancer-related mortality, as well as the use of targeted and personalized treatment approaches. For example, genomic markers such as EGFR mutations in lung cancer, BRAF V600E mutations in melanoma, and HER2 copy number amplification in breast cancer have been identified and used for targeted therapies [3].

K. Hadiloo
Department of Surgery, Velayat Clinical Research Development Unit, Qazvin University of Medical Sciences, Qazvin, Iran

S. Taremi · P. M. Yeganeh
School of Medicine, Zanjan University of Medical Sciences, Zanjan, Iran

K. Esmaeilzadeh
Department of Medical Nanotechnology, School of Advanced Medical Sciences, Tabriz University of Medical Sciences, Tabriz, Iran

F. Afshari
Tissue Engineering and Applied Cell Sciences Department, School of Advanced Medical Sciences and Technologies, Shiraz University of Medical Sciences, Shiraz, Iran

The completion of the Human Genome Project and the advent of advanced genomic technologies, particularly next-generation sequencing (NGS), have enabled the precise identification and characterization of cancer genome alterations, including somatic mutations, copy number variations (CNV), structural rearrangements, and epigenetic alterations [4]. In addition to identifying biomarkers predictive of response to treatment, this technology is also capable of detecting somatic and germline mutations, which aids in refining or confirming the diagnosis of cancer subtype in patients and provides physicians with valuable information about hereditary cancer risk, the likelihood of disease recurrence, and the risk of death. Despite its increasing utility and diminished cost, the development of new methods and identification of novel genomic biomarkers for diagnostic and therapeutic targets have enhanced the value of NGS-based profiling [5].

However, the widespread use of cancer genomics in the clinical setting still faces challenges. These include tumor heterogeneity, the existence of resistance mechanisms in cancer, the need for rigorous validation of genomic data, and the need for sophisticated computational approaches to interpret the data [6]. This chapter discusses the essential role of genomics and new technologies in this field as a bridge between basic science and clinical practice for advancing personalized cancer therapy. It is a comprehensive review of the importance and application of genomics in the diagnosis, treatment, and follow-up of cancer with a personalized medicine approach. It also provides an overview of emerging technologies such as liquid biopsy, single-cell sequencing, and the integration of genomics with other omics approaches. These advancements aim to develop personalized cancer treatment strategies that maximize efficacy while minimizing side effects. Additionally, it explores the future prospects of this field.

## 2 Types of Genomic Alterations in Cancer

A variety of genetic alterations have been observed in the genome of cancer cells, each promoting cancer progression and tumorigenesis through a distinct mechanism. Point mutations are among these changes, characterized by nucleotide substitutions. Missense mutations in oncogenes (such as KRAS G12V) and nonsense mutations in tumor suppressor genes (such as TP53 R175H) fall under this category [1]. A common mutational pattern in this group is the C > T transition in CpG regions. Small insertions and deletions (indels), usually less than 50 base pairs, can cause frameshifts in coding regions. One prominent clinical example is the deletion of exon 19 of EGFR, which influences sensitivity to tyrosine kinase inhibitors [7]. Another class of changes is copying number alterations (CNAs), which affect gene dosage. Prominent examples include HER2 in breast cancer and CDKN2A in glioma [8]. Chromosomal rearrangements such as inversions, translocations, and tandem duplications may give rise to new oncogene proteins (e.g., BCR-ABL1 in CML and EML4-ALK in NSCLC) [9].

One of the changes commonly seen in cancer cells is the gain or loss of entire chromosomes, known as aneuploidy. This change causes genomic instability.

Examples include trisomy 8 in AML and monosomy 17 in TP53-deficient cancers [10]. Epigenetic changes can also contribute to cancer progression. For example, hypermethylation of the MGMT promoter in glioma, which alters gene expression in the cell [11]. Oncogenic viruses such as HPV and HBV may also cause cancer by inserting their genomes near proto-oncogenes, as seen in the HPV E6/E7 fusion in cervical cancer [12]. Across various cancer types, these alterations tend to follow specific patterns. Typically, driver mutations confer a selective advantage during tumorigenesis, while passenger mutations accumulate during tumor development [1]. The use of novel genomic approaches, such as sequencing, enables the identification of variants, their functional interpretation, and the selection of targeted therapeutic strategies in personalized medicine.

## 3 Key Genes and Pathways Associated with Cancer

Cancer is a complex condition in which cells undergo genetic and molecular changes that affect important signaling pathways. These changes disrupt the normal processes of the cell, resulting in uncontrolled cell proliferation, resistance to apoptosis, and metastasis to other tissues. For this reason, this section reviews the important genes and pathways associated with cancer. Genes that contribute to cancer development or progression are known as driver genes. Driver genes are divided into two categories: oncogenes and tumor suppressor genes. Passenger mutations are genetic alterations that occur during cell proliferation but do not contribute to cancer development or progression.

### 3.1 Oncogenes

In the event of mutation, duplication, or chromosomal rearrangement, the proto-oncogene becomes an oncogene, and as a result, its activity increases, leading the cell to become cancerous. The most important key oncogenes include EGFR, KRAS, and MYC. EGFR is an epidermal growth factor receptor whose mutations and amplifications are associated with cancers such as glioblastoma and non-small cell lung cancer (NSCLC) [13]. Other molecules related to the tyrosine kinase signaling pathway whose mutations lead to cancer are ERBB2 (HER2) and BRAF, which cause breast cancer, gastrointestinal cancer, melanoma, and colon cancer, respectively, through activation of the MAPK/ERK signaling cascade. Mutations that cause activation of CDK4/6 and CCND1 lead to uncontrolled proliferation by disabling the G1/S checkpoint. The PIK3CA mutation has been reported in 30% of breast cancers. Metabolic signals and cell survival are also associated with AKT1 gene amplification [14].

KRAS, a member of the RAS gene family, is frequently mutated, and these alterations lead to persistent activation of downstream signaling pathways. Such mutations are commonly observed in lung, colorectal, and pancreatic cancers [15]. MYC is a transcription factor that is important in cell cycle, metabolism, and apoptosis.

According to studies, its mutation and disruption have been reported in prostate, breast, and hematological cancers [16].

## 3.2 Tumor Suppressor Genes in Cancer

Tumor suppressor genes play an important role in preventing tumor formation and regulating the cell cycle properly. Deletion mutations, epigenetic silencing, and loss-of-function mutations disrupt the function of these genes and lead to cancer progression. One of the most important of these genes is TP53, which is referred to as the guardian of the genome. This protein is involved in key processes of cell cycle control, apoptosis, and DNA repair. Mutations in this gene have been reported in more than 50% of human cancers [17]. Another gene in this group is the retinoblastoma gene (RB1), which controls the G1/S cell cycle checkpoint. Its inactivation leads to various cancers, including retinoblastoma [18]. Mutations in NF2 and LATS1/2 disrupt the Hippo pathway, impairing tissue growth control and contact inhibition, a mechanism commonly disrupted in sarcoma and mesothelioma [19]. Deletions of BAX and BAK1 also lead to cell escape from programmed cell death.

## 3.3 DNA Repair Genes

Inactivation of these genes contributes to genome instability and the occurrence of mutations in other genes and directly affects the DNA repair mechanism, cell cycle control, cell death pathways, and response to various drugs. For example, BRCA1/BRCA2 are genes involved in DNA repair through homologous recombination, mutations in which are associated with breast and ovarian cancers and confer sensitivity to PARP inhibitors [20]. MLH1/MSH2 defects cause microsatellite instability in Lynch syndrome. ERCC2 mutations also alter platinum response in bladder cancer (Fig. 1).

## 3.4 Signaling Pathways in Cancer

There are several signaling pathways whose disruption plays a significant role in cancer progression and tumorigenesis. The MAPK/ERK pathway is activated by the binding of growth factor receptors such as EGFR to ligands, which results in cell survival and increased proliferation. It is disrupted in melanoma and colorectal cancer. Another pathway that plays a role in regulating cell survival, growth, and metabolism is the PI3K/AKT/mTOR pathway. Disruption of this pathway has been reported in breast, prostate, and glioblastoma cancers [21]. The Wnt/β-catenin pathway is one of the pathways that is important in embryonic development and tissue homeostasis. However, its disruption is associated with colorectal cancer and hepatocellular carcinoma [22]. One of the important pathways in cancer is the TGF-β pathway, which can have a dual function depending on the stage of the cancer. In

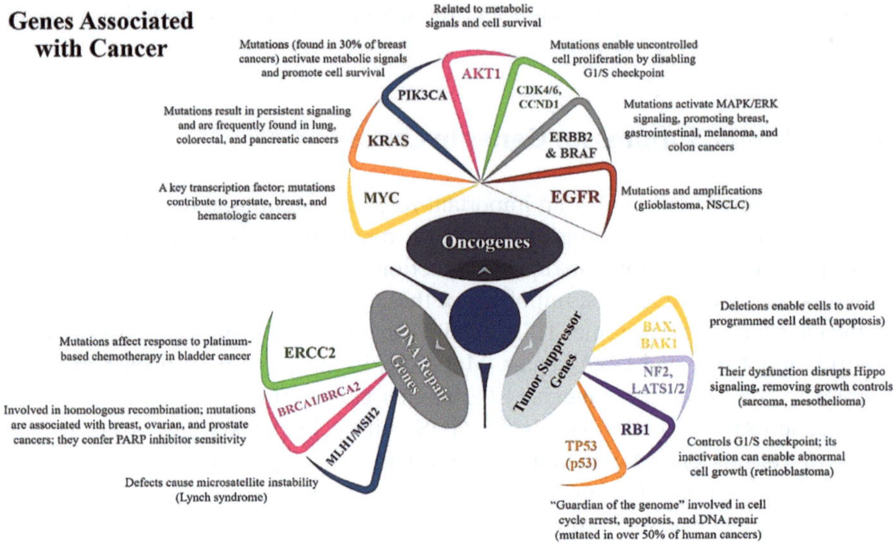


**Fig. 1** Key genes associated with cancer. (Created by Esmaeilzadeh et al.)

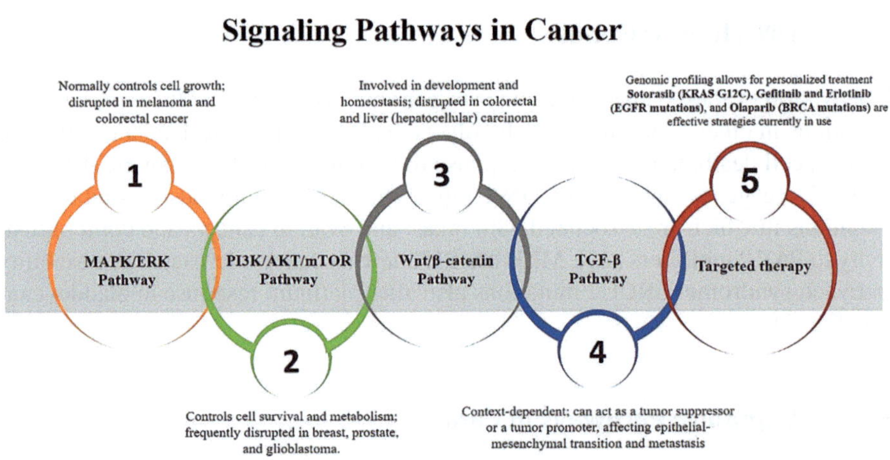


**Fig. 2** Signaling pathways in cancer. (Created by Esmaeilzadeh et al.)

fact, it can both inhibit tumor growth as a tumor suppressor and promote tumor progression as a tumor promoter. Epithelial-mesenchymal transition and metastasis are associated with disorders of this pathway [23]. Today, due to advances in cancer genomics and the use of genomic profiling, targeted therapy and personalized medicine for cancer patients have become possible. For example, the drug Sotorasib for lung cancer with KRAS G12C mutation [24], Gefitinib and Erlotinib for EGFR-mutated NSCLC [25], and Olaparib for BRCA-mutated breast/ovarian cancer are among the targeted therapies that have been developed so far [26] (Fig. 2).

## 4 Genomic Analysis Technologies in Cancer Research

In recent decades, significant advances in science and technology have provided tremendous power in enabling targeted and personalized cancer diagnosis and therapy. Among the most important genomic analysis technologies are next-generation sequencing (NGS) systems, which enable whole genome sequencing (WGS) or whole exome sequencing (WES) [27]. Additionally, targeted gene panel sequencing screens for mutations associated with various types of cancer. RNA sequencing helps identify gene expression profiles. Single-cell sequencing can also examine tumor heterogeneity [28]. Another category of genomic analysis technologies is high-throughput omics technologies that allow for the examination of DNA methylation arrays (Infinium EPIC), chromatin accessibility tests (ATAC-seq), proteomic profiling (mass spectrometry), and spatial transcriptomics (which retains spatial context within tissue samples) [29].

Liquid biopsy has emerged as a minimally invasive alternative to traditional tissue biopsy, using body fluids such as blood, urine, saliva, and serum to detect cancer biomarkers. Key analytes include circulating tumor DNA (ctDNA), extracellular vesicles (EVs), circulating tumor cells (CTCs), and extracellular RNA. EVs facilitate cell-to-cell communication by transferring proteins, lipids, RNA, and DNA, with cancer-derived EVs aiding tumor identification [30]. ctDNA, a subset of circulating cell-free DNA, carries tumor-specific genetic and epigenetic alterations useful for monitoring disease progression and predicting treatment response [40]. Although ctDNA has limitations for early cancer diagnosis, studies have shown its high potential in detecting mutations linked to various cancers, such as PIK3CA in breast cancer and KRAS/EGFR mutations in lung cancer [31]. Advanced computational tools, bioinformatics, and in silico, along with laboratory methods, have increased the efficiency, accuracy, sensitivity, and cost-effectiveness of research, diagnostic, and therapeutic methods in this field.

The most important applications of these tools include 3D genome reconstruction (Hi-C), variant calling with artificial intelligence (DeepVariant), drug response prediction modeling and signaling pathway analysis tools such as GSEA and OncoKB [32] (Table 1).

Functional genomic tools and platforms include CRISPR-Cas9 for screening and organoid models for testing personalized drugs [33]. Despite all these advances and capabilities, there are still challenges such as clinical interpretation of unknown variants, standardization of bioinformatics pipelines, and integration of bioinformatics databases [34]. However, the development and evolution of emerging genomic technologies such as digital spatial profiling and long-read sequencing methods (PacBio, Oxford Nanopore) will offer high-resolution molecular insights into cancer [35]. So expected to significantly advance personalized medicine in the near future (Fig. 3).

**Table 1** Some online resources for precision cancer medicine

| Name | Website | Supported by |
|---|---|---|
| Personalized cancer therapy | www.personalizedcancertherapy.org | The University of Texas MD Anderson |
| My cancer genome | www.mycancerGenome.org | Vanderbilt-Ingham Medical Center |
| OncoKb | OncoKb.org | Memorial Sloan Kettering Cancer |
| Drug gene interaction database | www.dgidb.genome.wustl.edu | Washington University in St. Louis |
| CIViC | www.civic.genome.wustl.edu | |
| JAX CKB | www.ckb.jax.org | Jackson Laboratory |
| PMKB | www.pmkb.weil.cornell.edu | Comell University |
| Precision cancer medicine | www.precisioncancermedicine.org | Dana Farber/Brigham and Woman's |

## 5 Key Alterations Across Cancer Types

Cancer is a highly heterogeneous disease characterized by genetic alterations that drive tumorigenesis. The classification of cancers mainly falls into three categories: solid tumors, hematological malignancies, and rare cancers. Each type exhibits distinct genomic alterations that influence treatment outcomes and patient prognosis. This section reviews key genetic changes across these cancer types and their implications for targeted therapies.

### 5.1 Genomic Alterations in Solid Tumors

Solid tumors such as lung, breast, colorectal, prostate, and gynecological cancers are largely driven by a complex interplay of somatic mutations, epigenetic alterations and chromosomal rearrangements. In non-small cell lung cancer (NSCLC), for instance, alterations in genes like KRAS, EGFR, BRAF (notably the V600E variant), and ALK are frequently observed [36]. EGFR mutations are particularly important, as they are often responsive to tyrosine kinase inhibitors (TKIs), which can significantly improve patient outcomes [37].

Breast cancer is often associated with germline mutations in BRCA1 and BRCA2, which increase hereditary cancer risk. Additionally, amplification of HER2 is commonly seen and serves as a common therapeutic target, especially for monoclonal antibody treatments such as trastuzumab [20]. In prostate cancer, mutations in BRCA1, BRCA2, TP53, and PMS2 are also frequently reported [17]. Colorectal cancer presents its distinct molecular profile, where microsatellite instability (MSI), often resulting from mutations in mismatch repair genes like MLH1 and MSH2, is a hallmark of Lynch syndrome (formerly known as hereditary nonpolyposis colorectal cancer, or HNPCC) [38].

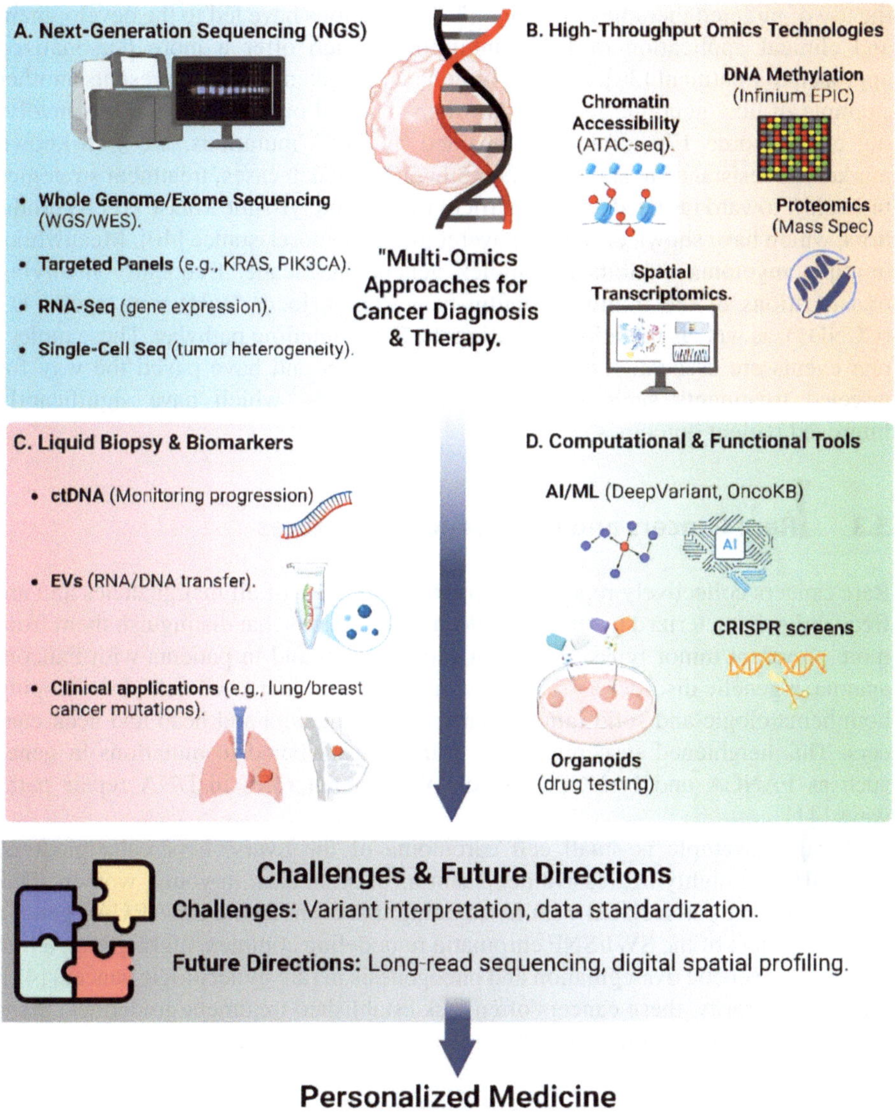


**Fig. 3** Genomic analysis technologies in cancer research. (Created by Esmaeilzadeh et al.)

## 5.2 Hematological Malignancies: Genetic Drivers

Hematologic malignancies, including leukemia, lymphoma, and multiple myeloma, are characterized by distinct genetic alterations that significantly impact both disease behavior and therapeutic strategies. In acute myeloid leukemia (AML), mutations in genes such as FLT3, NPM1, and DNMT3A are among the most frequently observed, with each playing a crucial role in determining prognosis and guiding the use of targeted therapies. Notably, FLT3 mutations have led to the development and clinical application of FLT3 inhibitors, which offer a more personalized approach to treatment [39]. Chronic lymphocytic leukemia (CLL) presents another example of how genomic features can shape clinical outcomes. Deletions involving chromosome 17p, often accompanied by TP53 mutations, are well-known markers of resistance to standard chemotherapy. In such cases, treatment strategies must shift toward targeted agents, particularly Bruton tyrosine kinase (BTK) inhibitors, which have shown efficacy in overcoming chemoresistance [40]. Meanwhile, multiple myeloma exhibits a complex genetic landscape, frequently involving translocations of the immunoglobulin heavy chain locus (IgH) with cyclin D1 (CCND1), as well as mutations affecting the RAS signaling pathway. These molecular events are associated with disease progression and have paved the way for targeted treatments such as proteasome inhibitors, which have significantly improved patient outcomes.

### 5.3 Rare Cancers and their Genomic Features

Rare cancers collectively represent approximately 20% of all malignancies and are frequently characterized by unique molecular alterations that distinguish them from more prevalent tumor types. One such example is found in patients with Fanconi anemia, a genetic disorder associated with a markedly increased risk of developing both hematologic and solid tumors, particularly leukemia and head and neck cancers. This heightened susceptibility is primarily attributed to mutations in genes such as FANCA and FANCD2, which play crucial roles in DNA repair pathways [41].

Another example is small cell carcinoma of the ovary, hypercalcemic type (SCCOHT), a highly aggressive neoplasm most often seen in young women. This tumor is strongly associated with inactivating mutations in the SMARCA4 gene, a key component of the SWI/SNF chromatin remodeling complex, highlighting a link between epigenetic dysregulation and oncogenesis in rare gynecologic cancers [42]. Due to their rarity, these cancers often lack established treatment guidelines, highlighting the need for further research.

## 6 Genomic Biomarker Discovery and Targeted Therapies

Biomarkers indicate physiological processes or treatment effects. In cancer research, factors such as genetic mutations, patterns of gene expression, levels of proteins, and various molecular indicators offer valuable information about tumor biology [5]. Alterations in epigenetic patterns, including DNA methylation and structural DNA changes, provide insights for biomarker discovery and personalized medicine [11]. Changes in epigenetic patterns frequently contribute to the onset, advancement, and resistance to treatments [43]. Biomarkers for lung cancer have been established for over a decade, resulting in a commercial NGS panel that identifies

15–21 genes for precision oncology, detecting structural variants on a single platform. NGS panels for various solid and liquid tumors enable precise detection of mutations across multiple cancer types. A comprehensive approach to gene testing in cancer enhances treatment effectiveness, reduces disease progression time, and ultimately improves quality of life, progression-free survival, and overall survival [44]. Biomarker-based strategies shift the focus from standard treatments to personalized plans designed for the specific molecular characteristics of each patient's tumor. The majority of genomic sequencing typically involves tissue samples, but there is a growing use of blood-derived circulating tumor DNA (ctDNA), circulating tumor cells, and exosomes [45].

## 7 The Role of Genomics in Immunotherapy

### 7.1 Genomic Influences on Immune Checkpoints

The dysregulation of immune-checkpoint proteins by tumors is a significant mechanism for immune resistance. In precision medicine, genomics and other technologies identify biomarkers for immunotherapy response. Key biomarkers predicting checkpoint inhibitor effectiveness include high tumor mutational burden, mismatch repair gene defects (MSI-H), PBRM1 gene alterations, and PDL1 amplification [45]. Despite the benefits of checkpoint inhibitors for various tumors, about 80% of patients do not experience positive outcomes.

### 7.2 Genomic Interactions with Immune Microenvironment

The interaction between genomic alterations in cancer cells and the tumor immune microenvironment is essential for developing precision cancer immunotherapies. Genomic changes drive tumor development and influence the immune landscape by affecting how the immune system recognizes and responds to tumors [5]. Tumor-specific neoantigens created by nonsynonymous mutations can make cancer cells more vulnerable to immune attacks, while loss-of-function mutations in genes responsible for antigen presentation can facilitate immune escape [1]. Furthermore, dysregulated pathways, such as WNT/β-catenin and PTEN/PI3K/AKT, can contribute to an immunosuppressive tumor microenvironment. Understanding these interactions is crucial for stratifying patients and designing therapies that combine targeted treatments with immune checkpoint inhibitors. Advanced technologies like single-cell sequencing are enabling researchers to explore these complex dynamics, paving the way for personalized immunotherapies [46].

### 7.3 Combining Genomics and Immunotherapy for Personalized Treatment

The success of immunotherapy varies among cancer types and patients due to heterogeneity in T cells and cancer cells, along with their complex interactions in the tumor microenvironment. Immunogenomics, leveraging NGS, helps profile the genetic makeup of cancer and immune cells. Single-cell RNA sequencing has revealed significant gene expression variability in glioblastoma, enhancing our understanding of tumor biology and potentially improving immunotherapy efficacy. Three leading immunotherapy approaches—checkpoint inhibitors, vaccine immunotherapy, and CAR T cell therapy—are advancing thanks to NGS and immunogenomics, promising improved cancer treatment outcomes [47]. NGS methods, such as exome sequencing and RNA-seq, can identify patients likely to respond to immunotherapy using checkpoint inhibitors, increasing efficacy and reducing side effects. Immunotherapy can involve developing vaccines targeting cancer neoantigens, training the immune system to fight cancer. NGS helps identify mutations in cancer genomes, allowing for the synthetic production of antigens for vaccination [47, 48]. A notable type of adoptive T-cell therapy is CAR-T-cell therapy, where T cells are genetically modified to express chimeric antigen receptors (CARs) targeting antigens on cancer cells. This modification enhances T-cell recognition and binding to tumor cells, resulting in effective treatment. However, challenges remain due to limited knowledge of cancer-specific antigens and T-cell heterogeneity. Recent advancements, such as single-cell RNA-seq, have identified various T-cell types in liver cancer, potentially aiding in the optimization of CAR T-cell therapies. Additionally, NGS and bioinformatics can help profile T cells and cancer cells for improved immunotherapy strategies [47].

## 8 Resistance Mechanisms and Genomic Adaptation

### 8.1 Genomic Basis of Therapy Resistance

Drug resistance is linked to various mechanisms, including increased drug efflux, alterations in drug metabolism, and mutations in drug targets. Additionally, epigenetic changes and the local tumor microenvironment play significant roles in chemoresistance [40]. For instance, mutations in the ROS1 kinase domain can cause lung adenocarcinoma to resist crizotinib, while ectodomain mutations in EGFR can lead to colorectal cancer's resistance to cetuximab. Similarly, alterations in genes like PIK3CA can induce resistance to EGFR inhibitors such as gefitinib and erlotinib. Resistance can also occur through pathway redundancy, where alternative receptors activate due to the primary inhibitor's action, or via pathway-independent mechanisms like epithelial–mesenchymal transition [40]. Moreover, the tumor microenvironment can shield cancer cells from treatments and induce innate resistance, exemplified by hepatocyte growth factor aiding RAF inhibitor resistance [46]. Resistance to immunotherapy is frequently linked to mutations in genes like beta-2-microglobulin (B2M), which disrupt antigen presentation, and JAK1/JAK2, which impair interferon signaling, reducing T-cell-mediated tumor destruction. Changes in the neoantigen landscape further affect immunotherapy effectiveness.

### 8.2 Strategies to Overcome Resistance Through Genomic Insights

Recent advancements in molecular biology have significantly enhanced our understanding of drug resistance mechanisms in cancer. Key developments include identifying specific ATP-binding cassette (ABC) transporters, such as P-glycoprotein (P-gp), multidrug resistance-associated protein (MRP), and breast cancer resistance protein (BCRP), which expel drugs from cells, contributing to multidrug resistance. Genomics and proteomics have also revealed genetic mutations in target genes, like tyrosine kinases and topoisomerases, that alter drug interactions and signaling pathways. Single-cell sequencing has unveiled tumor heterogeneity, identifying resistant cell populations [49].

## 9 Integrative Omics Approaches in Cancer Treatment

Advanced genome screening and mass spectrometry techniques now enable detailed profiling of DNA methylation, gene expression, proteins, and metabolites. While single-omics studies provide valuable insights, the complexity of cancer–host interactions requires integrative multi-omics approaches, including genomics, transcriptomics, proteomics, metabolomics, epigenomics, and radiomics, to fully understand cancer progression, develop novel therapies, and identify biomarkers [50]. Identifying biological characteristics and patient subgroups contributing to the pathophysiology of cancer is another goal of multi-omics techniques. Additionally, they help develop accurate prediction models that verify new medications and treatments. There is a concentrated effort to create computational and mathematical techniques that can unravel the complexity of cancer heterogeneity because genomic and epigenetic instability in tumors could affect intracellular responses and the tumorigenic process. Large omics datasets could be integrated to model tissue-level cellular behavior, which helps understand the pathophysiology of cancer and how tumor cells react to angiogenesis and drugs [51]. Novel deep learning models integrating multi-omics data have shown superior performance in predicting cancer drug responses compared to traditional methods. Deep learning approaches like moBRCA-net, using multi-omics datasets, have improved breast cancer subtype classification compared to single-omics data [52]. Collectively, multi-omics is a powerful but technically demanding approach. Key challenges include data heterogeneity, integration complexity, limited interpretability, and translational barriers. Overcoming these requires advanced computational methods, standardized practices, and robust validation to ensure meaningful biological and clinical insights.

## 10 Challenges and Ethical Considerations

Interpreting genomic data in cancer is complex, with sequencing results often challenging the somatic mutation theory. Integrating gene regulatory networks and tissue-level processes may better capture tumorigenesis's multifaceted nature [53]. Data privacy remains critical, as genomic information revealing cancer risks and hereditary conditions must be securely protected to prevent discrimination. Equitable access to genomic tools is limited, often excluding underserved populations, while incidental findings complicate informed consent, especially in pediatric cases [54]. Addressing these challenges requires standardized, user-friendly bioinformatics tools, non-invasive monitoring methods like circulating tumor DNA analysis, and policies prioritizing equitable access. Enhanced patient education and dynamic consent frameworks are essential to responsibly harness genomic data in cancer research.

## 11 Conclusion

The landscape of cancer research and treatment has been revolutionized by advances in genomics, providing a deeper understanding of the molecular underpinnings of tumor development, progression, and therapeutic response. This chapter has explored the scope of genomic alterations, from point mutations and copy number variations to epigenetic modifications and their role in shaping tumor behavior and heterogeneity. Key cancer-associated genes and pathways, including oncogenes, tumor suppressors, and DNA repair mechanisms, highlight critical targets for precision medicine. Cutting-edge technologies such as next-generation sequencing and single-cell genomics have enabled the characterization of complex cancer genomes and resistance mechanisms. The integration of genomics with immunotherapy, multi-omics approaches, and artificial intelligence holds immense promise for personalized oncology. However, challenges related to data interpretation, ethical considerations, and equitable access remain significant. Moving forward, genomics will continue to be central to early detection, prevention, and the development of individualized therapies, offering new hope for improved patient outcomes in cancer care.

**Declarations**

**Funding** The authors have no relevant financial or non-financial interests to disclose.

**Conflicts of Interest** There is no competing and conflict of interests.

## References


1. Hanahan D (2022) Hallmarks of cancer: new dimensions. Cancer Discov 12(1):31–46
2. Kamps R, Brandão RD, van den Bosch BJ, Paulussen ADC, Xanthoulea S, Blok MJ et al (2017) Next-generation sequencing in oncology: genetic diagnosis, risk prediction and cancer classification. Int J Mol Sci 18(2):308
3. Lynch TJ, Bell DW, Sordella R, Gurubhagavatula S, Okimoto RA, Brannigan BW et al (2004) Activating mutations in the epidermal growth factor receptor underlying responsiveness of non–small-cell lung cancer to gefitinib. N Engl J Med 350(21):2129–2139
4. Mardis ER (2013) Next-generation sequencing platforms. Annu Rev Anal Chem 6(1):287–303
5. Sameek R, Chinnaiyan AM (2014) Translating genomics for precision cancer medicine. Annu Rev Genomics Hum Genet 15(1):395–415
6. Tannock IF, Hickman JA (2016) Limits to personalized cancer medicine. N Engl J Med 375:1289–1294
7. Fujimoto N (2024) Subtype of EGFR exon 19 deletion mutations. Transl Lung Cancer Res 13(1):195
8. Beroukhim R, Mermel CH, Porter D, Wei G, Raychaudhuri S, Donovan J et al (2010) The landscape of somatic copy-number alteration across human cancers. Nature 463(7283):899–905
9. Mertens F, Johansson B, Fioretos T, Mitelman F (2015) The emerging complexity of gene fusions in cancer. Nat Rev Cancer 15(6):371–381
10. Ben-David U, Amon A (2020) Context is everything: aneuploidy in cancer. Nat Rev Genet 21(1):44–62
11. Esteller M (2008) Epigenetics in cancer. N Engl J Med 358(11):1148–1159
12. Moody CA, Laimins LA (2010) Human papillomavirus oncoproteins: pathways to transformation. Nat Rev Cancer 10(8):550–560
13. Zahonero C, Sánchez-Gómez P (2014) EGFR-dependent mechanisms in glioblastoma: towards a better therapeutic strategy. Cell Mol Life Sci 71:3465–3488
14. Miller TW, Balko JM, Arteaga CL (2011) Phosphatidylinositol 3-kinase and antiestrogen resistance in breast cancer. J Clin Oncol 29(33):4452–4461
15. Huang L, Guo Z, Wang F, Fu L (2021) KRAS mutation: from undruggable to druggable in cancer. Signal Transduct Target Ther 6(1):386
16. Dang CV (2012) MYC on the path to cancer. Cell 149(1):22–35
17. Kastenhuber ER, Lowe SW (2017) Putting p53 in context. Cell 170(6):1062–1078
18. Gordon GM, Du W (2011) Conserved RB functions in development and tumor suppression. Protein Cell [Internet] 2(11):864–878. Available from:. https://doi.org/10.1007/s13238-011-1117-z
19. Lv L, Zhou X (2023) Targeting hippo signaling in cancer: novel perspectives and therapeutic potential. MedComm 4(5):e375
20. Lord CJ, Ashworth A (2016) BRCAness revisited. Nat Rev Cancer 16(2):110–120
21. Castel P, Toska E, Engelman JA, Scaltriti M (2021) The present and future of PI3K inhibitors for cancer therapy. Nat Cancer 2(6):587–597
22. Qin K, Yu M, Fan J, Wang H, Zhao P, Zhao G et al (2024) Canonical and noncanonical Wnt signaling: multilayered mediators, signaling mechanisms and major signaling crosstalk. Genes Dis 11(1):103–134
23. Tzavlaki K, Moustakas A (2020) TGF-β signaling. Biomolecules 10(3):487
24. Canon J, Rex K, Saiki AY, Mohr C, Cooke K, Bagal D et al (2019) The clinical KRAS (G12C) inhibitor AMG 510 drives anti-tumour immunity. Nature 575(7781):217–223
25. Rosell R, Karachaliou N (2013) Maintenance therapy and precision medicine in NSCLC. Nat Rev Clin Oncol 10(10):549–550
26. Robson M, Im S-A, Senkus E, Xu B, Domchek SM, Masuda N et al (2017) Olaparib for metastatic breast cancer in patients with a germline BRCA mutation. N Engl J Med 377(6):523–533
27. Warburton PE, Sebra RP (2023) Long-read DNA sequencing: recent advances and remaining challenges. Annu Rev Genomics Hum Genet 24(1):109–132

28. Hwang B, Lee JH, Bang D (2018) Single-cell RNA sequencing technologies and bioinformatics pipelines. Exp Mol Med 50(8):1–14
29. Liu B, Li Y, Zhang L (2022) Analysis and visualization of spatial transcriptomic data. Front Genet 12:785290
30. Meng Y, Bian L, Zhang M, Bo F, Lu X, Li D (2020) Liquid biopsy and their application progress in head and neck cancer: focus on biomarkers CTCs, cfDNA, ctDNA and EVs. Biomark Med 14(14):1393–1404
31. Forshew T, Murtaza M, Parkinson C, Gale D, Tsui DWY, Kaper F et al (2012) Noninvasive identification and monitoring of cancer mutations by targeted deep sequencing of plasma DNA. Sci Transl Med 4(136):136ra68–136ra68
32. Huang J, Mao L, Lei Q, Guo A-Y (2024) Bioinformatics tools and resources for cancer and application. Chin Med J 137(17):2052–2064
33. Ford K, McDonald D, Mali P (2019) Functional genomics via CRISPR–Cas. J Mol Biol 431(1):48–65
34. Koboldt DC (2020) Best practices for variant calling in clinical sequencing. Genome Med 12(1):91
35. Logsdon GA, Vollger MR, Eichler EE (2020) Long-read human genome sequencing and its applications. Nat Rev Genet 21(10):597–614
36. Terbuch A, Konjic S, Schlintl V, Absenger G, Jost PJ, Lindenmann J et al (2024) Prognostic impact of targetable driver alterations in resected early-stage lung cancer. Transl Lung Cancer Res. 13(11):3096
37. Li J, Chen P, Wu Q, Guo L, Leong KW, Chan KI et al (2022) A novel combination treatment of antiADAM17 antibody and erlotinib to overcome acquired drug resistance in non-small cell lung cancer through the FOXO3a/FOXM1 axis. Cell Mol Life Sci 79(12):614
38. Wang X, Ni H, Zhu L, Huang H, Deng A, Hu J et al (2025) Analyzing pathogenic variants in mismatch repair genes: personalized prevention strategies for lynch syndrome in Chinese families. Front Med 12:1527249
39. Othman J, Potter N, Ivey A, Tazi Y, Papaemmanuil E, Jovanovic J et al (2024) Molecular, clinical, and therapeutic determinants of outcome in NPM1-mutated AML. Blood 144(7):714–728
40. Vasan N, Baselga J, Hyman DM (2019) A view on drug resistance in cancer. Nature 575(7782):299–309
41. Peake JD, Noguchi E (2022) Fanconi anemia: current insights regarding epidemiology, cancer, and DNA repair. Hum Genet 141(12):1811–1836
42. Gao Y, Zheng K, Tan H, Kang M, Lu B, Chen L et al (2025) Single-cell RNA sequencing reveals the intra-tumoral heterogeneity and immune microenvironment of small cell carcinoma of the ovary, hypercalcemic type. J Ovarian Res 18(1):76
43. Ahmad A, Imran M, Ahsan H (2023) Biomarkers as biomedical bioindicators: approaches and techniques for the detection, analysis, and validation of novel biomarkers of diseases. Pharmaceutics 15(6):1630
44. Satam H, Joshi K, Mangrolia U, Waghoo S, Zaidi G, Rawool S et al (2023) Next-generation sequencing technology: current trends and advancements. Biology (Basel) 12(7):997
45. Tsimberidou AM, Fountzilas E, Nikanjam M, Kurzrock R (2020) Review of precision cancer medicine: evolution of the treatment paradigm. Cancer Treat Rev 86:102019
46. Vitorakis N, Gargalionis AN, Papavassiliou KA, Adamopoulos C, Papavassiliou AG (2024) Precision targeting strategies in pancreatic cancer: the role of tumor microenvironment. Cancers (Basel) 16(16):2876
47. Mukherjee S (2019) Genomics-guided immunotherapy for precision medicine in cancer. Cancer Biother Radiopharm 34(8):487–497
48. Richard G, Princiotta MF, Bridon D, Martin WD, Steinberg GD, De Groot AS (2022) Neoantigen-based personalized cancer vaccines: the emergence of precision cancer immunotherapy. Expert Rev Vaccines 21(2):173–184
49. Garg P, Malhotra J, Kulkarni P, Horne D, Salgia R, Singhal SS (2024) Emerging therapeutic strategies to overcome drug resistance in cancer cells. Cancers (Basel) 16(13):2478

50. Zheng J, Lu W, Wang C, Chen S, Zhang Q, Su C (2024) Unfolding the mysteries of heterogeneity from a high-resolution perspective: integration analysis of single-cell multi-omics and spatial omics revealed functionally heterogeneous cancer cells in ccRCC. Aging (Albany NY) 16(13):10943
51. Heo YJ, Hwa C, Lee G-H, Park J-M, An J-Y (2021) Integrative multi-omics approaches in cancer research: from biological networks to clinical subtypes. Mol Cells 44(7):433–443
52. Choi JM, Chae H (2023) moBRCA-net: a breast cancer subtype classification framework based on multi-omics attention neural networks. BMC Bioinformatics 24(1):169
53. Huang S, Soto AM, Sonnenschein C (2025) The end of the genetic paradigm of cancer. PLoS Biol 23(3):e3003052
54. Droin-Mollard M, de Montgolfier S, Gimenez-Roqueplo A-P, Flahault C, Petit A, Bourdeaut F et al (2024) Psychological and ethical issues raised by genomic in paediatric care pathway, a qualitative analysis with parents and childhood cancer patients. Eur J Hum Genet 32(11):1446–1455

# Cancer Epigenomics in Precision Medicine

A. Esmaeilzadeh, M. Hassanzadeh Makoui, F. Ebrahimi, S. Taremi, P. M. Yeganeh, F. Afshari, K. Hadiloo, and K. Esmaeilzadeh

**Abstract**

Epigenomics investigates heritable modifications to DNA and associated proteins that regulate gene expression without altering the underlying nucleotide sequence. In cancer, aberrant epigenetic landscapes—primarily involving DNA methylation, histone modifications, and non-coding RNAs—disrupt normal transcriptional programs, thereby promoting tumorigenesis, progression, invasion, metastasis, and therapeutic resistance. DNA methylation, predominantly at CpG islands in promoter regions, frequently leads to hypermethylation and subsequent silencing of tumor suppressor genes, while global hypomethylation can activate oncogenes and contribute to genomic instability. Histone modifications, including acetylation, methylation, phosphorylation, and ubiquitination, dynamically alter chromatin structure and accessibility; for instance, aberrant hyperacetylation or altered methylation patterns (such as reduced H4K16 acetylation or H4K20 trimethylation) remodel chromatin to favor oncogenic gene expression or repress tumor-suppressive pathways. Non-coding RNAs, encompassing microRNAs, long non-coding RNAs, and others, further modulate epigenetic machinery by recruiting chromatin-modifying complexes, influencing DNA

A. Esmaeilzadeh (✉)
Pficell R&D Canadian Institution & Corporation, Profound Future Focused Innovative Cell and Gene Therapy, Toronto, Canada

Cancer Gene Therapy Research Center, Zanjan University of Medical Sciences, Zanjan, Iran

Cancer Immunology Project (CIP), Universal Scientific Education and Research Network (USERN), Tehran, Iran

M. Hassanzadeh Makoui
Student Research Committee, Zanjan University of Medical Sciences, Zanjan, Iran

F. Ebrahimi
Zanjan Pharmaceutical Biotechnology Research Center, Zanjan University of Medical Sciences, Zanjan, Iran



M. K. Ghosh, M. Basu (eds.), *Personalized Cancer Therapy*,
https://doi.org/10.1007/978-981-95-6845-1_4

methylation, histone marks, and chromatin remodeling, thus regulating both oncogenic and tumor-suppressive networks. A hallmark of epigenetic alterations is their reversibility, distinguishing them from irreversible genetic mutations. This property positions epigenetic changes as promising targets for therapeutic intervention and as potential biomarkers for early detection, prognosis, disease monitoring, and treatment response prediction. Epigenetic therapies, such as DNA methyltransferase inhibitors (e.g., 5-azacytidine) and histone deacetylase inhibitors, have gained regulatory approval for certain hematologic malignancies and are under active investigation in solid tumors, including combination regimens to enhance efficacy and overcome resistance. This chapter provides a comprehensive overview of epigenetic mechanisms in cancer, emphasizing their dysregulation, diagnostic and prognostic utility, and emerging therapeutic strategies to exploit reversible epigenetic reprogramming for improved clinical outcomes in oncology.

**Keyword**

Cancer · Epigenomics · Biomarkers · Precision medicine

## 1 Introduction

Epigenetic changes are widespread in cancer and influence both disease progression and clinical management. Noteworthy examples include hypermethylation of the BRCA1 promoter in breast and ovarian tumors, methylation of RASSF1A in lung cancer, and the CpG island methylator phenotype (CIMP), which involves silencing of the MLH1 gene and microsatellite instability in colorectal cancer [1–3]. By integrating epigenomic data with genomic information, we can achieve earlier detection, improved prognosis, and more precise, personalized treatment options.

S. Taremi · P. M. Yeganeh
School of Medicine, Zanjan University of Medical Sciences, Zanjan, Iran

F. Afshari
Tissue Engineering and Applied Cell Sciences Department, School of Advanced Medical Sciences and Technologies, Shiraz University of Medical Sciences, Shiraz, Iran

K. Hadiloo
Department of Surgery, Velayat Clinical Research Development Unit, Qazvin University of Medical Sciences, Qazvin, Iran

K. Esmaeilzadeh
Department of Medical Nanotechnology, School of Advanced Medical Sciences, Tabriz University of Medical Sciences, Tabriz, Iran

### 1.1 The Epigenetic Landscape of Cancer

Altered gene expression is a defining characteristic of cancer, resulting from both genetic mutations and epigenetic changes that modify transcription without altering the DNA sequence. The main mechanisms of epigenetic regulation include DNA methylation and histone modifications [4]. In cancer, there is often an increase in promoter hypermethylation of tumor suppressor genes, along with global hypomethylation. These patterns contribute to genomic instability and the activation of oncogenes, and they can also serve as biomarkers for diagnosis and monitoring [5]. Histone post-translational modifications—such as acetylation, methylation, phosphorylation, ubiquitination, and sumoylation—play a critical role in shaping chromatin accessibility. Aberrant states, such as the loss of H3K27me3, are associated with poor prognosis in certain cancers [6, 7].

Recent advancements in epigenomic profiling techniques, including whole-genome bisulfite sequencing and ChIP-seq, allow researchers to map methylation and histone landscapes across the genome. This work sheds light on regulatory networks and facilitates the discovery and classification of epigenetic biomarkers [8, 9]. These insights are crucial for precision oncology, as they help guide targeted treatments. Epigenetic therapies, such as DNA methyltransferase inhibitors and histone deacetylase inhibitors, aim to reactivate silenced tumor suppressor genes. When used in combination with other modalities, like immunotherapy, these treatments can enhance anti-tumor effectiveness [1, 2].

Non-coding RNAs add another regulatory layer. MicroRNAs regulate post-transcriptional gene expression; for instance, miR-21 is frequently overexpressed in glioblastoma and targets tumor suppressor genes [3]. Long non-coding RNAs (lncRNAs) can recruit chromatin-modifying complexes to specific locations, altering DNA methylation and histone states. Given their involvement in tumor development, lncRNAs are emerging as potential therapeutic targets [10].

In summary, cancer epigenetics reconfigures gene regulation through DNA methylation, histone modification, and non-coding RNA networks. A systematic investigation of these mechanisms is essential for identifying therapeutic targets and advancing personalized diagnosis and treatment.

## 2 Epigenetic Mechanisms in Tumorigenesis

Epigenetic modifications play a crucial role in the development of tumors, significantly affecting cancer initiation, progression, and response to treatment. Unlike genetic mutations, which involve changes in the DNA sequence, epigenetic changes alter gene expression through mechanisms such as DNA methylation, histone modifications, and the regulation of non-coding RNAs [10]. These changes can lead to the silencing of tumor suppressor genes, the activation of oncogenes, and an increase in tumor heterogeneity, which complicates treatment strategies. Understanding these mechanisms provides promising opportunities for developing targeted therapies in cancer care.

## 2.1 Epigenetic Silencing of Tumor Suppressor Genes

The silencing of tumor suppressor genes is a key characteristic of cancer [11]. Methylation of CpG islands in gene promoters leads to the recruitment of repressive histone marks and chromatin remodelers, which disrupt transcription. A well-known example of this process is the hypermethylation of the p16INK4a gene observed in various cancers [12]. Additionally, repressive histone modifications, particularly the H3K27me3 mark mediated by the PRC2 complex, further reinforce the silencing of tumor suppressor genes. This silencing contributes to unchecked cell proliferation and genomic instability [13].

## 2.2 Activation of Oncogenes Through Epigenetic Changes

Epigenetic alterations can activate oncogenes while also silencing tumor suppressor genes. Promoter hypomethylation leads to increased oncogene expression, which drives cellular proliferation and metastasis [14]. A notable example is the MYC gene, which is frequently upregulated due to reduced DNA methylation at its regulatory elements [15]. Additionally, histone acetylation by histone acetyltransferases (HATs) enhances chromatin accessibility and gene transcription. In acute myeloid leukemia (AML), acetylation at the HOXA9 locus promotes its overexpression, contributing to leukemogenesis [16]. Furthermore, non-coding RNAs play a role in regulating oncogenic programs; for instance, the long non-coding RNA HOTAIR recruits the Polycomb Repressive Complex 2 (PRC2) to reorganize chromatin and activate pathways that promote oncogenesis in breast and colorectal cancers [17].

## 2.3 Epigenetic Plasticity and Tumor Heterogeneity

Epigenetic plasticity is a key factor driving tumor heterogeneity, which enables tumors to adapt to stress, evade the immune system, and develop resistance to therapy [18]. Dynamic cellular states facilitate phenotypic switching and create diversity within tumors, particularly during the process of epithelial–mesenchymal transition (EMT). This reprogramming can endow cancer cells with stem-like characteristics and enhance their metastatic potential [19]. For instance, in melanoma, resistance to BRAF inhibitors is associated with increased levels of H3K27me3 at critical regulators that promote adaptive survival [20]. These findings highlight the potential of developing combination therapies that target epigenetic mechanisms to overcome drug resistance. Moreover, epigenetic alterations play a significant role in oncogenesis by silencing tumor suppressor genes, activating oncogenes, and maintaining cellular plasticity. Therapeutic strategies aimed at reversing these changes—such as using DNA methyltransferase inhibitors and histone deacetylase inhibitors—offer promising avenues for treatment and could lead to more precise cancer therapies [21].

# 3 Epigenetic Alterations in Different Cancer Types

## 3.1 Solid Tumors: Examples from Breast, Lung, and Colorectal Cancers

Epigenetic alterations, such as DNA methylation, histone modifications, non-coding RNAs, and chromatin remodeling, play a crucial role in the development of solid tumors. In breast cancer, abnormal methylation patterns and dysregulation of microRNAs contribute to drug resistance and the emergence of stem-like characteristics [22]. Similarly, lung cancer incorporates changes in DNA methylation and non-coding RNAs alongside oncogenic mutations to promote tumor development and resistance to chemotherapy [23]. In colorectal cancer, global hypomethylation, hypermethylation of promoters, and shifts in microRNA profiles interact with genetic alterations throughout the progression of the disease [24].

Since epigenetic states are reversible, they present valuable opportunities for targeted therapies. Although small-molecule epigenetic inhibitors have been approved for hematologic malignancies, their effectiveness in treating solid tumors remains limited, and the mechanisms behind responses and resistance are not fully understood [22, 25]. Emerging strategies include exploring the potential epigenetic effects of phytochemicals in lung cancer and utilizing miR-21 and XIST as diagnostic and prognostic biomarkers in colorectal cancer [24]. Key challenges include understanding the interplay between genetic and epigenetic factors to improve patient selection and develop rational treatment combinations. Important priorities involve identifying viable epigenetic targets, optimizing dosing and sequencing with standard therapies, and examining non-cell-autonomous interactions that propagate epigenetic dysregulation within tumor ecosystems [26]. Advancing these objectives is essential for accelerating the development of precision epigenetic therapies for breast, lung, and colorectal cancers.

## 3.2 Hematological Malignancies: Examples from Leukemia and Lymphoma

Epigenetic deregulation, which includes changes in DNA methylation, histone modification, and non-coding RNA control, plays a crucial role in the development of leukemia and lymphoma [27–29]. In acute lymphoblastic leukemia (ALL) and acute myeloid leukemia (AML), the hypermethylation of promoters leads to the silencing of tumor suppressor genes. Additionally, mutations in epigenetic regulators like DNMT3A and TET2 reprogram gene transcription, contributing to disease progression [30, 31]. Altered activity of histone-modifying enzymes and dysregulated non-coding RNAs—such as microRNAs and circular RNAs (e.g., circAF4 and circPVT1)—further change chromatin states and promote cancer advancement [32, 33]. In lymphoma, genetic alterations that affect epigenetic control—such as rearrangements of the BCL6 and IRF4 genes in B-cell lymphomas, as well as SOX11 overexpression in mantle cell lymphoma—add to the molecular diversity and pathogenesis of the disease [34, 35].

From a therapeutic standpoint, the reversible nature of epigenetic modifications supports the use of DNA methyltransferase inhibitors and histone deacetylase inhibitors. These treatments can restore gene expression patterns and induce cell death in leukemic and lymphomatous cells [29]. Despite the clinical potential, optimizing the efficacy and specificity of these treatments remains a critical goal. Integrating cytogenetic and multi-omics profiling is also essential for accurate patient stratification [35]. Key challenges include unraveling the complex genetic and epigenetic interactions that sustain these cancers and translating mechanistic insights into lasting treatment responses [27, 36]. Future research is focused on identifying viable epigenetic targets and developing tailored combination therapies that consider individual genomic and epigenomic profiles to enhance outcomes in leukemia and lymphoma [37, 38].

## 3.3 Rare Cancers and Their Unique Epigenetic Profiles

Epigenetic alterations—such as DNA methylation, histone modification, and chromatin remodeling—play a significant role in the initiation and progression of cancer, with impacts comparable to those of genetic mutations. These changes can silence tumor suppressor genes, activate oncogenes, promote chromosomal instability, and influence therapeutic responses. Key mechanisms involved include promoter hypermethylation and global or locus-specific hypomethylation. Additionally, the tumor microenvironment, which encompasses factors like hypoxia and interactions with the immune system, further influences these epigenetic programs. Insights gained from studying rare cancers, such as parathyroid cancer, provide valuable opportunities for biomarker discovery and targeted therapies [39–42]. Distinct epigenetic signatures can help differentiate cancer subtypes and predict outcomes. For instance, certain profiles are associated with poor prognosis in breast cancers, aiding in risk stratification and treatment selection. The reversible nature of epigenetic modifications supports the use of DNMT inhibitors and HDAC inhibitors, which can reactivate tumor suppressor networks and improve the effectiveness of combination treatments. However, intra-tumor epigenetic heterogeneity and the broad activity of currently available agents limit precision in treatment, highlighting the need for more selective options [43–45]. Future priorities should include identifying manageable epigenetic targets, developing highly specific inhibitors, and integrating genetic and epigenetic data to enhance personalized care. Additionally, the regulatory roles of non-coding RNAs, such as microRNAs and long non-coding RNAs, warrant further investigation as potential biomarkers and therapeutic targets, especially in less-studied rare cancers, to improve prevention, diagnosis, and outcomes [46, 47].

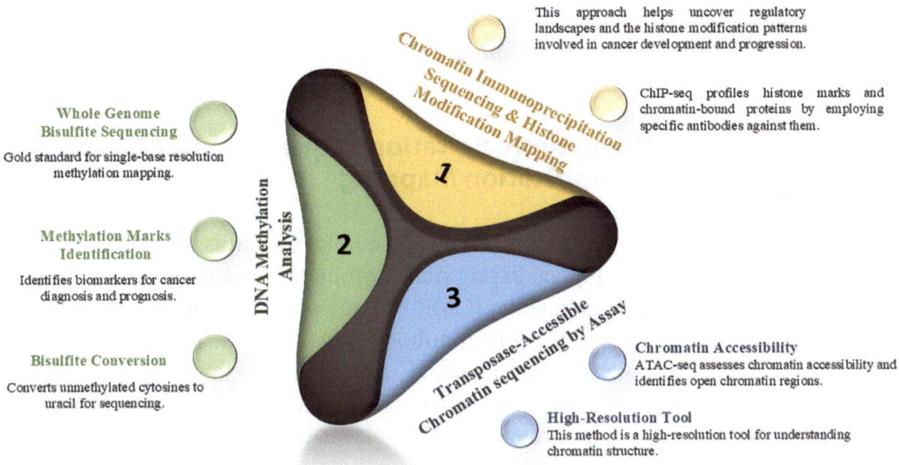


**Fig. 1** Key methodologies for epigenomic profiling in cancer. The diagram outlines three core approaches: (1) chromatin immunoprecipitation sequencing (ChIP-seq) and histone modification mapping—this method utilizes antibodies to map histone marks and chromatin-bound proteins, helping to elucidate the regulatory landscapes involved in tumor development; (2) DNA methylation analysis—this approach includes whole-genome bisulfite sequencing for single-base resolution, facilitating methylation biomarker discovery and enabling bisulfite conversion for the detection of cytosine methylation; (3) transposase-accessible chromatin sequencing (ATAC-seq)—this technique profiles open chromatin regions at high resolution, revealing chromatin accessibility patterns associated with oncogenesis

## 4 Technologies for Epigenomic Profiling

Recent advancements in epigenomic profiling technologies have significantly improved our understanding of gene regulation and chromatin dynamics, especially in the field of cancer biology. These techniques provide valuable insights into the epigenetic changes that drive tumor formation, which can aid in the discovery of biomarkers, the identification of therapeutic targets, and the development of precision medicine strategies. Figure 1 summarizes three principal experimental platforms used for epigenomic profiling in cancer research, providing an integrated view of how chromatin states and DNA methylation landscapes are interrogated at high resolution.

### 4.1 DNA Methylation Analysis

DNA methylation involves adding methyl groups to DNA and plays a crucial role in regulating gene expression. Whole-genome bisulfite sequencing (WGBS) is a technique that maps methylation at a single-base resolution by converting unmethylated cytosines into uracils, while preserving methylated cytosines. This gold-standard method has uncovered cancer-specific methylation patterns, facilitating the discovery of biomarkers for diagnosis and prognosis [8].

### 4.2 Chromatin Immunoprecipitation Sequencing (ChIP-seq) and Histone Modification Mapping

Histone modifications play a crucial role as epigenetic regulators, affecting chromatin structure and influencing gene expression. Chromatin immunoprecipitation followed by high-throughput sequencing (ChIP-seq) allows for genome-wide mapping of these modifications by using specific antibodies that target histone marks or transcription factors. This method has revealed the regulatory landscapes of cancer cells, uncovering histone modification patterns associated with oncogenic transcriptional programs and tumor progression [11].

### 4.3 Transposase-Accessible Chromatin Sequencing by Assay (ATAC-seq)

Assay for transposase-accessible chromatin using sequencing (ATAC-seq) is a highly effective and sensitive technique for profiling chromatin accessibility across the genome. This method allows researchers to identify open chromatin regions that are associated with active regulatory elements. ATAC-seq utilizes a hyperactive Tn5 transposase to insert sequencing adapters directly into accessible DNA, generating high-resolution maps of promoters, enhancers, and other cis-regulatory elements. Due to its minimal cell input requirements, streamlined workflow, and ability to capture dynamic chromatin states, ATAC-seq has become a preferred method in epigenomic research. In the field of oncology, it has played a crucial role in revealing tumor-specific regulatory landscapes, clarifying mechanisms of transcriptional reprogramming, and identifying biomarkers relevant to cancer progression and immune evasion [12].

### 4.4 Single-Cell Epigenomics

Single-cell epigenomics uncovers tumor heterogeneity that bulk assays often overlook. By using single-cell DNA methylation and scATAC-seq techniques, researchers can map cell-specific methylation patterns and accessibility throughout tumor evolution. This approach reveals the biological programs that drive tumor progression and drug resistance, providing valuable biomarkers and therapeutic targets for precision oncology [13]. However, the translation of these findings into clinical practice is hindered by technical complexity, the need for specialized equipment, and substantial computational requirements. Given the dynamic nature of epigenetics and the complexities of tumor microenvironments, longitudinal and

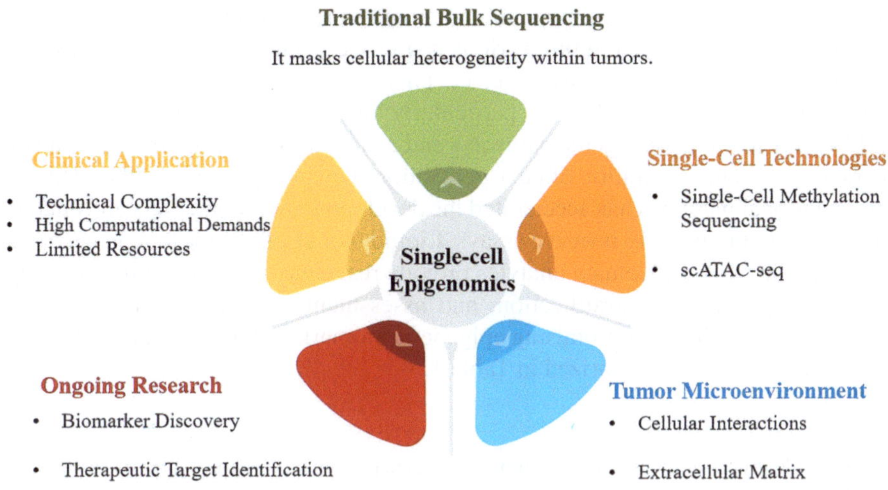


**Fig. 2** The conceptual framework of single-cell epigenomics in contrast to traditional bulk sequencing. This diagram highlights the limitations of bulk sequencing, which often obscures the heterogeneity within tumors, compared to single-cell epigenomic technologies that accurately depict cell-specific epigenetic landscapes. Key methodologies are featured, including single-cell DNA methylation sequencing and scATAC-seq, along with their applications in biomarker discovery, therapeutic target identification, and analysis of the tumor microenvironment. The figure also addresses the current challenges in clinical translation, such as technical complexity, high computational demands, and resource limitations

integrative studies are essential for achieving meaningful clinical outcomes. Figure 2 illustrates a comparison between bulk and single-cell epigenomic approaches, summarizing their applications, challenges, and relevance to tumor biology.

## 5 Epigenomic Biomarkers in Precision Medicine

### 5.1 Identification of Epigenetic Biomarkers for Cancer Detection and Prognosis

Epigenetic alterations, including DNA methylation, histone modifications, and noncoding RNAs, regulate gene transcription without altering the DNA sequence. These changes are closely linked to the initiation and progression of tumors. Their specific patterns in different cancer types make them valuable as biomarkers for early detection, prognosis, and treatment monitoring [48]. Advances in high-throughput sequencing and bioinformatics have facilitated systematic discovery and validation of these biomarkers across various malignancies, enhancing clinical applications. Promoter hypermethylation of tumor suppressor genes serves as an early indicator in several cancers. For example, the hypermethylation of BRCA1 in breast cancer and CDKN2A in lung cancer is noteworthy. Other loci, such as p16INK4a, APC, and MGMT, also provide diagnostic value across different tumor types [49]. In contrast, global hypo methylation can lead to genomic instability, and specific histone marks, including various lysine acetylation and methylation states, are associated with aggressive cancer phenotypes and poor outcomes [50]. Additionally, non-coding RNAs, particularly microRNAs and long non-coding RNAs, play a role in oncogenic pathways and are emerging as important prognostic markers [51]. The integration of next-generation sequencing, single-cell omics, and computational pipelines has accelerated the development of biomarkers and enabled the creation of minimally invasive assays. Liquid biopsy platforms that detect circulating tumor DNA and related analytes provide real-time, repeatable profiling, supporting diagnosis, risk stratification, and assessment of therapy responses [52]. Identification of the most promising epigenetic biomarkers for diagnosis in the most prevalent cancers is summarized in Table 1.

**Table 1** Identification of the most promising epigenetic biomarkers for diagnosis in the most prevalent cancers [53]

| Cancer | Biomarker | Sensitivity (%) | Specificity (%) | Positive predictive value (%) | Negative predictive value (%) | Area under curve (%) |
|---|---|---|---|---|---|---|
| Bladder | Methylation of GDF15; TMEFF2; VIM | 94 | 100 | 100 | 87 | 97.5 |
| Breast | Methylation (ITIH5, DKK3, and RASSF1A) | 67 | 69 | – | – | 69.7 |
| Colorectal | Methylation of VIM (ColoSure™) | 72.5 | 86.9 | – | – | – |
| | Methylation of SEPTIN9 (mSEPT9) | 90 | 50 | – | – | 79.3 |
| Kidney | miR-141/ miR-200b | 99.2 | 100 | 100 | 90.9 | 99.1 |
| | Methylation of APC, RASSF1A and RARβ2 | 94 | 94 | – | – | – |
| Lung | Methylation of SHOX2 (Epi proLung BL Reflex Assay®) | 78 | 96 | – | – | 94 |
| | MiRNA-based test: (Rosetta Lung Cancer Test®) | 93.7 | 98 | – | – | 94 |
| Prostate | PCA3 (PROGENSA™) | 82 | 89 | 75 | 90 | 87 |
| | Methylation of GSTP1, APC and RARβ2 (ProCaM assay™) | 69.4 | 81.6 | – | – | 77 |

### 5.2 Epigenetic Signatures for Early Detection and Risk Assessment

Epigenetic mechanisms—such as DNA methylation, chromatin remodeling, and non-coding RNAs—regulate gene expression without changing the DNA sequence itself. When these mechanisms are disrupted, it can lead to misregulation of transcription, contributing to the initiation and progression of tumors. Many tumors often remain asymptomatic until they have spread, which is why epigenetic biomarkers provide valuable opportunities for earlier detection, risk assessment, and prevention [54]. One significant mechanism is promoter CpG-island hypermethylation, which can silence tumor suppressor genes and is detectable through minimally invasive assays. For these biomarkers to be clinically useful, they should indicate early, cancer-specific events. Examples of such biomarkers include an 11-gene methylation panel that monitors the progression of cervical cancer, lung-associated loci like DAPK, RARB, and MGMT, and the prediagnostic detection of CDKN2A-p16(INK4A) and MGMT methylation, which can occur up to 3 years prior to the development of squamous cell carcinoma [55]. Key prognostic and predictive epigenetic biomarkers relevant to cancer diagnosis, progression, and treatment response are summarized in Table 2.

### 5.3 Liquid Biopsy and Circulating Epigenetic Markers (ctDNA)

Combining DNA methylation profiling with liquid biopsy allows for the detection of important circulating biomarkers. Analyzing the methylation patterns of circulating tumor DNA (ctDNA) can help identify the origin of tumors and support early diagnosis due to its chemical stability, prevalence in tumors, and detectability in preserved samples. Early-stage alterations are valuable for blood-based cancer screening and can assist in patient stratification for central nervous system (CNS) tumors and monitoring melanoma. Large-scale targeted methylation sequencing of cell-free DNA (cfDNA) has shown promise in detecting various cancers, including colorectal, lung, breast, and melanoma, by identifying hypermethylated CpG signatures [56]. Emerging methods also examine cfDNA fragmentation patterns, which reflect nucleosome positioning, as potential diagnostic indicators in cancer liquid biopsies. While these approaches are still in the proof-of-concept stage, they rely heavily on advanced sequencing technologies for accurate fragment mapping [57].

## 6 Epigenetic Therapies in Cancer

### 6.1 DNA Methyltransferase Inhibitors

DNA methyl transferase inhibitors (DNMTis) prevent the addition of methyl groups to cytosine at CpG sites, thereby reversing epigenetic silencing. At low doses, these inhibitors can reactivate tumor suppressor genes, while at higher doses, they induce

**Table 2** Some of the important prognostic and predictive epigenetic biomarkers of cancer [48]

| The biomarker | Being prognostic biomarker | Being predictive biomarker | Target cancer |
|---|---|---|---|
| **Histone modifications** | | | |
| cf-nucleosome epitope combination | + | – | Colorectal cancer |
| H3Cit | + | – | Advanced cancers |
| H3K4me3 and Wdr82 expression | + | – | Colorectal cancer |
| **Chromatin conformation** | | | |
| ARID1 (ARID1A and ARIDB) | + | – | Neuroblastoma tumors |
| Methylation status of CTCF locus | + | – | Colorectal cancer |
| SMARCA4/BRG1 | – | + | Non-small cell lung cancer |
| **Methylation** | | | |
| MLH1 Hypomethylation | + | – | Colorectal cancer |
| MGMT Hypermethylation | + | + | Glioblastoma |
| IDH1 p.R132H mutation and MGMT hypermethylation | + | – | Glioblastoma |
| RB1 hypermethylation | + | – | Retinoblastoma |
| GSTP1, RASSF1, APC methylation status | + | – | Prostate cancer |
| SEPT9 | + | – | Colorectal and lung cancers |
| MGMT-STP27 | – | + | Oligodendrogliomas and oligoastrocytomas |
| ESR1 | – | + | Breast cancer |
| ZNF331 | + | – | Colorectal cancer |
| SALL1 | – | + | Head and neck cancer |
| **miRNA** | | | |
| miR-21 | + | + | Multiple types of cancers |
| miR-30d, miR-21 | + | – | Non-small cell lung cancer |
| miR-31-3p | + | + | Colorectal cancer |
| miR-106a, miR125a-5p, miR-129-3p, miR-205, miR-21, miR-29b, miR-375, miR-7 | + | – | Non-small cell lung cancer |
| miR-29a, miR-92a | + | – | Colorectal cancer |
| miR-506, miR-4316 | + | – | Colorectal cancer |
| miR-126, miR-145, miR-210, miR-205-5p | + | – | Non-small cell lung cancer |
| miR-149-3p, miR-150-5p, miR-193a-3p | + | – | Melanoma |
| miR-200 family, miR-17 family | – | + | Prostate cancer |
| miR-17, miR-155 | + | – | Non-small cell lung cancer |

cytotoxicity. Although hepatotoxicity limits their use in solid tumors, DNMTis have shown effectiveness in hematological malignancies such as myelodysplastic syndrome (MDS), AML, and CML [58]. The nucleoside analogs azacitidine (AZA) and decitabine (DAC), discovered in the 1960s, are the most commonly used DNMTis. Once they are incorporated into DNA, they trap DNMTs, leading to the degradation of the enzyme. DAC specifically targets DNMT1 and DNMT3A, while AZA mainly inhibits DNMT1 and has the ability to integrate into RNA. Both agents are approved for treating MDS, but their activity in solid tumors is limited. Next-generation DNMTis include zebularine, which offers greater chemical stability, and guadecitabine, which demonstrates improved stability and activity in the treatment of AML. Additionally, novel analogs such as 4′-thio-2′-deoxycytidine are currently under evaluation. Non-nucleoside inhibitors, including procainamide and quinoline derivatives, work by demethylating CpG-rich promoters or promoting the degradation of DNMT1. Oligonucleotide-based inhibitors like MG98 and selective agents such as Nanaomycin A also show potential for reactivating tumor suppressor genes [58, 59].

### 6.2 Histone Deacetylase (HDAC) Inhibitors

HDAC inhibitors (HDACis) are both natural and synthetic compounds that specifically target classical HDAC enzymes. While histone hyperacetylation typically promotes gene transcription, the exposure to HDAC inhibitors does not uniformly lead to an increase in gene expression; approximately one-fifth of genes are affected, resulting in a balance of upregulation and downregulation. Four HDAC inhibitors—belinostat, vorinostat, panobinostat, and romidepsin—are approved by the FDA, with panobinostat also receiving approval from the EMA. Additionally, chidamide is approved in China. The various chemotypes of HDAC inhibitors include hydroxamic acids (such as vorinostat and trichostatin A), cyclic peptides (like romidepsin), benzamides (such as entinostat), and short-chain fatty acids. There is a growing interest in isotype-selective compounds, such as entinostat, which targets class I HDACs specifically. Clinically, HDAC inhibitors have shown the most effectiveness in treating hematologic malignancies, while their tolerability and efficacy in solid tumors have been more limited. Mechanistically, HDAC inhibitors remodel chromatin and reprogram transcription indirectly, leading to G1 cell cycle arrest through the activation of p21 and suppression of cyclin-dependent kinases (CDKs), which impacts the Rb/E2F pathway. They can also induce apoptosis by modulating death receptors, Bcl-2 family proteins, and the acetylation of p53. Furthermore, they inhibit tumor growth by influencing immune responses and angiogenesis, including the suppression of HIF-1α/VEGF under hypoxic conditions. For example, trichostatin A downregulates HIF-1α/VEGF while enhancing tumor-suppressive programs. These diverse mechanisms contribute to the selective anticancer activity of HDAC inhibitors and support ongoing development, particularly in the treatment of leukemia and lymphoma [60, 61].

## 6.3 Emerging Epigenetic Drugs

Targeted epigenetic therapies are transforming oncology, with bromodomain and extraterminal (BET) inhibitors and EZH2 inhibitors leading the way. BET proteins (BRD2, BRD3, BRD4, and BRDT) recognize acetyl-lysine histone marks to recruit transcriptional machinery. Inhibitors such as JQ1, OTX015, and ABBV-075 disrupt this interaction, thereby suppressing the expression of oncogenes like MYC and BCL2 in both hematologic malignancies and solid tumors. EZH2, the catalytic subunit of the Polycomb Repressive Complex 2 (PRC2), mediates gene silencing through the trimethylation of histone H3 at lysine 27 (H3K27me3). Dysregulation caused by gain-of-function mutations or overexpression can drive tumor formation. Small-molecule EZH2 inhibitors, such as tazemetostat and valemetostat, block its methyl transferase activity, restore tumor-suppressive transcriptional programs, and inhibit cancer cell proliferation. Clinically, these agents have shown efficacy in EZH2-mutated lymphomas and MYC-driven cancers and are promising candidates for rational combination regimens aimed at overcoming therapeutic resistance [62–64]. Notably, several of these epigenetic strategies are being evaluated in combination with immune checkpoint inhibitors that target the PD-L1/PD-1 axis, with the goal of enhancing antitumor immune responses—an approach summarized in Table 3.

## 6.4 Combination Therapies Involving Epigenetic Modulators

Combination regimens are increasingly preferred over monotherapy for complex diseases, as they aim to co-target complementary pathways for more durable control. In oncology, inhibitors of HDAC (histone deacetylase) and DNMT (DNA methyltransferase) often work synergistically. For instance, the combination of azacitidine and entinostat has shown activity in recurrent metastatic non-small cell lung cancer (NSCLC), while vorinostat combined with cladribine—a DNMT inhibitor that acts indirectly by inhibiting SAH hydrolase—has demonstrated synergy in NK-LGL leukemia. Rational triplet therapies, which include an HDAC inhibitor, a DNMT inhibitor, and either the lysine methyl transferase inhibitor DZNep or an EZH2 inhibitor, have led to enhanced responses in breast cancer models and AML. Additional combinations, such as vorinostat paired with the LSD1 inhibitor tranylcypromine, have been investigated in glioblastoma stem cells. BET inhibitors continue to show promise as partners in these regimens, despite some resistance. For example, JQ1 has been found to promote differentiation and inhibit growth in NUT midline carcinoma. Additionally, combinations of BET inhibitors and HDAC inhibitors are being studied in pancreatic ductal adenocarcinoma and other solid tumors. Overall, multi-agent epigenetic modulation presents a pathway to improved control of leukemia and a broader therapeutic reach [65, 66].

**Table 3** Epigenetic approaches work in conjunction with therapies that target the PD-L1/PD-1 pathway in cancer immunotherapy

| Epigenetic drug | Immune checkpoint inhibitors | Status/phase | Cancer type | Trial ID |
|---|---|---|---|---|
| Chidamide (HDACi) | Toripalimab | Recruiting/Ib/II | Advanced cervical cancer | NCT04651127 |
| Entinostat (HDACi) | Atezolizumab Bevacizumab | Suspended/I/II | Advanced renal cell carcinoma | NCT03024437 |
| Chidamide (HDACi) | Sintilimab | Not yet recruiting/II | Refractory and relapsed AITL | NCT04831710 |
| Chidamide (HDACi) | Sintilimab | Recruiting/II | Peripheral T-cell lymphoma | NCT04512534 |
| Chidamide (HDACi)<br>Decitabine (DNMTi) | Immune Checkpoint Inhibitors | Recruiting/I/II | R/R NHL and advanced solid tumors | NCT05320640 |
| Mocetinostat (HDACi) | Durvalumab | Terminated has results/I/II | Advanced solid tumors and NSCLC | NCT02805660 |
| Entinostat (HDACi) | Pembrolizumab | Active, not recruiting/Ib/II | NSCLC, melanoma, and colorectal cancer | NCT02437136 |
| Guadecitabine (DNMTi) | Atezolizumab | Active, not recruiting/I/II | Advanced MDS, refractory or relapsed CML | NCT02935361 |
| Decitabine (DNMT) | PDR001 MBG453 | Active, not recruiting/I | AML or high-risk MDS | NCT03066648 |
| Belinostat (HDACi) | Tremelimumab Durvalumab | Recruiting/I | ARID1A mutated metastatic or unresectable, locally advanced urothelial carcinoma | NCT05154994 |
| Entinostat (HDACi) | Nivolumab Ipilimumab | Active,not recruiting/I | Solid tumors that are metastatic or cannot be moved by surgery or locally advanced or metastatic HER2-negative breast cancer | NCT02453620 |
| Chidamide (HDACi) | Sintilimab | Completed/Ib/II | Refractory and relapsed ENKTCL | NCT03820596 |
| Romidepsin (HDACi) | Durvalumab | Recruiting/I/IIa | Lymphoma | NCT03161223 |
| ZEN003694 (BETi) | Nivolumab | Recruiting/I/Ib | Solid tumors | NCT04840589 |
| ZEN003694 (BETi) | Pembrolizumab | Recruiting/Ib | Advanced triple-negative breast | NCT05422794 |

# 7 Epigenomics in Immunotherapy

## 7.1 Impact of Epigenetic Regulation on Immune Response

Epigenetic programs play a crucial role in shaping the fate and function of immune cells by controlling access to lineage-defining genomic regions. In addition to transcription factors, various epigenetic mechanisms—including histone modifications, DNA methylation, and nucleosome positioning—coordinate the development of Th cells. For example, the enzyme DNMT3A influences the differentiation of Th1 and Th2 cells, while Th17 cells exhibit H3 acetylation and H3K4 trimethylation (H3K4me3) at the IL17 locus. The stability of Tregs is dependent on DNA methylation at the Foxp3 locus. Similarly, in B-cell development, the demethylation of the CD79a promoter leads to its activation, followed by chromatin remodeling. Epigenetic regulation is also essential for V(D)J recombination and the diversification of antibodies [67]. Innate immune responses are equally influenced by epigenetic modifications. TLR signaling facilitates the acetylation of histone lysines at inflammatory cytokine genes. This process is recognized by BET proteins, which then recruit positive transcription elongation factor b (P-TEFb) and RNA polymerase II to enhance transcription. Additionally, the polarization of macrophages is directed by epigenetic changes; for instance, the H3K27 demethylase Jmjd3, triggered by TLR–NF-κB signaling, is essential for M2 macrophage programs [68].

## 7.2 Role of Epigenetics in Modulating Immune Checkpoints

Immune checkpoints, such as PD-1 and CTLA-4, play a crucial role in maintaining immune tolerance. However, tumors can exploit these checkpoints to suppress antitumor immunity. Epigenetic mechanisms—including DNA methylation, histone modifications, and non-coding RNAs—regulate the expression of checkpoint genes and influence the tumor microenvironment. In various cancers, including melanoma, breast cancer, colorectal cancer, and lower-grade glioma, global DNA hypomethylation is associated with increased levels of PD-L1, PD-1, and CTLA-4. Additionally, a reduction in repressive histone marks (low levels of H3K9me3 and H3K27me3) is linked to the upregulation of these checkpoints, suggesting that lysine demethylases may serve as promising therapeutic targets. From a therapeutic standpoint, HDAC inhibitors have been shown to enhance PD-L1 expression and T-cell chemokine levels, which can improve the effectiveness of anti-PD-1/PD-L1 therapies. The protein HAT1 has been associated with poor prognosis in pancreatic cancer and plays a role in regulating PD-L1, indicating its potential as a drug target. Non-coding RNAs also influence checkpoint pathways. For instance, miR-138 and miR-424 are known to repress PD-1 and CTLA-4, while low levels of miR-34a are associated with PD-L1 overexpression and poor outcomes in non-small cell lung cancer (NSCLC) and AML. Combining immune checkpoint blockade with epigenetic therapies or miRNA-based agents presents a promising strategy to overcome resistance to treatment [69, 70].

## 7.3 Epigenetic Mechanisms Influencing Tumor Microenvironment

### 7.3.1 Epigenetic Regulation of Immune Cell Function and Tumor–Stroma Interactions

Cells within the tumor microenvironment, including stromal and immune cells, can acquire either pro-tumor or anti-tumor characteristics through epigenetic changes rather than permanent mutations. Modifications in chromatin structure and DNA methylation play a crucial role in regulating cytokine signaling, antigen transport, and immune cell infiltration. In cancer-associated fibroblasts, tumor-driven methylation—shaped by cytokines such as IL-1 and TGF-β—supports pro-tumor states. For instance, leukemia inhibitory factor (LIF) interacts with DNMT3B to reprogram fibroblasts, while reduced methylation in pancreatic ductal adenocarcinoma-associated fibroblasts increases the expression of factors like ADAM12. When EZH2 is lost and HDAC6 is gained, the ADAMTS1–STAT3/COX2 pathway is altered, leading to increased immunosuppression. Notably, inhibiting HDAC6 can reactivate T-cell functionality. Furthermore, overexpression of HMGA2 in stromal cells may even trigger the formation of epithelial neoplasia. Macrophage polarization is also directed by epigenetic mechanisms: DNMT1 and TET1 promote M1 polarization, which includes the production of TNF-α, while DNMT3B, in conjunction with PRMT1, SMYD3, and JMJD3, favors M2 polarization. Tumor-induced methylation of metabolic genes in pancreatic cancer (such as NQO-1 and ALDH1a3) can be reversed using decitabine. Additionally, enzymes like LSD1 and HDACs (including SIRT1/2 and HDAC7) connect metabolism to inflammatory responses, with histone lactylation playing a role in maintaining M1 homeostasis. In myeloid-derived suppressor cells (MDSCs), hypermethylation of genes such as S1PR4 and RUNX1 due to DNMT3A enhances their immunosuppressive function. SETD1B increases iNOS expression through H3K4 trimethylation. Moreover, HDAC11 and HDAC2 are involved in MDSC expansion and T-cell suppression; targeting HDACs or the bromodomains of CBP/EP300 can potentially reprogram MDSCs and slow tumor growth. T cells are also influenced by epigenetic regulation: DNMT1 and EZH2 silence chemokines CXCL9 and CXCL10, limiting T-cell infiltration. In contrast, hypomethylation marks tumor-reactive CD8+ T cells and is associated with the demethylation of exhaustion markers such as PDCD1 and CTLA4. The metabolite acetyl-CoA produced by LDHA enhances IFN-γ production through histone acetylation. HDAC inhibitors have the potential to instill memory-like properties in T cells, which include increased levels of CD28 and IL-21, while loss of Satb1 raises PD-1 expression. Ultimately, the interplay between chromatin flexibility and stability in tumor-specific T cells may predict their responsiveness to checkpoint blockade therapies and inform the design of epigenetic treatments [71–73].

### 7.3.2 Implications for Immunotherapy and Combination Therapies

Epigenetic agents have the potential to reprogram an immunosuppressive tumor microenvironment. DNMT inhibitors, such as azacitidine and decitabine, reduce the protumor activity of CAFs by downregulating DNMT1 and DNMT3A. They also help repolarize tumor-associated macrophages, decrease myeloid-derived suppressor cells (MDSCs), and work synergistically with PD-1 and CTLA-4 blockade. This combination limits CD8+ T-cell exhaustion and enhances type I interferon responses, major MHC expression, and chemokine signaling, thereby improving cytotoxic T-cell infiltration. Histone-directed therapies are context-dependent. For example, inhibiting EZH2 can shift Tregs towards pro-inflammatory states but may also impair CD8+ T-cell function due to Notch pathway disruption. Histone deacetylase (HDAC) inhibitors can vary in their effects, ranging from blocking the conversion of CAFs (as seen with Scriptaid) to partnering with DNMT inhibitors like entinostat to prevent the formation of metastatic niches [73]. Additionally, targeting chromatin readers with BET inhibitors can enhance antigen presentation (e.g., MHC-II) while suppressing CAF activity. Overall, strategically combining DNMT inhibitors, HDAC inhibitors, and BET inhibitors with checkpoint inhibitors presents a promising approach to overcoming resistance and improving long-lasting therapeutic responses [74].

## 8 Integration of Epigenomics and Multi-omics in Precision Medicine

Advances in next-generation sequencing have allowed for the large-scale generation of various types of data, including transcriptomic, proteomic, metabolomic, genomic, radiomic, and epigenomic information. This development has created opportunities to decode the complexity of cancer across different molecular layers [75]. Integrating multi-omics data can reveal the underlying pathways and regulatory networks associated with oncogenesis, disease progression, and therapeutic responses. However, this integration poses analytical challenges due to the scale, heterogeneity, and noise inherent in the data. Artificial intelligence, particularly through graph-based learning, helps address these challenges by modeling relationships within and across omics layers, thereby improving predictions and patient stratification [76, 77]. Graph neural network frameworks exemplify this progress. While earlier graph convolutional approaches, such as MOGONET and SUPREME, did not account for the weight of neighbor influence, attention-based models like MOGAT assign sample-specific attention coefficients and have demonstrated superior performance in predicting cancer subtypes. Integrative analytical pipelines applied to The Cancer Genome Atlas (TCGA) cohorts further highlight the clinical relevance of these methods. For instance, multi-omic clustering of lung adenocarcinoma (LUAD) identified two subtypes distinguished by distinct survival rates, immune infiltration, and patterns of copy number variations (CNV) and methylation. This study nominated CNTN4 and RFTN1 as potential biomarkers [77].

Deep learning techniques also help mitigate the challenges associated with having a small number of samples in relation to a large number of features. Variational autoencoders (VAEs) and maximum mean discrepancy variational autoencoders (MMD-VAEs) compress high-dimensional omics data into informative latent features that improve classification and survival modeling. In ovarian cancer studies,

VAE-derived representations have enabled accurate subtype classification and prognostication, with MMD-VAEs generally outperforming standard VAEs. Integrated multi-omics analyses have been shown to surpass or match the performance of single-omics approaches [78]. Overall, the combination of epigenomics with AI-driven multi-omics analysis holds the promise of achieving more precise cancer subtyping, risk prediction, and therapy selection. However, it is crucial to maintain a focus on benchmarking, reproducibility, and the development of clinically interpretable models.

## 9 Emerging Trends in Cancer Epigenomics

Single-cell technologies are transforming cancer epigenomics by uncovering cellular heterogeneity across genetic, transcriptional, and regulatory levels. Single-cell RNA sequencing (scRNA-seq) has mapped malignant ecosystems with unprecedented detail, while targeted single-cell DNA sequencing identifies recurrent mutations and copy-number variations to infer genotypes. However, transcriptional diversity often differs from clonal genetics, highlighting the necessity for single-cell epigenomic assays that can capture the specific regulatory states driving plasticity and resistance [79–82]. Despite rapid advancements, these methods encounter practical challenges, including low throughput, sparse coverage, high costs, amplification bias, and damage from processing. These issues underline the need for benchmarking and standardized approaches [83].

The complexity, sparsity, and noise of single-cell epigenomic data call for specialized computational methods. Analyses generally utilize dimensionality reduction and clustering techniques, along with imputation, pseudo-bulking, and genomic binning to address missing data. Integrative pipelines now facilitate multi-omic analyses across different modalities, allowing for the joint inference of chromatin states, transcription, and genotypes [84–88]. Spatial epigenomics is emerging as a field that incorporates regulation within native tissue architecture. Techniques like epigenomic MERFISH (Multiplexed Analysis of RNA and Protein in Single Cells) can map hundreds of genomic loci with optical readouts, showing region-specific enhancer and promoter landscapes, potential enhancer–promoter pairs, and subnuclear positioning in both developing and adult tissues [89]. Complementary methods such as Spatial-CUT&Tag enable de novo locus identification through sequencing, trading off spatial resolution for broader discovery. Together, these approaches create a framework for linking chromatin states to functional outcomes at the cellular level [90].

Epitranscriptomics introduces an additional dynamic regulatory layer that interfaces with epigenomic control. RNA modifications—including m6A, m1A, m5C, hm5C, m7G, m6Am, Nm, ac4C, and Ψ—regulate splicing, translation, stability, and export through specialized proteins that act as writers, erasers, and readers. These modifications have context-dependent roles that can be oncogenic or tumor-suppressive [91, 92]. For example, disease-specific modification patterns, such as changes in m6A distribution and burden in nasopharyngeal carcinoma compared to normal epithelium, highlight the diagnostic and mechanistic potential of integrating epitranscriptomic maps with genomic, epigenomic, and transcriptomic data. This integration is important for understanding the spatiotemporal heterogeneity of the tumor microenvironment [93].

## 10 Challenges in Cancer Epigenomics

Rapid advancements in cancer epigenomics are transforming our understanding of chromatin remodeling and regulation. However, this rapid pace presents challenges: methods are evolving, standards are lagging, and new layers of complexity are emerging as technologies develop. Both genetic and epigenetic variations contribute to heterogeneity in tumors, which can be classified as inter-tumoral (differences between tumors or patients) and intra-tumoral (variations within a single tumor). The intra-tumoral heterogeneity includes differences across spatial niches and changes over time, which drive tumor progression and resistance to therapy. Capturing these dynamics is crucial for developing durable and precise treatment strategies [94, 95]. Traditional bulk profiling methods can obscure rare but significant subpopulations within tumors. Single-cell epigenomic assays allow researchers to investigate clonal structures, interactions with the tumor microenvironment, cancer stem cells, circulating tumor cells, and mechanisms of resistance. However, since epigenomes function in conjunction with other molecular layers, single-modality profiling at the single-cell level yields an incomplete picture, highlighting the need for integrative single-cell multi-omics approaches [96].

The spatial organization of regulatory elements further complicates data interpretation. High-resolution spatial epigenomics techniques are necessary to resolve short enhancers and promoters while preserving their native context. New methods, including spatial ATAC-seq, spatial-CUT&Tag, and Epigenomic MERFISH, address histone marks and chromatin accessibility in situ, while imaging technologies such as chromatin tracing and Hi-M provide insights into 3D genome architecture [96, 97]. Linking epigenetic states to phenotypes remains a challenge. Marker levels fluctuate throughout life and are affected by drugs, comorbidities, and environmental factors. These markers can also influence prognosis and treatment responses. Thus, distinguishing causation from correlation requires careful study design and longitudinal validation [98].

Lifestyle and environmental exposures—such as diet, smoking, pollutants, endocrine disruptors, and stress—present complex translational and ethical dilemmas. Can epigenetic readouts inform risk modification, and how do we balance personal and societal responsibilities? Ensuring equitable access and avoiding stigmatization are crucial, as social and policy factors contribute to unequal exposure levels and may perpetuate "epigenetic disadvantage" in lower socioeconomic groups [99, 100].

In therapeutic contexts, the current epigenetic agents predominantly consist of DNMT and HDAC inhibitors. However, their broad effectiveness and context dependence impose limits on precision and durability, prompting efforts to refine therapeutic indications and combinations [101]. New targets, including histone methyltransferases and demethylases, are moving through clinical evaluation and may offer greater selectivity and a better fit for specific disease contexts [102]. It's important to note that epigenetic drugs often show reduced efficacy in many solid tumors due to factors such as slower proliferation rates, challenges related to stability and delivery, and the unique biology of these tumors compared to hematologic malignancies [103]. In summary, the central challenges facing the field—heterogeneity across different scales, spatial and temporal context, causal inference, ethical implementation, and therapeutic specificity—are becoming increasingly clear. Addressing these challenges will depend on standardization, the development of integrative single-cell and spatial platforms, rigorous longitudinal studies, and equitable clinical translation.

## 11 Conclusion and Future Directions

Recent advancements in cancer epigenomics have significantly improved our understanding of oncogenesis, tumor evolution, and therapeutic resistance. Key breakthroughs in DNA methylation, histone modifications, and chromatin structure have paved the way for the next phase of research. This phase will focus on single-cell epigenomics, which aims to uncover intratumoral heterogeneity; spatial epigenomics, which seeks to map the regulatory interactions within the tumor microenvironment; and multi-omics integration, which will help reconstruct gene regulatory networks influenced by epigenetics that underlie tumor plasticity and resistance. Artificial intelligence and advanced computational modeling will play a crucial role in analyzing these complex, high-dimensional datasets. They will facilitate robust pattern recognition, biomarker discovery, and predictive analytics. Clinical applications are progressing swiftly through liquid biopsy-based epigenomic profiling. Techniques that analyze ctDNA and exosomal RNAs allow for real-time, noninvasive monitoring to aid in early detection, assess minimal residual disease, and evaluate treatment responses. Strategic combinations, such as pairing epigenetic therapies with immunotherapy, have the potential to overcome the limitations of monotherapy and lead to longer-lasting remissions. For widespread adoption of these advancements, it's essential to establish standardized assay protocols, develop interoperable analysis pipelines, and ensure rigorous validation of biomarkers. Equally important are cross-disciplinary collaborations among molecular biologists, clinicians, and data scientists to effectively translate epigenomic insights into tangible patient benefits. Overall, the future looks very promising: epigenomics is set to enhance the accuracy, predictive capability, and personalization of cancer diagnosis, classification, monitoring, and treatment.

## References


1. Ceccon C, Borga C, Angerilli V, Bergamo F, Munari G, Sabbadin M et al (2025) MLH1 gene promoter methylation status partially overlaps with CpG methylator phenotype (CIMP) in colorectal adenocarcinoma. Pathol Res Pract 266:155786
2. Mohanad M, Hamza HM, Bahnassy AA, Shaarawy S, Ahmed O, El-Mezayen HA et al (2025) Molecular profiling of breast cancer methylation pattern in triple negative versus non- triple negative breast cancer. Sci Rep 15(1):6894
3. Jin Y, Lu R, Liu F, Jiang G, Wang R, Zheng M (2024) DNA methylation analysis in plasma for early diagnosis in lung adenocarcinoma. Medicine (Baltimore) 103(28):e38867
4. Rodenhiser D, Mann M (2006) Epigenetics and human disease: translating basic biology into clinical applications. CMAJ 174(3):341–348
5. Shi Z, Cao X, Ma Y, Li K, Wang X, Lin J et al (2025) RNA methyltransferase METTL16: from molecular mechanisms to therapeutic prospects in cancers. Cancer Lett 217698
6. Tsuchimochi S, Yamamoto Y, Taguchi A, Kawazu M, Sone K, Ikemura M et al (2025) BRCA1 promoter methylation in ovarian cancer: clinical relevance and a novel diagnostic approach using fragment analysis. Cancer Sci 116(7):1996–2007
7. Hosseini M, Feghhi-Najafabadi S, Azad M (2024) A review on the impact of aberrant methylation in breast cancer: diagnostic, prognostic, and therapeutic approaches. Iran J Biotechnol 22(4):e3897
8. Joo JE, Mahmood K, Clendenning M, Walker R, Georgeson P, Como J et al (2025) Causes of DNA mismatch repair deficiency in sebaceous skin lesions demonstrating loss of MLH1 protein expression: constitutional over somatic MLH1 promoter methylation. Fam Cancer 24(2):36
9. Ren X, Liu G, Zhou J (2025) Nuclear-activating miRNAs: unveiling the intricacies of subcellular miRNA function and regulation in cancer and immunity disease. Cancer Cell Int 25(1):147
10. Lauzon-Young C, Silva A, Sadikovic B (2025) Epigenomic insights and computational advances in hematologic malignancies. Mol Cytogenet 18(1):9
11. Kim DJ (2025) The role of the DNA methyltransferase family and the therapeutic potential of DNMT inhibitors in tumor treatment. Curr Oncol 32(2)
12. Ghimire B, Kurlberg G, Falk P, Singh Y, Wettergren Y (2024) Epigenetic differences in the tumor suppressor genes MLH1 and p16INK4a between Nepalese and Swedish patients with colorectal cancer. Innov Surg Sci 9(3):153–163
13. Chu L, Qu Y, An Y, Hou L, Li J, Li W et al (2022) Induction of senescence-associated secretory phenotype underlies the therapeutic efficacy of PRC2 inhibition in cancer. Cell Death Dis 13(2):155
14. Glaser S, Wagener R, Kretzmer H, López C, Baptista MJ, Bens S et al (2025) Subtyping Burkitt lymphoma by DNA methylation. Genes Chromosomes Cancer 64(4):e70042
15. Charalampopoulou S, Chapiro E, Nadeu F, Zenz T, Beà S, Martínez-Farran A et al (2024) Epigenetic features support the diagnosis of B-cell prolymphocytic leukemia and identify 2 clinicobiological subtypes. Blood Adv 8(24):6297–6307
16. Hayashi Y, Harada Y, Kagiyama Y, Nishikawa S, Ding Y, Imagawa J et al (2019) NUP98-HBO1-fusion generates phenotypically and genetically relevant chronic myelomonocytic leukemia pathogenesis. Blood Adv 3(7):1047–1060
17. Basu S, Nadhan R, Dhanasekaran DN (2025) Long non-coding RNAs in ovarian cancer: mechanistic insights and clinical applications. Cancers (Basel) 17(3)
18. Lucchini S, Constantinou M, Marino S (2024) Unravelling the mosaic: epigenetic diversity in glioblastoma. Mol Oncol 18(12):2871–2889
19. Wang Y, Liu Z, Lv Y, Long J, Lu Y, Huang P (2025) Mechanisms of radioresistance and radiosensitization strategies for Triple Negative Breast Cancer. Transl Oncol 55:102351
20. Jin SG, Xiong W, Wu X, Yang L, Pfeifer GP (2015) The DNA methylation landscape of human melanoma. Genomics 106(6):322–330

21. Haghir-Sharif-Zamini Y, Khosravi A, Hassan M, Zarrabi A, Vosough M (2025) c-FLIP/Ku70 complex; a potential molecular target for apoptosis induction in hepatocellular carcinoma. Arch Biochem Biophys 765:110306
22. Jiagge E, Jin DX, Newberg JY, Perea-Chamblee T, Pekala KR, Fong C et al (2023) Tumor sequencing of African ancestry reveals differences in clinically relevant alterations across common cancers. Cancer Cell 41(11):1963–71.e3
23. Fernández Rebollo I (2024) Targetable mechanisms of epigenetic age acceleration and rejuvenation in cancer. Universitat Politècnica de Catalunya
24. Moeinafshar A, Nouri M, Shokrollahi N, Masrour M, Behnam A, Tehrani Fateh S et al (2024) Non-coding RNAs as potential therapeutic targets for receptor tyrosine kinase signaling in solid tumors: current status and future directions. Cancer Cell Int 24(1):26
25. Rajavel T, Devi KP (2019) Chapter 13 - Phytochemicals as epigenetic modifiers for cancer management with special reference to lung cancer. In: Bishayee A, Bhatia D (eds) Epigenetics of cancer prevention. Academic Press, 8, pp 271–286
26. Prat A, Adamo B, Fan C, Vidal M, Galvan P, Felip E et al (2013) Molecular identification of basal-like breast cancer through genomic analyses across five cancer types. American Society of Clinical Oncology
27. Wajapeyee N, Gupta R (2021) Epigenetic alterations and mechanisms that drive resistance to targeted cancer therapies. Cancer Res 81(22):5589–5595
28. Blecua P, Martinez-Verbo L, Esteller M (2020) The DNA methylation landscape of hematological malignancies: an update. Mol Oncol 14(8):1616–1639
29. Hu D, Shilatifard A (2016) Epigenetics of hematopoiesis and hematological malignancies. Genes Dev 30(18):2021–2041
30. Wallace L, Obeng EA (2023) Noncoding rules of survival: epigenetic regulation of normal and malignant hematopoiesis. Front Mol Biosci 10:1273046
31. Grau M, López C, Martín-Subero JI, Beà S (2023) Cytogenomics of B-cell non-Hodgkin lymphomas: the "old" meets the "new". Best Pract Res Clin Haematol 36(4):101513
32. Fattizzo B, Rosa J, Giannotta JA, Baldini L (2020) The physiopathology of T-cell acute lymphoblastic leukemia: focus on molecular aspects. Front Oncol 10:273
33. Perez de Acha O, Rossi M, Gorospe M (2020) Circular RNAs in blood malignancies. Front Mol Biosci 7:109
34. Gonçalves MWA, Lavareze L, Chone CT, Egal ESA, Altemani A, Mariano FV et al (2024) Mantle cell lymphoma in the oral cavity: exploring the differential microscopic diagnosis of a rare case. Indian J Otolaryngol Head Neck Surg 76(5):4806–4809
35. Dufva O, Pölönen P, Brück O, Keränen MA, Klievink J, Mehtonen J et al (2020) Immunogenomic landscape of hematological malignancies. Cancer Cell 38(3):380–399. e13
36. Fathi E, Farahzadi R, Montazersaheb S, Bagheri Y (2021) Epigenetic modifications in acute lymphoblastic leukemia: from cellular mechanisms to therapeutics. Curr Gene Ther 21(1):60–71
37. Chi Y, Wang D, Wang J, Yu W, Yang J (2019) Long non-coding RNA in the pathogenesis of cancers. Cells 8(9):1015
38. del Pilar Navarrete-Meneses M, Pérez-Vera P (2017) Epigenetic alterations in acute lymphoblastic leukemia. Bol Med Hosp Infant Mex 74(4):243–264
39. Porten SP (2018) Epigenetic alterations in bladder cancer. Curr Urol 19:1–8
40. Verdelli C, Corbetta S (2017) Epigenetic alterations in parathyroid cancers. Int J Mol Sci 18(2):310
41. Yang J, Shay C, Saba NF, Teng Y (2024) Cancer metabolism and carcinogenesis. Exp Hematol Oncol 13(1):10
42. Maiuri AR, O'Hagan HM (2016) Interplay between inflammation and epigenetic changes in cancer. Prog Mol Biol Transl Sci 144:69–117
43. Plass C, Pfister SM, Lindroth AM, Bogatyrova O, Claus R, Lichter P (2013) Mutations in regulators of the epigenome and their connections to global chromatin patterns in cancer. Nat Rev Genet 14(11):765–780

44. Chakraborty S, Sharma G, Karmakar S, Banerjee S (2024) Multi-OMICS approaches in cancer biology: New era in cancer therapy. Biochim Biophys Acta Mol Basis Dis 1870(5):167120
45. Ballestar E, Esteller M (2008) Epigenetic gene regulation in cancer. Adv Genet 61:247–267
46. Wang S, Claret F-X, Wu W (2019) MicroRNAs as therapeutic targets in nasopharyngeal carcinoma. Front Oncol 9:756
47. Agirre X, Martínez-Climent JÁ, Odero M, Prosper F (2012) Epigenetic regulation of miRNA genes in acute leukemia. Leukemia 26(3):395–403
48. Kamińska K, Nalejska E, Kubiak M, Wojtysiak J, Żołna Ł, Kowalewski J et al (2019) Prognostic and predictive epigenetic biomarkers in oncology. Mol Diagn Ther 23(1):83–95
49. Heyn H, Esteller M (2012) DNA methylation profiling in the clinic: applications and challenges. Nat Rev Genet 13(10):679–692
50. Kanwal R, Gupta K, Gupta S (2015) Cancer epigenetics: an introduction. In: Verma M (ed) Cancer epigenetics: risk assessment, diagnosis, treatment, and prognosis. New York, NY, Springer, pp 3–25
51. Di Martino MT, Riillo C, Scionti F, Grillone K, Polerà N, Caracciolo D et al (2021) miRNAs and lncRNAs as novel therapeutic targets to improve cancer immunotherapy. Cancer 13(7):1587
52. Li L, Sun Y (2024) Circulating tumor DNA methylation detection as biomarker and its application in tumor liquid biopsy: advances and challenges. MedComm 5(11):e766
53. Costa-Pinheiro P, Diana M, Rui H, Jerónimo C (2015) Diagnostic and prognostic epigenetic biomarkers in cancer. Epigenomics 7(6):1003–1015
54. Herceg Z, Hainaut P (2007) Genetic and epigenetic alterations as biomarkers for cancer detection, diagnosis and prognosis. Mol Oncol 1(1):26–41
55. Verma M, Seminara D, Arena FJ, John C, Iwamoto K, Hartmuller V (2006) Genetic and epigenetic biomarkers in cancer. Mol Diagn Ther 10(1):1–15
56. Lianidou E (2021) Detection and relevance of epigenetic markers on ctDNA: recent advances and future outlook. Mol Oncol 15(6):1683–1700
57. Gai W, Sun K (2019) Epigenetic biomarkers in cell-free DNA and applications in liquid biopsy. Genes [Internet] 10(1)
58. Hu C, Liu X, Zeng Y, Liu J, Wu F (2021) DNA methyltransferase inhibitors combination therapy for the treatment of solid tumor: mechanism and clinical application. Clin Epigenetics 13(1):166
59. Pechalrieu D, Etievant C, Arimondo PB (2017) DNA methyltransferase inhibitors in cancer: from pharmacology to translational studies. Biochem Pharmacol 129:1–13
60. Ververis K, Alison H, Karagiannis TC, Licciardi PV (2013) Histone deacetylase inhibitors (HDACIs): multitargeted anticancer agents. Biologics 7:47–60
61. Shah RR (2019) Safety and tolerability of histone deacetylase (HDAC) inhibitors in oncology. Drug Saf 42(2):235–245
62. Genta S, Pirosa MC, Stathis A (2019) BET and EZH2 inhibitors: novel approaches for targeting cancer. Curr Oncol Rep 21(2):13
63. Doroshow DB, Eder JP, LoRusso PM (2017) BET inhibitors: a novel epigenetic approach. Ann Oncol 28(8):1776–1787
64. Adema V, Colla S (2022) EZH2 inhibitors: the unpacking revolution. Cancer Res 82(3):359–361
65. de Lera AR, Ganesan A (2016) Epigenetic polypharmacology: from combination therapy to multitargeted drugs. Clin Epigenetics 8(1):105
66. Morel D, Jeffery D, Aspeslagh S, Almouzni G, Postel-Vinay S (2020) Combining epigenetic drugs with other therapies for solid tumours—past lessons and future promise. Nat Rev Clin Oncol 17(2):91–107
67. Fernández-Morera JL, Calvanese V, Rodríguez-Rodero S, Menéndez-Torre E, Fraga MF (2010) Epigenetic regulation of the immune system in health and disease. Tissue Antigens 76(6):431–439
68. Obata Y, Furusawa Y, Hase K (2015) Epigenetic modifications of the immune system in health and disease. Immunol Cell Biol 93(3):226–232

69. Saleh R, Toor SM, Sasidharan Nair V, Elkord E (2020) Role of epigenetic modifications in inhibitory immune checkpoints in cancer development and progression. Front Immunol:11
70. Dai E, Zhu Z, Wahed S, Qu Z, Storkus WJ, Guo ZS (2021) Epigenetic modulation of antitumor immunity for improved cancer immunotherapy. Mol Cancer 20(1):171
71. Marks DL, Olson RO, Fernandez-Zapico ME (2016) Epigenetic control of the tumor microenvironment. Epigenomics 8(12):1671–1687
72. Pan X, Zheng L (2020) Epigenetics in modulating immune functions of stromal and immune cells in the tumor microenvironment. Cell Mol Immunol 17(9):940–953
73. Liu K, Li Y, Shen M, Xu W, Wu S, Yang X et al (2025) Epigenetic regulation of stromal and immune cells and therapeutic targets in the tumor microenvironment. Biomolecules [Internet] 15(1)
74. Liu Z, Ren Y, Weng S, Xu H, Li L, Han X (2022) A new trend in cancer treatment: the combination of epigenetics and immunotherapy. Front Immunol:13
75. Li L, Sun M, Wang J, Wan S (2024) Multi-omics based artificial intelligence for cancer research. Adv Cancer Res 163:303–356
76. Wu X, Li W, Tu H (2024) Big data and artificial intelligence in cancer research. Trends Cancer 10(2):147–160
77. Tanvir RB, Islam MM, Sobhan M, Luo D, Mondal AM (2024) MOGAT: a multi-omics integration framework using graph attention networks for cancer subtype prediction. Int J Mol Sci 25(5)
78. Hira MT, Razzaque MA, Angione C, Scrivens J, Sawan S, Sarker M (2021) Integrated multiomics analysis of ovarian cancer using variational autoencoders. Sci Rep 11(1):6265
79. Casado-Pelaez M, Bueno-Costa A, Esteller M (2022) Single cell cancer epigenetics. Trends Cancer 8(10):820–838
80. Costa A, Kieffer Y, Scholer-Dahirel A, Pelon F, Bourachot B, Cardon M et al (2018) Fibroblast heterogeneity and immunosuppressive environment in human breast cancer. Cancer Cell 33(3):463–79.e10
81. Aoki T, Chong LC, Takata K, Milne K, Hav M, Colombo A et al (2020) Single-cell transcriptome analysis reveals disease-defining T-cell subsets in the tumor microenvironment of classic Hodgkin lymphoma. Cancer Discov 10(3):406–421
82. Miles LA, Bowman RL, Merlinsky TR, Csete IS, Ooi AT, Durruthy-Durruthy R et al (2020) Single-cell mutation analysis of clonal evolution in myeloid malignancies. Nature 587(7834):477–482
83. Mehrmohamadi M, Sepehri MH, Nazer N, Norouzi MR (2021) A comparative overview of epigenomic profiling methods. Front Cell Dev Biol 9:714687
84. Danese A, Richter ML, Chaichoompu K, Fischer DS, Theis FJ, Colomé-Tatché M (2021) EpiScanpy: integrated single-cell epigenomic analysis. Nat Commun 12(1):5228
85. Wang C, Sun D, Huang X, Wan C, Li Z, Han Y et al (2020) Integrative analyses of single-cell transcriptome and regulome using MAESTRO. Genome Biol 21(1):198
86. Wu KE, Yost KE, Chang HY, Zou J (2021) BABEL enables cross-modality translation between multiomic profiles at single-cell resolution. Proc Natl Acad Sci 118(15):e2023070118
87. Li G, Fu S, Wang S, Zhu C, Duan B, Tang C et al (2022) A deep generative model for multiview profiling of single-cell RNA-seq and ATAC-seq data. Genome Biol 23(1):20
88. Argelaguet R, Velten B, Arnol D, Dietrich S, Zenz T, Marioni JC et al (2018) Multi-Omics Factor Analysis-a framework for unsupervised integration of multi-omics data sets. Mol Syst Biol 14(6):e8124
89. Lu T, Ang CE, Zhuang X (2022) Spatially resolved epigenomic profiling of single cells in complex tissues. Cell 185(23):4448–64.e17
90. Deng Y, Bartosovic M, Kukanja P, Zhang D, Liu Y, Su G et al (2022) Spatial-CUT&Tag: spatially resolved chromatin modification profiling at the cellular level. Science 375(6581):681–686
91. Wiener D, Schwartz S (2021) The epitranscriptome beyond m(6)A. Nat Rev Genet 22(2):119–131

92. Liu Z, Gao L, Cheng L, Lv G, Sun B, Wang G et al (2023) The roles of N6-methyladenosine and its target regulatory noncoding RNAs in tumors: classification, mechanisms, and potential therapeutic implications. Exp Mol Med 55(3):487–501
93. Zhao Y, Li J, Dian M, Bie Y, Peng Z, Zhou Y et al (2024) Role of N6-methyladenosine methylation in nasopharyngeal carcinoma: current insights and future prospective. Cell Death Discov 10(1):490
94. Dentro SC, Leshchiner I, Haase K, Tarabichi M, Wintersinger J, Deshwar AG et al (2021) Characterizing genetic intra-tumor heterogeneity across 2,658 human cancer genomes. Cell 184(8):2239–54.e39
95. Guo M, Peng Y, Gao A, Du C, Herman JG (2019) Epigenetic heterogeneity in cancer. Biomark Res 7:23
96. Hu Y, Shen F, Yang X, Han T, Long Z, Wen J et al (2023) Single-cell sequencing technology applied to epigenetics for the study of tumor heterogeneity. Clin Epigenetics 15(1):161
97. Farzad N, Enninful A, Bao S, Zhang D, Deng Y, Fan R (2024) Spatially resolved epigenome sequencing via Tn5 transposition and deterministic DNA barcoding in tissue. Nat Protoc 19(11):3389–3425
98. Smith DA, Sadler MC, Altman RB (2023) Promises and challenges in pharmacoepigenetics. Camb Prism Precis Med 1:e18
99. Santaló J, Berdasco M (2022) Ethical implications of epigenetics in the era of personalized medicine. Clin Epigenetics 14(1):44
100. World Medical Association Declaration of Helsinki (2013) Ethical principles for medical research involving human subjects. JAMA 310(20):2191–2194
101. Patnaik S, Anupriya. (2019) Drugs targeting epigenetic modifications and plausible therapeutic strategies against colorectal cancer. Front Pharmacol 10:588
102. Cheng Y, He C, Wang M, Ma X, Mo F, Yang S et al (2019) Targeting epigenetic regulators for cancer therapy: mechanisms and advances in clinical trials. Signal Transduct Target Ther 4:62
103. Piekarz RL, Frye R, Turner M, Wright JJ, Allen SL, Kirschbaum MH et al (2009) Phase II multi-institutional trial of the histone deacetylase inhibitor romidepsin as monotherapy for patients with cutaneous T-cell lymphoma. J Clin Oncol 27(32):5410–5417

# Pharmacogenomics in Personalised Treatment

Sayani Ghosh, Sunny Kumar, and Malini Basu

**Abstract**

Predicting the clinical results of various treatments in various patients and offering tailored care are the two main objectives of personalised medicine. One of the fundamental components of personalised medicine is pharmacogenomics. By customising drug selection and dosage to each patient's unique genetic profile, the study of pharmacogenomics—which studies how a person's genetic composition affects how they react to medications. It plays a critical part in personalised medicine. By maximising medication efficacy and minimising side effects, this strategy seeks to provide safer and more effective treatments. The fundamental idea is that a variety of factors, such as genetics, epigenomics, the environment, and a patient's attributes, such as gender, age, and concurrent medicine, can result in interindividual variability in drug response. By using this information, medical professionals can maximise treatment results by selecting the best medication and dosage for each patient. Pharmacogenomics can enhance treatment effectiveness and lower the risk of treatment failure by customising medication therapy to each patient's unique genetic makeup. Although genetic variants in drug-metabolising enzymes were revealed to change medication response 30 years ago, more than 90% of pharmaceuticals currently used in clinical practice lack reliable and predictive biomarkers for therapeutic effects to prevent serious side effects. In recent years, pharmacogenomics has revolutionised our understanding of disease susceptibility and pathophysiology by utilising a new generation of technologies known as "omics" methods. This has greatly

S. Ghosh · S. Kumar (✉)
Cancer Biology and Inflammatory Disorder Division, Council of Scientific and Industrial Research-Indian Institute of Chemical Biology (CSIR-IICB), TRUE Campus, Kolkata, India

M. Basu (✉)
Department of Microbiology, Dhruba Chand Halder (DCH) College, University of Calcutta, Kolkata, West Bengal, India



M. K. Ghosh, M. Basu (eds.), *Personalized Cancer Therapy*,
https://doi.org/10.1007/978-981-95-6845-1_5

increased the potential for new therapeutic approaches. Pharmacogenomics unquestionably aids in the creation of targeted treatments.

**Keyword**

Pharmacogenomics · Pharmacokinetics · Pharmacodynamics · Proteomics · Next-generation sequencing

## 1 Introduction

The foundation of personalised medicine is the use of an individual's genetic makeup to forecast their optimum therapeutic decision. Personal variables such as age, sex, disease state, body weight, socioeconomic situation, and common diagnostic measures like hypertension, blood sugar and cholesterol levels etc. governs whether the individuals will benefit from a given medication or experience severe side effects. Personalised medicine is the therapeutic choice based on empirical facts. It acknowledges the individuality of patients rather than evidence-based medications. Optimising treatments catered to certain patient subgroups or even to each patient individually has been possible with the recent scientific and technology advancements, particularly in the field of genomics. Simmons et al. defined personalised medicine as personalised technologies that cater to provide patients with predictive care through integrated service [1]. The key component of personalised medicine is Genomic medicine or precision medicine that targets each patient individually. However, implementing genomics in clinical practice has been challenging due to complicated genetic linkages and practical difficulties. Genomics now includes proteomics, metabolomics, and other fields that provide biomarkers for therapeutic purposes. Pharmacogenomics is a key driver of personalised medicine, with many drugs now in clinical use. However, comparable challenges remain. This document explores the potential and limitations of pharmacogenomics in personalized treatment [2, 3].

## 2 Pharmacogenomics Fundamentals

Pharmacogenomics is the study of individual medication reactions using genomic information. It may be possible to make clinical decisions based on genetics. Modulation of drug or dose is possible when a patient's gene variant is linked to a specific drug response. Scientists evaluate genetic variations by influencing a person's drug reaction by first finding genetic loci linked to known drug responses, and then assessing people whose response is unknown. Whole-genome single-nucleotide polymorphism (SNP) profiles and multigene analysis are examples of contemporary methods that are only now beginning to be used in clinical settings for medication development and discovery. Researchers concentrate on two main factors when

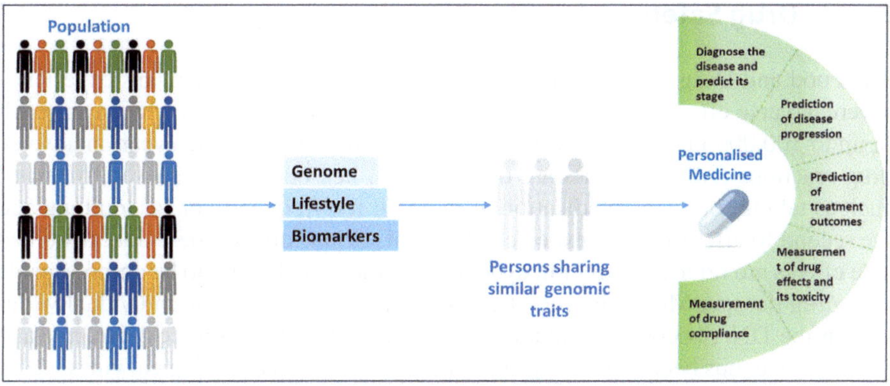


**Fig. 1** Demonstration defining personalised medicines and their functions

examining drug activity in humans i.e. pharmacokinetics and pharmacodynamics. These two factors are important in the study of pharmacogenomics (Fig. 1).

## 2.1 Pharmacokinetics

Pharmacokinetics is the amount of medication required to reach its target in the body. The four processes that make up pharmacokinetics are absorption, distribution, metabolism, and excretion acronymed as ADME. Typically, absorption describes how a medication enters the bloodstream via a tablet or inhalation. Intravenous injection is the exception because it avoids absorption by administering a medication straight into the bloodstream. Distribution explains how much of the medicine reaches the target site after absorption. Several medications, for instance, are unable to cross the blood-brain barrier. The term "metabolism" describes the process by which a drug is broken down in the body. This can occur instantly due to stomach enzyme activity, or it can occasionally entail end products that have pharmacologic effects of their own. Lastly, excretion refers to the process by which medications are eliminated from the body, including through bile, urine, and occasionally exhalation.

## 2.2 Pharmacodynamics

Pharmacodynamics is the degree to which the target cells react to the drug. It is the molecular action of a drug on its target, be it a receptor, an ion channel, an enzyme, a regulatory protein, or a cell surface target. For instance, patients respond positively to beta-agonists, which include polymorphisms. The beta-2 adrenergic receptor, which is used to treat asthma, and beta-blockers, which are used to treat hypertension [4].

## 3 Drug Safety

The Food and Drug Administration reports that two million Americans experience severe adverse effects from pharmaceutical medications annually, which can result in up to 100,000 fatalities. An estimated $136 billion is spent each year on adverse drug reactions (ADRs). ADRs are of two types, i.e., one type is caused by abuse, such as taking too much medication or taking it too frequently or for too long and the second focuses on the unexpected effects of various drugs. These unaccustomed side effects are referred to as "idiosyncratic" as they are thought to be unforeseeable with drug dosage. Most idiosyncratic effects are generated due to individual variation in the DNA encoded. Individual differences in pharmacological potency and toxicity have thus been linked to genetic variance in genes encoding enzymes which metabolise drugs, drug receptors, and transporters. Hypersensitivity reactions, where the body mounts a quick, aggressive immune response that can result in not only a rash but also breathing difficulties and oedema to the point of cardiovascular collapse, are also genetically based in patients who are allergic to specific medications, such as penicillin. Pharmacogenomic research is focused on predicting adverse drug reactions. For instance, several antidepressants, antiarrhythmic, and antipsychotic medications are disintegrated, and their effects are stopped by the liver's CYP2D6 enzyme, which is a member of a class of drug-metabolising enzymes. More than 70 variant alleles have been identified by molecular cloning and characterisation investigations of the gene encoding this enzyme [5]. While some of these alleles entail gene deletions and duplications that can result in increased enzyme activity, these alleles contain single or more point mutations, only some of which have an impact on enzyme activity.

## 4 History of Pharmacogenetics

Archibald Garrod postulated in 1909 that when consumed in large doses, a drug turns to poison. Garrod also suggested that metabolic defects at birth of an individual determine the genetic basis for some diseases. Arthur L. Fox's 1932 discovery demonstrated that some people cannot taste the bitter taste of phenylthiocarbamide, which is found in broccoli. It shows that genes are responsible for xenobiotic response. Fox also identified this as an inherited trait, which was later linked to polymorphisms in TAS2R38, i.e. bitter-taste receptor, which is involved in food acceptance (Keller and Adise 2016). Following the identification of genetic defects in N-acetyl transferase, glucose-6-phosphate dehydrogenase deficiency, and plasma cholinesterase deficiency, Friedrich Vogel first used the term "pharmacogenetics" in 1959. Next was the CYP2D6 deficiency, which had severe negative effects on debrisoquin and sparteine [6]. This set of findings led to a thorough investigation into the role of genetics in medication response, including the cloning of the genes in question and the identification of the genetic variants that cause it. Many drug-metabolising enzymes and drug transporters were found in later research. The genes cloned and the mutations identified led to influencing the activities of the encrypted protein, with possible therapeutic significance. For instance, it was discovered that a CYP2D6 mutation abrogates debrisoquine metabolism, resulting in serious negative consequences. The aim of pharmacogenetics is to predict medication exposure and drug response using genetic biomarkers, which were made possible by growing understanding of the underlying genetic reasons. To maximise personal therapies, any such biomarker must be incorporated into a comprehensive decision framework, because many factors influence medication response.

Blood drug levels can act as a transitional phenotype that combines genetic influences with all other elements, such as renal functioning, enzyme induction and inhibition, and more. Therapeutic drug monitoring, which has been used for decades as a gauge of drug exposure, necessitates pharmacokinetic modelling, frequently with limited data. Drug-level monitoring is still important today thanks to novel drug-level tests and data analytics, which serve as useful biomarkers in numerous fields, such as depression, inflammatory illnesses, and cancer.

The identification of drug response receptors followed a similar historical trajectory to these gene clusters that encode the G protein-coupled receptors continue to be the primary target of medications in use today. Pharmacodynamics, which aims to maximise therapeutic response across various patient groups, is used to evaluate drug effects. However, the multidimensional structure of complicated biological response pathways complicates the notion that genetic polymorphisms in the genes of drug-receptor would be useful for therapy optimisation. Although pharmacological efficacy is typically polygenic, significant adverse drug reactions (ADRs) are less common and more likely to occur due to specific genetic variables.

According to the WHO, ADRs from routine medication, excluding overdoses, are a leading cause of hospitalisations and mortality. Numerous studies are being conducted to identify genetic factors that contribute to ADR susceptibility [7]. ADRs are classified as type A (warfarin and thiopurines) and type B (idiosyncratic drug reactions). Research suggests that genetics has a large role in ADRs and that using known pharmacogenetic variations can help prevent them. Gene variations that encode drug-metabolising enzymes were found to have strong relationships with type A ADRs, with activity levels directly related to drug levels in the body. Drug-induced prolongation of QT raises the possibility of "torsade de pointes," a fatal ventricular arrhythmia associated with various drug groups such as antibiotics, antipsychotics, cardiovascular, and anticancer treatments.

Pharmacogenetics has traditionally focused on gene interactions by a single drug, which effect therapeutic decisions. It has a few examples of multi-gene interactions [8, 9]. However, the complexity of diagnostic experiments is quickly rising. Human physiology and response to therapy are influenced by many variables other than genes. The integration of these complicated elements has become a popular trend, driven by breakthroughs in genomics. This approach reflects the complete genome rather than just individual genes.

# 5 Pharmacogenomics

Rapid developments in genomic sciences have resulted from the 2001 sequencing of the human genome, which has allowed us to concentrate on the ensemble of all genes and genomic functions rather than the genes of an individual. The National Human Genome Research Institute defines "pharmacogenomics" as "a branch of pharmacology concerned with using DNA and amino acid sequence data to inform drug development and testing." This more expansive perspective gave rise to the name.

## 5.1 Transition to Pharmacogenomics

Constitutional mutations in drug transporters and in enzymes metabolising drugs have a significant function in response to drugs, when the therapeutic window is limited. Targeted cancer therapies mostly rely on somatic mutations in oncogenes [8, 9]. The complete genome, protein-coding exome, transcriptome, methylome, and epigenome can now be examined to predict disease risk and treatment response [10]. However, the complicated landscape created by such massive data sets makes it difficult to identify strong genome-phenome connections responsible for challenges in clinical application [11]. Though few examples exist for treatments, the potential of genomics in therapeutic applications is frequently recognised.

The field of pharmacogenomics is further elaborated by new treatment techniques that go beyond conventional medications [12]. In addition to radiation therapy, nutrition, and physiotherapy, novel therapies include proteins, living cells, tissues, bacteria, viruses, RNA and DNA molecules, gene therapy, gene editing, and more. Thus, to reflect the extremely complex therapeutic environment, treatment tactics are tailored to individual variables. Tribute to genomics insights, biomarkers, and RNA medicines which are effective in vaccination and treatments for infections, hereditary illnesses, cancer, and heart disease [13]. Furthermore, as a substitute for or addition to medical treatments, the National Institutes of Health (NIH) advocates for "precision nutrition," which targets the microbiome. It is now clear that microbiota affects immunological response and cancer [14, 15]. In the development of the most individualised treatments and preventative measures, these characteristics come together in the field of personalised medicine, which necessitates the integration of various datasets.

## 5.2 Molecular Genomics of Pharmacogenes

At least 26 pharmacogenes have frequent variations that primarily act as biomarkers for treatment. It consists of over 90 drugs. Clinical value is supported by knowledge of genetic mechanisms, interactions between many variants per gene locus, and ethnogeographic variations in frequencies. Polygenic SNP panels, which primarily use surrogate markers for unknown causal variations, are used to evaluate the many other variants that may contribute to the overall genetic effect [16]. SNPs, insertions/deletions, variations i.e., in gene copy number and in gene structural are among the majority of currently clinically relevant variants found in transcribed regions. These variants alter the amino acid sequence, disrupt splicing, shift open reading frames, and alter the start or stop codons during translation, all of which have an impact on the function of encoded proteins. GWAS, however, show that around 90% of SNPs linked to traits or diseases are noncoding variants [17], with gene expression modulation being a key protocol.

According to Ingelman-Sundberg [18], genetic variations in CYP genes can change the metabolism of about 60% of clinically used medications. Other uncommon coding variations have been found as a result of exon sequencing and GWAS [19]. Furthermore, many variations outside of transcribed areas linked to CYP activity are shown by whole genome sequencing. Numerous variations linked to GWAS are still being investigated and may contribute to the "missing heritability". Although CYP3A4, the significant drug-metabolising enzyme, exhibits significant interindividual variation in expression and a significant genetic effect. The known CYP3A4 variations only explain a small percentage of population variance. However, new medications are frequently made to be metabolised by CYP3A4.

The identification of distant regulatory elements is made easier by new functional genomics tools. The Encyclopaedia of DNA elements and other publicly available genomic databases demonstrate that enhancer–promoter communication does not always follow the rule of proximity. Numerous distal enhancers target gene promoters through long-range DNA looping, interacting spanning 100 kb to 1 Mb [20]. Rapid identification of interacting chromatin regions is made possible by chromatin conformation capture technologies [21, 22]. These technologies also reveal intricate connections of the human genome, which include multiple interactions among distal enhancers and promoters [23].

## 5.3 Potential Therapeutics Derived from the Human Genome Project

The conclusion of the Human Genome Project sparked expectations that genetic information would greatly improve medicine, increase the predictability of side effects, and allow people to be checked for possible drug sensitivities. Development has been much slower thus far than it was during the early excitement. A large amount of this delay can be explained by the fact that an individual's response to drugs is multifactorial, resulting from several gene and environmental interactions [24]. While the realm of knowledge is expanding, scientists also recognise that the requirement of therapeutic knowledge is valid evidence for a specific disease and medication combination, prior treatment may be customised for the genotype of a particular. Therefore, before personalised medicine can reach its full potential, a lot more work needs to be done.

## 6 Pharmacogenomics-Based Cancer Treatment

Aggressive cytotoxic chemotherapies and surgery are two traditional cancer treatments that have numerous negative side effects. Since the 1980s, the search for molecularly focused therapies—defined as interventions based on biologic outcomes like protein overexpression or pathway activation—has gained attention.

### 6.1 Early Molecular Markers

Many years after its effectiveness against ER-positive breast carcinoma was suggested, tamoxifen—regarded as the first targeted cancer medication—was used as an adjuvant therapy in women with ER-positive breast cancer in the 1990s [25]. Since then, the pace of drug-biomarker pair discovery has skyrocketed, and hundreds of targeted medicines have been approved for almost all types of cancer [26, 27]. Using immunohistochemistry, fluorescence in situ hybridisation, or karyotyping, proteins or genes exhibiting amplification, translocation, or overexpression were examined in tumours as one of the earliest molecular cancer biomarkers. Through abnormal protein activation or enhanced activity, these genes and proteins mostly encourage the formation of tumours or aggressive behaviour. The discovery in a small category of breast tumours that amplification of the HER2 gene leading to greater expression of protein, corresponds with retarded effects. Due to its increased effectiveness when combined with the biomarker test, the FDA approved trastuzumab (Herceptin), an antibody against the HER2/neu protein, in 1998 for the treatment of HER2/neu-positive breast tumours exclusively [27].

### 6.2 Proteomics and Next-Generation Sequencing

Standard-of-care targeted sequencing panels, which identify overexpressed genes, constitutive and somatic variants, or biomarkers like tumour mutational burden (TMB) and microsatellite instability (MSI), have been made possible by the widespread use of NGS [28].

Gene amplification, translocation, oncogene activating mutations, mutational signatures, and TMB are examples of genetic/genomic biomarkers. More patients are undergoing tumour testing using tissue-agnostic cancer gene panels due to reduction of NGS costs [26]. Although there are few therapy options for many tumours with somatic mutations, targeted treatments have demonstrated clinical success in haematologic and solid malignancies. The oncogenes EGFR, ALK, BRAF, MET, and KIT are among the targets; these proteins lack conventional drug adherence sites, and lately, KRAS (p.Gly12Cys) encodes a protein that was thought to be resistant to pharmacological inhibition. Most heterozygous carriers have inactivating germline mutations in tumour-suppressor genes, which cause cancer when the wild-type allele is somatically lost [29]. To combine mass spectrometry-based proteomics with genomic sequence data, large-scale protein sequencing has evolved [30]. As molecular targets or indicators of therapeutic response, proteogenomics identifies protein fusions, new splice forms, and neo-antigens [31].

## 6.3 Molecular Signatures and Pathways

Molecular signals of disrupted cancer pathways may serve as possible biomarkers due to genetic and chromosomal variation. MSI is one such biomarker. Random insertions or deletions in repeated DNA sequences, like homopolymers or microsatellite repeats, are found in tumours with inadequate mismatch repair. This results in heritable mutations of important cancer-causing genes. Lynch syndrome, a hereditary cancer brought on by germline pathogenic mutations in mismatch repair genes. MSI was initially identified in endometrial and colorectal tumours that developed in patients with Lynch syndrome. MSI appears to be a promising biomarker for immune-based checkpoint inhibitor response, according to recent research [32].

When the wild-type allele is somatically lost, tumours that develop in people with germline BRCA1 or BRCA2 pathogenic alleles usually exhibit a lack of homology-directed repair (HDR) capability, which is crucial for double-strand break repair. As a synthetic lethality strategy, poly-adenosine diphosphate-ribose polymerase (PARP) inhibitors (PARPi) were created to take advantage of the dependence of BRCA-deficient tumours on base excision and single-strand break repair, which are carried out by PARPs. Cell death and irreversible double-strand breaks in BRCA-deficient tumours result from the failure to fix single-strand breaks in tumours treated with PARPi [33].

PARPi was first approved by the FDA in 2014 for advanced BRCA-deficient ovarian cancers with mutations in BRCA1 or BRCA2. BRCA-deficient metastatic breast, pancreatic, and prostate tumours were the next to receive PARPi [34]. The goal of ongoing research is to apply the PARPi method to more HDR genes, including CHEK1, ATR, RAD51C/D, and PLB2. Assessing "BRCAness" phenotypes and HDR genomic scoring techniques, such as differential gene expression profiling, DNA copy-number alterations, chromosomal rearrangements, genomic instability scores, and mutations, have been developed to increase the use of PARP inhibitor therapies. Research is being conducted to evaluate PARPi in tumours that have somatic mutations in any DNA damage repair genes [35].

Additionally, mutational signatures function as biomarkers of treatment resistance. The dysregulated activity of the apolipoprotein b mRNA editing enzyme, catalytic polypeptide-like (APOBEC), results in a common mutational signature [36]. APOBEC dysregulated tumours exhibit a distinctive mutational pattern linked to treatment resistance, resulting in the deamination of cytidines at particular sequence motifs. For instance, tumours treated with osimertinib, which targets EGFR-mutant lung cancers and receptor-tyrosine kinase-driven lung malignancies, both commonly have APOBEC mutational signatures [37].

## 6.4 Rising Role of "Basket Trials"

Initially, targeted treatments were evaluated for tissue-specific malignancies; however, the design of clinical trials was hindered by the lack of patients with a target mutation. "Basket trials" are used for tissue-agnostic therapy of tumours that has the same molecular biomarker because molecular targets can appear in multiple tumour types, presuming that the tumour driver mutation is the primary determinant of drug efficacy. Imatinib was tested in one of the earliest basket trials to treat tumours that have activating mutations or high expression of receptor tyrosine kinases such as PDGFRA, PDGFRB, or KIT. Six cancers out of the more than 40 tumour types examined exhibited signs of clinical responses. However, due to pathway-escape mechanisms, some treatments are more effective against melanoma and non-small cell lung cancer rather than colorectal cancer, while others are more effective against a variety of tumour types [38].

## 6.5 Molecular Biomarkers for Response to Onco-Immune Therapy

By identifying foreign tumor-specific antigens and proteins that are not typically produced by mature cells, the immune system plays a crucial role in controlling malignancies. Cytotoxic immune cells may eradicate tumour cells because of this immunosurveillance. By overexpression of immune checkpoint proteins such PD-1, PD-L1, or CTLA-4, which compete for T-cell interaction to B7 molecules, cancer cells might evade death. By preventing inhibitory tumour tissue interaction with T-cells and facilitating the identification and destruction of tumour cells, antibodies against these immune checkpoint proteins, either alone or in combination, have been successful in the treatment of melanoma and non-small cell lung carcinoma.

Only a small percentage of patients who overexpress these proteins respond to immune checkpoint-based antibody treatments. Molecular biomarkers that predict responsiveness and successful combination therapy are sought after. Because more neoantigens are produced, tumours with a high mutational burden (TMB) typically react well to checkpoint inhibitors. But according to certain studies, TMB's predictive value is not very strong. The evaluation of immune gene signatures, selective polyadenylation signals, changes in somatic copy number CD274, and polymorphisms in CD274 and PDCD1 that promise improved prediction of responses to onco-immune therapies are all ongoing efforts to improve biomarkers for immune checkpoint inhibitor response.

## 6.6 Acquired Resistance to Targeted Therapy and Counterstrategies

### 6.6.1 Somatic Mutations of Genes in Parallel Pathway

Although targeted cancer treatments might produce striking initial results, tumours often become resistant. For instance, vismodegib, which blocks the Hedgehog pathway, was first authorised for use in treating patients with PTCH1 mutations who had locally or metastatic basal cell carcinoma. According to early trial results, after 6–12 months of treatment, both primary basal cell carcinomas and metastatic lesions were dramatically eliminated [39]. However, 20% of patients experienced tumour recurrence within a year. Furthermore, a sizable portion of patients experience significant adverse effects. By amplification of GLI genes or activating mutations in SMO, the latest mutations in other areas of the Hedgehog pathway allow tumours to evade inhibition. Targeted therapy resistance is frequently caused by secondary mutations in other pathway genes. Triggering mutations in ESR1 in oestrogen receptor-positive breast tumours treated with aromatase inhibitors or oestrogen receptor ligands like tamoxifen are just one of several other instances of secondary mutations that result in therapeutic resistance [40]. Since ESR1 mutations are uncommon in tumours that have not responded to treatment, it is likely that the therapy's selection pressure is the main cause of their formation. Clinical trials assessing intermittent dosing or combination therapy that may lessen adverse effects and resistance have been called for because of tumour resistance to targeted therapies [41].

### 6.6.2 Reverse Mutations

Breast tumours with germline pathogenic variations in BRCA1 or BRCA2 may be resistant to PARPi due to reversion mutations that fix the initial defect [42]. Rarely, tumours have secondary mutations that "correct" the original germline abnormality, such as restoring the gene's reading frame. Platinum-resistant or PARPi-refractory tumours often have reversion mutations in BRCA1 and BRCA2, which repair DNA damage caused by drugs [43].

### 6.6.3 Tumour Heterogeneity

Tumour heterogeneity is caused by genetic and nongenetic variations. It corresponds with awful outcomes and a greater incidence of therapy resistance. Tumours with significant heterogeneity and neoantigen variety is less susceptible to immune checkpoint treatments [44]. Cancer therapy techniques are shifting from focussing on specific biological targets to incorporating combinatorial medicines to address potential tumour resistance pathways. Molecularly targeted and immunological checkpoint medicines have been effective in treating many tumours. Liquid biopsies, a non-invasive method for monitoring tumours, can detect resistance in real-time before clinical presentation. Tumour DNA is delivered into the bloodstream when tumour cells undergo apoptosis. Circulating DNA may be analysed by NGS to identify clonal tumour diversity, targetable mutations, therapy response, minimal residual disease after surgery, and novel mutations that may lead to therapy resistance. These technologies may replace tumour imaging modalities, particularly in tumours with greater ctDNA levels. The FDA has approved ctDNA monitoring, which has demonstrated success in clinical trials for lung and colorectal malignancies as well as identifying reversion mutations in PARPi-treated tumours [45]. Detection of tumour evasion in near real time can inform therapy tactics.

## 7 Pharmacogenomics in Drug Discovery and Development

Genomics now plays a crucial role in medication discovery and development, guiding innovative treatments to clinical trials. Integrating genetic sequence, transcriptomes in affected tissues, proteome, and metabolome can help detect appropriate therapeutic targets [46]. Genetic research has identified PCSK9 deficiency as a protective factor against cardiovascular disease by lowering cholesterol levels. This has led to the development of successful medicines utilising PCSK9 antibodies or RNA blockers. Many medications in clinical usage today require a biomarker to ensure efficacy or prevent harm. According to Jacobs et al., Herceptin would not have been approved without the use of a biomarker [27]. Kim et al. reported that theranostics now account for 75% of biomarker drug pairings requiring pharmacogenomic testing in cancer therapy. Radiolabelled medicines bind to specific molecular targets, allowing for disease imaging and therapy with higher-energy molecules. Membrane antigens that are Prostate-specific can be visualised using gallium-68-labeled positron emission tomography (PET) scans and treated with 177Lu-PSMA-617. The biomarker has the same value as the medication if mandated by regulatory bodies, i.e., the FDA. The table provides instances of recent personalised drugs (Table 1). The Personalised Medicine Coalition provides annual reports on FDA-approved "personalised therapeutics," which make up about 30% of all new drug approvals.

Discovery of drugs in the pharmaceutical business has typically focused on creating medicines for prevalent illnesses with high economic returns. Bringing "one-size-fits-all" medications to clinics is challenging and expensive. In 1983, the United States Congress created the Orphan Drug Act to encourage the evolution of medicines for rare diseases. The legislation has resulted in a significant increase of approved orphan medications, largely due to genomics and innovative drug discovery technologies, as well as support for gene therapy. Precision medications often require biomarker testing to determine patient eligibility. Upto 2019, 88% of all precision medications were cancer-targeting orphan pharmaceuticals, with 44% of those therapies focussing on this disease. Biomarkers can identify subgroups of cancer patients with shared mutations, leading to personalised therapy and orphan drug classification, despite the disease's widespread prevalence. The combination of orphan drug designation and pharmacogenetic testing is being used to address rare illnesses. For example, numerous mutations in the CFTR gene cause cystic fibrosis, which affects protein folding, channel function, and transport to the plasma

**Table 1** Pharmacogenetics-based medications in oncology

| FDA Drug | Biomarker | Pharmacogenomic role |
|---|---|---|
| Abemaciclib | ERBB2 (HER2) | Indications and usage, adverse reactions, clinical studies |
| Adagrasib | KRAS | Indications and usage, dosage and administration, adverse reactions, clinical pharmacology, clinical studies |
| Ado-Trastuzumab Emtansine | ERBB2 (HER2) | Usage, dosage, and administration, adverse reactions, clinical pharmacology, clinical studies |
| Afatinib | EGFR | Indications and usage, dosage and administration, adverse reactions, clinical studies |
| Alectinib | ALK | Indications and usage, dosage and administration, adverse reactions, clinical pharmacology, clinical studies |
| Allopurinol | HLA-B | Warnings |
| Alpelisib | ERBB2 (HER2), ESR, PIK3CA | Indications and usage, dosage and administration, adverse reactions, clinical studies |
| Amivantamab | EGFR | Indications and usage, dosage and administration, adverse reactions, clinical studies |
| Arsenic trioxide | PML-RARA | Indications and usage, clinical studies |
| Bosutinib | BCR-ABL1 | Indications and usage, dosage and administration, warnings and precautions, adverse reactions, use in specific |
| Brigatinib | ALK | Indications and usage, dosage and administration, adverse reactions, clinical studies |
| Cetuximab | EGFR, RAS, BRAF | Indications and usage, dosage and administration, adverse reactions, use in specific populations, clinical studies |
| Dacomitinib | EGFR | Indications and usage, dosage and administration, adverse reactions, use in specific populations, clinical studies |
| Docetaxel | ESR, PGR | Clinical studies |
| Fluorouracil | DPYD | Warnings and precautions, clinical pharmacology, patient counseling information |
| Gefitinib | EGFR, CYP2D6 | Indications and usage, dosage and administration, clinical studies |
| Imatinib | KIT, BCR-ABL1, PDGFRB, FIP1L1-PDGFRA | Indications and usage, dosage and administration, clinical studies |
| Lapatinib | ESR, PGR | Indications and usage, dosage and administration, adverse reactions, use in specific populations, clinical studies |

(continued)

**Table 1** (continued)

| FDA Drug | Biomarker | Pharmacogenomic role |
|---|---|---|
| Mercaptopurine | TPMT, NUDT15 | Dosage and administration, warnings and precautions, clinical pharmacology |
| Neratinib | ERBB2 (HER2) | Indications and usage, adverse reactions, clinical studies |
| Olaparib | BRCA, ERBB2, ESR, PGR, PP2R2A | Indications and usage, dosage and administration, warnings and precautions, adverse reactions, clinical studies |
| Panitumumab | EGFR, RAS | Indications and usage, dosage and administration, warnings and precautions, adverse reactions, clinical studies |
| Pazopanib | UGT1A1, HLA-B | Clinical pharmacology |
| Quizartinib | FLT3 | Indications and usage, dosage and administration, warnings and precautions, adverse reactions, clinical studies |
| Rucaparib | BRCA, CYP2D6 | Indications and usage, dosage and administration, adverse reactions, clinical studies |
| Sotorasib | KRAS | Indication and usage, dosage and administration, adverse reactions, clinical pharmacology, clinical studies |
| Tamoxifen | F5, F2, CYP2D6, ESR, PGR | Warnings and precautions, clinical pharmacology, indications and usage, adverse reactions, clinical pharmacology, clinical studies |
| Trastuzumab | ESR, PGR, ERBB2 | Clinical studies, indications and usage, dosage and administration, clinical pharmacology, clinical studies |
| Tretinoin | PML-RARA | Indications and usage, dosage and administration, warnings and precautions, patient counselling, information |
| Tucatinib | ERBB2, RAS | Indications and usage, dosage and administration, adverse reactions, clinical studies |
| Vemurafenib | RAS, chromosome 17p & 11q, IDH1, IDH2, TP53, IGHV, NPM1 | Warnings and precautions, adverse reactions, clinical studies |
| Vincristine | BCR-ABL1 | Indications and usage, adverse reactions, clinical studies |
| Zanubrutinib | MYD88 chromosome 17p TP53 | Adverse reactions, clinical studies |

membrane. CFTR mutations affect CFTR function differently; therefore, different medications may be most effective in treating the defect. Compilation of tezacaftor, ivacaftor, and elaxacaftor effectively restores D508 CFTR function and may be effective for additional CFTR mutations [47].

## 8 Conclusion

The use of genetic data raises concerns about confidentiality, abuse by other parties, and potential consequences for patients' families if harmful mutations are detected. In clinical trials, patients' rights to know or be protected from detrimental variations should be clarified through signed consent. Personalised medicine involves integrating medical, genetic, personal, cultural, and socioeconomic characteristics into easily available electronic health records [48]. While it is still unclear if this strategy can replace experienced healthcare workers' intuition, it is important to prioritise beneficence, justice, and equality. To advance personalised health care, it's necessary to integrate several disciplines and technology to identify interconnected and influencing factors. Artificial intelligence and machine learning have the potential to help understand such complexities in the future. To improve clinical translation, basic genomics research should combine critical features of personalised medicine, as pharmacogenomics plays a crucial role.

**Declarations**

**Author's Contributions** Conceived the idea and review structure: SG, SK, MKG; Wrote the manuscript: SG; Revised and edited the manuscript: SG, SK, MKG. The authors have read, agreed, and approved the final draft of the manuscript.

**Funding** This work is jointly supported by the Department of Science and Technology (EMR/2017/000992), Focused Basic Research (FBR), HCT, and OLP-121, Govt. of India.

**Data Availability** The software Biorender (https://www.biorender.com/) is used to draw some elements of the figures and, hence, the authors are happy to acknowledge Biorender.

**Conflicts of Interests** There is no competing and conflict of interests.

## References

1. Simmons LA, Dinan MA, Robinson TJ, Snyderman R (2012) Personalized medicine is more than genomic medicine: confusion over terminology impedes progress towards personalized healthcare. Pers Med 9(1):85–91. https://doi.org/10.2217/pme.11.86
2. Sadee W, Wang D, Hartmann K, Toland AE (2023) Pharmacogenomics: driving personalized medicine. Pharmacol Rev 75(4):789–814
3. Cecchin E, Stocco G (2020) Pharmacogenomics and personalized medicine. Genes 11(6):679. https://www.mdpi.com/2073-4425/11/6/679. Accessed 23 July 2025
4. Goldstein DB, Tate SK, Sisodiya SM (2003) Pharmacogenetics goes genomic. Nat Rev Genet 4(12):937–947
5. Meyer UA (2000) Pharmacogenetics and adverse drug reactions. Lancet 356(9242):1667–1671
6. Mueller CM, Rao GR, Miller Needleman KI (2019) Precision medicines' impact on orphan drug designation. Clin Transl Sci 12(6):633–640. https://doi.org/10.1111/cts.12667
7. Cacabelos R, Cacabelos N, Carril JC (2019) The role of pharmacogenomics in adverse drug reactions. Expert Rev Clin Pharmacol 12(5):407–442. https://doi.org/10.1080/17512433.2019.1597706

8. Miteva-Marcheva NN, Ivanov HY, Dimitrov DK, Stoyanova VK (2020) Application of pharmacogenetics in oncology. Biomark Res 8(1). https://doi.org/10.1186/s40364-020-00213-4
9. Miteva-Marcheva N, Ivanov H, Linev A et al (2023) Pharmacogenetic markers associated with drug metabolism in patients with oncological diseases. Bulg Chem Commun 55:45
10. Rabbani B, Nakaoka H, Akhondzadeh S, Tekin M, Mahdieh N (2016) Next generation sequencing: implications in personalized medicine and pharmacogenomics. Mol BioSyst 12(6):1818–1830
11. Ji X, Niu X, Qian J et al (2018) A phenome-wide association study uncovers a role for autoimmunity in the development of chronic obstructive pulmonary disease. Am J Respir Cell Mol Biol 58(6):777–779. https://doi.org/10.1165/rcmb.2017-0409le
12. Brown DG, Wobst HJ (2021) A decade of FDA-approved drugs (2010–2019): trends and future directions. J Med Chem 64(5):2312–2338. https://doi.org/10.1021/acs.jmedchem.0c01516
13. Servick K (2020) 'Huge hole' in testing data blurs racial, ethnic disparities. Science 369(6501):237–238. https://doi.org/10.1126/science.369.6501.237
14. Mager LF, Burkhard R, Pett N et al (2020) Microbiome-derived inosine modulates response to checkpoint inhibitor immunotherapy. Science 369(6510):1481–1489. https://doi.org/10.1126/science.abc3421
15. Woelk CH, Snyder A (2021) Modulating gut microbiota to treat cancer. Science 371(6529):573–574. https://doi.org/10.1126/science.abg2904
16. Zhou Y, Lauschke VM (2022) Population pharmacogenomics: an update on ethnogeographic differences and opportunities for precision public health. Hum Genet 141(6):1113–1136. https://doi.org/10.1007/s00439-021-02385-x
17. Hindorff LA, Sethupathy P, Junkins HA et al (2009) Potential etiologic and functional implications of genome-wide association loci for human diseases and traits. Proc Natl Acad Sci 106(23):9362–9367. https://doi.org/10.1073/pnas.0903103106
18. Ingelman-Sundberg M, Nebert DW, Lauschke VM (2023) Emerging trends in pharmacogenomics: from common variant associations toward comprehensive genomic profiling. Hum Genomics 17(1):105. https://doi.org/10.1186/s40246-023-00554-9
19. Loukola A, Buchwald J, Gupta R et al (2015) A genome-wide association study of a biomarker of nicotine metabolism. PLoS Genet 11(9):e1005498
20. Heidari N, Phanstiel DH, He C et al (2014) Genome-wide map of regulatory interactions in the human genome. Genome Res 24(12):1905–1917
21. Dekker J, Rippe K, Dekker M, Kleckner N (2002) Capturing chromosome conformation. Science 295(5558):1306–1311. https://doi.org/10.1126/science.1067799
22. Tolhuis B, Palstra RJ, Splinter E, Grosveld F, De Laat W (2002) Looping and interaction between hypersensitive sites in the active β-globin locus. Mol Cell 10(6):1453–1465
23. Sanyal A, Lajoie BR, Jain G, Dekker J (2012) The long-range interaction landscape of gene promoters. Nature 489(7414):109–113
24. Rich EC, Burke W, Heaton CJ et al (2004) Reconsidering the family history in primary care. J Gen Intern Med 19(3):273–280. https://doi.org/10.1111/j.1525-1497.2004.30401.x
25. Jordan VC (1994) The development of tamoxifen for breast cancer therapy. Long-term Tamoxifen therapy for the treatment of breast cancer. Cancer Control 3:26
26. Aleksakhina SN, Imyanitov EN (2021) Cancer therapy guided by mutation tests: current status and perspectives. Int J Mol Sci 22(20):10931
27. Jacobs AT, Castaneda-Cruz DM, Rose MM, Connelly L (2022) Targeted therapy for breast cancer: an overview of drug classes and outcomes. Biochem Pharmacol 204:115209
28. Mosele F, Remon J, Mateo J et al (2020) Recommendations for the use of next-generation sequencing (NGS) for patients with metastatic cancers: a report from the ESMO precision medicine working group. Ann Oncol 31(11):1491–1505
29. Gilson P, Merlin JL, Harlé A (2021) Detection of microsatellite instability: state of the art and future applications in circulating tumour DNA (ctDNA). Cancer 13(7):1491
30. Mani DR, Krug K, Zhang B et al (2022) Cancer proteogenomics: current impact and future prospects. Nat Rev Cancer 22(5):298–313

31. Subbannayya Y, Di Fiore R, Urru SAM, Calleja-Agius J (2021) The role of omics approaches to characterize molecular mechanisms of rare ovarian cancers: recent advances and future perspectives. Biomedicine 9(10):1481
32. Cercek A, Lumish M, Sinopoli J et al (2022) PD-1 blockade in mismatch repair–deficient, locally advanced rectal cancer. N Engl J Med 386(25):2363–2376. https://doi.org/10.1056/NEJMoa2201445
33. Chen A (2011) PARP inhibitors: its role in treatment of cancer. Chin J Cancer 30(7):463
34. Sigorski D, Iżycka-Świeszewska E, Bodnar L (2020) Poly(ADP-ribose) polymerase inhibitors in prostate cancer: molecular mechanisms, and preclinical and clinical data. Target Oncol 15(6):709–722. https://doi.org/10.1007/s11523-020-00756-4
35. Zhang D, Xu X, Wei Y et al (2022) Prognostic role of DNA damage response genes mutations and their association with the sensitivity of Olaparib in prostate cancer patients. Cancer Control 29:10732748221129451. https://doi.org/10.1177/10732748221129451
36. Mertz TM, Collins CD, Dennis M, Coxon M, Roberts SA (2022) APOBEC-induced mutagenesis in cancer. Annu Rev Genet 56(1):229–252. https://doi.org/10.1146/annurev-genet-072920-035840
37. Selenica P, Marra A, Choudhury NJ et al (2022) APOBEC mutagenesis, kataegis, chromothripsis in EGFR-mutant osimertinib-resistant lung adenocarcinomas. Ann Oncol 33(12):1284–1295
38. Robert C, Grob JJ, Stroyakovskiy D et al (2019) Five-year outcomes with Dabrafenib plus Trametinib in metastatic melanoma. N Engl J Med 381(7):626–636. https://doi.org/10.1056/NEJMoa1904059
39. Dessinioti C, Plaka M, Soura E et al (2019) A practical guide for the follow-up of patients with advanced basal cell carcinoma during treatment with hedgehog pathway inhibitors. Oncologist 24(8):e755–e764
40. Brett JO, Spring LM, Bardia A, Wander SA (2021) ESR1 mutation as an emerging clinical biomarker in metastatic hormone receptor-positive breast cancer. Breast Cancer Res 23(1):85. https://doi.org/10.1186/s13058-021-01462-3
41. Leavitt E, Lask G, Martin S (2019) Sonic hedgehog pathway inhibition in the treatment of advanced basal cell carcinoma. Curr Treat Options in Oncol 20(11):84. https://doi.org/10.1007/s11864-019-0683-9
42. Zong H, Zhang J, Xu Z et al (2022) Comprehensive analysis of somatic reversion mutations in homologous recombination repair (HRR) genes in a large cohort of Chinese pan-cancer patients. J Cancer 13(4):1119
43. Pettitt SJ, Frankum JR, Punta M et al (2020) Clinical BRCA1/2 reversion analysis identifies hotspot mutations and predicted neoantigens associated with therapy resistance. Cancer Discov 10(10):1475–1488
44. Craig DJ, Bailey MM, Noe OB et al (2022) Subclonal landscape of cancer drives resistance to immune therapy. Cancer Treat Res Commun 30:100507
45. Malla M, Loree JM, Kasi PM, Parikh AR (2022) Using circulating tumor DNA in colorectal cancer: current and evolving practices. J Clin Oncol 40(24):2846–2857. https://doi.org/10.1200/JCO.21.02615
46. Plenge RM (2019) Priority index for human genetics and drug discovery. Nat Genet 51(7):1073–1075
47. Fiedorczuk K, Chen J (2022) Molecular structures reveal synergistic rescue of Δ508 CFTR by Trikafta modulators. Science 378(6617):284–290. https://doi.org/10.1126/science.ade2216
48. Hoffman JM, Dunnenberger HM, Kevin Hicks J et al (2016) Developing knowledge resources to support precision medicine: principles from the clinical pharmacogenetics implementation consortium (CPIC). J Am Med Inform Assoc 23(4):796–801

# Precision Cancer Medicine in Clinical Trial Design

Arijit Bhowmik, Souvik Das. Sandip Ghosh, and Biswarup Basu

**Abstract**

Precision oncology in medicine treats a cancer patient based on the molecular characteristics. Traditionally, cancer therapy relied on histological classifications and site of origin, but this paradigm fails to take into account tumor heterogeneity. Using immense analytical NGS assays and/or multiomic environments and functional assays, the tumor has undergone deeper profiling for the detection of actionable biomarkers and patient-tailored therapy selection. The practice of using cohort clinical trial designs such as basket, umbrella, platform, and N-of-1 trials arose in response to the molecular complexity of cancers, aiming to fast-track the evaluation of targeted therapies. Due to the limited durability of monotherapies, drug combinations and adaptive dosing are increasingly employed. Artificial intelligence with advanced machine learning finds real-time application in sifting through enormous molecular data for drug match and trial optimization. Decentralized as well as community-engaged trials keep widening the horizon of therapeutics, especially for the needy areas, while real-world evidence, synthetic control arms, and patient-reported outcomes enrich evidence generation. Trials of I-PREDICT, WINTHER, NCI-MATCH, TAPUR, and TRACK exemplify how these approaches are applied across tumor types and settings. Ethical issues, such as informed consent, data privacy, equity in access, etc., are essential for the future of precision oncology. Clinical implementation is advancing, yet challenges persist due to resistance mechanisms, hurdles in data interpretation, and disparities in access to genomic testing and targeted therapies. However, with innovations in diagnosis and trial methods, combined with an integration of data, precision medicine promises to change cancer care—a step toward more effective and individualized treatment.

A. Bhowmik (✉) · S. D. S. Ghosh · B. Basu (✉)
Department of Neuroendocrinology and Experimental Hematology, Chittaranjan National Cancer Institute, Kolkata, West Bengal, India

 
M. K. Ghosh, M. Basu (eds.), *Personalized Cancer Therapy*,
https://doi.org/10.1007/978-981-95-6845-1_6

## 1 Introduction

Treatment for cancer is rapidly changing, driven by rapid improvements in molecular technologies and enhanced tumor biology knowledge. Cancers have historically been characterized and treated on the basis of tissue of origin. Yet next-generation sequencing (NGS) and multiomic profiling have revealed an immense degree of inter- and intra-tumoral heterogeneity, which proves that cancers, even those occurring in the same tissue, commonly possess exceedingly diverse genomic and molecular features [1–3]. These findings have prompted a move away from the "one-size-fits-all" approach to precision medicine strategies that align therapeutic interventions with the distinct molecular signatures of each patient's cancer. Molecularly guided therapies, particularly those targeting specific genomic alterations, have yielded notable clinical successes. Nonetheless, monotherapies directed at a single biomarker frequently fail to produce durable responses, especially in advanced or metastatic settings, where tumors tend to harbor complex constellations of co-occurring mutations [3]. As a result, personalized combination therapies, tailored to each patient's unique tumor biology, are increasingly being recognized as a necessary strategy for achieving better clinical outcomes [1, 4]. Precision oncology has also fueled an attendant reinvention of clinical trial designs. The conventional phased model of oncology drug development (phase 1, 2, and 3 trials) is wasteful and time-consuming, taking years to make absolute decisions. In contrast, new trial designs have been conceived that are more agile, faster, and better at leveraging molecular information. These include umbrella trials (multiple therapies tested within a single cancer type, stratified by biomarker), basket trials (one treatment tested across multiple cancer types with a shared biomarker), and adaptive designs allowing real-time protocol modification as data are accrued [1, 5–7]. Most recently, patient-centered N-of-1 trials have come into prominence, in which individual patients receive treatments based on the molecular profile of their tumor. These trials assess the clinical utility of the drug-matching algorithm, but not the activity of a single treatment on a single cancer type [8–11]. Multiomic technologies—such as transcriptomics, proteomics, epigenomics, metabolomics, and immunomics—now complement genomic analysis to provide a more complete understanding of tumor behavior and therapeutic vulnerabilities [1, 4, 12]. These data sources are essential for identifying cancer-driving alterations that may not be captured by DNA sequencing alone. Furthermore, emerging platforms such as organoid-based functional assays and single-cell sequencing, and spatial information are beginning to play a role in preclinical therapeutic testing and biomarker discovery [13–15]. Machine learning (ML) and artificial intelligence (AI) are increasingly being applied to analyze such vast, high-dimensional data sets. With over 1 trillion combinations of genomic alterations and millions of drug pairs, computational strategies must be applied to accelerate therapy choice and hasten drug development [1]. They can also supply real-time information and rapid therapy adaptation when the tumor evolves or develops resistance mechanisms. This transition to molecularly defined trials also fits with trial logistics and access modifications. Decentralized and virtual clinical trial models are being investigated to access a larger population of patients, particularly in underserved or rural populations [1, 16]. These models have the potential to decrease barriers to trial entry and to precision medicine equity. Another step forward is the application of real-world data (RWD), synthetic control arms, and patient-reported outcomes (PROs) to clinical trial designs. Each of these can potentially speed the timeline and reduce the cost of trials, while enhancing the salience and comprehensiveness of clinical results [1, 17]. Taken together, these represent a departure from traditional, rigid, protocol-driven trials toward flexible, patient-focused studies more in line with the recording of the subtleties of modern cancer treatment.

In this chapter, the discussion is about how clinical trial design is changing in the face of the molecular reclassification of cancer. The technologies and methods that are facilitating this change, from multiomic integration and functional assays to AI-based decision-making and decentralized study models, are also identified. The need for equitable access to precision medicine is also discussed and offered a patient's view of the promise and pitfalls of these new methodologies.

## 2 Evolution of Clinical Trial Design

### 2.1 Introduction to Drug-Centered Approaches

Traditionally, for many decades, classic oncology clinical trials have been drug-centered, with emphasis placed on developing and proving proof-of-principle of novel treatments [18]. With this design, patient groups were selected to fit the trial design, which was focused on assessing pre-clinically proven drugs. All-inclusive, based primarily on cancer histologic type and organ of origin, defined this paradigm. While valuable in certain contexts, this design is constrained in the precision oncology era that recognizes the molecular heterogeneity of cancer within and between tumor types [3]. Drug-specific trials span a spectrum of study designs from the classic phase 1, 2, and 3 trials to those that focus on specific tumor histologies or specific genomic deficiencies. Increasingly, over the past few years, these trials have been accompanied by tumor-agnostic basket trials that enroll patients on the basis of shared molecular abnormalities, e.g., NTRK fusions, regardless of tumor type [19]. Even in the face of growing individualization, drug-based trials remain the mainstay of oncology research and still provide valuable data on safety, efficacy, and mechanism of action (Fig. 1). More and more, however, they are supplemented—and in some cases, replaced—by more nimble and more individualized methods [20].

### 2.2 Traditional Drug-Centric Trial Designs

#### 2.2.1 Phase 1: First-in-Human Studies

Phase 1 trials are the initial human experience with a therapy. They are performed primarily to assess the safety, tolerability, and pharmacokinetics of a new agent,

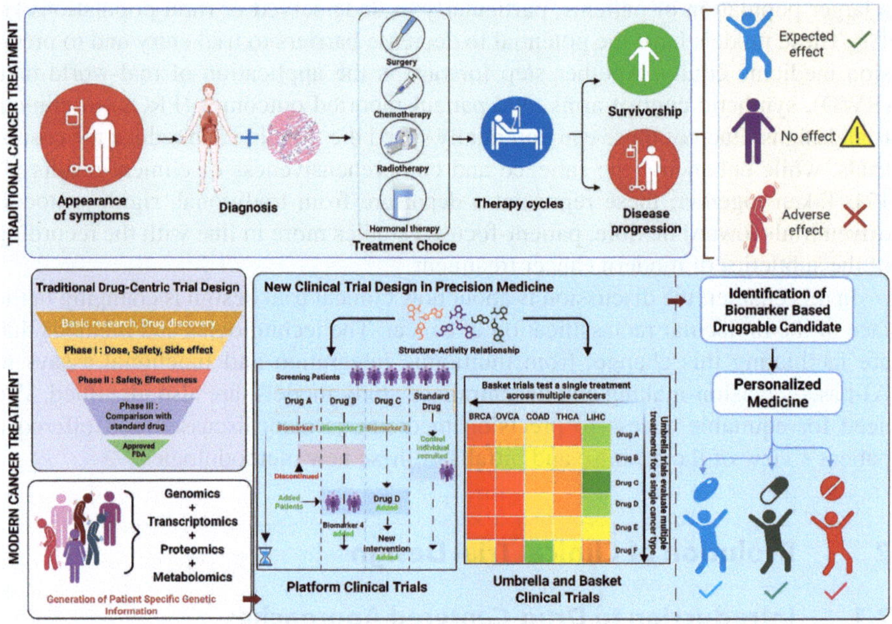


**Fig. 1** Overview of traditional and modern clinical trial designs in cancer treatment. The top panel illustrates the traditional cancer treatment pathway beginning with the appearance of symptoms, followed by diagnosis and selection of treatment choices (surgery, chemotherapy, radiotherapy, hormonal therapy, etc.). Patients undergo chosen therapy cycles that result in either survivorship or disease progression, with outcomes as favorable or unfavorable. Traditional trials proceed in sequential phases (preclinical research, phase 1–3 trials) and culminate in regulatory approval. Compared to traditional drug-centric trials, and to further address the modern precision medicine approach, modern trial frameworks—enabled by omics technologies (genomics, transcriptomics, proteomics, metabolomics), and integration of advanced in silico data—support the generation of patient-specific genetic profile-based therapeutics. These enable adaptive, biomarker-driven trials such as platform, umbrella, and basket trials. Platform trials dynamically add or remove arms based on real-time data. Umbrella trials test multiple treatments within one cancer type stratified by biomarkers, while basket trials assess a single treatment across different cancer types sharing a common molecular alteration. These innovative designs facilitate personalized medicine by linking identified druggable biomarkers to tailored therapies

typically establishing the MTD and RP2D [21]. The traditional 3 + 3 dose-escalation design is most widely practiced, in which small cohorts of patients receive progressively increasing doses to identify DLTs. More recently evolved, more accelerated phase 1 designs have been created to accelerate development, but at the cost of ignoring toxicities with reduced sample sizes. In the past, patients with very advanced, treatment-refractory cancers have been enrolled in phase 1 trials. More recently, enrolling larger populations of patients in phase 1 trials has become more common. Phase 1 trials have even yielded significant efficacy signals, leading to regulatory approval in some instances [21].

### 2.2.2 Phase 2: Targeted Populations Effectiveness

Phase 2 trials assess the RP2D established in phase 1 in a larger, often histologically characterized patient population. Phase 2 trials are intended to determine whether a drug demonstrates sufficient activity to warrant further evaluation in a phase 3 trial. In contemporary trial designs, phase 2 trials may target biomarker-defined subpopulations within a single tumor type or increasingly treat patients with more than one tumor type based on a shared molecular profile [20].

### 2.2.3 Phase 3: Randomized Controlled Trials

Phase 3 trials are the gold standard for the assessment of new therapies. Large, frequently multi-center trials compare the new therapy to the standard of care in a randomized, controlled environment. Double-blinded in most instances, phase 3 trials are usually superiority or non-inferiority trials. Although methodological sophistication is unparalleled, phase 3 trials are frequently costly, time-consuming, and rigid—factors that can delay access to promising therapies and constrain responsiveness to new scientific findings [22].

## 2.3 Innovative Drug-Centered Trial Designs

With the acknowledgment of the shortcomings of the previous models, a number of new trial designs have been developed. These models are created to provide greater flexibility, efficiency, and applicability in the age of molecular oncology.

### 2.3.1 Broad Phase 2 Trials

Used almost exclusively during the 1970s and 1980s, phase 2 trials with a broad label enrolled patients with a wide variety of solid and hematologic cancers. A typical example was the first trial of doxorubicin performed by the Southwest Oncology Group, which treated over 470 patients with a variety of tumors [23]. In this single trial, not only the dosage and toxicity patterns were defined but also the activity against a variety of tumor types was demonstrated. Despite their success, large phase 2 trials fell out of favor because, increasingly, there was a suspicion that tumor histology was the greatest determinant of drug response. This led to an explosion of single-disease, single-agent studies, which in retrospect were less successful than the larger earlier designs.

### 2.3.2 Bayesian Adaptive Trials

Bayesian adaptive designs enable adaptation on the basis of interim data analysis. Bayesian trials provide real-time flexibility with the possibility of early stopping of inferior arms, selective patient entry to promising treatment, and adding or dropping arms. Bayesian trials are randomizable and are less demanding on the patient count, accelerate discovery while maintaining statistical power. They are not particularly suited to investigate toxicity and are equally cumbersome to execute and interpret [20].

### 2.3.3 Umbrella and Platform Trials

Umbrella trials enroll patients with one tumor type but randomize by molecular features to treatment arms. For instance, patients with non-small-cell lung cancer are randomized to arms for treatment of EGFR, ALK, or KRAS. Platform trials take a step further in having multiple tumor types and treatment regimens in one protocol. They are adaptive and new arms can be added while others are dropped as evidence accumulates. They provide operational efficiency and are optimal when comparing multiple therapies simultaneously in multiple settings [24].

### 2.3.4 Master Protocols

Master protocols unify umbrella, basket, and platform trial approaches in one overarching design. Master protocols enable simultaneous testing of several interventions across several subgroups and diseases. A master protocol was used by the DART trial (Dual Anti–CTLA-4 and Anti–PD-1 blockade in Rare Tumors) to test immunotherapy on 53 rare tumor cohorts [19].

### 2.3.5 Seamless Telescoping Trials

Telescope trials merge phase 1, 2, and 3 stages into a single trial. Patients go through different stages according to the same protocol, saving administrative expenses and accelerating the length of time for overall development. These trials provide a consistent data structure that can support both exploratory and confirmatory endpoints. However, their susceptibility to changes in the standard of care with time and design complexity is the key limitations [20].

### 2.3.6 Basket Trials: A Tumor-Agnostic Development

Basket trials enroll patients with heterogenous cancer subtypes based on the occurrence of a shared molecular abnormality, e.g., BRAF mutations and NTRK fusions [20]. Trial design follows the molecular, but not the histologic, stratification of disease. Examples of successful ones include the NCI-MATCH, MyPathway, and TAPUR trials. They have led to FDA tissue-agnostic approvals and have resulted in a redefinition of cancer by genomic signature rather than organ of origin. Basket trials are particularly valuable for rare alterations since they can facilitate successful accrual across different cancers [20, 24]. They are challenging in molecular screening and must overcome tumor microenvironment heterogeneity and comorbidities between disease subtypes.

Strengths and weaknesses of drug-centric trials and drug-focused trials yield invaluable information for drug licensure and regulation (Table 1). The stringent methodology of drug-focused trials ensures that treatments are extremely safe and effective. The trials, however, do not take into account heterogeneity of individual cancers and might exclude the application of promising treatment for patients who do not meet restrictive trial inclusion criteria. Limitations of this are long timelines. Extensive, sequential trials take years to finish. Along with this, experiments are resource-intensive and logistically intensive. Trial procedures are rigid and routinized, limiting flexibility to new information. Other than these, the majority of trials employ pre-set regimens that may not reflect the molecular heterogeneity of

**Table 1** Clinical trial evaluating the application of precision medicine approaches in the treatment of cancer (https://clinicaltrials.gov/)

| Trial name | Institute/collaborator | Cancer type | Study details |
|---|---|---|---|
| NCT02827370 | Sidney Kimmel Cancer Center at Thomas Jefferson University | Breast cancer | The study aims to evaluate whether a targeted dietary intervention can enhance the effectiveness of neoadjuvant chemotherapy. |
| NCT02664350 | University of Florida | Solid tumor with or without metastasis | Genetic testing for CYP2D6 metabolic pathway at the baseline and OPRM1 after 8 weeks |
| NCT05090605 | Asfendiyarov Kazakh National Medical University | Breast cancer | Identification of potential candidate genes associated with breast cancer through the analysis of single-nucleotide polymorphisms (SNPs) in a cohort of breast cancer patients. |
| NCT04067960 | Mayo Clinic | Stage III-IV Cancer | A one-time saliva sample is collected for pharmacogenomic testing. Patients also complete a quality-of-life assessment at baseline and again 3 months after the testing. |
| NCT03521141 | Vanderbilt University Medical Canter | Lung cancer | Nicotine patch (Nicotine Replacement Therapy) and varenicline (Chantix) are FDA-approved smoking cessation pharmacotherapies, while the NCI's "Clearing the Air" is a booklet-led behavioural intervention designed to support individuals at any stage of their quitting journey. |
| NCT05189171 | Xilis, Inc. | Colorectal cancer | This study aims to assess the feasibility of generating MicroOrganoSpheres (MOS) from liver-metastatic colorectal cancer biopsies and completing a drug screen within 14 days using standard treatments: oxaliplatin, irinotecan, 5-FU/capecitabine (Xeloda), bevacizumab, panitumumab, cetuximab, trifluridine/tipiracil (Lonsurf), regorafenib, and pembrolizumab or nivolumab. |
| NCT05697198 | Labcorp Corporation of America Holdings, Inc | Stage III and IV cancers | This study aims to collect and analyze blood and/or tissue biopsies from individuals with advanced solid tumors to compare liquid biopsy results with next-generation sequencing of matched or archival tissue. Samples from stage III/IV cancers (e.g., lung, breast, GI, gynaecologic, prostate, head and neck, soft tissue, melanoma) will be used for diagnostic and research purposes by Labcorp/ OmniSeq or designated third parties, with de-identified clinical data maintained for future research. |

(continued)

**Table 1** (continued)

| Trial name | Institute/collaborator | Cancer type | Study details |
|---|---|---|---|
| NCT04951921 | Chang Gung Memorial Hospital | Cervical cancer | This single-centre study will assess advanced imaging methods (MRF, CEST, DNP-MRI) in 30 LACC patients undergoing CCRT. Participants will undergo standard imaging, biopsies, and two investigational scans, including DNP-MRI with hyperpolarized pyruvate (^13C) to evaluate tumor metabolism. |
| NCT05772767 | Fondazione Policlinico Universitario Agostino Gemelli IRCCS | Glioblastoma stem cells | This study explores cilium-related gene expression in patient-derived glioblastoma stem cells and examines the effects of targeting cilia components in vitro, in vivo, and in brain organoids. It also tests cilia disassembly inhibitors and evaluates a cilium-based gene signature for prognostic value. |
| NCT03056599 | Presage Biosciences | Soft Tissue Sarcoma | This pilot study tests the feasibility of using the CIVO system to assess in situ responses of soft tissue sarcoma to microdoses of anti-cancer drugs injected before surgery. Tumor samples collected during surgery will be analyzed to evaluate drug effects, with no impact on patient care. |
| NCT02274025 | Fore Biotherapeutics | Lung cancer | NovellusDx uses a functional assay to detect abnormal movement of mutated signaling proteins to the nucleus, helping identify active driver mutations in tumors, even if they are previously uncharacterized or unlinked to specific cancer types |
| NCT01726582 | Medical College of Wisconsin | Pancreas cancer | This trial uses molecular profiling of pancreatic adenocarcinoma to guide personalized treatment. A biopsy is tested to select pre-surgery chemo and radiation, and the tumor is tested again after surgery to tailor post-surgery therapy. All treatments are already approved; the goal is to match the best option to each patient's tumor. |
| NCT06782022 | Jan Gerard Maring | Gastrointestinal cancer | This study compares 5-FU clearance in patients with common DPYD variants versus wild-type and assesses toxicity and safety of reduced dosing in DPD deficiency. It also aims to improve dosing accuracy and minimise patient discomfort during monitoring. |

| | | | |
|---|---|---|---|
| NCT03871790 | CENTOGENE GmbH Rostock | Colon and pancreas carcinoma | This study aims to identify tumor-specific molecular features and neoepitopes in colorectal and pancreatic cancer patients undergoing surgery, to support the development of personalized, neoepitope-based cancer vaccines. |
| NCT03470402 | Columbia University | Breast and ovarian cancer | This study aims to increase genetic testing for hereditary breast and ovarian cancer in high-risk women by using an EHR-embedded BNAV tool for primary care referrals and a web-based decision aid, RealRisks, to address patient barriers and support informed decision-making. |
| NCT02666885 | University Hospital Tuebingen | Head and neck cancer | This trial evaluates the use of pre-treatment FDG-PET/MRI in guiding adjuvant radiochemotherapy for head and neck cancer and aims to develop a personalized decision support system using imaging, γH2AX analysis, and genetic data. |
| NCT04105582 | Universidad Nacional de Colombia | Breast cancer | This study is the first in Colombia to evaluate a personalized vaccine for triple-negative breast cancer using tumor-specific synthetic peptides presented by autologous dendritic cells, aiming to assess safety and immunogenicity and validate a workflow for identifying patient-specific tumor epitopes. |
| NCT04065555 | Presage Biosciences | Head and neck cancer | This multi-center, single-arm Phase 0 trial uses the CIVO device to deliver microdoses of TAK-981 alone or with cetuximab or avelumab directly into tumors to study their localized biological effects in the tumor microenvironment. |
| NCT04015466 | Fundación para la Investigación del Hospital Clínico de Valencia | Gastric cancer | This study aims to characterize a multi-centric cohort of advanced gastric cancer (GC) patients from EU and CELAC regions using a multi-omic approach—proteomics, genomics, transcriptomics, microbiome, and exposome—to identify disease determinants, regional risk differences, and potential biomarkers for developing a therapy-guiding algorithm. |

(continued)

**Table 1** (continued)

| Trial name | Institute/collaborator | Cancer type | Study details |
|---|---|---|---|
| NCT02142036 | Oslo University Hospital | Metastatic cancer | The MetAction study treats patients based on molecular abnormalities in metastatic lesions, rather than tumor type, recognizing that metastases can differ from the primary tumor due to heterogeneity, clonal selection, and prior treatments. Therapy is guided by actionable target identification (ATI). |
| NCT02752893 | Dartmouth-Hitchcock Medical Center | Breast cancer | This study aims to establish a platform at Dartmouth-Hitchcock Medical Center to create new models of ER+/HER2- breast cancer, enabling research into treatment strategies and the signaling pathways behind drug sensitivity and resistance. |
| NCT06062602 | Presage Biosciences | HNSCC | This multi-center Phase 0 substudy evaluates the localized pharmacodynamics of TAK-676 alone or with Carboplatin, 5-FU, or Paclitaxel in head and neck squamous cell carcinoma using microdose intratumoral delivery via the CIVO device prior to scheduled surgery. It is part of the PBI-MST-01 Master Protocol. |

individual tumors. Moving towards Patient-Centered Paradigms. While drug-targeting trials remain relevant, the field is trending toward patient-targeted approaches, like N-of-1 trials that adapt treatment to each patient's individual molecular profile. In such trials, the emphasis increasingly shifts from assessing the effectiveness of a single treatment and to identifying the effectiveness of a treatment algorithm that matches patients with optimal treatments [20, 24]. This is in line with the growing understanding that cancer is not a monolithic disease, but a group of many, each with its own molecular profile.

## 3 Ethical and Regulatory Considerations

The application of precision cancer medicine in the clinical setting is linked with many ethical and regulatory issues. The basic challenges in scientific and technical aspects arise from usage of complex genomic data, customization of treatments, and the ever-changing landscape of clinical trials.

### 3.1 Informed Consent and Genetic Privacy

In cases concerning precision oncological therapeutics, a patient may undergo extensive genetic testing that may reveal sensitive information regarding the present health status of the person, as well as future concerns, and possible threats to the family members. Informed consent, therefore, ought to be more than what is generally explained in clinical situations [22]. More importantly, patients should actually comprehend the possible outcomes of genomic analyses, namely, incidental findings like germline mutations that may constitute hereditary cancers [22]. They should also be informed of how their data would be used, stored, and shared, especially in cases of research or sharing across databases. Protecting genetic privacy is paramount; thus, the concerned institutions must ensure data security through data protection laws such as the General Data Protection Regulation (GDPR), Europe, or HIPAA, USA [22].

### 3.2 Equity of Access

Equity in oncology is an ethical concern as advanced diagnostics and personalized therapies are accessed inequitably. For instance, in less resourceful settings, genomic sequencing, targeted therapies, or even clinical trials remain out of reach of the patients, and hence there is variation in care [21]. For palliative implementation of precision medicine, we require strategies that democratize access to molecular testing and insurance coverage for the biomarker tests that have been checked through validated tests. Private and public healthcare systems ought to work against the geographic, socioeconomic, and racial inequalities existing in cancer care.

### 3.3 Regulatory Hurdles

From a regulatory standpoint, the accelerated development of targeted therapies and companion diagnostics often surpasses existing frameworks for approval. Regulatory bodies including the FDA and EMA, have, in recent years, sought implementation of accelerated pathways for approval, breakthrough therapy designations, and oncology reviews in real-time to maximize earlier patient access to promising treatments [24]. However, these accelerated pathways still need to ensure the application of rigorous standards relating to safety and efficacy. Moreover, adaptive trials and basket trials create interesting new challenges in regulatory scrutiny because they do not fit traditional drug development models. Regulators must weigh innovation and patient safety and ensure that trial designs are well-situated to be ethical, statistically sound, and interpretable.

### 3.4 Sharing of Data, Data Ownership

Finally contemplates the ethical question concerning data ownership in the world of precision oncology. Large-scale data sharing may improve cancer outcome research and the design of targeted therapies so that it fine-tunes its use, yet it must be balanced against the individual's right to custodianship over their health data [22]. A new set of policies concerning data sharing, de-identification, and informed consent might give greater consideration to the said parties and preserve the trust of the health-care system.

## 4 Case Study: Precision Medicine in Action

To provide a clinical example of the applicability of precision oncology, the case of a 58-year-old man with metastatic colorectal cancer (mCRC) is considered [25]. Symptoms usually presented by the patient included weight loss, a change in bowel habits, and fatigue. Colonoscopy showed a mass in the sigmoid, and biopsy demonstrated a moderately differentiated adenocarcinoma of the colon. The whole-body CT scan evidence pinpointed the presence of hepatic and pulmonary metastases, stage IV disease.

### 4.1 Molecular Probing and Personalization of Treatment Strategy

As per precision medicine guidelines, the patient's tumor underwent NGS to search for actionable mutations. Molecular profiling showed wild-type KRAS and NRAS genes, BRAF V600E negative, and Microsatellite stable (MSS) tumors with low tumor mutational burden (TMB). All these factors consisted of the patient being eligible for anti-EGFR treatment [26]. The patient was negative for HER2 amplification as well as PD-L1 expression. Germline testing for Lynch syndrome and other inherited cancer syndromes was such that it came back negative as well, placing more limits on possible treatment targets.

### 4.2 Initiation of Targeted Therapy and Monitoring

Based on these molecular results, the patient was started on first-line FOLFIRI (5-fluorouracil, leucovorin, and irinotecan) plus cetuximab, an anti-EGFR monoclonal antibody [25]. Following four cycles, clinical and radiologic improvement were noted in the form of partial regression of liver metastases and relief of systemic symptoms.

### 4.3 Emergence of Resistance and Adaptation of Therapy

One of the strategies used for treatment efficacy and resistance monitoring involved ctDNA liquid biopsies. They allowed tumor evolution to be monitored in real time without the need for invasive and repeated tissue biopsies [27]. By 6 months of treatment, ctDNA testing detected the emergence of a KRAS G12D mutation, as expected for clonal evolution and resistance to EGFR inhibition [28]. This molecular alteration was not present at diagnosis and suggested the expansion of a resistant minor subclone under therapeutic pressure. Cetuximab was therefore discontinued. Therapy was switched to FOLFOX (5-FU, leucovorin, oxaliplatin) plus bevacizumab, a monoclonal antibody against VEGF, useful in RAS-mutant CRC patients [29]. The disease remained stable for 3 months, emphasizing the clinical utility of molecularly guided therapy shifts.

### 4.4 Clinical Implications

This case proves that precision medicine allows cancer treatment in a responsive, evidence-based manner. The clinical team was able to choose an appropriate targeted therapy by integrating thorough molecular profiling during diagnosis according to the tumor biology of the patient. The application of liquid biopsy allowed real-time tracking of tumor evolution, enabling early detection of resistance mechanisms before clinical progression had occurred.

This active surveillance enabled timely conversion of therapy, hence guaranteeing disease control. What is also interesting is that the individualized approach focuses on the need for multidisciplinary interaction between oncologists, pathologists, molecular biologists, and genetic counselors in decision-making. With dynamically changing therapy based on genomic information, the patient was able to achieve a better quality of life and prolonged disease stability. In general, this case demonstrates the potential of precision oncology to revolutionize the optimization of results by coordination of therapeutic decisions with evolving tumor biology (Table 2).

**Table 2** Timeline of precision oncology treatment

| Time frame | Interference | Key molecular finding | Therapeutic verdict |
|---|---|---|---|
| Week 0 | Diagnosis | mCRC, KRAS/NRAS wild-type | Eligible for anti-EGFR |
| Week 1 | NGS & baseline ctDNA | MSS, BRAF wild-type | Cetuximab + FOLFIRI |
| Week 12 | Interim scan | Partial response | Continue same regimen |
| Week 24 | ctDNA analysis | KRAS G12D mutation | Switch to FOLFOX + bevacizumab |
| Week 36 | Follow-up imaging | Stable disease | Maintain adjusted therapy |

## 5 Transforming Oncology Trials Through Precision and Personalization

Cancer clinical trial systems are in the process of fundamental change, shifting methods from conventional "one-size-fits-all" toward therapeutic approaches with a molecular-profile- and personalized-therapy-based angle. In the pre-genomic era, almost all oncology trials were fixture around tumor histology. After a drug had cleared Phase I, it would almost always be shoved into trials in patients suffering a named cancer type, say lung, breast, or colorectal reflectively without any consideration for molecular heterogeneity on the part of the patients [30]. New program of trials have been propelled into existence by the burgeoning genomically-informed medicine paradigm. These early trials, still tissue-specific, began targeting defined molecular subgroups-for example, EGFR-mutant lung cancers being treated with EGFR inhibitors [31]. More recently, however, dramatic development has taken place toward tissue-agnostic trial designs. These studies enroll patients solely based on the presence of certain molecular aberrations, without consideration for the origin of the tumor. Regulatory approvals have quite followed. The FDA approval of drugs such as pembrolizumab (for MSI-H tumors), larotrectinib, and entrectinib (NTRK fusions), and selpercatinib (RET fusions) has thus highlighted the increase in acceptance toward this strategy [32]. NGS is a strategy that has helped to identify molecular drivers that are shared by different malignancies and the limitations of single-target inhibitors. However, once indicted that it is rarely that inhibition of one pathway produces a durable response because of the complexity of tumors and the redundancy of pathways, the attention shifted to rational drug combinations [33]. ComboMATCH, among other trials, now tests multitherapy regimens for molecular alterations that account for both primary and adaptive resistance. These studies take into consideration the variability in drug-target matching and consist of both targeted therapies and conventional agents, often following an adaptive or platform trial design [34]. The very complexity of cancer biology is a huge problem. The theoretical number of three-drug combinations is over 4.5 million, given by an average of five alterations per tumor, and over 700 genes known to be oncogenic, in addition to more than 300 oncology drugs. Due to the innumerable permutations, a tedium of classical trials for each case is practically impossible. To address this, the personalized N-of-1 trials such as I-PREDICT and WINTHER arose. These studies offer an individualized treatment plan, combining multiple agents per patient's unique tumor profile. Importantly, dose titrated to effect or side effect does not follow the standard phase I approach but rather intrapatient dose adjustment to achieve a balance between safety and efficacy [4]. This method of adaptation thereby enables investigators to offer hitherto untried combinations to their single patient. The closest that N-of-1 trials can come to a fossilized formalism is the provision of a molecular matching score to quantify how much a given therapy matches with a patient's tumor profile. Higher matching scores have been strongly correlated with better outcomes such as disease control, prolonged progression-free survival (PFS) and overall survival (OS) in resistant and even treatment-naive patients [4]. In the process, these results reverse the direction of oncology, from drug-centered, targeting a single pathway, to dynamic, patient-centered approaches that integrate multiomic data, combinatorial treatment, and real-time dose adaptation. With entering this new age for oncology trials, precision medicine is no longer an unreachable goal but now is being erected as the very foundation on which all cancer therapy stands.

## 6 Special Considerations for Precision Oncology Trials

### 6.1 Enhancing Immunotherapy Through Precision Biomarkers

Immune checkpoint inhibitors are currently revolutionizing therapy for many cancers, especially those with immunologically active profiles. In the world of precision medicine, however, there is still much room to fine-tune the art of patient selection for immunotherapy. In fact, tumors generally deemed immunologically "cold" often do not respond, while some labeled "hot" fall prey to complete resistance.

Well-established biomarkers like PD-L1 expression, MsI-H, and high TMB—are the ones helping clinicians to pinpoint those who might respond to PD-1/PD-L1 blockade, but they make up just a piece of a far larger puzzle [35]. Genomic profiling has proven that each tumor has its own molecular and immunologic landscape, yet much of immunotherapy research continues to ignore this complexity. To forward the field, researchers really need to look into other factors, such as neoantigen load, antigen presentation, T-cell receptor diversity, and immune evasion mechanisms [35]. Including any one or all of these in clinical trial design could foster results in immunotherapy and decrease exposure to ineffective therapies.

### 6.2 Personalized Dosing Strategies for Combination Therapies

Personalized therapy does not end with drug selection; it has to include dosing, too. Today, in many cases, especially concerning novel drug combinations, the currently existing trial data seems to be insufficient for properly guiding this choice. According to one large meta-analysis, when used in combination therapy, monoclonal antibodies generally present less toxicity than small molecules. For new combinations, this suggests that a conservative starting dose, often 50% of the monotherapy dose, should be administered [36].

Indeed, this approach was employed in the I-PREDICT trial, where dose titration was done in patients according to real-time determinations of safety and efficacy parameters [4]. Dosing adaptations were made on an individual basis, considering parameters such as age, organ function, co-morbidities, and drug-drug interactions. Most importantly, this proved to be a safe and effective strategy, and the results support the need for a flexible dosing method that is patient- and case-specific. Growing concerns regarding dose precision led to initiating Project Optimus by the FDA, with the goal of reforming dose determination for oncology drugs [37]. Whereas I-PREDICT utilizes intrapatient dose adaptation, the focal point of Project Optimus is seeking a consistent dose biologically effective across populations [37]. Both initiatives stand as signals of a paradigm shift, oncology dosing is no longer about finding the maximum tolerated dose; rather, it is about finding the right dose for the right patient.

## 6.3 The Case for Early Integration of Precision Medicine

In fact, much of precision oncology has been dealt with at the later stages of disease, when patients become treatment-resistant. Experience, however, points to the opposite: the earlier, the greater the therapeutic benefit. In chronic myelogenous leukemia (CML), early treatment with BCR-ABL inhibitors has made a fatal disease treatable, providing almost normal life expectancy; these drugs are much less effective when applied in crisis [38].

The same reasoning goes for solid tumors. Advanced cancers gather more mutations and biological complexity, and thus become more difficult to treat. Early administration of selective therapies may intercept the disease well before it grows widely heterogeneous. An example being a recent study of MSI-H rectal cancer patients treated with PD-1 blockade (dostarlimab) at diagnosis, following which all patients underwent complete remission, thus there was no need for surgery or chemo or radiation [39]. Such results should robustly encourage consideration of positioning precision medicine approaches at the very front line, more so in rare or aggressive cancers.

## 6.4 Real-Time Precision Medicine: The Next Horizon

With the ongoing advancement of AI, multiomics, and synthetic biology, the possibility for the real-time development of truly patient-specific therapies moves ever so much closer. Tumor profiles will then be rapidly analyzed, and matched or even custom-designed therapies will be generated through an AI-guided platform and 3D printing technologies. Instead of approving static drugs, regulators in the future may approve a platform that designs and delivers a unique drug for each patient's tumor.

Though quite theoretical at the moment, this personalized drug manufacturing mode, if ever realized, may very well revolutionize how treatment delivery works and under what terms oncology drugs are approved.

## 6.5 Real-World Evidence in Modern Oncology Practice

Real-world data offer an alternative perspective to clinical trial results by looking at outcomes in different patient populations that often are underrepresented. RWD are considered important, as they come from electronic health records, insurance claims, and disease registries and include variables that explain therapy response for patients in real-life scenarios, rather than in highly controlled environments of trials [40].

RWD helps to bridge research and clinical practice. The FDA Real-World Evidence program accepts these various data types into regulatory decisions, for label expansions, safety evaluations, etc. [40]. In the situations of early-phase trials or rare cancers, RWD can also be employed as a synthetic comparator arm to strengthen early efficacy signals.

## 6.6 Registry-Based Evidence and the Power of Outliers

Structured registry protocols such as the ROOT study (NCT04028479) collect clinical, molecular, and outcomes data in large patient populations [41, 42]. Synthesizing AI analytics, registries can track outliers to determine response patterns and strategies for future treatment. Exceptional responders are paramount as they represent a unique set of patients who experience a great level of benefit from treatments that do not work in general. Profiling exceptional responders may unveil hidden biomarkers or mechanisms of drug action. These outliers hardly ever present themselves, yet they might be incredibly useful in identifying which subpopulations are best able to benefit from drugs that do not perform well.

## 6.7 Elevating the Patient Voice in Clinical Research

Patient-reported outcomes (PROs) describe the processes by which patients assess treatment in actuality. These PROs include symptoms, side effects, and quality of life, and depending on the trial, could be gathered through survey administration and electronic design or even telephonic interviews. Use of PROs in clinical trials would help with making prompt dosage adjustments, improving patient outcomes, and earlier hospitalizations [43].

Despite being suggested very strongly by regulatory bodies, PROs seldom form part of drug approval decisions. Giving much more voice to patients throughout the clinical trial process would bring in additional dimensions from patient experience that are presently lacking compared to the existing focus on clinical endpoints.

### 6.8 Rethinking Control Groups: The Role of Synthetic Arms

Synthetic control arms are the use of external datasets to form comparator groups for clinical studies [44]. Synthetic control arms are primarily applicable for rare cancers or very aggressive diseases where recruiting patients for both the control and experimental arms is both unfeasible and unethical.Desirably, these controls would closely represent the population to be studied to allow for valid comparison as recommended by the FDA. The FDA is slowly endorsing synthetic arms as an increasing accepted and effective mechanism to study new treatments in circumstances where one would ordinarily need randomized controlled trials.

## 7 Correcting Inequities and Access Disparities Among Under-Represented/Underserved Populations

While precision medicine holds great promise to improve cancer therapy and/or survivorship, achieving equitable access to these trials continues to be problematic. Patients in rural areas or access-challenged populations regularly come in contact with several issues, including geographic distance, financial barriers, and logistical challenges to participate in a clinical study. In the scenario where patients are expected to travel frequently to large academic centers for access to investigational treatments, these barriers are insurmountable for the majority of patients with limited resources. To alleviate these access challenges, some decentralized trial designs or trials that co-enroll several sites, like those created by the Cancer Cooperative Groups, could expand access to clinical trials [45].

### 7.1 Home-Based Decentralized Trials

Home-run or decentralized trials allow patients to receive investigational therapies locally and potentially at home. This type of study design minimizes travel burden, expands access, and/or improves access to investigational therapy, especially for patients with advanced cancer. The DART (Dual Anti–CTLA-4 and Anti–PD-1 Blockade in Rare Tumors) study is a great exemplar since immunotherapy was being tested in patients with 53 distinct rare cancer types under a single protocol [46]. DART enrolled patients at over 1000 sites across the U.S. via the SWOG network. Similarly (but not the same), the NCI-MATCH trial matched investigational therapies to tumors based upon specific genomic alterations and ultimately enrolled over 1200 patients at 1117 sites [46]. Both of these studies illustrate that national-scale precision trials are feasible. The TAPUR study, sponsored by ASCO, assesses FDA-approved targeted therapies in patients who have actionable genomic alterations, with enrollment through more than 260 sites in the USA. It provides quick assessments of efficacy and safety across tumor types [47]. The TRACK study, a fully remote N-of-1 precision trial for rare cancers, is representative of an emerging model [48]. This method incorporates remote consent, mobile phlebotomy, virtual tumor boards, and recommends therapies based on molecular profiling. This model allows patients to access individualized treatment options regardless of geographic limitations.

### 7.2 Community-Based Clinical Trial Networks

Regional community oncology networks, like the U.S. Oncology Network, are essential in enabling the trial to be available in a local community. The U.S. Oncology Network has activated more than 600 sites, approximately more than 2500 physicians at participating sites, participated in over 2000 trials, and has aided in contributing to nearly 100 therapies that have received FDA approval [49]. These networks increase the extent to which trials can be available outside of an academic center, as well as the diversity of patients that can be included throughout the USA.

### 7.3 Advancing Diversity, Equity, and Inclusion (DEI)

Improving DEI for clinical trials will include intentional approaches at every stage. Important approaches will include DEI-based goals, diversity in research teams, engaging with communities as early as possible, and developing flexible trial procedures that incorporate technology. Continuous monitoring to minimize biases, promoting patient involvement, and establishing a long-term relationship with underrepresented communities are key. These efforts could lead to better participation and ensure the generalizability of findings among diverse populations [50].

As oncology transitions to molecular-based treatment, widespread use of molecular diagnostics and matching therapies will become common practice. High-quality FDA-approved next-generation sequencing (NGS) technologies are becoming more available, which will facilitate widespread use in community practices. To sustain progress, oncology training should include teaching molecular profiling and therapy selection, so clinicians in every setting can provide up-to-date care.

## 8 Patient Experience and Perspectives Related to Precision Oncology

The patient perspective is critically important in precision oncology. The I-PREDICT trial at UC San Diego's Moores Cancer Center is a powerful example that balances promise in the field by providing genomic testing and expert molecular tumor board recommendations to align patients with individualized therapies [4, 10]. In 2016, one patient was running out of possibilities. After a managed tumor for years in remission, it was back, and was currently identified as stage IV with triple-negative pleomorphic lobular breast cancer [4, 10]. Her tumor markers were dangerously high, and available standard therapies were not working. She was not given hope as the likelihood of hospice was high due to an impending bowel obstruction that was being caused by the cancer. She was referred to surgery at UCSD, and learned about their clinical trial called I-PREDICT. Ample patient education was provided to ensure that this was not a usual clinical trial as it afforded her the opportunity of tailored therapy based on her tumor's genomic profile; there was no risk of placebo. She relayed the experience like being in a very high-stakes chess match where each trial decision would greatly impact her quality of life. Her tumor presented with very high tumor mutational burden (76 mutations/megabase), strongly indicating a potential response to immunotherapy, even though this was not yet standard for breast cancer. Ultimately, coverage from her insurance was denied. However, acceptance for compassionate use of a PD-1 inhibitor was approved [35]. In two infusions, her tumor markers were down 75%, and after 8 weeks, the scans showed a complete response. Now, over 7 years cancer-free, she attributes her life to precision medicine and the scientists working in that field. She encourages everyone she comes across to get genomic testing and believes the most expensive treatment is the treatment that doesn't work. For her, every day is a gift, and when she wakes up each morning, she is joyful and grateful.

## 9 Challenges and Future Directions

Despite the potential for revolution by precision oncology, a number of barriers stand against its complete integration into the realm of clinical practice. Key among these barriers is the not-so-broad availability of genomic testing and targeted therapy, particularly in the low- and middle-income nations, causing disparities in treatment [51]. In addition to this, tumor heterogeneity and clonal evolution pose a challenge to therapy because resistant subclones can expand when treated, making the long-term response decline [52]. The other major issue is the interpretation of complex genomic data. The majority of the changes that are discovered are variants of uncertain significance (VUS), and it is challenging for clinicians to determine their treatment implications [53]. Furthermore, the integration of big data and artificial intelligence into practice is still in its initial stages, with challenges in the standardization of data, data privacy, and model validation [54, 55].

## 10 Conclusion

Precision cancer medicine marks a new era of tailored oncologic therapy, with huge gains in efficacy and patient benefit. But to achieve its full potential, there needs to be ongoing innovation in clinical trial design, global availability, and strong embedding of new technologies such as AI. In the future, oncology will be very individualized, dynamic, and molecular-driven, transforming cancer care across the board.

**Declarations**

**Author's contributions** Conceived the idea and review structure: AB, BB; Wrote the manuscript: AB, SD, SG; Revised and edited the manuscript: AB, BB. Authors have read, agreed, and approved the final draft of the manuscript.

**Funding** AB and BB would like to thank the Director, CNCI, for institutional support. This work is further supported by funding from Department of Biotechnology, Govt. of India (Sanction No. BT/PR38493/TRM/120/465/2020), Indian Council of Medical Research, Govt. of India (File No. 17X (3)/Adhoc/87/2022-ITR) and West Bengal Department of Science Technology and Biotechnology, Govt. of West Bengal, India (Sanction No. 2402 (Sanc.)/STBT-11012 (12)/2/2022-ST SEC).

**Data Availability** The software Biorender (https://www.biorender.com/) is used to draw the figure.

**Conflicts of Interests** There is no competing and conflict of interests.

## References

1. Nikanjam M, Kato S, Allen T, Sicklick JK, Kurzrock R (2025) Novel clinical trial designs emerging from the molecular reclassification of cancer. CA Cancer J Clin [Internet] [cited 2025 May 19] 75(3):243–267. Available from: https://onlinelibrary.wiley.com/doi/abs/10.3322/caac.21880
2. Swanton C (2012) Intratumor heterogeneity: evolution through space and time. Cancer Res 72(19):4875–4882
3. Bedard PL, Hansen AR, Ratain MJ, Siu LL (2013) Tumour heterogeneity in the clinic. Nature 501(7467):355–364
4. Sicklick JK, Kato S, Okamura R, Schwaederle M, Hahn ME, Williams CB et al (2019) Molecular profiling of cancer patients enables personalized combination therapy: the I-PREDICT study. Nat Med 25(5):744–750
5. Kim ES, Herbst RS, Wistuba II, Lee JJ, Blumenschein GR, Tsao A et al (2011) The BATTLE trial: personalizing therapy for lung cancer. Cancer Discov 1(1):44–53
6. Kris MG, Johnson BE, Berry LD, Kwiatkowski DJ, Iafrate AJ, Wistuba II et al (2014) Using multiplexed assays of oncogenic drivers in lung cancers to select targeted drugs. JAMA 311(19):1998–2006
7. Park JW, Liu MC, Yee D, Yau C, van't Veer LJ, Symmans WF et al (2016) Adaptive randomization of Neratinib in early breast cancer. N Engl J Med 375(1):11–22
8. Rodon J, Soria JC, Berger R, Miller WH, Rubin E, Kugel A et al (2019) Genomic and transcriptomic profiling expands precision cancer medicine: the WINTHER trial. Nat Med 25(5):751–758
9. Tsimberidou AM, Iskander NG, Hong DS, Wheler JJ, Falchook GS, Fu S et al (2012) Personalized medicine in a phase I clinical trials program: the MD Anderson Cancer Center initiative. Clin Cancer Res 18(22):6373–6383
10. Sicklick JK, Kato S, Okamura R, Patel H, Nikanjam M, Fanta PT et al (2021) Molecular profiling of advanced malignancies guides first-line N-of-1 treatments in the I-PREDICT treatment-naïve study. Genome Med [Internet] [cited 2025 May 19] 13(1):155. Available from: https://doi.org/10.1186/s13073-021-00969-w
11. Coyne GO, Takebe N, Chen AP (2017) Defining precision: the precision medicine initiative trials NCI-MPACT and NCI-MATCH. Curr Probl Cancer 41(3):182–193

12. Hasin Y, Seldin M, Lusis A (2017) Multi-omics approaches to disease. Genome Biol 18(1):83
13. Vlachogiannis G, Hedayat S, Vatsiou A, Jamin Y, Fernández-Mateos J, Khan K et al (2018) Patient-derived organoids model treatment response of metastatic gastrointestinal cancers. Science 359(6378):920–926
14. Sachs N, de Ligt J, Kopper O, Gogola E, Bounova G, Weeber F et al (2018) A living biobank of breast cancer organoids captures disease heterogeneity. Cell 172(1–2):373–386.e10
15. Ooft SN, Weeber F, Dijkstra KK, McLean CM, Kaing S, van Werkhoven E et al (2019) Patient-derived organoids can predict response to chemotherapy in metastatic colorectal cancer patients. Sci Transl Med 11(513):eaay2574
16. Dorsey ER, Topol EJ (2016) State of telehealth. N Engl J Med 375(2):154–161
17. Khozin S, Blumenthal GM, Pazdur R (2017) Real-world data for clinical evidence generation in oncology. J Natl Cancer Inst 109(11)
18. Sharma MR, Schilsky RL (2011) Role of randomized phase III trials in an era of effective targeted therapies. Nat Rev Clin Oncol 9(4):208–214
19. Chabner BA (2014) Approval after phase I: Ceritinib runs the three-minute mile. Oncologist [Internet] [cited 2025 May 19] 19(6):577–8. Available from: https://www.ncbi.nlm.nih.gov/pmc/articles/PMC4041677/
20. Lu C (Cindy), Li X (Nicole), Broglio K, Bycott P, Jiang Q, Li X et al (2021) Practical considerations and recommendations for master protocol framework: basket, umbrella and platform trials. Ther Innov Regul Sci [Internet] [cited 2025 May 19] 55(6):1145–1154. Available from: https://www.ncbi.nlm.nih.gov/pmc/articles/PMC8220876/
21. Wages NA, Slingluff CL, Bullock TN, Petroni GR (2019) Tailoring early-phase clinical trial design to address multiple research objectives. Cancer Immunol Immunother CII [Internet] [cited 2025 May 19] 69(1):95–102. Available from: https://www.ncbi.nlm.nih.gov/pmc/articles/PMC6952569/
22. Rego S, Grove ME, Cho MK, Ormond KE (2020) Informed consent in the genomics era. Cold Spring Harb Perspect Med [Internet] [cited 2025 May 19] 10(8):a036582. Available from: https://www.ncbi.nlm.nih.gov/pmc/articles/PMC7397836/
23. O'Bryan RM, Luce JK, Talley RW, Gottlieb JA, Baker LH, Bonadonna G (1973) Phase II evaluation of adriamycin in human neoplasia. Cancer 32(1):1–8
24. Woodcock J, LaVange LM (2017) Master protocols to study multiple therapies, multiple diseases, or both. N Engl J Med [Internet] [cited 2025 May 19] 377(1):62–70. Available from: https://www.nejm.org/doi/full/10.1056/NEJMra1510062
25. Cutsem EV, Köhne CH, Hitre E, Zaluski J, Chien CRC, Makhson A et al (2009) Cetuximab and chemotherapy as initial treatment for metastatic colorectal cancer. N Engl J Med [Internet] [cited 2025 May 19] 360(14):1408–1417. Available from: https://www.nejm.org/doi/full/10.1056/NEJMoa0805019
26. Tathineni P, Joshi N, Jelinek MJ (2023) Current state and future directions of EGFR-directed therapy in head and neck cancer. Curr Treat Options in Oncol 24(6):680–692
27. Diaz LA, Bardelli A (2014) Liquid biopsies: genotyping circulating tumor DNA. J Clin Oncol Off J Am Soc Clin Oncol 32(6):579–586
28. Misale S, Yaeger R, Hobor S, Scala E, Janakiraman M, Liska D et al (2012) Emergence of KRAS mutations and acquired resistance to anti-EGFR therapy in colorectal cancer. Nature 486(7404):532–536
29. Hurwitz H, Fehrenbacher L, Novotny W, Cartwright T, Hainsworth J, Heim W et al (2004) Bevacizumab plus irinotecan, fluorouracil, and leucovorin for metastatic colorectal cancer. N Engl J Med 350(23):2335–2342
30. Schilsky RL (2010) Personalized medicine in oncology: the future is now. Nat Rev Drug Discov 9(5):363–366
31. Mok TS, Wu YL, Thongprasert S, Yang CH, Chu DT, Saijo N et al (2009) Gefitinib or carboplatin-paclitaxel in pulmonary adenocarcinoma. N Engl J Med 361(10):947–957
32. Thein KZ, Myat YM, Park BS, Panigrahi K, Kummar S (2024) Target-driven tissue-agnostic drug approvals—a new path of drug development. Cancers [Internet] [cited 2025 Jun 16];16(14):2529. Available from: https://www.ncbi.nlm.nih.gov/pmc/articles/PMC11274498/

33. Remon J, Dienstmann R (2018) Precision oncology: separating the wheat from the chaff. ESMO Open [Internet]. [cited 2025 Jun 16] 3(6):e000446. Available from: https://www.ncbi.nlm.nih.gov/pmc/articles/PMC6212683/
34. ComboMATCH Overview [Internet]. ECOG-ACRIN Cancer Research Group. [cited 2025 Jun 16]. Available from: https://ecog-acrin.org/clinical-trials/eay191-combomatch/overview/
35. Yarchoan M, Hopkins A, Jaffee EM (2017) Tumor mutational burden and response rate to PD-1 inhibition. N Engl J Med 377(25):2500–2501
36. Wages NA, Chiuzan C, Panageas KS (2018) Design considerations for early-phase clinical trials of immune-oncology agents. J Immunother Cancer [Internet] [cited 2025 Jun 16] 6(1):81. Available from: https://doi.org/10.1186/s40425-018-0389-8
37. Murphy R, Halford S, Symeonides SN (2023) Project Optimus, an FDA initiative: considerations for cancer drug development internationally, from an academic perspective. Front Oncol 13:1144056
38. Druker BJ, Guilhot F, O'Brien SG, Gathmann I, Kantarjian H, Gattermann N et al (2006) Five-year follow-up of patients receiving imatinib for chronic myeloid leukemia. N Engl J Med 355(23):2408–2417
39. Cercek A, Lumish M, Sinopoli J, Weiss J, Shia J, Lamendola-Essel M et al (2022) PD-1 blockade in mismatch repair–deficient, locally advanced rectal cancer. N Engl J Med [Internet] [cited 2025 Jun 16] 386(25):2363–2376. Available from: https://www.nejm.org/doi/full/10.1056/NEJMoa2201445
40. Sherman RE, Anderson SA, Dal Pan GJ, Gray GW, Gross T, Hunter NL et al (2016) Real-world evidence - what is it and what can it tell us? N Engl J Med 375(23):2293–2297
41. Dickson D, Johnson J, Bergan R, Owens R, Subbiah V, Kurzrock R (2020) The master observational trial: a new class of master protocol to advance precision nedicine. Cell [Internet] [cited 2025 Jun 16] 180(1):9–14. Available from: https://www.sciencedirect.com/science/article/pii/S0092867419313716
42. Dickson D, Johnson J, Bergan R, Owens R, Subbiah V, Kurzrock R (2020) Snapshot: trial types in precision medicine. Cell [Internet] [cited 2025 Jun 16] 181(1):208–208.e1. Available from: https://www.sciencedirect.com/science/article/pii/S0092867420302099
43. Basch E, Deal AM, Dueck AC, Scher HI, Kris MG, Hudis C et al (2017) Overall survival results of a trial assessing patient-reported outcomes for symptom monitoring during routine cancer treatment. JAMA [Internet] [cited 2025 Jun 16] 318(2):197–198. Available from: https://doi.org/10.1001/jama.2017.7156
44. Nuño MM, Pugh SL, Ji L, Piao J, Dignam JJ, Steingrimsson JA (2025) On the use of external controls in clinical trials. J Natl Cancer Inst Monogr 2025(68):30–34
45. Daly B, Brawley OW, Gospodarowicz MK, Olopade OI, Fashoyin-Aje L, Smart VW et al (2024) Remote monitoring and data collection for decentralized clinical trials. JAMA Netw Open [Internet] [cited 2025 Jun 16] 7(4):e246228. Available from: https://doi.org/10.1001/jamanetworkopen.2024.6228
46. Adashek JJ, Kurzrock R (2023) Home-run trials for rare cancers: giving the right drug(s) to the right patients at the right time and in the right place. Npj Precis Oncol [Internet] [cited 2025 Jun 16] 7(1):129. Available from: https://www.nature.com/articles/s41698-023-00487-5
47. Targeted Agent & Profiling Utilization Registry (TAPUR) Study - ASCO [Internet] [cited 2025 Jun 16]. Available from: https://www.asco.org/research-data/tapur-study
48. Palma J, Kato S, Nikanjam M, Shreenivas AV, Sicklick JK, Adashek JJ et al (2024) On the right TRACK: providing comprehensive genomic profiling (CGP) and molecular tumor board (MTB) for patients (pts) with rare cancers. J Clin Oncol [Internet] [cited 2025 Jun 16] 42(16_suppl):3127–3127. Available from: https://ascopubs.org/doi/10.1200/JCO.2024.42.16_suppl.3127
49. Hortman K (2023) The Network is growing—now more than 2,000 providers strong [Internet]. The US Oncology Network. [cited 2025 Jun 16]. Available from: https://usoncology.com/news/the-network-is-growing-now-more-than-2000-providers-strong/
50. Washington V, Franklin JB, Huang ES, Mega JL, Abernethy AP (2023) Diversity, equity, and inclusion in clinical research: a path toward precision health for everyone. Clin Pharmacol

Ther [Internet] [cited 2025 Jun 16] 113(3):575–584. Available from: https://onlinelibrary.wiley.com/doi/abs/10.1002/cpt.2804
51. Pramesh CS, Badwe RA, Borthakur BB, Chandra M, Raj EH, Kannan T et al (2014) Delivery of affordable and equitable cancer care in India. Lancet Oncol 15(6):e223–e233
52. McGranahan N, Swanton C (2017) Clonal heterogeneity and tumor evolution: past, present, and the future. Cell 168(4):613–628
53. Dienstmann R, Rodon J, Barretina J, Tabernero J (2013) Genomic medicine frontier in human solid tumors: prospects and challenges. J Clin Oncol Off J Am Soc Clin Oncol 31(15):1874–1884
54. Topol EJ (2019) High-performance medicine: the convergence of human and artificial intelligence. Nat Med [Internet] [cited 2025 May 19] 25(1):44–56. Available from: https://www.nature.com/articles/s41591-018-0300-7
55. Das S, Mazumder S, Alam N, Vernekar M, Dam A, Bhowmick AK et al (2024) Precision oncology in the era of genomics and artificial intelligence. J Curr Oncol Trends [Internet] [cited 2025 May 21] 1(1):22. Available from: https://journals.lww.com/jcot/fulltext/2024/01010/precision_oncology_in_the_era_of_genomics_and.5.aspx

# Clinical Implementation and Outcomes of Personalized Cancer Therapy

Kirat Kumar Ganguly

**Abstract**

Cancer's intrinsic heterogeneity, marked by diverse genetic and epigenetic alterations, contributes significantly to its complexity and resistance to conventional therapies. Traditional treatments such as chemotherapy and radiation often lack specificity, resulting in varied efficacy and substantial systemic toxicity. Personalized cancer therapy aims to overcome these limitations by utilizing detailed molecular profiling to identify actionable genomic alterations tailored to each patient's unique tumor biology. Advancements in high-throughput sequencing technologies and bioinformatics have facilitated the discovery of critical oncogenic drivers like EGFR mutations in non-small cell lung cancer (NSCLC) and BRAF V600E mutations in melanoma, enabling the development of targeted therapies such as tyrosine kinase inhibitors and monoclonal antibodies. Integrating genomic, transcriptomic, proteomic, and immunophenotyping data provides a comprehensive understanding of tumor biology and the tumor microenvironment, crucial for precise therapeutic interventions. Clinically, personalized approaches have demonstrated improved progression-free survival and overall survival, as seen with agents like crizotinib for ALK-positive NSCLC and trastuzumab for HER2-positive breast cancer. Immunotherapy combined with biomarker-guided patient selection, exemplified by PD-1/PD-L1 inhibitors in mismatch repair-deficient tumors, further illustrates personalized therapy's potential. However, challenges including tumor heterogeneity, therapeutic resistance driven by on-target mutations and bypass signaling pathways, economic and access barriers to advanced diagnostics and treatments, and regulatory inconsistencies limit widespread adoption. Ongoing developments in molecular diagnostics, immunotherapy, nanotechnology, and digital pathology are poised

K. K. Ganguly (✉)
Michael Madhusudan Memorial College, Kabiguru Sarani,
City Centre Durgapur, West Bengal, India

 
M. K. Ghosh, M. Basu (eds.), *Personalized Cancer Therapy*,
https://doi.org/10.1007/978-981-95-6845-1_7

to enhance the precision and efficacy of cancer treatments. Addressing infrastructural, economic, and regulatory challenges remains critical to fully realize personalized oncology's promise and to ensure equitable patient access to precision therapies.

**Keywords**

Cancer · Biomarker · Precision medicine · Clinical implementation

## 1 Introduction

Cancer is a highly heterogeneous disease characterized by diverse genetic and epigenetic alterations that vary extensively among individual patients, influencing tumor behavior, progression, and therapeutic response. This heterogeneity underpins the complexity of cancer and presents significant challenges to conventional therapies. Traditional treatments such as chemotherapy and radiation often exhibit limited specificity, targeting rapidly dividing cells indiscriminately and resulting in systemic toxicities alongside variable efficacy across patients. These limitations highlight the urgent necessity for more precise interventions. Personalized cancer therapy has emerged to address this critical need by leveraging molecular insights to identify actionable genomic alterations—mutations or aberrations that drive oncogenesis and are susceptible to targeted therapy—thus enabling the rational selection of tailored treatments optimized for each patient's unique tumor profile [1, 2].

The launch of high-throughput sequencing technologies alongside advancements in bioinformatics has revolutionized the identification of key oncogenic drivers across various cancer types. For instance, mutations in the epidermal growth factor receptor (EGFR) gene have been identified as critical drivers in non-small cell lung cancer (NSCLC), with clinical trials showcasing the efficacy of EGFR tyrosine kinase inhibitors leading to significant improvements in patient survival and quality of life. Similarly, activating mutations in the BRAF gene, particularly the V600E variant, have been implicated in melanoma, with the advent of selective BRAF inhibitors markedly transforming the therapeutic landscape for these patients. These breakthroughs have facilitated the development and FDA approval of targeted agents such as imatinib, a BCR-ABL tyrosine kinase inhibitor effective in chronic myeloid leukemia by specifically inhibiting the fusion oncoprotein driving the malignancy, and trastuzumab, a monoclonal antibody directed against HER2-overexpressing breast cancers that has markedly improved outcomes when combined with chemotherapy [3–6].

This review critically examines the current state of clinical implementation of personalized cancer therapy, highlighting its transformative impact on patient outcomes by increasing response rates, progression-free survival, and reducing treatment-related toxicity through precise molecular targeting. Equally important are the persistent challenges that hamper broader adoption—including tumor heterogeneity leading to therapeutic resistance, access and affordability issues related to advanced diagnostics and costly targeted agents, and regulatory complexities surrounding biomarker validation and standardization. Addressing these challenges will be essential to unlock the full potential of precision oncology and to ensure equitable delivery of personalized therapies.

## 2 Biomarker Discovery and Molecular Profiling

### 2.1 Genomic and Transcriptomic Profiling

Genomic and transcriptomic profiling have emerged as cornerstone techniques in personalized cancer therapy, revolutionizing biomarker discovery by enabling in-depth interrogation of the genetic and gene expression landscapes within tumors. These methodologies allow identification of unique molecular signatures linked to cancer development, progression, and therapeutic response, thereby facilitating the customization of treatment strategies tailored to the individual patient's tumor biology. By integrating genomic and transcriptomic datasets, researchers gain a holistic understanding of tumor heterogeneity and the underlying biological mechanisms, which enhances the discovery of clinically relevant biomarkers for diagnosis, prognosis, and therapeutic targeting.

Next-generation sequencing (NGS) technologies have dramatically expanded the scope of genomic profiling, permitting comprehensive detection of somatic mutations, copy number variations, gene fusions, and other alterations critical for cancer pathogenesis. Large-scale projects like The Cancer Genome Atlas (TCGA) and the International Cancer Genome Consortium (ICGC) have systematically catalogued such molecular aberrations across diverse cancer types, serving as invaluable resources for pinpointing biomarkers relevant to personalized treatment approaches [7, 8]. Complementing genomic data, liquid biopsy techniques analyzing circulating tumor DNA (ctDNA) offer minimally invasive, real-time monitoring of tumor evolution, including the detection of resistance mutations such as the EGFR T790M variant in non-small cell lung cancer (NSCLC) patients, guiding adaptive therapy selection [9, 10].

Genomic profiling primarily characterizes DNA-level alterations—single nucleotide polymorphisms (SNPs), somatic mutations, structural variants—that influence tumor behavior and therapeutic vulnerabilities. Techniques such as whole-exome and whole-genome sequencing enable the identification of driver mutations, facilitating molecular stratification of patients to appropriate targeted therapies. Such profiling also unveils complex mutational landscapes reflecting intratumoral heterogeneity, informing precision medicine [11, 12].

Transcriptomic profiling offers a complementary dimension by quantifying RNA expression patterns, revealing the functional consequences of genomic changes and uncovering dynamic gene regulatory alterations. RNA sequencing (RNA-seq) and microarray technologies facilitate detection of differentially expressed genes and noncoding RNAs, including microRNAs (miRNAs), which regulate gene expression and impact tumor progression. For example, specific miRNA signatures have prognostic significance in adrenocortical tumors, reflecting transcriptomics' value in refining molecular subtyping and biomarker discovery [13].

One notable challenge is the integration of heterogeneous high-throughput data derived from various platforms, complicated by batch effects and technical variability. Advanced bioinformatics pipelines such as the Gene Expression Data Integration (GEDI) package have been developed to harmonize transcriptomic datasets, enhancing statistical power and reproducibility in biomarker detection [14]. Further, computational tools like RegTools integrate genomic variants with transcriptomic splicing data, identifying splice-associated mutations that could serve as novel cancer biomarkers.

The synergy of genomic and transcriptomic data is particularly powerful in elucidating complex tumor biology, highlighting dysregulated pathways and networks involved in tumorigenesis and therapy resistance, thereby accelerating precision oncology approaches. Resources like Astraea compile expansive biomarker databases integrating multi-omics and tumor microenvironment features, supporting clinicians in evidence-based therapy selection prioritized by molecular biomarkers [16].

As sequencing technologies advance and costs decrease, the capacity for comprehensive biomarker discovery expands accordingly. Machine learning techniques are increasingly harnessed to manage high-dimensional multi-omics data, enabling refined feature selection and robust predictive modeling, enhancing the accuracy of patient stratification and therapeutic decision-making [17].

In summary, integrating genomic and transcriptomic profiling constitutes an essential paradigm in personalized cancer therapy. The continual evolution of high-throughput sequencing and computational analytics underpins detailed molecular characterization and biomarker discovery, which are critical for tailored cancer diagnosis, prognosis, and treatment decisions in precision oncology (Fig. 1).

## 2.2 Proteomic and Immunophenotyping

Proteomics and immunophenotyping are pivotal components in advancing precision medicine by providing detailed insights into cancer biology beyond genetic information. Proteomics, the large-scale study of proteins—the functional effectors within cells—enables comprehensive characterization of the dynamic protein landscape of tumors and their microenvironment. By quantifying and profiling protein expression, post-translational modifications, and protein–protein interactions, proteomics reveals the real-time molecular processes driving cancer progression, therapeutic resistance, and tumor heterogeneity. This depth of information is essential for identifying novel biomarkers, understanding disease mechanisms, and discovering therapeutic targets within the complex and variable tumor microenvironment (TME), which plays a critical role in tumor growth and immune evasion. Hanash et al. [18] highlight how proteomic technologies complement genomic data,

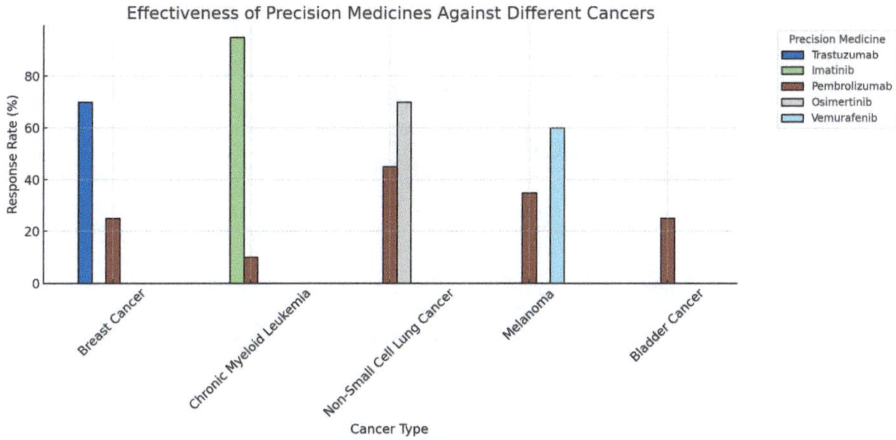


**Fig. 1** Comparative effectiveness of selected precision medicines across different cancer types. This bar graph illustrates the therapeutic response rates (in percentage) of five FDA-approved precision medicines—Trastuzumab, Imatinib, Pembrolizumab, Osimertinib, and Vemurafenib—across five cancer types: Breast Cancer, Chronic Myeloid Leukemia (CML), Non-Small Cell Lung Cancer (NSCLC), Melanoma, and Bladder Cancer. Data represent approximate response rates from published clinical trials and treatment outcomes, reflecting the targeted efficacy of each agent based on tumor-specific molecular aberrations. Notably, Imatinib shows high efficacy in CML due to BCR-ABL targeting, Trastuzumab in HER2+ breast cancer, and Osimertinib in EGFR-mutant NSCLC. Pembrolizumab demonstrates moderate efficacy across several cancers through immune checkpoint inhibition, while Vemurafenib is effective in BRAF V600E-mutant melanoma. The chart highlights the importance of molecular profiling in guiding cancer treatment decisions through precision medicine

refining personalized therapy through precise molecular phenotyping that accounts for both tumor-intrinsic factors and the surrounding milieu [18] (Fig. 2).

Immunophenotyping, the detailed analysis of the immune cell composition and functional states within the TME, further enhances precision oncology by elucidating the immunological landscape that influences cancer behavior and response to immunotherapies. Techniques such as immunohistochemistry (IHC), flow cytometry, and multiplex immunofluorescence allow for the detection and quantification of immune checkpoint molecules, immune cell subsets, and activation markers. Among these, PD-L1 expression on tumor or immune cells measured by immunohistochemistry has emerged as an actionable biomarker that guides the selection of immune checkpoint inhibitors (ICIs), including PD-1/PD-L1 blockers. Topalian et al. [19] demonstrated that tumors with higher PD-L1 expression are more likely to respond to PD-1 pathway blockade, thus making PD-L1 IHC a critical diagnostic tool for personalized immunotherapy decision-making [19]. The integration of immunophenotyping data with proteomic profiles now enables a multidimensional understanding of tumor-immune interactions, facilitating the stratification of patients who will benefit most from immunotherapy and informing combinational strategies to overcome resistance.

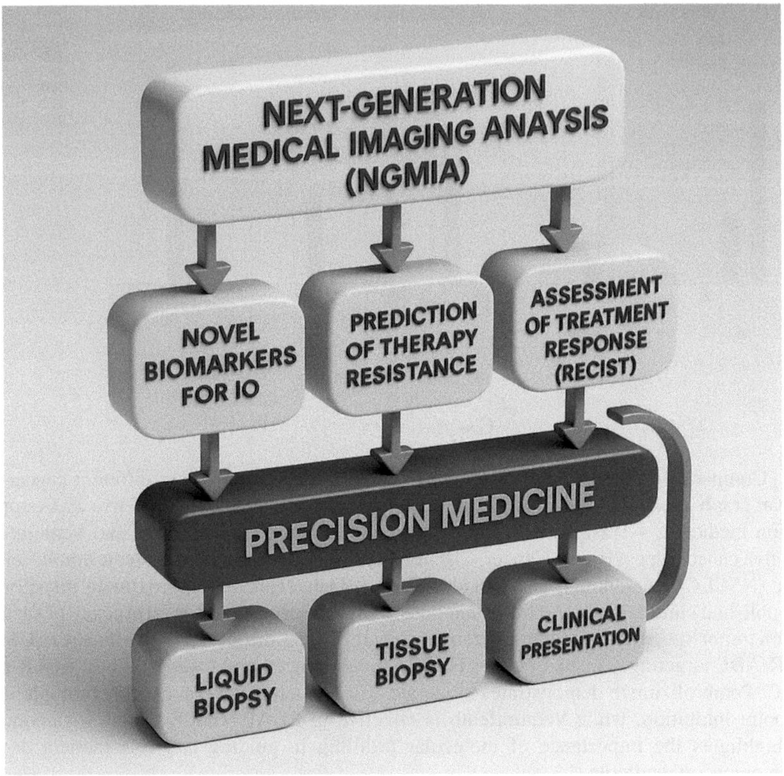


**Fig. 2** This diagram depicts the integration of Next-Generation Medical Imaging and Analysis (NGMIA) into the framework of precision medicine. NGMIA contributes to the development of personalized treatment strategies by identifying novel biomarkers for immuno-oncology (IO), predicting resistance to therapy, and assessing treatment response using RECIST (Response Evaluation Criteria in Solid Tumors). These insights complement traditional diagnostic inputs, including liquid biopsy, tissue biopsy, and clinical presentation, to guide tailored therapeutic interventions for individual patients. The feedback loop highlights how advanced imaging and data analytics synergize with clinical data to optimize cancer care

In essence, proteomics and immunophenotyping synergize to provide a refined molecular and cellular framework for precision medicine. By assessing protein-level alterations and immune context within the TME, these approaches permit the tailoring of therapeutic regimens that are more effective and individualized, advancing the goal of precise, patient-centric cancer care.

# 3 Clinical Implementation of Precision Oncology

## 3.1 Targeted Therapies

Targeted therapies have become a cornerstone in the modern treatment of various cancers by focusing on specific molecular abnormalities driving tumor growth, leading to improved clinical outcomes.

**EGFR Inhibitors in EGFR-Mutant NSCLC**
Epidermal growth factor receptor (EGFR) inhibitors such as gefitinib and osimertinib have significantly enhanced progression-free survival (PFS) in patients with EGFR-mutant non-small cell lung cancer (NSCLC). Mok et al. [22] demonstrated in a landmark randomized trial that first-line gefitinib improved PFS compared to standard chemotherapy in East Asian patients with NSCLC harboring EGFR mutations, marking a paradigm shift in treatment [22]. More recently, Soria et al. [23] reported that osimertinib, a third-generation irreversible EGFR tyrosine kinase inhibitor (TKI), further extended PFS and demonstrated a favorable safety profile compared to earlier generation TKIs in treatment-naïve EGFR-mutant NSCLC patients, establishing osimertinib as the new standard of care [23] (Fig. 3).

**BRAF/MEK Inhibitors in BRAF V600E Melanoma**
For patients with metastatic melanoma harboring the BRAF V600E mutation, the combination of dabrafenib (a BRAF inhibitor) and trametinib (a MEK inhibitor) has shown superior efficacy compared to monotherapy. Long et al. [24] conducted a phase 3 trial demonstrating that this combination significantly improved progression-free survival, overall survival, and objective response rates, while concurrently reducing some toxicities typically associated with BRAF inhibitor monotherapy [24]. This combination therapy thus represents a major advancement in targeted melanoma treatment.

**PARP Inhibitors in BRCA-Mutated Ovarian Cancer**
PARP inhibitors such as olaparib have transformed the therapeutic landscape for patients with BRCA-mutated ovarian cancer by exploiting defects in DNA repair mechanisms. Moore et al. [25] showed in a pivotal phase 3 trial that maintenance therapy with olaparib significantly prolonged progression-free survival compared to placebo in patients with newly diagnosed, advanced BRCA-mutated ovarian cancer following platinum-based chemotherapy [25]. These results established olaparib as an essential component of personalized treatment in ovarian cancer. Collectively, these examples highlight the crucial role of targeted therapies in improving patient outcomes by addressing specific genetic aberrations within tumors.

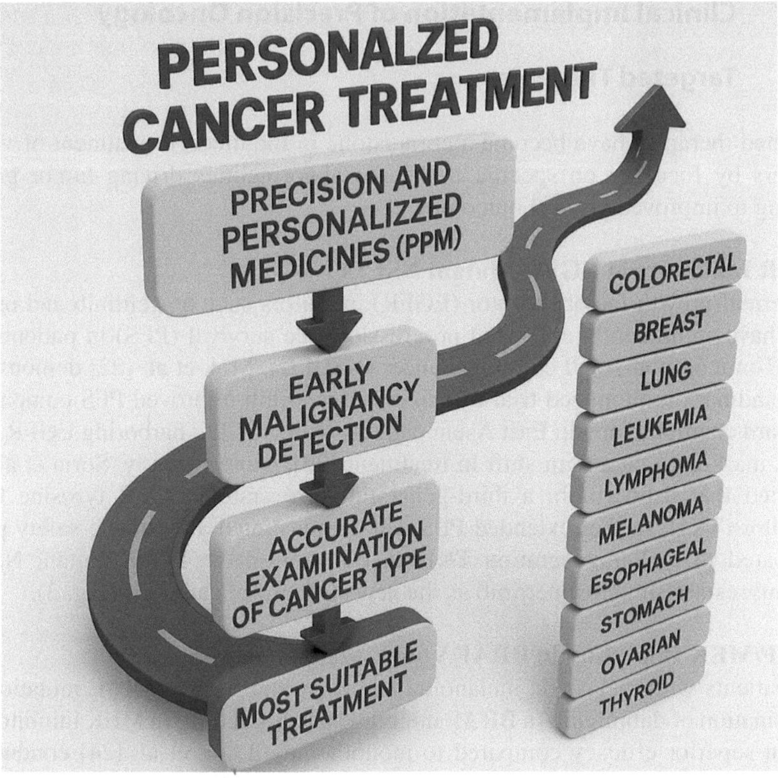


**Fig. 3** This infographic illustrates the concept and application of personalized cancer treatment (Precision and Personalized Medicines, PPM). It outlines how PPM tailors medical care based on a patient's unique genetic makeup and specific disease characteristics. The flowchart describes the process beginning with identifying individuals at high risk, early malignancy detection, and accurate examination of cancer types, culminating in the selection of the most suitable treatment. The infographic also lists various cancer types—such as colorectal, breast, lung, leukemia, lymphoma, melanoma, esophageal, stomach, ovarian, and thyroid cancers—where personalized treatment approaches are effectively implemented

## 3.2 Immunotherapy and Biomarker Selection

Immunotherapy has revolutionized cancer treatment by harnessing the patient's immune system to recognize and eliminate tumor cells. Among immunotherapeutic agents, inhibitors targeting the programmed death-1 (PD-1)/programmed death-ligand 1 (PD-L1) axis have shown remarkable clinical benefits across multiple cancer types, especially in tumors characterized by specific molecular biomarkers.

**PD-1/PD-L1 Inhibitors in Mismatch Repair-Deficient Tumors**

Mismatch repair-deficient (dMMR) tumors exhibit a high frequency of mutations due to defective DNA repair mechanisms, leading to increased neoantigen formation and enhanced immunogenicity. Le et al. [20] conducted a seminal study demonstrating that PD-1 blockade with pembrolizumab significantly improved survival and induced durable responses in patients with dMMR tumors across various cancer types compared to mismatch repair-proficient tumors. This study not only established dMMR status as a predictive biomarker for response to immune checkpoint inhibitors (ICIs) but also led to the FDA's landmark tissue-agnostic approval of pembrolizumab for dMMR or microsatellite instability-high (MSI-H) solid tumors [20].

**Tumor Mutational Burden as a Predictor of Immune Checkpoint Inhibitor Response**

Tumor mutational burden (TMB), defined as the total number of somatic mutations per coding area of a tumor genome, has emerged as another robust biomarker predictive of response to ICIs. Samstein et al. [21] performed a comprehensive pan-cancer analysis of over 1600 patients treated with ICIs and found that higher TMB was associated with improved overall survival across multiple tumor types. Their findings supported the use of TMB as a tissue-agnostic biomarker to stratify patients likely to benefit from immune checkpoint blockade therapy, although the optimal TMB cutoffs varied by cancer histology [21].

Together, PD-1/PD-L1 inhibitors and molecular biomarkers such as dMMR status and TMB enable personalized immunotherapy, allowing clinicians to better select and tailor treatments to achieve maximal clinical benefit while minimizing unnecessary toxicity (Fig. 4).

## 3.3 Challenges in Implementation

The implementation of personalized cancer therapy faces several formidable challenges, primarily stemming from biological complexity, economic constraints, and regulatory inconsistencies. These challenges collectively hinder the widespread integration of tailored treatments into routine clinical practice despite their promising potential.

**Tumor Heterogeneity and Clonal Evolution Leading to Resistance**

A significant biological challenge in personalized cancer therapy is tumor heterogeneity, characterized by genetic, phenotypic, and functional diversity within and between tumor lesions in the same patient. This heterogeneity arises through clonal evolution, a Darwinian process where tumor cell populations acquire genetic mutations over time, leading to branched evolutionary patterns. As a result, some subclones harbor resistance-conferring mutations while others remain sensitive to targeted therapies. This dynamic diversity undermines the efficacy of treatments aimed at specific molecular targets, as resistant clones can survive and proliferate despite therapy, resulting in disease progression or relapse. The intricate landscape of intratumoral heterogeneity complicates the identification of actionable targets and necessitates ongoing monitoring to adapt treatment strategies according to tumor evolution.

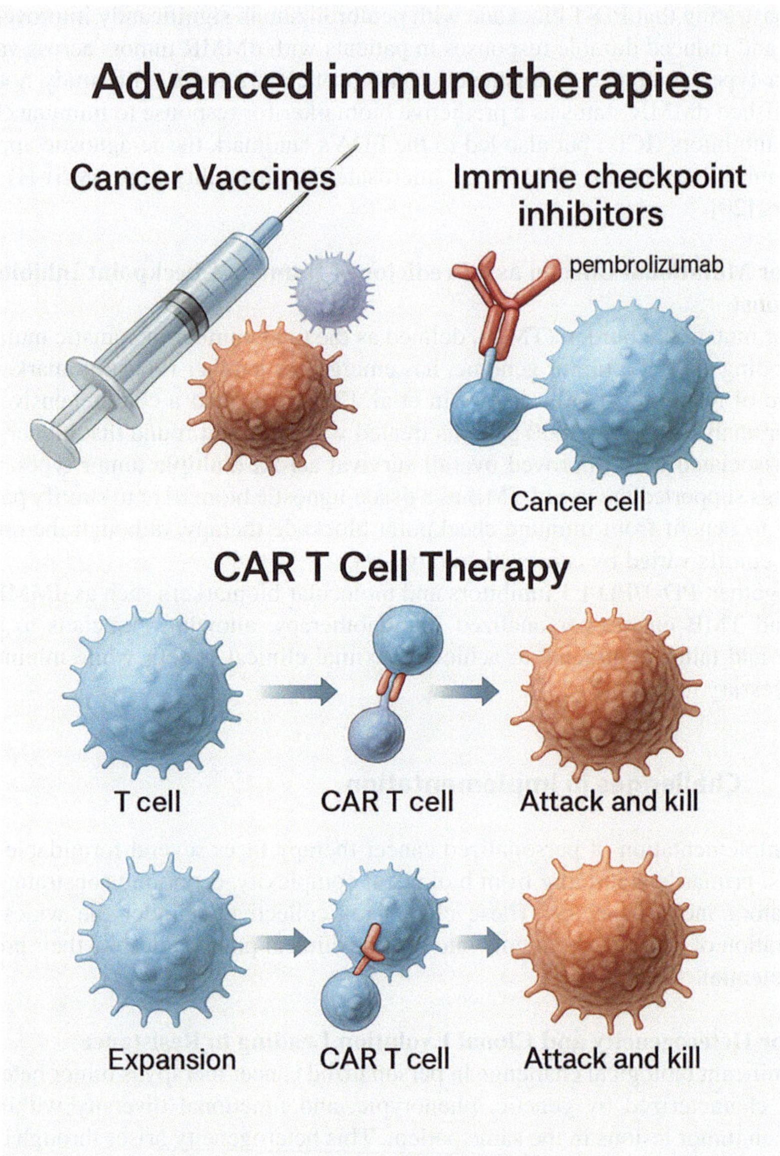


**Fig. 4** This rendered infographic illustrates three advanced immunotherapy strategies used in precision medicine for cancer treatment. On the top left, Cancer Vaccines are depicted, highlighting their role in enhancing the immune system's response to cancer cells for both prevention and targeted therapy. On the top right, Immune Checkpoint Inhibitors such as pembrolizumab are shown, which block immune checkpoints to restore immune activity against cancer cells. At the bottom, CAR T Cell Therapy is visualized, representing the process of engineering a patient's T cells to target cancer-specific antigens, expanding them in the lab, and reinfusing them to attack and kill cancer cells. These approaches demonstrate how personalized immunotherapies are revolutionizing oncology by tailoring treatments to individual patients' immune profiles and tumor characteristics

### Access and Cost Barriers Associated with Next-Generation Sequencing (NGS) and Targeted Therapies

Despite the remarkable strides in genomic technologies such as next-generation sequencing (NGS), which enable comprehensive molecular profiling essential for personalized therapy, access to these technologies remains uneven, particularly in low- and middle-income countries. The high costs of NGS assays, coupled with the expensive nature of targeted drugs, limit availability to many patients globally. Infrastructure deficits, including a lack of technical expertise and laboratory capacity, further exacerbate disparities in molecular testing. Additionally, targeted therapies often entail chronic administration with substantial financial burdens, making affordability and reimbursement significant obstacles. The persistence of elevated prices for both sequencing and therapeutics, despite some reductions in sequencing costs, presents a critical barrier to equitable access and widespread clinical application [43] (Table 1).

### Regulatory Hurdles from Variability in Biomarker Testing Standards

The clinical implementation of personalized cancer therapy is also hampered by regulatory challenges, notably the variability in biomarker testing methodologies and standards across institutions and regions. Biomarkers serve as crucial indicators

**Table 1** Summarizes ten highly effective precision medicines in cancer therapy, detailing their mechanisms, approved clinical indications, and references documenting their use and efficacy. Each reference corresponds to high-quality evidence from clinical studies and reviews that support their role as established precision oncology treatments

| Precision medicine | Mechanism of action | Clinical indications |
|---|---|---|
| Gefitinib/ Osimertinib | EGFR tyrosine kinase inhibitors targeting EGFR mutations | EGFR-mutant non-small cell lung cancer (NSCLC) |
| Dabrafenib + Trametinib | BRAF and MEK inhibitors targeting BRAF V600E mutation | BRAF V600E-mutant metastatic melanoma |
| Olaparib | PARP inhibitor exploiting defects in DNA repair (BRCA mutations) | BRCA-mutated ovarian and breast cancer |
| Trastuzumab | Monoclonal antibody against HER2 receptor overexpression | HER2-positive breast and gastric cancers |
| Crizotinib | ALK and ROS1 tyrosine kinase inhibitor | ALK or ROS1 fusion-positive NSCLC |
| Sotorasib/ Adagrasib | KRAS G12C inhibitors locking KRAS in inactive GDP-bound state | KRAS G12C-mutant NSCLC, pancreatic, colorectal cancers |
| Pembrolizumab | PD-1 immune checkpoint inhibitor guided by biomarkers (e.g., dMMR, MSI-H, TMB) | Various advanced cancers, including MSI-high tumors |
| Alpelisib | PI3K alpha-selective inhibitor targeting PIK3CA mutations | PIK3CA-mutant hormone receptor-positive, HER2-negative breast cancer |
| Talazoparib | PARP inhibitor targeting homologous recombination repair deficiencies | BRCA-mutated breast and prostate cancers |
| Larotrectinib/ Entrectinib | TRK fusion inhibitors acting on NTRK gene fusions | NTRK fusion-positive solid tumors (tumor-agnostic) |

for selecting appropriate therapies, but their reliability is contingent upon standardized, validated assays. Divergent regulatory frameworks result in inconsistent biomarker assessment, complicating the interpretation and comparability of molecular testing results and delaying therapeutic decisions. Moreover, complex approval processes for multiplex testing platforms and companion diagnostics increase the time and cost required for clinical integration. The lack of unified regulatory guidance and reimbursement policies further impedes the scalability of personalized approaches. Overcoming these hurdles necessitates coordinated efforts to harmonize testing standards, ensure assay accuracy, and streamline regulatory pathways to foster the adoption of precision oncology in clinical settings [34, 36, 37, 40].

In summary, the clinical advancement of personalized cancer therapy is challenged by the biological complexity of tumor heterogeneity and evolution, as well as the economic and infrastructural barriers to accessing sophisticated genomic tools and therapies. Additionally, regulatory inconsistencies affect biomarker testing. Addressing these challenges through innovative research, improved healthcare infrastructure, and streamlined regulatory policies is essential to realizing the full potential of personalized oncology.

## 4 Clinical Outcomes of Personalized Therapy

### 4.1 Survival Benefits

Personalized cancer therapy has demonstrated significant survival benefits across multiple cancer types by targeting specific molecular alterations unique to patients' tumors.

**Survival Benefits in ALK-Positive NSCLC with Crizotinib**
In non-small cell lung cancer (NSCLC) patients harboring ALK rearrangements, treatment with the targeted ALK inhibitor crizotinib has shown marked improvement in progression-free survival (PFS) compared to conventional chemotherapy. Shaw et al. [26] reported that crizotinib-treated patients achieved a median PFS of approximately 10.9 months as a first-line therapy and 7.7 months in previously treated patients, significantly outperforming chemotherapy [26]. This improved PFS was accompanied by higher objective response rates and better quality of life, establishing crizotinib as a standard treatment for ALK-positive NSCLC. Their findings underscore the clinical efficacy of targeting oncogenic drivers with precision therapies to prolong disease control and potentially survival in molecularly defined patient subgroups.

**Survival Benefits in HER2-Positive Breast Cancer with Targeted Therapies**
For HER2-positive breast cancer, the introduction of targeted HER2 therapies such as trastuzumab has revolutionized patient outcomes by significantly reducing cancer recurrence and improving overall survival. Piccart-Gebhart et al. [27] conducted a pivotal randomized controlled trial showing that adding trastuzumab after

adjuvant chemotherapy significantly improved disease-free survival and overall survival in early-stage HER2-positive breast cancer patients compared to chemotherapy alone [27]. This study established the integration of HER2-targeted agents into standard breast cancer care, highlighting profound survival advantages through precision medicine approaches.

In summary, personalized cancer therapy exemplified by crizotinib in ALK-positive NSCLC and trastuzumab in HER2-positive breast cancer delivers substantial survival benefits by tailoring treatment based on molecular characteristics, improving progression-free and overall survival outcomes.

## 4.2 Limitations and Resistance Mechanisms

Personalized cancer therapy has advanced significantly, yet its clinical implementation faces critical limitations, primarily due to resistance mechanisms that reduce long-term efficacy. These resistance mechanisms can be broadly categorized into on-target and off-target resistance, each posing distinct challenges.

**On-Target Resistance: Secondary Mutations**

On-target resistance occurs when mutations arise within the drug target itself, diminishing therapeutic efficacy. A well-documented example is the EGFR C797S mutation in non-small cell lung cancer (NSCLC). This tertiary mutation interferes with the covalent binding of third-generation EGFR tyrosine kinase inhibitors (TKIs) like osimertinib, leading to resistance [28]. Studies have shown that the C797S mutation abolishes the ability of these inhibitors to effectively suppress EGFR signaling, causing treatment failure [29]. The allelic context of C797S (whether it occurs in cis or trans with other mutations such as T790M) critically influences subsequent treatment options and drug sensitivity [29]. Furthermore, cases have been reported where the C797S mutation arises alongside other secondary mutations or oncogenic fusions, adding complexity to resistance management [30]. Consequently, fourth-generation EGFR TKIs and novel therapeutic strategies such as PROTACs are under development to overcome C797S-mediated resistance, demonstrating promising preclinical and early clinical results [31].

**Off-Target Resistance: Activation of Bypass Pathways**

Off-target resistance involves the activation of alternative signaling pathways that circumvent the inhibited target. This bypass leads to sustained oncogenic signaling despite the presence of targeted inhibitors. For instance, activation or amplification of receptor tyrosine kinases (RTKs) such as MET, HER2, and AXL, or downstream effectors like RAS/RAF/MEK/ERK and PI3K/AKT/mTOR pathways, has been implicated in resistance to EGFR TKIs [32]. Increased expression of EGFR ligands such as TGF$\alpha$ can activate parallel pathways, contributing to resistance against ALK inhibitors in lung cancer [33]. Additionally, histological transformations, including small cell lung cancer (SCLC) phenotype shifts and epithelial-to-mesenchymal transition (EMT), are associated with resistance via off-target mechanisms. The heterogeneity of these bypass pathways often requires combination therapies that concurrently target the primary oncogenic driver and compensatory signaling routes to overcome resistance.

**Limitations in Clinical Implementation**

Beyond biological resistance, clinical adoption of personalized cancer therapy faces hurdles including cost and access to next-generation sequencing (NGS) and targeted agents, variable regulatory standards for biomarker testing, and the need for multidisciplinary expertise to interpret complex genomic data [34, 40]. Moreover, intratumoral heterogeneity and clonal evolution during treatment complicate persistent tumor control, necessitating real-time monitoring strategies such as liquid biopsies to dynamically adapt therapy [35]. These challenges underscore the importance of continuous development of novel inhibitors, combinatorial treatment regimens, and standardized clinical frameworks to realize the full potential of precision oncology.

In summary, both on-target mutations like EGFR C797S and off-target bypass mechanisms significantly impede personalized therapy effectiveness. Addressing these resistance mechanisms through innovative drug development, combination therapies, and improved clinical implementation strategies is essential to enhance patient outcomes in cancer precision medicine.

## 5 Future Directions

Personalized cancer therapy represents a transformative frontier in oncology, offering tailored treatments based on individual patient tumor characteristics, genetics, and molecular profiles. The future directions of clinical implementation and outcomes for personalized cancer therapy are driven by advances in genomics, biomarker identification, targeted therapeutics, immunotherapy, and emerging technologies such as next-generation sequencing and nanotechnology.

One critical direction is the integration of comprehensive genomic profiling in clinical practice, enabling precise identification of actionable mutations and guiding treatment selection. Studies demonstrate that matching therapeutics to patient-specific molecular markers improves response rates, progression-free survival, and overall survival across cancer types, highlighting the efficacy of personalized strategies over traditional approaches. The development of robust predictive and prognostic biomarkers, including protein and genetic signatures, is central to optimizing patient selection and monitoring therapeutic response, thus enhancing clinical outcomes and reducing toxicity.

Immunotherapy, particularly immune checkpoint inhibitors and chimeric antigen receptor (CAR) T-cell therapies, is a promising pillar within personalized cancer treatment. These approaches harness the patient's immune system to target cancer cells specifically, with ongoing research focusing on overcoming resistance, refining immune targets, and developing combination therapies to maximize efficacy. The advent of novel cancer vaccines, including mRNA and neoantigen-based vaccines, holds significant potential for personalized immunotherapy, though challenges remain in vaccine delivery and immune response variability.

Technological advancements such as digital pathology, liquid biopsies, and quantitative phase imaging facilitate real-time tumor characterization and dynamic monitoring, thus enabling adaptive and precisely tailored therapeutic interventions. Moreover, nanotechnology platforms offer multifunctional theranostic capabilities, integrating targeted drug delivery with imaging to improve treatment precision and patient stratification.

Despite these advancements, challenges in clinical implementation persist, including infrastructure limitations, high costs, regulatory considerations, and the need for multidisciplinary collaborations and education to ensure widespread adoption. Future efforts are aimed at overcoming these barriers, standardizing biomarker assays, and conducting adaptive clinical trials to expedite the translation of personalized cancer therapies into routine care.

In conclusion, the future of personalized cancer therapy lies in the seamless integration of molecular diagnostics, immunotherapy, and innovative technologies, fostering improved patient outcomes through precision medicine approaches that are continually refined by emerging research and clinical evidence.

## 6 Conclusion

Personalized cancer therapy has fundamentally redefined the oncology landscape by shifting away from the traditional one-size-fits-all paradigm toward precision interventions tailored to individual tumor molecular profiles. This transformation leverages comprehensive molecular profiling technologies—including genomic, transcriptomic, proteomic, and immunophenotyping approaches—that together unravel tumor heterogeneity and delineate actionable targets enabling rational therapy selection. Clinical successes with targeted agents such as EGFR inhibitors in NSCLC, BRAF/MEK inhibitors in melanoma, and PARP inhibitors in BRCA-mutated ovarian cancer have evidenced clear survival benefits and improved quality of life, confirming the clinical value of matching treatments to specific molecular aberrations. Moreover, integration of immunotherapy guided by biomarkers such as mismatch repair deficiency or tumor mutational burden has broadened therapeutic efficacy across diverse tumor types and advanced personalized immuno-oncology.

Nevertheless, personalized therapy faces persistent challenges that complicate its clinical implementation and limit access. Tumor heterogeneity and clonal evolution generate resistance via on-target mutations (e.g., EGFR C797S) and activation of bypass pathways (e.g., MET or HER2 amplification), undermining durable responses and necessitating continual adaptation of therapeutic regimens. Economic constraints and infrastructure disparities impede equitable access to next-generation sequencing and costly targeted agents, disproportionately affecting underserved populations. Regulatory fragmentation in biomarker testing and approval processes introduces variability and delays, underscoring the need for harmonized standards and streamlined pathways to foster clinical integration.

Future progress in personalized oncology hinges on multidimensional advances: expanding comprehensive genomic and transcriptomic profiling facilitated by machine learning to manage complex datasets; harnessing proteomic and immunophenotyping insights to refine tumor microenvironment characterization; innovating targeted and immunotherapeutic strategies, including novel vaccines and CAR-T cell therapies; and developing nanotechnology-enabled theranostics for real-time tumor monitoring and delivery precision. Furthermore, overcoming economic and regulatory barriers through coordinated policy initiatives, infrastructure investment, education, and multidisciplinary collaboration is imperative to ensure widespread adoption and access.

In conclusion, personalized cancer therapy represents a paradigm shift toward patient-centric, biology-driven oncology with the potential to maximize therapeutic efficacy, minimize toxicity, and improve survival outcomes. Realizing this promise requires sustained efforts to surmount biological, economic, and regulatory challenges, coupled with the integration of emerging technologies and biomarkers into adaptive clinical frameworks. As molecular diagnostics and therapeutic options continue to evolve rapidly, personalized oncology will increasingly become the standard of care, transforming cancer management into a dynamic, precise discipline tailored to the unique molecular attributes of each patient's disease.

**Declarations**

**Author's Contributions** Conceived the idea and review structure: KKG; Wrote the manuscript: KKG; Revised and edited the manuscript: KKG. The author has read, agreed, and approved the final draft of the manuscript.

**Funding** None

**Data Availability** The software MS PowerPoint is used to draw some elements of the few figures.

**Conflicts of Interest** There is no competing and conflict of interest.

## References

1. Vogelstein B, Papadopoulos N, Velculescu VE, Zhou S, Diaz LA, Kinzler KW (2013) Science 339(6127):1546–1558
2. Collins FS, Varmus H (2015) New Engl J Med 372(9):793–795
3. Chapman PB, Hauschild A, Robert C et al (2011) New Engl J Med 364(26):2507–2516
4. Flaherty KT, Robert C, Hersey P et al (2010) New Engl J Med 367(2):107–114
5. Druker BJ, Talpaz M, Resta DJ et al (2001) New Engl J Med 344(14):1031–1037
6. Slamon DJ, Leyland-Jones B, Shak S et al (2001) New Engl J Med 344(11):783–792
7. Mardis ER (2017) Annu Rev Genomics Hum Genet 6:287–303
8. Weinstein JN et al (2013) The cancer genome atlas pan-cancer analysis project. Nat Genet. https://doi.org/10.1038/ng.2764
9. Diehl F et al (2008) Circulating mutant DNA to assess tumor dynamics. Nat Med. https://doi.org/10.1038/nm.1789

10. Oxnard GR, Paweletz CP, Kuang Y, Mach SL, O'Connell A, Messineo MM, Luke JJ, Butaney M, Kirschmeier P, Jackman DM, Jänne PA (2014) Noninvasive detection of response and resistance in EGFR-mutant lung cancer using quantitative next-generation genotyping of cell-free plasma DNA. Clin Cancer Res, 20(6):1698–1705. https://doi.org/10.1158/1078-0432.CCR-13-2482. PMID: 24429876; PMCID: PMC3959249
11. Pankiw M, Brezden-Masley C, Charames GS (2023) Comprehensive genomic profiling for oncological advancements by precision medicine. Med Oncol, 41(1):1. https://doi.org/10.1007/s12032-023-02228-x. PMID: 37993657
12. Fertig E, Slebos R, Chung C (2012) Application of genomic technologies in oncology. ASCO Educational Book
13. Mock A, Braun M, Scholl C et al (2023) Transcriptome profiling for precision cancer medicine using shallow nanopore cDNA sequencing. Sci Rep, 13:2378. https://doi.org/10.1038/s41598-023-29550-8
14. Stokholm MN, Rabaglino MB, Kadarmideen HN (2024) GEDI: An R Package for Integration of Transcriptomic Data from Multiple Platforms for Bioinformatics Applications. Curr Protoc, 4(10):e70046. https://doi.org/10.1002/cpz1.70046. PMID: 39453042
15. Gafurov A, Mamichev I, Vasileva EV, Sagaradze GD, Shitova MS, Nos G, Fowler N (2021) Astraea: A first-in-class biomarker database integrating genomic, transcriptomic, and tumor microenvironment properties for precision oncology. Cancer Res, 81(13_Supplement):205–205. https://doi.org/10.1158/1538-7445.AM2021-205
16. Reel PS, Reel S, Pearson E, Trucco E, Jefferson E. Using machine learning approaches for multi-omics data analysis: A review. Biotechnol Adv. 2021 Jul-Aug;49:107739. https://doi.org/10.1016/j.biotechadv.2021.107739. Epub 2021 Mar 29. PMID: 33794304.
17. Hanash SM, Pitteri SJ, Faca VM (2011) Mining the plasma proteome for cancer biomarkers. Genome Med 3(11):53
18. Topalian SL, Hodi FS, Brahmer JR et al (2012) Safety, activity, and immune correlates of anti–PD-1 antibody in cancer. New Engl J Med 366(26):2443–2454
19. Le DT, Durham JN, Smith KN et al (2017) Mismatch repair deficiency predicts response of solid tumors to PD-1 blockade. New Engl J Med 377(26):2509–2520
20. Samstein RM, Lee CH, Shoushtari AN et al (2019) Tumor mutational load predicts survival after immunotherapy across multiple cancer types. Nat Genet 51(2):202–206
21. Mok TS, Wu YL, Thongprasert S et al (2009) Gefitinib or carboplatin–paclitaxel in pulmonary adenocarcinoma. New Engl J Med 361(10):947–957
22. Soria JC, Ohe Y, Vansteenkiste J et al (2018) Osimertinib in untreated EGFR-mutated advanced non–small-cell lung cancer. New Engl J Med 378(2):113–125
23. Long GV, Hauschild A, Santinami M et al (2017) Adjuvant dabrafenib plus trametinib in stage III BRAF-mutated melanoma. New Engl J Med 377(19):1813–1823
24. Moore KN, Secord AA, Geller MA et al (2018) Maintenance olaparib in patients with newly diagnosed advanced ovarian cancer. New Engl J Med 379(26):2495–2505
25. Shaw AT, Kim DW, Nakagawa K et al (2013) Crizotinib versus chemotherapy in advanced ALK-positive lung cancer. New Engl J Med 368(25):2385–2394
26. Piccart-Gebhart MJ, Procter M, Leyland-Jones B et al (2005) Trastuzumab after adjuvant chemotherapy in HER2-positive breast cancer. New Engl J Med 353(16):1659–1672
27. Yu H et al (2015) Acquired resistance of EGFR-mutant lung cancer to a T790M-Specific EGFR inhibitor: emergence of a third mutation (C797S) in the EGFR tyrosine kinase domain. JAMA Oncol. https://doi.org/10.1001/jamaoncol.2015.1066
28. Niederst MJ et al (2015) The Allelic Context of the C797S mutation acquired upon treatment with third-generation EGFR inhibitors impacts sensitivity to subsequent treatment strategies. Clin Cancer Res. https://doi.org/10.1158/1078-0432.CCR-15-0560
29. Wang LS et al (2022) Acquired EML4-ALK fusion and EGFR C797S in cis mutation as resistance mechanisms to osimertinib in a non-small cell lung cancer patient with EGFR L858R/T790M. Anti-Cancer Drugs. https://doi.org/10.1097/CAD.0000000000001489

30. Du Y et al (2022) HJM-561, a potent, selective and orally bioavailable EGFR PROTAC that overcomes osimertinib-resistant EGFR triple mutations. Mol Cancer Ther. https://doi.org/10.1158/1535-7163.MCT-21-0835
31. Ou X et al (2024) Mechanisms of resistance to tyrosine kinase inhibitor-targeted therapy and overcoming strategies. MedComm (2020). https://doi.org/10.1002/mco2.694
32. Tani T et al (2015) Activation of EGFR bypass signaling by TGFα overexpression induces acquired resistance to Alectinib in ALK-translocated lung cancer cells. Mol Cancer Ther. https://doi.org/10.1158/1535-7163.MCT-15-0084
33. Meric-Bernstam F, Mills GB (2012a) Overcoming implementation challenges of personalized cancer therapy. Nat Rev Clin Oncol. https://doi.org/10.1038/nrclinonc.2012.127
34. Zhu S, Xie Z, Chen Y, Liu S, Kwan YW, Zeng S, Yuan W, Ho HP (2022) Real-Time Detection of Circulating Tumor Cells in Bloodstream Using Plasmonic Fiber Sensors. Biosensors 12(11):968. https://doi.org/10.3390/bios12110968
35. Glickman A, Weiner J (2018) Opportunities for precision cancer: reflections from the Gant consortium. https://repository.upenn.edu/handle/20.500.14332/38101
36. Godman B, Frost I, Harrington R (2016) Personalising medicine: feasibility and future implications from a payers' perspective. https://pureportal.strath.ac.uk/en/publications/personalising-medicine-feasibility-and-future-implications-from-a
37. Caswell DR, Swanton C (2017) The role of tumour heterogeneity and clonal cooperativity in metastasis, immune evasion and clinical outcome. BMC Med. https://link.springer.com/article/10.1186/s12916-017-0900-y
38. Hirsch F, Zarić B, Rabea A, Thongprasert S, Lertprasertsuke N, Dalurzo M, Varella-Garcia M (2017) Biomarker testing for personalized therapy in lung cancer in low – and middle-income countries. In: American Society of Clinical Oncology educational book. American Society of Clinical Oncology. Annual Meeting. https://ascopubs.org/doi/10.1200/EDBK_175243
39. Meric-Bernstam F, Mills GB (2012b) Overcoming implementation challenges of personalized cancer therapy. Nat Rev Clin Oncol. https://www.nature.com/articles/nrclinonc.2012.127
40. Dagogo-Jack I, Shaw AT (2018) Tumour heterogeneity and resistance to cancer therapies. Nat Rev Clin Oncol. https://www.nature.com/articles/nrclinonc.2017.166
41. Tannock IF (2022) Clinical research in oncology: in memory of Professor Gordon McVie. ecancermedicalscience. https://pmc.ncbi.nlm.nih.gov/articles/PMC8831109/
42. Wistuba II, Gelovani JG, Jacoby JJ, Davis SE (2011) Methodological and practical challenges for personalized cancer therapies. Nat Rev Clin Oncol 8:135. https://www.nature.com/articles/nrclinonc.2011.2

# Cancer Specific Molecular Profiling, Diagnosis and Treatment

# Advanced Personalized Treatment Strategies for Glioblastoma

Sunny Kumar and Mrinal K. Ghosh

**Abstract**

Glioblastoma (GBM) presents the biggest therapeutic challenge among gliomas, which are some of the most aggressive and resistant to treatment brain tumors. Due to the limited survival benefits of traditional treatment modalities, including chemotherapy, radiation, and surgery, precision medicine has become necessary. Using molecular profiling, next-generation sequencing, and sophisticated computational methods, precision medicine in glioma therapy aims to customize treatments according to the genetic and epigenetic makeup of each tumor. Important biomarkers that direct therapy selection include key molecular changes like EGFR and TP53 pathway changes, MGMT promoter methylation, and IDH mutations. Glioma treatment is being revolutionized by new approaches such as immunotherapy, targeted medicines, and innovative drug delivery methods. The state of precision medicine for gliomas is examined in this chapter, with particular attention paid to developments in molecular diagnostics, individualized treatment plans, and the application of artificial intelligence to therapy optimization. Precision medicine has the ability to greatly enhance patient outcomes and open the door to more efficient, customized glioma treatments by incorporating these developments into clinical practice.

**Keywords**

Glioma · Precision medicine · Molecular profiling · Immunotherapy · Targeted therapy

S. Kumar · M. K. Ghosh (✉)
Cancer Biology and Inflammatory Disorder Division, Council of Scientific and Industrial Research-Indian Institute of Chemical Biology (CSIR-IICB), TRUE Campus, Jadavpur, Kolkata, West Bengal, India
e-mail: mrinalghosh@iicb.res.in

 
M. K. Ghosh, M. Basu (eds.), *Personalized Cancer Therapy*,
https://doi.org/10.1007/978-981-95-6845-1_8

## 1 Introduction

WHO divides gliomas into four classes according to patient survival rates and aggressiveness. With an overall survival (OS) of less than a year, grade IV GBM is the most aggressive. The survival duration for anaplastic astrocytoma (Grade III) is 2–3 years, that of diffuse astrocytoma (Grade II) is 7–8 years, and that of pilocytic astrocytoma (Grade I) is 8–10 years. Different genetic changes are linked to different glioma grades, which affect how the tumor progresses and how well a treatment works. Mutations across glioma grades highlight the genetic variety of these tumors. There are main and secondary forms of glioblastoma; primary glioblastoma is more aggressive and has a poorer prognosis. EGFR amplifications (40–60%), p53 mutations (30%), PTEN mutations (25%), and chromosome 10 deletion (70%), are typical characteristics of primary GBM. Treatment responsiveness is impacted by the low rate of MGMT methylation (36%). EGFR amplifications (10%), p53 mutations (65%), IDH1 mutations (70%), and increased MGMT methylation (75%), on the other hand, are characteristics of secondary GBM, which develops from lower-grade gliomas (Grades I/II). While primary GBM primarily affects older people (~60 years), secondary GBM usually affects younger patients (~40 years) [1, 2].

The most prevalent primary tumors of the central nervous system are gliomas, which are a diverse collection of neoplasms with unique histological and molecular characteristics. With a median lifespan of only around 15 months after standard treatment. GBM, the most aggressive type, nevertheless has a poor prognosis despite advancements in surgery, radiation, and chemotherapy. This difficulty has prompted the need for a paradigm change in the treatment of gliomas, one that abandons the conventional "one-size-fits-all" method in favor of a more individualized approach based on the distinct molecular and genetic makeup of every tumor [2].

Glioma therapy precision medicine uses advances in proteomics, transcriptomics, and genomics to customize treatments for each patient. Clinicians can maximize treatment interventions, reduce off-target effects, and enhance patient outcomes by identifying important genetic mutations, signaling pathways, and epigenetic alterations. Glioma care has started to change as a result of advancements in immunotherapy, targeted treatments, and biomarker-driven treatment selection, providing fresh hope in an otherwise difficult environment [1]. This chapter explores the evolving field of precision medicine in glioma therapy, covering the molecular underpinnings of gliomas, advances in biomarker identification, targeted therapies, and the role of artificial intelligence in treatment selection. The integration of precision medicine into clinical practice holds great promise for enhancing therapeutic efficacy and overcoming the resistance mechanisms that have historically hindered progress in glioma treatment.

## 2 Molecular Profiling

Gliomas are the most common primary brain tumors, characterized by significant genetic and epigenetic heterogeneity. WHO classifies gliomas based on histological and molecular features, incorporating key genetic alterations that influence prognosis and therapeutic strategies. Advances in next-generation sequencing (NGS), transcriptomic analysis, and proteomics have led to a refined molecular classification, distinguishing gliomas into distinct subtypes with specific driver mutations. Understanding these molecular characteristics has enabled the development of targeted therapies and personalized treatment strategies [3, 4].

Mutations in the IDH genes (IDH1 and IDH2) define a major subset of gliomas with distinct metabolic and prognostic implications. IDH-mutant gliomas are predominantly found in lower-grade gliomas (WHO grade 2 and 3) and secondary glioblastomas. These mutations result in the production of the oncometabolite D-2-hydroxyglutarate (D-2HG), which alters DNA and histone methylation, leading to widespread epigenetic dysregulation and tumorigenesis. Clinically, IDH-mutant gliomas exhibit a better prognosis compared to IDH-wildtype gliomas, responding more favorably to therapy. IDH inhibitors, such as ivosidenib (AG-120) and vorasidenib, are being explored as potential targeted therapies to reverse the effects of mutant IDH activity. Furthermore, the co-deletion of chromosome arms 1p and 19q is a hallmark of oligodendrogliomas, distinguishing them from astrocytomas. This genetic alteration is strongly associated with enhanced responsiveness to chemotherapy (e.g., procarbazine, lomustine, and vincristine [PCV] regimen and temozolomide) and prolonged survival. Oligodendrogliomas with 1p/19q co-deletion and IDH mutation exhibit a favorable prognosis, with a median survival exceeding 15 years in some cases. Molecular testing for 1p/19q status is now essential in glioma classification to guide treatment decisions. Furthermore, the MGMT gene encodes a DNA repair enzyme that counteracts the effects of alkylating chemotherapeutic agents, such as temozolomide (TMZ), by repairing O6-methylguanine lesions. Methylation of the MGMT promoter silences gene expression, reducing the repair activity and making glioma cells more susceptible to alkylating agents. MGMT promoter methylation serves as a predictive biomarker for the effectiveness of TMZ, particularly in GBM patients. In MGMT-unmethylated tumors, resistance to TMZ is common, necessitating alternative therapeutic strategies, such as PARP inhibitors, immunotherapy, and tumor-treating fields (TTF). Additionally, EGFR amplifications and mutations are frequently observed in glioblastoma, particularly in the classical GBM subtype. One of the most common alterations is EGFR-v-III, a truncated, constitutively active mutant that drives tumor growth, invasion, and resistance to apoptosis. EGFR-v-III-expressing gliomas are highly aggressive and often resistant to standard therapies. Several targeted approaches, including EGFR inhibitors (e.g., afatinib, osimertinib) and monoclonal antibodies (e.g., depatuxizumab, mafodotin), have been investigated, but clinical efficacy remains limited due to adaptive resistance mechanisms. Newer approaches, such as bispecific antibodies and CAR-T cell therapy targeting EGFR-v-III, are under development to improve therapeutic outcomes. Moreover, mutations in the TERT promoter are prevalent in high-grade gliomas, including glioblastomas and anaplastic astrocytomas. These mutations lead to increased telomerase activity, promoting cellular immortality and tumor progression. TERT promoter mutations often co-occur with EGFR amplifications and PTEN loss, defining a subgroup of IDH-wildtype glioblastomas with poor prognosis. Given their role in glioma pathogenesis, TERT-targeted therapies, such as telomerase inhibitors and G-quadruplex stabilizers, are being explored as potential therapeutic strategies. Furthermore, the diffuse midline gliomas (DMGs), including diffuse intrinsic pontine gliomas (DIPGs), are aggressive pediatric and young adult tumors characterized by mutations in histone H3 variants (H3K27M mutations). This mutation leads to global epigenetic dysregulation, silencing tumor suppressor genes and promoting a highly malignant phenotype. H3K27M-mutant gliomas exhibit extremely poor prognosis, with median survival often less than 1 year. EZH2 inhibitors, HDAC inhibitors, and immunotherapy approaches are being investigated to target this epigenetic dysregulation [1, 3–6].

Numerous significantly mutated genes in GBM have been found by whole-exome sequencing analysis from The Cancer Genome Atlas (TCGA). These findings confirm previously reported key driver alterations, including EGFR, PDGFRA, PIK3CA, PTEN, NF1, RB1, and TP53, while also revealing novel genetic alterations and pathway disruptions, particularly in chromatin remodeling genes. Amplifications in EGFR, MET, and CDK6 on chromosome 7, CDK4 and MDM2 on chromosome 12, and PDGFRA on chromosome 4 were among the most commonly seen genomic gains and losses. Deletions in chromosome 9's CDKN2A/CDKN2B were also frequent. Oncogenic RNA fusion events in GBM have also been found by sequencing investigations, and they could be targets for treatment. FGFR1-TACC1 and FGFR3-TACC3 in-frame fusions and EGFR-SEPT14 are noteworthy examples, as they have shown in vitro sensitivity to FGFR and EGFR inhibitors, respectively [3, 6]. Even though our knowledge of the genetic landscape of GBM is expanding, these findings have not yet resulted in appreciable improvements in patient outcomes. To close the gap between molecular understanding and efficient, focused treatments for GBM, more investigation is required.

## 3 Imaging-Based Diagnosis of Glioma

Choosing the best course of treatment for brain tumors requires an accurate diagnosis. The usual imaging method for glioblastoma is MRI, which evaluates variables such as water diffusion, cerebral blood volume, blood flow rate, and transit time. High glucose levels in brain tissue render FDG-PET useless, whereas FET-PET provides superior resolution for characterizing tumors. While magnetic resonance spectroscopy imaging (MRSI) tracks metabolic indicators like lactate, choline, and N-acetyl aspartate, MRI in conjunction with NMR spectroscopy has demonstrated promise in identifying mesenchymal GBM subtypes. MRSI is becoming a useful method for tracking clonal populations and detecting tumor heterogeneity [7].

5-aminolevulinic acid (5-ALA), a fluorescent marker, has become a promising technique for improving glioblastoma surgical resection. This method makes use of 5-ALA, which is given orally. 5-ALA is selectively absorbed by tumor cells and fluoresces when exposed to filtered light, allowing for more accurate tumor excision. There is little absorption in healthy brain tissue, blood vessels, and olfactory areas, and peak fluorescence happens 6–8 hours after treatment. 5-ALA fluorescence-guided surgery was approved by the FDA in 2017 and has shown promise in increasing the extent of resection (EOR) and progression-free survival (PFS), both of which will ultimately improve patient outcomes [7].

## 4 Conventional Therapies

A multimodal treatment approach that includes radiation therapy, chemotherapy, and maximally safe surgical resection is the current standard of care for GBM. The first-line treatment for tumor debulking is still surgery, which aims to lower intracranial pressure and lessen tumor burden. However, total surgical excision of glioblastoma is nearly impossible because of its extremely infiltrative nature. It can be difficult to remove the entire mass without impairing neurological function since tumor cells frequently infiltrate nearby healthy brain tissue. Because of this, leftover tumor cells often remain after surgery, which makes tumor recurrence inevitable [1] (Table 1).

The prognosis of the patient is significantly influenced by the extent of resection (EOR). Even while a higher EOR is linked to better survival rates, it is frequently constrained by the tumor's closeness to expressive brain areas. Radiation and surgery were the main therapeutic options for GBM for a long time, although they only slightly increased survival. When the FDA approved TMZ, an oral alkylating agent, as the first chemotherapeutic medication for glioblastoma in 2005, it was a major breakthrough. Particularly in cells with compromised DNA repair systems, including mismatch repair (MMR), base excision repair (BER), and most importantly, O6-methylguanine-DNA methyltransferase (MGMT), TMZ causes DNA damage that leads to apoptosis by methylating DNA at the O6 position of guanine [7].

Despite its potential, tumors' innate capacity to repair TMZ-induced DNA lesions through MGMT activity frequently undermines TMZ's effectiveness. By removing alkyl groups from guanine's O6 position, the DNA repair enzyme encoded by the MGMT gene effectively counteracts the cytotoxicity caused by TMZ. Resistance to TMZ is conferred in glioblastomas because the MGMT promoter is unmethylated and MGMT expression is still active. On the other hand, tumors that have a methylation MGMT promoter show less MGMT expression, which makes them more vulnerable to TMZ therapy. The MGMT promoter's epigenetic state has become a crucial prognostic biomarker for TMZ responsiveness and is frequently utilized to inform treatment choices. Interestingly, secondary glioblastomas are more likely than primary glioblastomas to contain methylated MGMT promoters, which increases their sensitivity to TMZ treatment [8].

**Table 1** List of several FDA-approved treatment options for high-grade gliomas (HGGs)

| S No. | Drug | Approval year | Mode of action | Use | Route of administration | Survival (Weeks) | Progression free survival |
|---|---|---|---|---|---|---|---|
| 1 | TMZ | 2005 | DNA methylation | HGG | Oral | 63.51 | 53.90% |
| 2 | CCNU | 1976 | Crosslinking of DNA and RNA of dividing cells | Recurrent-HGG | Oral | 49.97 | – |
| 3 | BCNU | 1977 | Crosslinking of DNA and RNA dividing cells | Recurrent-HGG | IV | 51.05 | – |
| 4 | BCNU wafers | 2003 | Crosslinking of DNA and RNA dividing cells | Recurrent-HGG | Implanted directly to brain | 60.39 | – |
| 5 | BVZ | 2009 | Anti-VEGF monoclonal Ab | Recurrent-HGG | IV | 40.41 | 36% |
| 6 | TTF | 2015 | Disrupt mitosis | Recurrent-HGG | On scalp | 89.16 | 56% |

Abbreviations: *TMZ* Temozolomide, *CCNU* Lomustine, *BCNU* Carmustine, *BVZ* Bevacizumab, *TTF* Tumor-treating fields, *FDA* Food and Drug Administration, and *IV* Intravenous

The goal of ongoing research is to find therapeutic drugs that can either reduce MGMT expression or induce methylation of the MGMT promoter, making resistant tumors more sensitive to TMZ. Finding these adjuvants has the potential to improve treatment outcomes, particularly for patients whose tumors have unmethylated MGMT promoters. In conclusion, overcoming TMZ resistance, especially through targeted modulation of MGMT expression, remains a crucial area of therapeutic development in the pursuit of more potent glioblastoma treatments, even though surgery, radiation, and TMZ constitute the current standard of care.

## 5 Precision Medicine in Glioma

Precision medicine in glioma aims to tailor treatment based on individual tumor genetics and molecular profiles. By identifying specific mutations and biomarkers, it enables the use of targeted therapies and immunotherapies that improve efficacy and reduce toxicity. This section highlights key advancements in targeted therapies and immunotherapeutic approaches shaping personalized glioma treatment (Table 2).

**Table 2** List of several molecules under development for glioma therapy

| S No. | Drug | Target | Use | Route | Phase | Survival (Weeks) |
|---|---|---|---|---|---|---|
| 1 | Nivolumab | PD-1 | Recurrent-GBM | IV | III | 41.27 |
| 2 | Gefitinib | EGFR-TKI | Recurrent-GBM | Oral | II | – |
| 3 | Durvalumab | PD-1/2 & CD80 | Recurrent-GBM | IV | II | |
| 4 | DCVax-L | Inhibits recurrence | Recurrent-GBM | Direct | II | >52.14 |
| 5 | Afatinib | EGFR & v-III | Recurrent-GBM | Oral | I/II | – |
| 6 | Tremelimumab & Durvalumab | PD-1 & CTLA-4 | HGG | IV | II | – |
| 7 | AEE788 | EGFR, VEGFR-TKI | Recurrent-GBM | Oral | I | – |
| 8 | Dacomitinib | EGFR-TKI | Recurrent-GBM | Oral | II | – |
| 9 | Neratinib | EGFR-TKI | Recurrent-GBM | Oral | II | – |
| 10 | Nivolumab + Bevacizumab | PD-1, VEGF | Recurrent-GBM | IV | III | 43.45 |
| 11 | Pembrolizumab | PD-1 | Recurrent-GBM | Oral | II | 26.07 |
| 12 | Enasidenib | Mutant IDH2 | All types | Oral | II | – |
| 13 | Ivosidenib | Mutant IDH1 | All types | Oral | I | 56.49 |
| 14 | Erlotinib | EGFR-v-III | Recurrent-GBM | Oral | II | 52.14 |
| 15 | Rindopepimut | EGFR-v-III | Early-glioma | | III | 87.34 |
| 16 | Asunercept | CD95 | Recurrent-GBM | | II | – |
| 17 | Depatux-M | EGFR | Early-glioma | | III | In-effective |

## 5.1 Targeted Therapies

Targeted therapies in glioma aim to inhibit key molecular drivers identified through tumor profiling. This section covers therapies against IDH mutations, EGFR alterations, DNA repair defects (PARP inhibitors), and BRAF V600E mutations, all of which offer promising avenues for personalized treatment in glioma.

***IDH Inhibitors*** Mutations particularly in IDH 1&2 are key molecular drivers in lower-grade gliomas and secondary glioblastomas, and they have become important targets in the era of precision medicine. These mutations result in a neomorphic enzyme activity that converts α-ketoglutarate to the oncometabolite D-2-hydroxyglutarate (D-2HG), which contributes to tumorigenesis by promoting epigenetic dysregulation, blocking cellular differentiation, and impairing DNA repair.

As a result, mutant IDH enzymes present an attractive therapeutic target. Several small-molecule inhibitors have been developed to specifically target these mutations. Ivosidenib (AG-120), an oral selective IDH1 inhibitor, has shown clinical activity in patients with IDH1-mutant gliomas by reducing intra-tumoral D-2HG levels and delaying disease progression. Similarly, enasidenib (AG-221), which targets IDH2 mutations, has been evaluated in preclinical models and other IDH2-mutant malignancies, with potential application in gliomas [9–11] (Box 1).

**Box 1: Glioma Precision Medicine**

A. **Targeted Therapies**
   IDH inhibitors
   EGFR targeting
   PARP inhibitors
   BRAF and MEK Inhibitors

B. **Immunotherapies**
   Immune checkpoint inhibitors
   CAR-T cell therapy
   Dendritic cell (DC) vaccines

Vorasidenib (AG-881), a dual IDH1/2 inhibitor with excellent brain penetration, has shown particular promise in glioma. In the INDIGO phase III trial, vorasidenib significantly improved progression-free survival in patients with non-enhancing, recurrent IDH-mutant low-grade gliomas, with a favourable safety profile. Another dual inhibitor, olutasidenib (FT-2102), is also under investigation and has shown preliminary evidence of clinical activity in IDH1-mutant gliomas. These inhibitors are crucial components of personalized treatment strategies, not only due to their tumor-targeted activity but also because IDH mutation status serves as a prognostic and predictive biomarker. Furthermore, ongoing research is exploring the combination of IDH inhibitors with other therapeutic approaches, such as immune checkpoint inhibitors, radiotherapy, or agents targeting the epigenome, to overcome resistance and maximize therapeutic benefit. Collectively, IDH inhibitors represent a cornerstone in the development of precision medicine for glioma, providing more tailored and potentially less toxic treatment options for patients with IDH-mutant tumors [12–14].

***EGFR Targeting*** EGFR amplification and mutations, including the constitutively active EGFR-v-III variant, are among the most prevalent genetic alterations in glioma, making EGFR a key target for precision therapy. Several EGFR-targeted strategies have been explored, though clinical success remains modest due to challenges like intra-tumoral heterogeneity, adaptive resistance, and limited drug delivery across the BBB. Tyrosine kinase inhibitors (TKIs) such as erlotinib and gefitinib were among the first targeted agents tested in EGFR-amplified glioma. However, their clinical efficacy has been inconsistent, largely due to poor BBB penetration, redundancy in downstream signaling pathways, and compensatory survival mechanisms within tumor cells. Monoclonal antibodies like cetuximab, which directly bind the extracellular domain of EGFR, have demonstrated limited therapeutic benefit in GBM clinical trials, partially due to the receptor's mutated or truncated forms (e.g., EGFR-v-III) that escape antibody recognition [15, 16].

To address this, vaccines targeting the tumor-specific EGFR-v-III mutation, such as rindopepimut (CDX-110), have been developed. Rindopepimut is a peptide vaccine that stimulates an immune response against EGFR-v-III-expressing tumor cells. Early-phase clinical trials (e.g., ACT III) showed encouraging results with prolonged progression-free and overall survival in patients with EGFR-v-III-positive glioma, especially when combined with standard therapies like temozolomide. However, later trials such as the ACT IV phase III study failed to meet their primary endpoint, likely due to the loss of EGFR-v-III expression in recurrent tumors and immunoediting. Despite these setbacks, EGFR-targeted therapies remain a focus of research, with newer approaches involving bispecific antibodies, antibody-drug conjugates, and CAR T-cell therapies designed to overcome previous limitations [17, 18]. Collectively, these strategies underscore the complexity of EGFR signaling in glioma and the need for combinatorial or multi-targeted approaches to achieve durable therapeutic responses.

***PARP Inhibitors*** Emerged as promising agents in glioma precision medicine by targeting tumor-specific vulnerabilities in DNA repair pathways. PARP enzymes play a critical role in the repair of single-strand DNA breaks; inhibition of PARP leads to the accumulation of DNA damage, ultimately resulting in synthetic lethality in tumor cells with homologous recombination deficiencies (HRD), such as those with BRCA1/2 mutations or defects in other DNA repair genes. Agents like olaparib and niraparib have demonstrated efficacy in preclinical glioma models, particularly in tumors exhibiting HRD or PTEN loss, which impairs homologous recombination. Notably, these inhibitors show synergistic effects when combined with TMZ, the standard chemotherapeutic for glioblastoma, and radiotherapy, enhancing tumor cell death while potentially sparing normal tissue. Early-phase clinical trials have investigated olaparib in recurrent glioblastoma and in combination with TMZ or radiation, showing tolerable safety profiles and preliminary signs of efficacy. Moreover, PARP inhibitors can potentiate the effects of DNA-damaging agents in MGMT-methylated gliomas, which are more sensitive to alkylating agents. Ongoing research focuses on identifying reliable biomarkers of response and optimizing combination regimens, positioning PARP inhibitors as valuable components in personalized glioma therapy strategies [19, 20].

***BRAF and MEK Inhibitors*** BRAF V600E mutations, although relatively rare, are present in a distinct subset of gliomas, particularly in pleomorphic xanthoastrocytomas, gangliogliomas, and some pediatric low-grade gliomas, making them actionable targets in precision medicine. The aberrant activation of the MAPK signaling pathway driven by this mutation has been successfully targeted using BRAF inhibitors such as vemurafenib and dabrafenib, often in combination with MEK inhibitors like trametinib, to enhance efficacy and prevent resistance. Clinical studies have shown that BRAF/MEK inhibitor combinations induce significant and durable tumor regression, with improved progression-free and overall survival in BRAF V600E-mutant gliomas compared to standard therapies. Notably, the ROAR (Rare Oncology Agnostic Research) basket trial demonstrated encouraging response rates in high-grade glioma patients treated with dabrafenib and trametinib, leading to the FDA approval of this combination for BRAF V600E-mutant solid tumors, including gliomas. These therapies are generally well-tolerated and have become a paradigm of targeted treatment in molecularly defined glioma subsets, emphasizing the importance of genomic profiling in guiding individualized treatment strategies [21–23].

## 5.2 Immunotherapeutic Approaches

With a low survival rate in malignant glioma, traditional chemotherapy offers limited success. Immunotherapy-based strategies in clinical trials show promising results, offering hope for more effective, targeted treatments in precision medicine for glioma.

***Immune Checkpoint Inhibitors*** PD-1/PD-L1 and CTLA-4 inhibitors, like pembrolizumab, nivolumab, and ipilimumab, are being tested in clinical trials for glioma. While pembrolizumab and nivolumab show similar median OS in recurrent glioma (around 9.8 months), nivolumab has a shorter PFS (1.5 months) compared to bevacizumab (3.5 months). Nivolumab alone or with ipilimumab is well tolerated by some patients, but combination therapy has limited tolerance. Retrospective studies indicate modest OS improvement with pembrolizumab (4–6.6 months), and neo-adjuvant pembrolizumab before surgery shows better OS (13.7 months) than post-surgery treatment (7.5 months). Neo-adjuvant nivolumab shows no significant OS improvement. (Fig. 1) The failure of immunotherapy strategies in GBM is

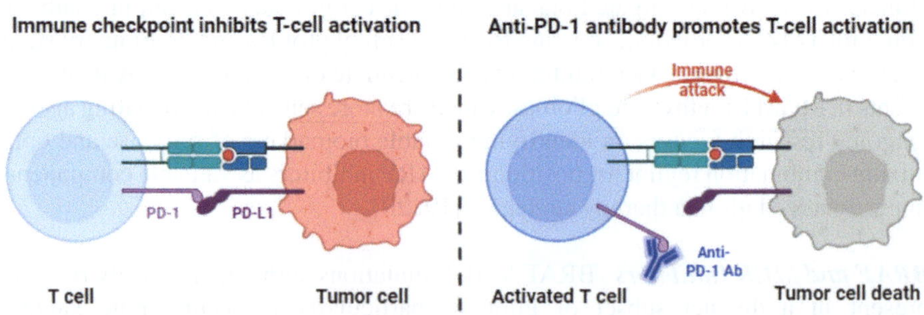


**Fig. 1** Schematic illustration of PD-1/PD-L1 immune checkpoint pathway and its blockade. Diagram showing the interaction between PD-1 on T cells and PD-L1 on tumor cells, leading to T cell exhaustion. Inhibition of PD-1 restores T-cell activity, promoting anti-tumor immunity

linked to factors like the lack of immunological biomarkers, variable PD-L1 expression, and poor T-cell infiltration. In the Checkmate-143 trial, only 27% of patients had >10% PD-L1 expression, while 32% had <1%. TMB is a key biomarker, with higher TMB associated with improved responses to nivolumab and pembrolizumab. However, typical GBM patients exhibit low TMB, reducing ICB efficacy. Other factors, such as the BBB and high corticosteroid use, also limit ICB effectiveness. Patient grouping in clinical trials is crucial for optimizing therapeutic responses [1].

***CAR-T-Cell Therapy*** Glioma treatment by targeting antigens like EGFR-v-III and IL-13Rα2. In a pilot trial, anti-EGFR-v-III CAR T-cells improved survival in 9/10 recurrent GBM patients, with some surviving for 3 years. However, challenges like antigen loss, immunosuppressive factor overexpression, and Treg cell recruitment were noted. Anti-IL-13Rα2 CAR T-cells showed lower immune responses in recurrent GBM patients. Local and intra-ventricular CAR T-cell administration improved X-ray changes and delayed disease recurrence for up to 7.5 months. In pre-clinical models, HER-2 targeting CAR-T cells achieved a median OS of 11.1 months. Single-target CAR T-cell therapy faces challenges due to tumor heterogeneity and antigen resistance. Multi-target approaches, such as combining HER2, IL13Rα2, and EphA2-directed CAR T-cells or targeting EGFR and EGFR-v-III, help address antigen escape and improve efficacy by overcoming tumor variability [24, 25] (Fig. 2).

***Dendritic Cell (DC) Vaccines*** DC-based vaccines, primed ex vivo against tumor-specific antigens, offer a promising alternative to peptide vaccines by effectively activating cytotoxic T-cells and enhancing anti-tumor responses. ICT-107, targeting

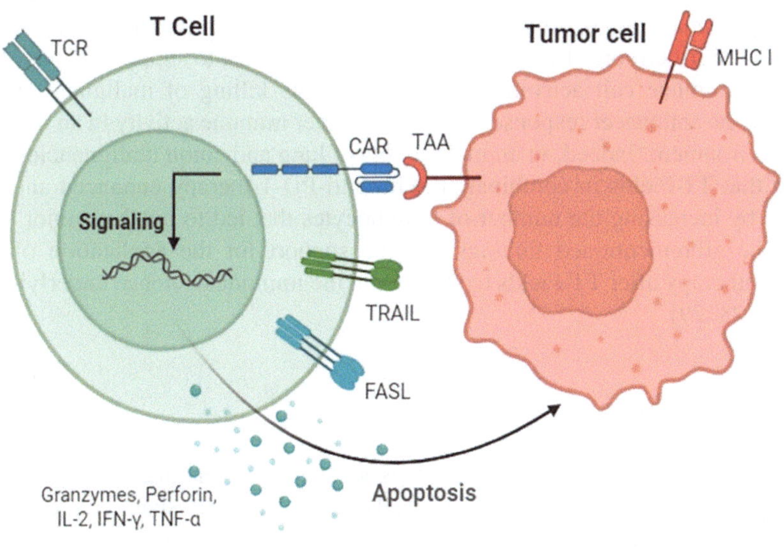


**Fig. 2** Schematic illustration of CAR-T-cell therapy

multiple antigens (TRP-2, AIM-2, HER2, IL-13Rα2, gp100, MAGE1), showed a median OS of 38.4 months in early trials but failed to replicate results in a phase II study. Combining DCVax-L with standard therapy improved OS to 23.1 months compared to 17 months with standard therapy alone. Personalized DC-peptide vaccine combos also showed benefit, raising OS to 21 months. Fusion of patient-derived GBM cells with autologous DCs led to OS of 25.3 months in newly diagnosed and 12.6 months in recurrent GBM cases. Targeting CMV, found in GBM but not healthy brains, with pp65 RNA-pulsed DCs improved survival, achieving a PFS of 25.3 months and OS of 41.1 months in recent trials. DC vaccines offer key advantages over peptide vaccines, including ease of antigen selection, personalization, and cost-effectiveness [26, 27].

## 6 Novel Approaches in Glioma Precision Medicine

Several innovative approaches are being explored to enhance current therapies of glioma, the most common and aggressive neuroepithelial tumor with limited treatment options. Key strategies are discussed below:

### 6.1 Tumor-Treating Fields (TT-Fields)

TT-Fields has become the fourth cancer therapy technique by applying low-intensity, intermediate-frequency shifting electric fields in the direction of the tumor. The FDA recently approved the TT-Fields treatment for newly diagnosed GBM 198 after a phase-III clinical trial showed that the median OS rose from 16.0 months in the TMZ-only group to 20.9 months in the TT-Fields-plus-TMZ group. The genetic basis of gliomas can also enhance the impacts of tumor death in numerous TT-Field activities. Furthermore, TT-Fields has sufficient evidence to support DC maturation in vitro, immune cell activation in vivo, and DC killing of malignant cells. To increase the anticancer response, it appears to alter immune activity in an inflammatory environment. Indeed, in animal models of lung and colon malignancies, it was found that TT-Fields in combination with anti-PD-1 therapy enhanced anticancer effects by increasing the number of lymphocytes that led to notable tumor shrinkage. The aforementioned findings provide support for the application of GBM immunotherapy after TT-Fields to strengthen the immune response underlying cancer [17, 28, 29].

### 6.2 CAR-NK Cells

Even without HLA matching (human leukocyte antigen), naturally occurring resistance effectors, such as natural killer cells with a short lifespan, can kill targets; nonetheless, germline-coded receptors engage with ligands on target cells to provide cytotoxic effects. Because they reduce the relapse or resistance caused by the loss of CAR-targeted antigens instead of producing cytokine release syndrome, graft-versus-host disease, or other long-term negative effects, NK cells are now a desirable alternative to T-cells for CAR engineering. The ErbB2-specific CAR expressed by the human NK-92 NK cell line exhibited potent anticancer action and greatly prolonged the mice's manifestation-free life. In NK-92, CARs with regions for cell-attaching EGFR and EGFR-v-III also controlled EGFR wild-type or mutant clones. Because it targets both CAR-NK cells, adaptive immunotherapy increases longevity without resulting in an immunological rescue right away. NK-92 cells must undergo radioactive treatment prior to being put into patients due to chromosomal abnormalities and the potential for cancer. Collecting, filtering, and transforming primary NK cells is extremely difficult. Better antitumor effects have been demonstrated in vivo by both CAR-engineered NK cells and human iPSC-derived NK cells. Thus, the creation of future anti-GBM immunotherapy could be facilitated by iPSC-NK cells with distinct CAR designs [30–32].

## 6.3 CSF Biopsies

The creation of the immunotherapy approach is guided by the diverse character of tumors and the capacity of antigenic epitopes to alter over the course of the disease, which is why tumor samples are repeatedly analyzed before and after treatment. Liquid biopsies derived from circulating tumor DNA (ctDNA) provide a less invasive and inexpensive way to genotype cancers. It has been demonstrated that the molecular profiles of gliomas can be detected more accurately by sequencing ctDNA recovered from CSF than by sequencing ctDNA isolated from blood. More importantly, because the genetic profile of tumor biopsies and gliomas in the CSF closely resembles each other, it can monitor the development of EGFR signaling pathways. These findings underscore the utility of CSF biopsies for assessing TMB in GBM, enabling identification of patients likely to benefit from ICBs and monitoring genomic evolution to detect immunotherapy resistance and adjust treatment strategies accordingly [33–35].

## 6.4 GAMs

Glioma-associated macrophages (GAMs) are key players in HGG, with their antitumor activity suppressed by tumor-derived cytokines. Depleting Nrp1 in bone marrow-derived macrophages or microglia reprograms them toward a more antitumorigenic phenotype. This leads to reduced tumor growth, lower vascularity, enhanced CTL infiltration, and diminished immunosuppression. These findings highlight Nrp1 as a critical regulator of GAM function and a promising therapeutic target in glioma [36, 37].

IFN-α, a potent immunomodulatory and anti-angiogenic cytokine, has limited therapeutic use due to its toxicity and short half-life. To overcome this, monocytes transduced with lentiviral vectors were engineered to express IFN-α under the

GBM-specific Tie2 promoter. These cells are home to the tumor, enabling sustained, localized IFN-α release within the TME. This targeted approach activates immune cells like DCs, macrophages, and T-cells, suppresses angiogenesis, and effectively impedes tumor growth and metastasis while minimizing systemic toxicity. Engineered human macrophages can secrete BiTEs targeting EGFRvIII in GBM, activating T-cells and releasing IL-12. In vitro and xenograft models showed these GEMs enhance T-cell recruitment and reshape the immunosuppressive TME, offering a promising strategy to boost antitumor immunity [38].

### 6.5 Organoids

Organoids are three-dimensional stem cell formations. They perform better than existing cell lines and almost perfectly mimic tumor variability and the 3D stem cell-derived models that closely mimic GBM tumor heterogeneity and the TME, offering a faster, cost-effective alternative to xenografts. Using the air–liquid interface (ALI) technique, patient-derived tumor organoids retain native TILs and T-cell receptor diversity. ALI organoids effectively model PD-1-mediated immune checkpoints, enabling personalized testing of PD-1 inhibitors. Co-culturing tumor organoids with patient lymphocytes helps identify tumor-reactive T-cells that spare healthy tissue. Paired samples before and after relapse offer insights into resistance mechanisms, guiding the development of therapies to improve treatment response [39, 40].

## 7 Challenges

A significant challenge in the use of precision medicine is the classification and prioritization of variants discovered by integrated genomic research. When an alteration is well-defined and previously published, existing literature can provide valuable insights into its functional significance and potential drug response links. However, the vast majority of alterations discovered through whole-genome or exome sequencing have not yet been explained, and it is unclear what their possible therapeutic uses and functional importance are. These genetic alterations range widely, from well-characterized variants with established clinical significance to novel mutations with unknown functional and therapeutic implications. The subset of significantly mutated genes found in The Cancer Genome Atlas (TCGA) GBM dataset includes several that are direct targets of FDA-approved therapies (EGFR, PDGFRA, BRAF) or are linked to pathways affected by approved drugs or investigational agents (PTEN, PIK3CA, NF1, TP53). Numerous techniques have been created to help with variation annotation and prioritizing in order to handle this complexity. By combining a number of variables, such as mutation type (e.g., frameshift, nonsense, or missense), location within functional domains (e.g., kinase domains), mutant allele fraction, corresponding gene expression from RNA sequencing, and previous reports in cancer databases like TCGA and COSMIC

(Catalogue of Somatic Mutations in Cancer), computational tools and functional predictive models are essential for ranking mutations. Evidence-based knowledge repositories that connect genetic variants to drug responses can be used to map these prioritized mutations to potential therapeutic alternatives in order to find clinically significant and potentially actionable changes. It is essential to take into account drug–gene connections from other tumor types since glioblastoma lacks prognostic biomarkers. The strength of supporting evidence, such as whether the mutation is identified as an FDA-defined pharmacogenomic biomarker in glioblastoma or another cancer type, and whether clinical or preclinical data indicate a correlation between the alteration and drug response, can be used to further categorize these therapeutic associations [41–43].

For instance, we classified somatic mutations found in the TCGA glioblastoma dataset according to their therapeutic relevance using a strength-of-evidence grading system. A number of changes found in GBM have been linked to positive treatment outcomes in other malignancies. The FDA-designated pharmacogenomic biomarker for vemurafenib, dabrafenib, and trametinib in BRAF V600E-mutant melanoma is one prominent example, which is the BRAF V600E mutation. The single level 1 mutation in this dataset is this change, which was discovered in a sample of GBM patients. We also found level 2 and level 3 mutations, which have been previously described and associated with medication response in preclinical or clinical investigations. These consist of nonsense mutations in NF1 and PTEN, EGFR A289V, PIK3CA E545K, and H1047R. Furthermore, there is no direct literature support for a drug-response association for a portion of changes that map to important functional domains or were previously discovered in cancer but are still functionally uncharacterized. Incorporating these ideas into the treatment of glioblastoma will need improving annotation techniques and growing the database of drug–mutation correlations [7, 44].

Even though therapeutic targets for GBM have been identified, it is still difficult to turn these discoveries into efficient treatments. These challenges are highlighted by the poor success of EGFR inhibitors in clinical trials. Even in tumors with EGFR and PTEN positive, EGFR tyrosine kinase inhibitors have not demonstrated much therapeutic effect, despite the fact that EGFR changes, including mutations, amplifications, and fusions, are seen in almost half of GBM patients. This failure highlights important obstacles to precision medicine in GBM, such as tumor heterogeneity and problems with medication efficacy and delivery. Intratumor heterogeneity is a major barrier to effective GBM treatment. Tumor regions vary in oxygenation, proliferation, vascularization, and infiltration, reflecting underlying genotypic diversity. A subset of GBMs exhibits mosaic amplification, where key oncogenes like EGFR and PDGFRA are amplified in distinct tumor subpopulations. In such cases, dual inhibition of both targets is necessary to suppress signaling pathways effectively. Beyond receptor tyrosine kinase heterogeneity, genome-wide copy number analyses reveal extensive genetic variability across spatially distinct tumor regions. Treating these regional differences is challenging due to limitations in routine imaging and the impracticality of repeated biopsies. Beyond intratumor heterogeneity, treatment itself can drive clonal evolution by selecting for preexisting resistant subclones or inducing new mutations. TMZ, for instance, promotes a mutator phenotype, contributing to glioblastoma progression through mutations in the retinoblastoma and Akt-mTOR pathways. Secondary mutations further drive resistance to targeted therapies, as seen with EGFR T790M in non-small cell lung cancer, which emerges at low levels before treatment and expands under selective pressure. In contrast, resistant mutations in BRAF-mutant melanoma appear to arise de novo during BRAF/MEK inhibitor therapy. Consequently, repeat biopsies at recurrence are crucial for guiding precision treatment strategies. Extensive intratumor heterogeneity and molecular complexity of GBM, effective treatment will likely require combination therapy targeting multiple pathways. Single-cell clonal analysis has identified preexisting TMZ-resistant subclones within untreated GBM, each with distinct genomic alterations and drug sensitivities. This highlights the need for multi-agent approaches to address diverse driver events within aggressive clones. Preclinical studies suggest that targeting parallel or redundant pathways, such as EGFR and PI3K/mTOR, can be effective. However, clinical trials, including a phase I/II study of erlotinib and temsirolimus, have faced challenges due to dose-limiting toxicity and lack of efficacy. Optimizing combination therapy requires careful management of toxicity and drug interactions [1, 40].

GBM's inherent resistance to therapy, combining treatments with chemosensitizing agents, is a promising strategy. Autophagy inhibitors like chloroquine have enhanced TMZ sensitivity in xenograft models and shown clinical activity in combination with TMZ and radiation. However, hydroxychloroquine exhibited dose-limiting toxicity and inconsistent autophagy inhibition in a phase I/II trial with standard GBM therapy [45–47]. Further research into novel autophagy inhibitors and optimized regimens is needed. Immunotherapy also holds potential, with early clinical trials showing promise. While its role in combination therapy is still being explored, evidence suggests synergy with radiation and chemotherapy, warranting further investigation.

## 8 Future Directions

One of the most significant hurdles in glioma therapy is the development of resistance to targeted treatments. Gliomas are highly heterogeneous tumors, with multiple genetic and epigenetic alterations that allow them to evade therapeutic interventions. Resistance mechanisms include activation of compensatory signaling pathways, mutations in drug targets, and tumor microenvironment-mediated resistance. For instance, glioblastomas frequently exhibit redundancy in receptor tyrosine kinase (RTK) signaling, allowing them to bypass inhibition of a single pathway. Additionally, GSCs contribute to resistance by promoting tumor plasticity and survival under therapeutic stress. To counteract these resistance mechanisms, combination therapies that target multiple signaling pathways simultaneously have shown promise. By inhibiting compensatory mechanisms, researchers aim to prevent tumor adaptation and enhance treatment efficacy. For example, combining RTK inhibitors with downstream pathway inhibitors, such as PI3K/AKT/mTOR inhibitors, may prevent resistance and prolong tumor control. Furthermore, epigenetic modulators, such as HDAC inhibitors, are being explored to reprogram resistant glioma cells and sensitize them to existing treatments. Overcoming resistance will require a deeper understanding of tumor evolution and the development of adaptive therapeutic strategies tailored to individual patients [1].

The integration of multi-omics data—genomic, transcriptomic, proteomic, and metabolomic—into personalized treatment algorithms represents a crucial advancement in glioma precision medicine. Each glioma exhibits a unique molecular profile that dictates its response to therapy. Traditional treatment approaches based on histological classification are being replaced by molecularly informed decision-making, allowing clinicians to tailor therapies to individual patients based on their tumor-specific alterations. One major challenge in implementing personalized algorithms is the complexity and variability of glioma molecular profiles. Comprehensive genomic sequencing has revealed a wide range of mutations, such as IDH1/2 mutations, TERT promoter mutations, and MGMT promoter methylation, which influence prognosis and therapy selection. However, translating this molecular knowledge into actionable treatment plans remains difficult due to intertumoral and intratumoral heterogeneity. Future advancements will rely on the development of predictive computational models that integrate diverse patient data to recommend optimal therapeutic regimens. Additionally, real-time monitoring of treatment response through liquid biopsies and circulating tumor DNA (ctDNA) analysis may help refine treatment decisions and allow for early intervention when resistance emerges [1, 47].

Artificial intelligence (AI) is revolutionizing glioma therapy by enabling the analysis of vast amounts of patient data to identify patterns and predict treatment responses. Machine learning algorithms can process complex datasets from imaging, histopathology, and molecular profiling to assist in diagnosis, prognosis prediction, and therapy selection. AI-driven platforms have shown promise in distinguishing glioma subtypes based on radiographic features and predicting patient outcomes with higher accuracy than conventional methods. In precision medicine, AI can aid in drug discovery by identifying novel therapeutic targets and repurposing existing drugs for glioma treatment. By analyzing gene expression data and protein interaction networks, AI can uncover previously unrecognized vulnerabilities in glioma cells. Furthermore, AI-powered decision-support systems can optimize clinical trial designs by identifying patient populations most likely to benefit from specific therapies, thereby accelerating the development of new treatments. Despite its potential, the integration of AI into clinical practice faces several challenges. The reliability of AI predictions depends on the quality and diversity of training datasets, which must include data from a broad range of patient populations to ensure generalizability. Additionally, the interpretability of AI models remains a concern, as complex deep learning algorithms often function as "black boxes" with limited transparency regarding their decision-making processes. Overcoming these challenges will require collaboration between clinicians, data scientists, and regulatory agencies to develop robust AI-driven tools that can be safely and effectively implemented in glioma therapy [1].

## 9 Conclusion

An important development in the battle against these aggressive brain tumors is the incorporation of precision medicine into glioma therapy. Clinicians can shift from traditional treatment paradigms to more individualized and successful therapeutic approaches by comprehending the genomic heterogeneity of gliomas and identifying important hereditary causes. Glioma treatment is already changing due to developments in immunotherapy, targeted medicines, and molecular diagnostics, giving patients who had few alternative options new hope. Drug resistance, intra-tumoral heterogeneity, and the requirement for more thorough clinical validation of new biomarkers are some of the major obstacles that still need to be overcome. There is also potential for improving treatment selection and response prediction with the application of AI and machine learning in precision oncology. Glioma therapy's future depends on a multidisciplinary, data-driven, and patient-centered strategy as research continues to identify new molecular targets and improve therapeutic strategies. Precision medicine has the potential to transform glioma treatment and enhance long-term survival results by implementing these scientific discoveries in clinical settings.

## References

1. Ghosh MK, Kumar S, Begam S, Ghosh S, Basu M (2024) GBM immunotherapy: Exploring molecular and clinical frontiers. Life Sci 356:123018
2. Wen PY, Reardon DA (2016) Progress in glioma diagnosis, classification and treatment. Nat Rev Neurol 12(2):69–70
3. Nicholson JG, Fine HA (2021) Diffuse glioma heterogeneity and its therapeutic implications. Cancer Discov 11(3):575–590
4. Aum DJ, Kim DH, Beaumont TL, Leuthardt EC, Dunn GP, Kim AH (2014) Molecular and cellular heterogeneity: the hallmark of glioblastoma. Neurosurg Focus 37(6):E11
5. Lo Greco MC, Marano G, La Rocca M et al (2025) Latest advancements in the management of H3K27M-mutant diffuse intrinsic pontine glioma: a narrative review. Cancer 17(3):420
6. Lee P, Murphy B, Miller R et al (2015) Mechanisms and clinical significance of histone deacetylase inhibitors: epigenetic glioblastoma therapy. Anticancer Res 35(2):615–625
7. Iyer K, Saini S, Bhadra S, Kulavi S, Bandyopadhyay J (2023) Precision medicine advancements in glioblastoma: a systematic review. Biomedicine 13(2):1
8. Sarkar S, Kumar S, Saha G, Basu M, Ghosh MK (2024) Glioma nanotherapy: unleashing the synergy of dual-loaded DIM and TMZ. Int J Pharm 665:124697
9. Rudà R, Horbinski C, van den Bent M, Preusser M, Soffietti R (2024) IDH inhibition in gliomas: from preclinical models to clinical trials. Nat Rev Neurol 20(7):395–407
10. Gatto L, Franceschi E, Tosoni A et al (2021) IDH inhibitors and beyond: the cornerstone of targeted glioma treatment. Mol Diagn Ther 25:457–473
11. Turkalp Z, Karamchandani J, Das S (2014) IDH mutation in glioma: new insights and promises for the future. JAMA Neurol 71(10):1319–1325
12. Han S, Liu Y, Cai SJ et al (2020) IDH mutation in glioma: molecular mechanisms and potential therapeutic targets. Br J Cancer 122(11):1580–1589
13. Nakhate V, Lasica AB, Wen PY (2024) The role of mutant IDH inhibitors in the treatment of glioma. Curr Neurol Neurosci Rep 24(12):631–643

14. Giacobetti SA, Fine HA (2024) Shifting the landscape: the role of IDH inhibitors in glioma cell fate. Cancer Cell 42(5):741–743
15. Westphal M, Maire CL, Lamszus K (2017) EGFR as a target for glioblastoma treatment: an unfulfilled promise. CNS Drugs 31(9):723–735. https://doi.org/10.1007/s40263-017-0456-6
16. Lo HW (2010) EGFR-targeted therapy in malignant glioma: novel aspects and mechanisms of drug resistance. Curr Mol Pharmacol 3(1):37–52
17. Padfield E, Ellis HP, Kurian KM (2015) Current therapeutic advances targeting EGFR and EGFRvIII in glioblastoma. Front Oncol 5:5
18. Oprita A, Baloi SC, Staicu GA et al (2021) Updated insights on EGFR signaling pathways in glioma. Int J Mol Sci 22(2):587
19. Sim HW, Galanis E, Khasraw M (2022) PARP inhibitors in glioma: a review of therapeutic opportunities. Cancer 14(4):1003
20. Bisht P, Kumar VU, Pandey R, Velayutham R, Kumar N (2022) Role of PARP inhibitors in glioblastoma and perceiving challenges as well as strategies for successful clinical development. Front Pharmacol 13:939570
21. Schreck KC, Grossman SA, Pratilas CA (2019) BRAF mutations and the utility of RAF and MEK inhibitors in primary brain tumors. Cancer 11(9):1262
22. Fusco MJ, Piña Y, Macaulay RJ et al (2021) Durable progression-free survival with the use of BRAF and MEK inhibitors in four cases with BRAF V600E-mutated gliomas. Cancer Control 28:10732748211040013. https://doi.org/10.1177/10732748211040013
23. Rizzo D, Ruggiero A, Amato M, Maurizi P, Riccardi R (2016) BRAF and MEK inhibitors in pediatric glioma: new therapeutic strategies, new toxicities. Expert Opin Drug Metab Toxicol 12(12):1397–1405. https://doi.org/10.1080/17425255.2016.1214710
24. Bagley SJ, Desai AS, Linette GP, June CH, O'Rourke DM (2018) CAR T-cell therapy for glioblastoma: recent clinical advances and future challenges. Neuro Oncol 20(11):1429–1438
25. Maggs L, Cattaneo G, Dal AE, Moghaddam AS, Ferrone S (2021) CAR T cell-based immunotherapy for the treatment of glioblastoma. Front Neurosci 15:662064
26. Qian D, Liu Y, Zheng J, Cai J (2024) Dendritic cell therapy for neurospoagioma: immunomodulation mediated by tumor vaccine. Cell Death Discov 10(1):11
27. Wong CE, Chang Y, Chen PW et al (2024) Dendritic cell vaccine for glioblastoma: an updated meta-analysis and trial sequential analysis. J Neuro-Oncol 170(2):253–263
28. Wu H, Zhou F, Gao W et al (2024) Current status and research progress of minimally invasive treatment of glioma. Front Oncol 14:1383958
29. Guo Z, Su Z, Wei Y, Zhang X, Hong X (2024) Pyroptosis in glioma: current management and future application. Immunol Rev 321(1):152–168. https://doi.org/10.1111/imr.13294
30. Strassheimer F, Elleringmann P, Ludmirski G et al (2025) CAR-NK cell therapy combined with checkpoint inhibition induces an NKT cell response in glioblastoma. Br J Cancer 132:1–12
31. Xiong Q, Zhu J, Zhang Y, Deng H (2023) CAR-NK cell therapy for glioblastoma: what to do next? Front Oncol 13:1192128
32. Ma R, Lu T, Li Z et al (2021) An oncolytic virus expressing IL15/IL15Rα combined with off-the-shelf EGFR-CAR NK cells targets glioblastoma. Cancer Res 81(13):3635–3648
33. Friedman JS, Hertz CAJ, Karajannis MA, Miller AM (2022) Tapping into the genome: the role of CSF ctDNA liquid biopsy in glioma. Neuro Oncol Adv 4(Supplement_2):ii33–ii40
34. Eibl RH, Schneemann M (2023) Liquid biopsy and glioblastoma. Explor Target Antitumor Ther 4(1):28
35. Otsuji R, Fujioka Y, Hata N et al (2024) Liquid biopsy for glioma using cell-free DNA in cerebrospinal fluid. Cancer 16(5):1009
36. Lin C, Wang N, Xu C (2023) Glioma-associated microglia/macrophages (GAMs) in glioblastoma: immune function in the tumor microenvironment and implications for immunotherapy. Front Immunol 14:1123853
37. Xu C, Xiao M, Li X et al (2022) Origin, activation, and targeted therapy of glioma-associated macrophages. Front Immunol 13:974996

38. Matsuzaki H, Pan C, Komohara Y et al (2025) The roles of glioma-associated macrophages/microglia and potential targets for anti-glioma therapy. Immunol Med 48(1):24–32. https://doi.org/10.1080/25785826.2024.2411035
39. Mariappan A, Goranci-Buzhala G, Ricci-Vitiani L, Pallini R, Gopalakrishnan J (2021) Trends and challenges in modeling glioma using 3D human brain organoids. Cell Death Differ 28(1):15–23
40. Xu X, Li L, Luo L et al (2021) Opportunities and challenges of glioma organoids. Cell Commun Signal 19(1):102. https://doi.org/10.1186/s12964-021-00777-0
41. Fathi Kazerooni A, Bagley SJ, Akbari H et al (2021) Applications of radiomics and radiogenomics in high-grade gliomas in the era of precision medicine. Cancer 13(23):5921
42. Naithani N, Atal AT, Tilak T, Vasudevan B, Misra P, Sinha S (2021) Precision medicine: uses and challenges. Med J Armed Forces India 77(3):258–265
43. Yan RE, Greenfield JP (2024) Challenges and outlooks in precision medicine: expectations versus reality. World Neurosurg 190:573–581
44. Ghiaseddin A, Hoang Minh LB, Janiszewska M et al (2021) Adult precision medicine: learning from the past to enhance the future. Neuro Oncol Adv 3(1):vdaa145
45. Kumar S, Basu M, Ghosh MK (2024) E3 ubiquitin ligases and deubiquitinases in colorectal cancer: emerging molecular insights and therapeutic opportunities. Biochim Biophys Acta BBA-Mol Cell Res 1871:119827
46. Kumar S, Basu M, Ghosh MK (2022) Chaperone-assisted E3 ligase CHIP: a double agent in cancer. Genes Dis 9(6):1521–1555
47. Kumar S, Chatterjee M, Ghosh P, Ganguly KK, Basu M, Ghosh MK (2023) Targeting PD-1/PD-L1 in cancer immunotherapy: an effective strategy for treatment of triple-negative breast cancer (TNBC) patients. Genes Dis 10(4):1318–1350

# Personalized Treatment Options in Head and Neck Cancer

Sourav Sanyal, Arundhaty Pal,
Sumita Bandyopadhyay Sengupta,
and Tapas Kumar Sengupta

**Abstract**

Head and neck cancer is the seventh most common cancer globally, with approximately 900,000 new cases every year and a 50% mortality rate. Major risk factors for head and neck cancers include tobacco, alcohol, viral infections like human papillomavirus and Epstein-Barr virus, and genetic alterations. Despite the recent advancements in current treatment options like surgery, radiation therapy, chemotherapies, targeted therapies, and immunotherapies, the clinical and molecular heterogeneity of head and neck cancers makes any treatment challenging and often found to be associated with considerable side effects. In this context, personalized medicine has emerged as an effective therapeutic choice that enables medical professionals to customize treatments depending on genetic variations of an individual, resulting in better patient outcomes and fewer side effects. Recent advances in next-generation sequencing and molecular profiling, along with the developments in diagnostic methods, have revolutionized the identification of specific genetic traits/mutations and/or epigenetic biomarkers to customize medical choices and treatment procedures for individual patients. This chapter explores the recent advancements in personalized medicine for head and neck cancer treatment, highlighting targeted therapies, immunotherapies, gene therapies, and stem cell therapies. Furthermore, the prospects and challenges of using personalized medicine in head and neck cancer are also discussed.

S. Sanyal · A. Pal · T. K. Sengupta (✉)
Department of Biological Sciences, Indian Institute of Science Education and Research Kolkata, Haringhata Farm, West Bengal, India
e-mail: senguptk@iiserkol.ac.in

S. B. Sengupta
Department of Biophysics, Molecular Biology and Bioinformatics, University of Calcutta, Kolkata, West Bengal, India



M. K. Ghosh, M. Basu (eds.), *Personalized Cancer Therapy*,
https://doi.org/10.1007/978-981-95-6845-1_9

**Keywords**

Head and neck cancer · Personalized treatment · Precision medicine · Epigenetics · Targeted therapy

## Abbreviations

| | |
|---|---|
| AKAP9 | A-kinase anchoring protein 9 |
| BCL2 | B-cell leukemia/lymphoma 2 |
| BMF | Bcl-2-modifying factor |
| CDKN2A | Cyclin-dependent kinase inhibitor 2A |
| CASP8 | Caspase 8 |
| CD | Cluster of Differentiation |
| CGH | Comparative Genomic Hybridization |
| cfDNA | Cell-free DNA |
| CRISPR | Clustered regularly interspaced short palindromic repeats |
| Cas | CRISPR-associated protein |
| CCND1 | Cyclin D1 |
| CDK | cyclin-dependent kinase |
| CDKN2A | cyclin-dependent kinase inhibitor 2A |
| c-MET | Mesenchymal-Epithelial Transition Factor |
| COX2 | Cyclooxygenase-2 |
| CSC | Cancer stem cell |
| CTC | Circulating tumour cell |
| ctDNA | Circulating tumour DNA |
| CTLA-4 | Cytotoxic T-lymphocyte antigen 4 |
| CT | Computed Tomography |
| DAPK | Death-Associated Protein Kinase |
| DNA | Deoxyribonucleic Acid |
| DNMT3B | DNA methyltransferase 3 beta |
| E2F | Early region 2 binding factor |
| EBV | Epstein-Barr virus |
| E-cadherin | Epithelial cadherin |
| EGFR | Epidermal growth factor receptor |
| EMT | Epithelial-Mesenchymal Transition |
| ErbB | Erythroblastic leukaemia viral oncogene homologue |
| ERK | Extracellular Signal-Regulated Kinase |
| FAK | Focal adhesion kinase |
| FDA | Food and Drug Administration |
| FGFR | Fibroblast Growth Factor Receptor |
| FIH | Factor-inhibiting hypoxia-inducible factor |
| GLOBOCAN | Global Cancer Observatory |
| HDAC | Histone Deacetylase |
| HER | Human epidermal growth factor receptor |
| HNC | Head and Neck Cancer |

| | |
|---|---|
| HNSCC | Head and Neck Squamous Cell Carcinoma |
| HPV | Human papillomavirus |
| HRAS | Harvey rat sarcoma viral oncogene homolog |
| IDO-1 | Indoleamine 2, 3-dioxygenase 1 |
| IGF-1R | Insulin-Like Growth Factor 1 Receptor |
| INK4a | Inhibitor of cyclin-dependent kinase 4a |
| JAK | Janus Kinase |
| KIT | Proto-oncogene, Receptor tyrosine kinase |
| LAG-3 | Lymphocyte-activation gene 3 |
| LA-HNSCC | Locally advanced- HNSCC |
| LATS2 | Large tumor suppressor kinase 2 |
| MEK | Mitogen-activated protein kinase |
| MET | Mesenchymal-Epithelial Transition factor |
| miRNA | MicroRNA |
| MRI | Magnetic Resonance Imaging |
| MSP | Methylation-specific polymerase chain reaction |
| MTC | Medullary thyroid cancer |
| mTOR | Mammalian target of rapamycin |
| NGS | Next Generation Sequencing |
| NPC | Nasopharyngeal cancer |
| NOTCH1 | Neurogenic locus notch homolog protein 1 |
| OPC | Oropharyngeal cancer |
| p110α | phosphatidylinositol-4,5-bisphosphate 3-kinase, catalytic subunit alpha |
| PET | Positron Emission Tomography |
| PD-1 | Programmed cell death protein 1 |
| PDCD | Programmed Cell Death |
| PDGFR-α | Platelet-derived Growth Factor Receptor alpha |
| PD-L1 | Programmed cell death ligand 1 |
| PI3K | Phosphoinositide 3-kinase |
| PIK3CA | Phosphatidylinositol-4, 5-bisphosphate 3-kinase catalytic subunit alpha |
| PTEN | Phosphatase and TENsin homolog |
| RAS | Rat sarcoma viral oncogene homologue |
| RASSF1A | Ras association domain family protein 1, isoform A |
| RAF | Rapidly Accelerated Fibrosarcoma |
| RET | REarranged during Transfection |
| R/M | Recurrent and Metastatic |
| RNA | Ribonucleic acid |
| RNAi | RNA interference |
| RT | Radiotherapy |
| RTK | Receptor Tyrosine Kinase |
| shRNA | Short hairpin RNA |
| siRNA | Small Interfering RNA |
| SOCS1 | Suppressor of Cytokine Signaling 1 |

| | |
|---|---|
| SOX2 | SRY-box transcription factor 2 |
| STAT | Signal Transducer and Activator of Transcription |
| SUV | Standardized uptake value |
| TNM | Tumour, Nodes, Metastasis |
| TP53 | Tumour Protein 53 |
| TPF | Taxanes, cisplatin and 5-FU |
| TPM1 | Tropomyosin 1 |
| TSG | Tumour Suppressor Gene |
| TS-miR | Tumour suppressor microRNA |
| VEGFR | Vascular Endothelial Growth Factor Receptor |
| WWOX | WW domain-containing oxidoreductase |
| ZEB | Zinc finger E-box binding homeobox |

## 1 Introduction

Head and neck cancer (HNC) is a combined term for a variety of cancers originating from the squamous cells that line the mucosal surfaces of the head and neck region, particularly the oral cavity, pharynx (or throat), and larynx (or voice box). These malignancies are generally referred to as head and neck squamous cell carcinoma (HNSCC) [1]. Although far less frequent than the HNSCC, HNC can also arise from the nasal cavity and paranasal sinuses, salivary glands, and nerves or muscles in the head and neck. Cancers arising from the skin of the head and neck, eye, brain, thyroid gland, and oesophagus are not generally considered as HNC [2]. Use of tobacco, alcohol consumption, betel nut chewing, human papillomavirus (HPV) infection (especially infection with HPV types 16 and 18), Epstein-Barr virus (EBV) infection, and occupational exposure to nickel dust, asbestos dust, wood dust, and formaldehyde are the major risk factors associated with HNC [3]. According to the data published by GLOBOCAN in 2020, HNSCC is the seventh most frequent cancer worldwide, with ~900,000 new cases (approximately 4.5% of all cancer diagnoses worldwide) and ~450,000 annual deaths (about 4.6% of all cancer deaths worldwide), with an overall mortality rate of ~50% [2].

The conventional treatment methods for HNC include a combination of surgery, radiation therapy, and chemotherapy [2]. Other than that, Cetuximab (an anti-EGFR antibody) is the sole FDA-approved medication used as targeted therapy to treat HNSCC [4]. However, these traditional treatment methods have drawbacks because they do not take into account the genetic variations of an individual. As a result, these treatments often lead to considerable side effects, chemoresistance, and variable patient outcomes [5]. To address these limitations, the paradigm for treating HNC is progressively moving towards personalized treatment options or precision medicines, which adapt therapeutic approaches based on individual patient characteristics such as lifestyle, molecular and genetic profile, and the presence of certain biomarkers. The rationale for using personalized medicines in HNC is rooted in the fact that it is a group of cancers that come with significant clinical and biological complexity and heterogeneity. Molecular and genetic traits also differ from one person to another, further highlighting the need for more customized treatment strategies [2]. Therefore, the ultimate goal of personalized medicine in HNC is to increase the treatment efficacy, reduce side effects and drug resistance, and eventually enhance the prognostic outcomes and overall survival.

This chapter aims to discuss major molecular alterations, diagnostic tools for precision medicine, and current personalized treatment options in HNC, along with a hint toward future directions for personalized treatment approaches. By understanding the current landscape and prospects of personalized medicine in HNC, this chapter intends to provide insights into improving patient outcomes and advancing the field of HNC.

## 2 Key Molecular Mechanisms Involved in Head and Neck Cancer

Genetic alterations, deregulations of the cell cycle and signalling pathways, epigenetic changes, viral infections, and dysregulations of the immune checkpoint pathways are the major molecular mechanisms that aid in the development and progression of cancer. Understanding these molecular mechanisms is crucial for molecular profiling, identifying diagnostic and prognostic biomarkers, and developing diagnostic tools and personalized treatment options for head and neck cancer.

### 2.1 Genetic Alterations

#### 2.1.1 Epidermal Growth Factor Receptor (EGFR)

EGFR is a transmembrane glycoprotein and a cell surface receptor of the HER/ErbB receptor tyrosine kinase (RTK) family. Although rarely amplified or mutated, EGFR remains upregulated in ~80% of HNC patients and is strongly linked to poor prognostic outcomes, increased recurrence rates, and decreased overall survival [6, 7]. Overexpression of EGFR hyperactivates the signalling pathways like RAS/RAF/MEK/ERK and PI3K/AKT/mTOR. These hyperactivated signalling cascades promote cell proliferation, invasion, migration, angiogenesis, and inhibit apoptosis [5, 7]. Molecular profiling of EGFR in HNC patients can help in precision prediction and diagnosis. Moreover, personalized treatments like targeted therapy, gene disruption therapy following RNA interference (RNAi) technique, and targeted radionuclide therapy can be utilized to inhibit or downregulate EGFR for treating HNC [5]. Apart from that, several reports have showed that inhibition of EGFR can improve the effectiveness of radiation and chemotherapy in HNC patients. Targeting EGFR can also improve patient response to therapy and raise survival rates [7] (Fig. 1).

#### 2.1.2 Tumour Protein 53 (TP53)

The TP53 gene encodes the p53 tumour suppressor protein, which is instrumental in repairing DNA, maintaining genomic stability, cell cycle, apoptosis, and senescence

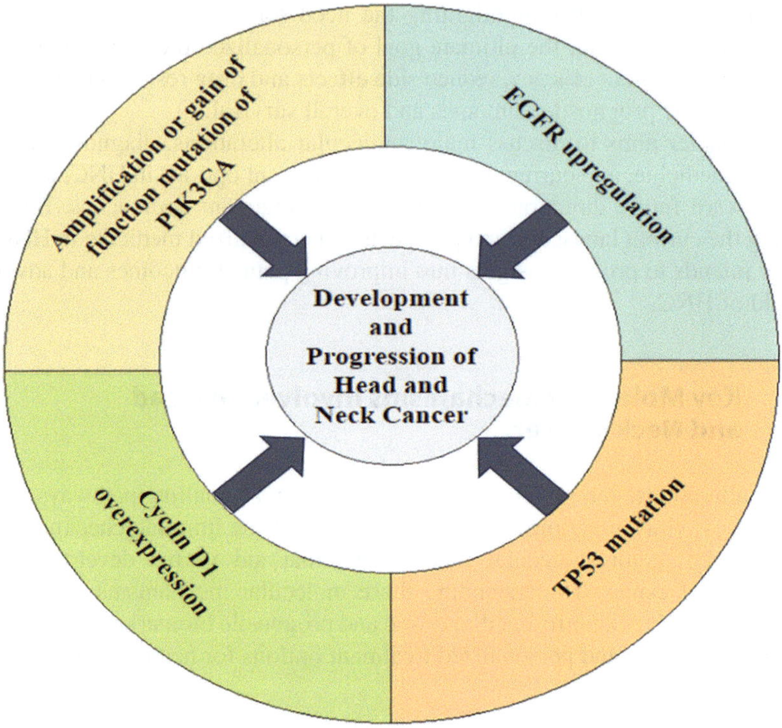


**Fig. 1** Key genetic alterations involved in head and neck cancer. (Created by Microsoft Paint and Microsoft PowerPoint)

[6]. The mutation of the TP53 gene is seen in ~50% of HNC patients [5]. Loss-of-function mutations in the TP53 gene render it unable to stop cell proliferation and tumour growth, and help the cells with damaged DNA to continue dividing and aiding in the spread of tumours [7]. Other than that, several reports have found that the mutated p53 protein plays oncogenic roles and helps in tumour growth, metastasis, and drug resistance [8]. Molecular profiling of the p53 protein in HNC patients can aid in diagnosis and precision prediction [5]. Additionally, p53 mutation in HNSCC patients can be overcome following gene addition therapy by directly injecting an adenoviral vector expressing the wild-type TP53 gene [9]. Apart from that, the diagnosis of p53 mutation in HNC patients can also help in precision surgery [5] (Fig. 1).

### 2.1.3 Cyclin D1

The CCND1 gene encodes the Cyclin D1 protein, which generally forms a complex with either CDK4 or CDK6 and promotes the transition of the cells from the G1 phase to the S phase of the cell cycle. The overexpression of Cyclin D1 protein is often seen in HNC, mainly due to the amplification of the CCND1 gene [7]. Other than that, the overexpression of nuclear EGFR can also increase the expression of

Cyclin D1 [6]. The upregulation of Cyclin D1 protein leads to uncontrolled cell proliferation and tumour growth. Researchers are now trying to develop CDK inhibitors to target Cyclin D1-CDK complexes, and this could result in the impediment of tumour growth and a favourable patient outcome [7]. Gene disruption therapy following the RNAi technique can also be employed to inhibit or downregulate Cyclin D1 for treating HNC [5] (Fig. 1).

#### 2.1.4 PI3K/AKT/mTOR Pathway

The PI3K/AKT/mTOR pathway is one of the pivotal pathways that is dysregulated in ~30% of HNSCC patient samples [10]. This pathway plays an important role in cell growth, proliferation, differentiation, survival, and metabolism [5, 11]. The PIK3CA gene encodes the protein p110α, the main catalytic subunit of PI3K. The amplification or gain-of-function mutation of the PIK3CA gene makes this pathway always active [6, 7]. On the other hand, one of the major downstream regulators of this pathway is the tumour suppressor gene PTEN, which inhibits this pathway. Although PTEN is rarely mutated in HNC, PTEN gene copy number loss and PTEN deficiency are frequent in HNC, and this in turn activates the PI3K/AKT/mTOR pathway [6]. This persistent signalling cascade promotes HNSCC development and progression, and the cells often become resistant to radiation therapy and chemotherapy [7]. Researchers are constantly searching for targeted therapeutic agents to inhibit or downregulate PI3K, AKT, and mTOR. Additionally, the development of molecular profiling techniques can help to identify this dysregulated signalling pathway, which will eventually enable clinicians to overcome the resistance to conventional therapies and improve the prognostic outcomes (Fig. 1).

### 2.2 Epigenetic Alterations

Other than genetic alterations, epigenetic modifications such as DNA methylation, histone modifications, and non-coding RNA-mediated gene regulation also play vital roles in the development and progression of HNC. Epigenetic alterations can activate oncogenes or deactivate tumour suppressor genes without changing the underlying DNA sequence [7]. Non-coding RNAs and chemical modifications of DNA and histone proteins can also be used as biomarkers for diagnosis, prognosis, and precision prediction of HNC. These biomarkers can also help clinicians to decide treatment strategies [2].

Hypermethylation of the promoter or first exon region of a gene by DNA methyltransferase enzymes can suppress the expression of that gene at the transcriptional level. Hypermethylation of various tumour suppressor genes such as CDKN2A (p16INK4a), RASSF1A, and DAPK has been reported in HNSCC [2, 12]. p16 is a cell cycle regulator that inhibits the transition of the cells from G1 phase to S phase of the cell cycle by inhibiting CDK4 and CDK6 cyclin-dependent kinases [13]. Molecular techniques such as methylation-specific polymerase chain reaction (MSP) can identify the hypermethylation of the p16 gene, and it is often used as a biomarker for the early diagnosis and prognosis of HNSCC [2, 12]. Silencing of

RASSF1A and DAPK genes by promoter hypermethylation leads to increased cell proliferation, survival, aggressive tumour phenotype, and poor prognostic outcomes in HNSCC patients [2, 12]. Determining the methylation status of RASSF1A and DAPK may be utilized as biomarkers in HNSCC tumour samples and may have diagnostic and prognostic consequences for HNSCC patients.

MicroRNAs (miRNAs) are 18–25 nucleotide-long non-coding RNAs. miRNAs regulate gene expression at the post-transcriptional level and play a central role in various biological processes such as cell proliferation, survival, migration, invasion, and therapeutic response. miRNAs can function as tumour-suppressors (TS-miR) or oncogenes (oncomiR) [14]. Some well-studied miRNAs that play a crucial role in HNC are discussed in Table 1. These miRNAs are promising candidates as biomarkers for diagnosis and prognosis, and for tracking therapeutic response in HNC patients.

## 2.3 Oncogenesis by Viral Infection

Human papillomavirus (HPV) infection (especially infection with HPV types 16 and 18) is the main risk factor for oropharyngeal cancer (OPC). In contrast, Epstein-Barr virus (EBV) infection is primarily associated with nasopharyngeal cancer (NPC) [3].

### 2.3.1 HPV Infection

HPV contains E6 and E7, two main oncoproteins that are responsible for the development of OPC. E6 generally forms a complex with the E6-associated protein (E6AP). This complex then ubiquitinates the p53 protein and promotes the degradation of p53 protein by the ubiquitin-proteasome system. This degradation of the tumour-suppressor p53 protein leads to uncontrolled cell proliferation, inhibition of apoptosis, and helps the cells to grow with damaged DNA, which further aids in the tumour growth [23]. On the other hand, E7 protein binds to the 'pocket domain' (the sequence which is crucial for the tumour suppressive role of Rb protein) of the Rb protein (pRb). This E7/pRb binding prevents the interaction between pRb and E2F transcription factors. As a result, E2F transcription factors get released, leading to the increased expression of genes responsible for the uncontrolled progression of the cells from the G1 phase to the S phase of the cell cycle [23]. The development of tailored treatment and preventive HPV vaccines has been made easier by the identification of these E6 and E7 viral markers. Several reports also suggest that HPV-positive OPCs exhibit improved prognoses and treatment responses compared to HPV-negative OPCs [14].

### 2.3.2 EBV Infection

Various mechanisms of action of EBV have been reported in causing NPC. EBV-encoded viral oncoproteins and non-coding RNAs play the central role in persistent infection and tumorigenesis. Viral oncoproteins such as EBNA1, BALF3, LMP1, and non-coding RNAs like miR-BART5-5p, 7-3p, 9-3p, 14-3p induce genomic

**Table 1** Role of miRNAs in head and neck cancer

| miRNA | Role and mechanism | Status and implications in HNC | References |
|---|---|---|---|
| miR-21 | oncomiR; targets PTEN, TPM1, and PDCD4 TSGs | Upregulation; promotes cell proliferation, survival, and invasion | [14, 15] |
| miR-29b | TS-miR; targets DNMT3B, which suppresses the expression of E-cadherin by promoter hypermethylation | Downregulation; induction of EMT, invasion, and metastasis | [16] |
| MiR-31 | oncomiR; targets LATS2 and FIH, which are involved in Hippo signalling pathway and hypoxia response, respectively | Upregulation; promotes proliferation, invasion, and angiogenesis | [14, 15] |
| miR-34a | TS-miR; targets BCL2, MET, and CDK6 oncogenes | Downregulation; inhibition of apoptosis, and induction of cell proliferation and metastasis | [14, 15] |
| miR-125a | oncomiR; suppresses the production of p53, which is a tumour suppressor protein | Upregulation; increased cell proliferation and migration | [2, 15] |
| miR-134 | oncomiR; targets WWOX and PDCD7 TSGs | Upregulation; increased cell division and metastasis | [17, 18] |
| miR-155 | oncomiR; targets SOCS1, which negatively regulates JAK/STAT pathway | Upregulation; increased cell growth and cell survival | [14] |
| miR-196a | oncomiR; targets annexin A1, which is involved in cell differentiation and apoptosis | Upregulation; promotes cell proliferation, migration, and invasion | [19] |
| miR-200c | TS-miR; inhibition of ZEB1 and ZEB2 transcription factors that promote EMT and metastasis | Downregulation; induction of invasion and metastasis | [14, 20, 21] |
| miR-221/222 cluster | oncomiR; targets CDK inhibitor p27Kip1 and proapoptotic protein BMF | Upregulation; increased cell cycle progression, survival, and chemoresistance | [14] |
| miR-375 | TS-miR; targets IGF-1R, which is involved in cell cycle regulation | Downregulation; increased cell proliferation and aggressive tumour behaviour | [22] |

instability in NPC, EBNA1 and LMP1 along with other viral oncoproteins like LMP2A, LMP2B, and BZLF1/Zta and non-coding RNAs such as miR-BART6p, miR-BART11, 17-3p help in the immune evasion, EBERs, EBNA1, and miR-BART1-5p induce angiogenesis, BRLF1, EBNA1, LMP1, LMP2A, miR-BART7-3p, miR-BART8-3p, miR-BART9, miR-BART10-3p, miR-BART13-3p, and miR-BART22 promote invasion and metastasis, and LMP1, LMP2A, and miR-BART1-5p aid in metabolic reprogramming in NPC [24]. Identifying these viral markers and understanding the molecular mechanisms behind EBV-infected NPC would facilitate the development of targeted therapies and preventive vaccines.

### 2.4 Immune Checkpoint Pathways

Head and neck cancer cells use different strategies to evade the body's immune surveillance mechanisms. PD-1/PD-L1 and CTLA-4/B7 are the two major immune checkpoint pathways that are manipulated by cancer cells for immune evasion. Cancer cells express a high level of PD-L1, which binds to the PD-1 receptor present on the T-cells. This interaction stimulates a cascade of intracellular signalling in the T-cell, leading to the inhibition of T-cell activation and function. Cancer cells also upregulate the expression of the immune checkpoint ligand B7. In order to successfully inhibit T-cells from attacking the cancer cells, B7 binds to the CTLA-4 receptor present on the T-cells and signals the T-cells to cease or reduce their activity [25]. Nivolumab and Pembrolizumab are the two major FDA-approved immune checkpoint inhibitors that target the PD-1/PD-L1 pathway, and several others are presently in the pre-clinical or clinical phase [26].

## 3 Diagnostic Tools for Precision Treatment

With the rise of head and neck cancer (HNC) in the past few years, the diagnostic tools for the better characterization of the disease have also evolved. The rapid increase in the field of genomics and the analysis of genomic data has revolutionized cancer treatment, as patients with the same cancer may require different treatments based on their genomic characteristics. The genomic era started with the introduction of comparative genomic hybridization (CGH) and microarray techniques in the last decade. Later, next-generation sequencing (NGS) replaced these techniques. With the recent advancement of diagnostic tools and considering the heterogeneous tumour environment and limited access to deeper tissues, liquid biopsies have aided in the identification of circulating DNAs in blood. Other techniques like CT, MRI, and PET have helped in the quantitative diagnosis of the disease.

### 3.1 Genetic Testing

In the past decade introduction of comparative genomic hybridization (CGH) has allowed to identify the alteration of DNA copy number and aberrations in chromosomes by screening the whole genome [27], and it was further upgraded to array-CGH, and this marked the beginning of the 'array era'. This development paves the path for the introduction of microarray techniques that target genomic features like DNA methylation, nucleotide polymorphism, microRNA expression, and DNA binding [28]. Lately, next generation sequencing technique (NGS) has become popular for its capability of sequencing and storing huge DNA data at a lower sequencing cost and user-friendly data analysis software [29]. NGS-based methods have been applied to characterize the molecular background of different cancers, including HNC. As already described, a large number of HNSCC cases are caused by

HPV infection, and NGS has played a remarkable role in identifying different molecular markers and signalling pathways. Analysis of NGS data unveiled that genes like TP53 (67.5% mutated cases), CDKN2A (16.7% mutation, 27% deletion cases), and PIK3CA (16.5% mutated, 16% amplified cases) have the highest mutation rate, along with NOTCH mutation with NOTCH1 gene with the third highest mutation rate. Some of the apoptotic genes, like CASP8 and HRAS, are also mutated in the population with copy number alteration [29]. NGS is also a very convenient method for the identification of the HPV+ population of HNSCC, as it has been observed that there is a remarkable difference in molecular characteristics between HPV+ and HPV- cancers. HPV+ tumours have fewer mutations than HPV- cancers [30]. HPV- cancers are rich in TP53 [31] mutations, whereas HPV+ cancers show gain-of-function mutation of PIK3CA [32]. These different molecular profiles play important roles in the prognosis. In conclusion, NGS has brought a new perception to HNC biology and has unveiled the complex genetic nature of HNC.

## 3.2 Biopsies: Liquid Biopsies and ctDNA

Despite different techniques for the identification of HNC, liquid biopsies have gained attention due to their ability to identify molecular markers from the body fluid and for being cost-effective. Moreover, it has also helped to overcome the limitations like time, patient discomfort, and inaccuracy of conventional techniques. In 1948, the presence of nucleic acid in blood plasma was reported for the first time by Mandel and Metais [33]. Cell-free DNAs (cfDNA) are fragments of nucleic acids of different sizes in bodily fluid, independent of cells [34]. cfDNAs can circulate either freely in the blood and plasma or can be found as macromolecules attached to protein and lipids and inside the vesicles [34]. DNAs are released in the body fluids through processes like apoptosis, necrosis, NETosis, etc., and can result in cell damage and death. Due to this, cfDNAs are used as a marker for inflammation and disease [35]. Circulating tumour DNA (ctDNA) is cfDNA originating from tumours, and it has been observed that patients with different types of cancers have higher cfDNA levels [35]. ctDNAs are also released either by apoptosis and necrosis or through active secretion. It ranges in size from 150 bp to 250 bp [33]. During certain treatment procedures like surgery, radiotherapy, and chemotherapy, cell death is triggered, and that damaged tissue accumulates in the surrounding area as a result of faulty phagocytosis by the immune cells, leading to the release of ctDNAs [35]. Hence, the analysis of ctDNA plays a crucial role in identifying patients' disease, other clinical features like gene mutation, methylation status for early cancer detection, treatment, cancer origin, stage, and also gives an idea about tumour burden, growth, and localization [35]. Liquid biopsy allows the examination of ctDNA and different biomarkers present in bodily fluids like saliva, blood, urine, pleural, peritoneal, and cerebrospinal fluids [36]. Initially, liquid biopsy was introduced in 2010 for the study of circulating tumour cells (CTC), but later its application was extended to the analysis of ctDNA. In HNSCC, which is the most reported HNC, has been observed with the help of liquid biopsy that the TP53 gene is the most mutated in

HPV- negative HNC. In a study involving 93 HNSCC patients, it was observed that in HPV positive cancer, HPV DNA was detected in both saliva and plasma samples [37]. It was also observed that the abundance of ctDNA in saliva and plasma is also related to cancer stage. Early-stage cancer has more ctDNA in saliva compared to late-stage cancer, and the opposite trend was observed in the case of plasma [37]. Not only that, but examination of ctDNA also revealed that plasma ctDNA carries 97% of the mutations present in metastatic tumour sites. Mostly mutated genes in HNSCC were found to be PIK3CA (30%), TP53 (50%), BRAF (10%), and AKAP9 (10%) [37]. These applications make liquid biopsy and ctDNA a promising tool for HNC diagnosis, but they also have some limitations and challenges, like cost and complex data analysis, inadequate materials, and storage of patients' samples, etc. These limitations may affect the liability and accuracy of the technique. Other challenges involve disease, treatment, and data collection-related issues, and false positive/negative results from ctDNA using liquid biopsy that affect the treatment and increase stress among the patients. Despite having these limitations, liquid biopsy is still a valuable tool for early diagnosis of the disease [35].

## 3.3 Imaging Techniques

In recent years, there has been a remarkable advancement in diagnostic imaging with the introduction of new technologies. Ultrasound, computed tomography (CT), magnetic resonance imaging (MRI), and nuclear medicine techniques like positron emission tomography (PET) are widely used imaging modalities for diagnosis, staging and management of HNC in recent years. All these techniques have a strong impact on the treatment of the disease. Hence, understanding each of them is important for accurate staging and characterization of the HNC.

### 3.3.1 Ultrasound

Ultrasound has been the preferred diagnostic tool for the preliminary identification of cervical mass, and also helps in clinical decision-making [38]. This technique is not only one of the easily accessible but also provides excellent characterization of non-invasive soft tissue during assessment of nodal basins and superficial primary sites, e.g., thyroid nodules, neck lumps, lesions of the salivary gland, and morphology of regional lymph nodes [39]. Ultrasound also helps to assess the size, composition, location, and relation to the great vessels [38]. However, some limitations of the technique include dependency on technicians' skills and incompetence to go deep into the tissue, restricted penetration in the soft tissue, air, and overlying bone artefacts [38].

### 3.3.2 Computed Tomography (CT)

Computed tomography is one of the useful techniques for HNC assessment as it helps in identifying the primary site, nodal disease, and staging of HNC, as per the TNM system [38]. CT is widely available and gives good bone details, and images can be obtained in a very short time, making it ideal for HNC patients who have problems lying for a long time due to the disease [40]. Some disadvantages of CT include being inappropriate for soft tissue, exposure to radiation, risk of allergic reactions, and contrast-induced nephropathy, etc. [38].

#### 3.3.3 Magnetic Resonance Imaging (MRI)

MRI is the technique of choice for the characterization of the suprahyoid neck, including oral cavity, sinuses, nasopharynx, and oropharynx [38]. It provides an excellent definition of soft tissues compared to other modalities and provides a clear assessment of lesions in the head and neck region [40]. MRI is influenced by proton density, constituents of blood, and melanin content; different MRI sequences and weightings are applied to distinguish pathology from normal tissue, e.g., T1-weighted images for fat planes and T2-weighted images for highlighting oedema and therefore pathology [40]. One of the disadvantages of the process is that it takes more time than CT, and dental amalgam can reduce the image quality [40].

#### 3.3.4 Positron Emission Tomography Combined with CT (PET-CT)

PET-CT is used to identify the primary tumour of unidentified origin and is generally used for cancer patients with a chance of having recurrent HNC [40]. Radiolabelled glucose is used in the technique to identify regions of high metabolic activity as it is a feature of tumours, including normal structures like brain, heart, and urinary tract [40]. It is represented as a standardized uptake value (SUV), with advanced tumours having a high SUV [40]. However, PET-CT is expensive, and the radiation dose is also high [40].

## 4 Personalized Treatment Options in Head and Neck Cancer

### 4.1 Targeted Therapies

Targeted therapy involves the detailed understanding of the molecular mechanisms and identification of signalling pathways in cancers. Identification of unique proteins, genetic markers that help in the progression of the disease are general target of the therapy.

#### 4.1.1 Immunotherapy

Immune cells of our body can target different antigens expressed by the cancer cells and destroy them in the same way as pathogenic infections are dealt with. Programmed cell death protein 1 (PD-1), a member of the CD28/cytotoxic T-lymphocyte antigen 4 (CTLA-4) family of T cell regulators and Programmed cell death Ligand-1 (PD-L1) axis is the best studied pathway and a target for immune therapy in many cancers. PD-1 is expressed on immune and cancer cells and inhibits self-destruction by prohibiting the immune system from destroying cancer cells and subduing the inflammatory activity of T cells [41]. PD1, PD-L1, and CTLA-4 are the common immunotherapy drug targets, recently used for the treatment of multiple cancer types, including head and neck cancers (HNC). Targeting PD1/PD-L1 restores the immune activity of CD8+ T cells and has been found to be useful for HPV positive HNC [26]. Besides this, anti-LAG-3, anti-CTLA4, and anti-TIM3 drugs have also been introduced for HNC treatment [26]. In HNSCC, the anti-LAG-3 antibody, BMS 986016, in combination with nivolumab, has shown promising results. A soluble LAG-3 protein, eftilagimod alpha (efti), has been examined in combination with pembrolizumab, and TACTI-002 has shown results in a phase II treatment study for PD-L1 unselected metastatic HNSCC patients [42]. IDO-1 enzyme was another target of choice for HNSCC, although application of IDO-1 inhibitors was halted after failure in phase III trials in melanoma [42]. Besides these, some other targets for immunotherapy include EGFR, VEGFR, FGFR, PDGFR-α, RET, KIT, and HDAC inhibitors like vorinostat combined with pembrolizumab, showing promising results [42].

#### 4.1.2 Gene Therapy

Gene therapies aim to find novel targets/critical genes to suppress the growth of neoplasms. This can be achieved by RNA interference (RNAi) or post-transcriptional gene silencing (PTGS), which temporarily knockdown targeted genes and CRISPR/Cas9, a revolutionary gene-editing technology, to particularly target genetic alterations involved in tumour growth.

RNAi has been employed for downregulating E6 and/or E7 proteins and upregulation of p53 and pRb tumour suppressor genes in HNSCC associated with high-risk human papillomavirus (HR-HPV). However, RNAi against HPV was traditionally applied in cervical cancer cell lines and animal models, but shRNA and RNAi for targeting E6/E7 have shown positive results against HR-HPV HNSCC cells both in vitro and in vivo conditions [2]. Lipid nanoparticles coated with anti-EGFR antibodies have been studied for effective distribution of anti E6/E7 siRNA [2].

E6/E7 is also the target of interest for CRISPR/Cas9-mediated gene editing. Guide RNA has been designed to specifically target E6/E7 [2]. However, the delivery method for CRISPR/Cas9 is a challenge. Lipid nanoparticle-mediated delivery has been used for targeting HPV E6/E7 in cervical cancer cells, and similar possibilities exist for HNC as well. To optimize the system and to minimize the off-target delivery, modified Cas9 or related enzymes like Cas13 have also been explored [2].

### 4.2 Stem Cell Therapies

Stem cell therapy involves the identification of specific markers and downstream pathways of cancer stem cells (CSCs) and other therapeutic targets. CD44, Bmi-1, FoxM1, and ALDH are some unique markers in CSCs that provide potential targets for therapy. Scientists have developed therapies and drugs that can target these markers to inhibit CSCs, reducing self-renewal, and to sensitize CSCs for other therapeutic approaches. The focal adhesion kinase (FAK) pathway can act as a target for CSCs to enhance sensitivity to conventional therapies. Some of the examples

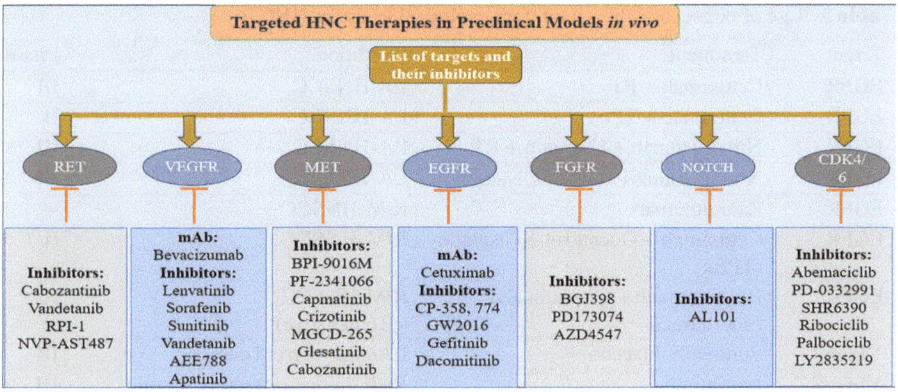


**Fig. 2** List of targets and their inhibitors in preclinical studies. (Created by Microsoft PowerPoint)

include: Antibody-drug conjugates, like MEDI0641, which targets antigens like 5T4, HDAC inhibitors, which have reverted cisplatin resistance [2]. Transcription factors like brachyury or SOX2, which play important roles in CSCs regulation in certain HNSCC, can be explored to develop targeted therapies for suppressing CSCs to enhance sensitivity to chemotherapy or radiotherapy. Besides these, EGFR, IGF-1R, STAT3, Wnt/β-catenin, and COX-2 are a few other relevant biomarkers involved with CSC activity [2]. Moreover, the identification of specific genetic alterations and mutations in the tumour can help guide treatment decisions [2].

### 4.3 Targeted Therapies in Pre-clinical and Clinical Trials

Several molecules have been developed for targeting different proteins and pathways in HNC, and many of those are under pre-clinical (Fig. 2) [6] and clinical trials.

A number of them have been employed in combination with conventional chemotherapy as well. Some of such examples include: inhibitors of EGFR, VEGF, FGFR, AKT, mTOR, c-MET, and many more, which are listed in Table 2.

## 5 Conclusion

Recent advances in personalized medicine have the potential to bring a revolution in cancer therapy, including Head and Neck cancer. The application of personalized medicine has improved the quality of patients' lives globally. In this chapter, current approaches and techniques in HNC have been summarized. Identification of different biomarkers and signalling pathways that play a major role in the development of HNC has paved the path for developing personalized medicines. Diagnostic tools like NGS, liquid biopsy have provided excellent in-depth data analysis for HNC and also helped in identifying early and late-stage disease and the presence of markers in bodily fluids. Imaging tools like ultrasound, CT, MRI, and PET-CT have helped

**Table 2** List of targets and their inhibitors in clinical trials for HNC [6]

| Target | Treatment | Condition | Phase |
|---|---|---|---|
| EGFR | Cetuximab + RT | LA-HNSCC | III |
| EGFR | Cetuximab + TPF | LA-HNSCC | II |
| EGFR | Nimotuzumab + Cisplatin + RT | LA-HNSCC | III |
| EGFR | Nimotuzumab+ RT +/– Cisplatin | LA-HNSCC | III |
| EGFR | Zalutumumab | R/M-HNSCC | – |
| EGFR | Cetuximab + Docetaxel + Cisplatin (TPEx) | R/M-HNSCC | II |
| EGFR | Panitumumab + Cisplatin and Fluorouracil | R/M-HNSCC (p16 negative) | III |
| VEGFR | Sorafenib, Placebo | LA/RM- thyroid cancer | III |
| VEGFR | Lenvatinib | Radioiodine-refractory thyroid cancer | III |
| VEGFR + EGFR | Bevacizumab + Cetuximab/ Erlotinib | R/M-HNSCC | II |
| PI3K | Buparlisib + Cetuximab | R/M-HNSCC | II |
| PI3K | Buparlisib + Paclitaxel | R/M-HNSCC | II |
| PI3K | Alplisib | LA-HNSCC | I |
| AKT | MK-2206, Perifosine, Ipatasertib | HNC | II |
| mTOR | A. Everolimus + Erlotinib<br>B. Everolimus + Cisplatin and Radiotherapy | Aggressive radioiodine-refractory or advanced follicular-derived thyroid cancer | II |
| mTOR | A. Temsirolimus<br>B. Aemsirolimus + Low dose chemotherapy (Carboplatin and Paclitaxel) | R/M-HNSCC | II |
| mTOR + EGFR | Temsirolimus + Cetuximab | Cetuximab-resistant R/M-HNSCC | II |
| C-MET | Cabozantinib | Metastatic MTC | III |
| RET | Selpercatinib (LOXO-292) | RET-altered thyroid cancer | I/II |
| RET | Pralsetinib | RET altered thyroid | I/II |

identify tumour mass, size, and cancer of unknown origin. Based on the identified targets, different targeted therapies have been developed, and a number of them are already in clinical trials, and many are under preclinical studies. A combination of targeted therapies with traditional treatment like chemotherapy and radiation therapy has improved the overall survival of cancer patients and reduced cancer burden. Hence, the detailed study, understanding, and application of personalized therapy will improve the patient outcome and life expectancy of HNC patients.

**Declaration**

**Acknowledgement** SS and AP acknowledge the Department of Biotechnology, Govt. of India. SS, AP, and TKS acknowledge the Indian Institute of Science Education and Research Kolkata (IISER Kolkata) and SBS acknowledges the University of Calcutta for providing research facilities.

**Authors' Contributions** Conceived the idea and review structure: SS, AP, SBS, TKS; Wrote the manuscript: SS, AP, SBS, TKS; Edited the manuscript: SS, AP, SBS, TKS. All authors have read, agreed, and approved the final draft of the manuscript.

**Funding** This work is supported by fellowship to SS and AP from the Department of Biotechnology, Govt. of India. Indian Institute of Science Education and Research Kolkata (IISER Kolkata) provided funds to TKS, and the University of Calcutta provided funds to SBS.

**Data Availability** Microsoft Paint and Microsoft PowerPoint software programs are used to draw all the figures.

**Conflicts of Interests** There is no competing or conflict of interests.

## References

1. Johnson DE, Burtness B, Leemans CR, Lui VWY, Bauman JE, Grandis JR (2020) Head and neck squamous cell carcinoma. Nat Rev Dis Primers 6(1). https://doi.org/10.1038/s41572-020-00224-3
2. Jayawickrama SM, Ranaweera PM, Pradeep RGGR et al (2024) Developments and future prospects of personalized medicine in head and neck squamous cell carcinoma diagnoses and treatments. Cancer Rep 7(3). https://doi.org/10.1002/CNR2.2045
3. Barsouk A, Aluru JS, Rawla P, Saginala K, Barsouk A (2023) Epidemiology, risk factors, and prevention of head and neck squamous cell carcinoma. Med Sci 11(2):42. https://doi.org/10.3390/medsci11020042
4. Solis RN, Silverman DA, Birkeland AC (2022) Current trends in precision medicine and next-generation sequencing in head and neck cancer. Curr Treat Options in Oncol 23(2):254–267. https://doi.org/10.1007/s11864-022-00942-8
5. Gong W, Xiao Y, Wei Z et al (2017) Toward the use of precision medicine for the treatment of head and neck squamous cell carcinoma. Oncotarget 8:2141. www.impactjournals.com/oncotarget/
6. Li Q, Tie Y, Alu A, Ma X, Shi H (2023) Targeted therapy for head and neck cancer: signaling pathways and clinical studies. Signal Transduct Target Ther 8(1). https://doi.org/10.1038/s41392-022-01297-0
7. Sakthi Sanjana D, Elizabeth Rani E, Madhumitha S, Yuvaraj M, Karthikeyan E (2024) Precision oncology in head and neck cancers: from molecular mechanisms to novel therapeutic strategies. Oral Oncol Rep 12. https://doi.org/10.1016/j.oor.2024.100669
8. Parrales A, Iwakuma T (2015) Targeting oncogenic mutant p53 for cancer therapy. Front Oncologia 5. https://doi.org/10.3389/fonc.2015.00288
9. Chen GX, Zhang S, He X hua, Liu SY, Ma C, Zou XP. Clinical utility of recombinant adenoviral human p53 gene therapy: current perspectives. Onco Targets Ther 2014;7:1901–1909. https://doi.org/10.2147/OTT.S50483
10. Aung KL, Siu LL (2016) Genomically personalized therapy in head and neck cancer. Cancers Head Neck 1(1). https://doi.org/10.1186/s41199-016-0004-y
11. Polverini PJ, D'Silva NJ, Lei YL (2018) Precision therapy of head and neck squamous cell carcinoma. J Dent Res 97(6):614–621. https://doi.org/10.1177/0022034518769645
12. Kumari S, Mishra S, Anand N, Hadi R, Rastogi M, Husain N (2023) Circulating free DNA integrity index and promoter methylation of tumor suppressor gene P16, DAPK and RASSF1A as a biomarker for oropharyngeal squamous cell carcinoma. Pathol Res Pract 246. https://doi.org/10.1016/j.prp.2023.154489

13. Liggett WH, Sidransky D (1998) Role of the P16 tumor suppressor gene in cancer. Biol Neoplasia 16:1197
14. Vakili S, Behrooz AB, Whichelo R et al (2024) Progress in precision medicine for head and neck cancer. Cancers (Basel) 16(21). https://doi.org/10.3390/cancers16213716
15. Thomaidou AC, Batsaki P, Adamaki M et al (2022) Promising biomarkers in head and neck cancer: the Most clinically important miRNAs. Int J Mol Sci 23(15). https://doi.org/10.3390/ijms23158257
16. Chen LH, Hsu WL, Tseng YJ, Liu DW, Weng CF (2016) Involvement of DNMT 3B promotes epithelial-mesenchymal transition and gene expression profile of invasive head and neck squamous cell carcinomas cell lines. BMC Cancer 16(1). https://doi.org/10.1186/s12885-016-2468-x
17. Liu CJ, Shen WG, Peng SY et al (2014) MiR-134 induces oncogenicity and metastasis in head and neck carcinoma through targeting WWOX gene. Int J Cancer 134(4):811–821. https://doi.org/10.1002/ijc.28358
18. Peng SY, Tu HF, Yang CC et al (2018) miR-134 targets PDCD7 to reduce E-cadherin expression and enhance oral cancer progression. Int J Cancer 143(11):2892–2904. https://doi.org/10.1002/ijc.31638
19. Suh YE, Raulf N, Gäken J et al (2015) MicroRNA-196a promotes an oncogenic effect in head and neck cancer cells by suppressing annexin A1 and enhancing radioresistance. Int J Cancer 137(5):1021–1034. https://doi.org/10.1002/ijc.29397
20. Perdigão-Henriques R, Petrocca F, Altschuler G et al (2016) MiR-200 promotes the mesenchymal to epithelial transition by suppressing multiple members of the Zeb2 and Snail1 transcriptional repressor complexes. Oncogene 35(2):158–172. https://doi.org/10.1038/onc.2015.69
21. Park SM, Gaur AB, Lengyel E, Peter ME (2008) The miR-200 family determines the epithelial phenotype of cancer cells by targeting the E-cadherin repressors ZEB1 and ZEB2. Genes Dev 22(7):894–907. https://doi.org/10.1101/gad.1640608
22. Zhang B, Li Y, Hou D, Shi Q, Yang S, Li Q (2017) MicroRNA-375 inhibits growth and enhances Radiosensitivity in Oral squamous cell carcinoma by targeting insulin like growth factor 1 receptor. Cell Physiol Biochem 42(5):2105–2117. https://doi.org/10.1159/000479913
23. Yim EK, Park JS (2005) The role of HPV E6 and E7 oncoproteins in HPV-associated cervical carcinogenesis. Cancer Res Treat 37:319
24. Su ZY, Siak PY, Leong CO, Cheah SC (2023) The role of Epstein–Barr virus in nasopharyngeal carcinoma. Front Microbiol 14. https://doi.org/10.3389/fmicb.2023.1116143
25. Kaidar-Person O, Gil Z, Billan S (2018) Precision medicine in head and neck cancer. Drug Resist Updat 40:13–16. https://doi.org/10.1016/j.drup.2018.09.001
26. Fasano M, Corte CMD, Di Liello R et al (2022) Immunotherapy for head and neck cancer: present and future. Crit Rev Oncol Hematol 174:103679. https://doi.org/10.1016/J.CRITREVONC.2022.103679
27. Agrawal N, Frederick MJ, Pickering CR et al (2011) Exome sequencing of head and neck squamous cell carcinoma reveals inactivating mutations in NOTCH1. Science 333(6046):1154–1157. https://doi.org/10.1126/SCIENCE.1206923
28. Aziz N, Zhao Q, Bry L et al (2015) College of American pathologists' laboratory standards for next-generation sequencing clinical tests. Arch Pathol Lab Med 139(4):481–493. https://doi.org/10.5858/ARPA.2014-0250-CP
29. Giefing M, Wierzbicka M, Szyfter K et al (2016) Moving towards personalized therapy in head and neck squamous cell carcinoma through analysis of next generation sequencing data. Eur J Cancer 55:147. https://doi.org/10.1016/J.EJCA.2015.10.070
30. Jiang X, Zhou J, Giobbie-Hurder A, Wargo J, Hodi FS (2013) The activation of MAPK in melanoma cells resistant to BRAF inhibition promotes PD-L1 expression that is reversible by MEK and PI3K inhibition. Clin Cancer Res 19(3):598–609. https://doi.org/10.1158/1078-0432.CCR-12-2731
31. Lui VWY, Hedberg ML, Li H et al (2013) Frequent mutation of the PI3K pathway in head and neck cancer defines predictive biomarkers. Cancer Discov 3(7):761–769. https://doi.org/10.1158/2159-8290.CD-13-0103

32. Li G, Qiu Y, Su Z et al (2013) Genome-wide analyses of Radioresistance-associated miRNA expression profile in nasopharyngeal carcinoma using next generation deep sequencing. PLoS One 8(12):e84486. https://doi.org/10.1371/JOURNAL.PONE.0084486
33. Huang X, Duijf PHG, Sriram S et al (2023) Circulating tumour DNA alterations: emerging biomarker in head and neck squamous cell carcinoma. J Biomed Sci 30(1). https://doi.org/10.1186/S12929-023-00953-Z
34. Pös O, Biró O, Szemes T, Nagy B (2018) Circulating cell-free nucleic acids: characteristics and applications. Eur J Hum Genet 26(7):937–945. https://doi.org/10.1038/S41431-018-0132-4
35. Nassar SI, Suk A, Nguyen SA, Adilbay D, Pang J, Nathan CAO (2024) The role of ctDNA and liquid biopsy in the diagnosis and monitoring of head and neck cancer: towards precision medicine. Cancers (Basel) 16(18):3129. https://doi.org/10.3390/CANCERS16183129
36. Santos V, Freitas C, Fernandes MGO et al (2022) Liquid biopsy: the value of different bodily fluids. Biomark Med 16(2):127–145. https://doi.org/10.2217/BMM-2021-0370
37. Economopoulou P, Kotsantis I, Kyrodimos E, Lianidou ES, Psyrri A (2017) Liquid biopsy: an emerging prognostic and predictive tool in Head and Neck Squamous Cell Carcinoma (HNSCC). Focus on Circulating Tumor Cells (CTCs). Oral Oncol 74:83–89. https://doi.org/10.1016/J.ORALONCOLOGY.2017.09.012
38. López F, Mäkitie A, de Bree R et al (2021) Qualitative and quantitative diagnosis in head and neck cancer. Diagnostics 11(9). https://doi.org/10.3390/DIAGNOSTICS11091526
39. Vishwanath V, Jafarieh S, Rembielak A (2020) The role of imaging in head and neck cancer: an overview of different imaging modalities in primary diagnosis and staging of the disease. J Contemp Brachyther 12(5):512–518. https://doi.org/10.5114/JCB.2020.100386
40. Batuwitage BT, Hanlon R, Charters P (2021) Imaging in head and neck cancers. BJA Educ 21(1):2–9. https://doi.org/10.1016/J.BJAE.2020.08.006
41. Seidel JA, Otsuka A, Kabashima K (2018) Anti-PD-1 and anti-CTLA-4 therapies in cancer: mechanisms of action, efficacy, and limitations. Front Oncol 8. https://doi.org/10.3389/FONC.2018.00086
42. Jayawickrama SM, Ranaweera PM, Pradeep RGGR et al (2024) Developments and future prospects of personalized medicine in head and neck squamous cell carcinoma diagnoses and treatments. Cancer Rep 7(3). https://doi.org/10.1002/cnr2.2045

# Breast Cancer: Targeted Therapy and Neoantigen Vaccine

Shaonlee Ghosh, Gourab Das, and Abhijit De

**Abstract**

Breast cancer is a global burden that continues to show the highest incidence among women. The disease is well characterized, and clear recommendations on available therapies on the basis of pathological staging and molecular diagnosis of patients are available. For the majority of breast cancer disease subtypes, personalized treatment options in the form of patient-specific molecularly targeted medicines are also available, showing improvement in survival outcomes. Advancements in computational biology have greatly enhanced this effort of druggable target identification, thereby shaping the targeted therapy landscape available today. However, irrespective of enriched treatment options available for breast cancer patients, drug resistance remained a major clinical desolation for this disease. As a continued global research effort, new clinical trials are now exploring the prospect of various new drugs and novel combinatorial treatment approaches. More recently, patient tumor-specific neoantigens are in the limelight, for their unique presence in the cancer cells, as well as the way neoantigen-based vaccines can engage the patient's own immune system to combat the disease. Hence, the present chapter epitomizes current and future advancements in targeted therapy regimens, including prospective neoantigen vaccine-based immunotherapy approaches developed for breast cancer management.

S. Ghosh · A. De (✉)
Molecular Functional Imaging Laboratory, Advanced Centre for Treatment, Research and Education in Cancer (ACTREC), Tata Memorial Centre, Navi Mumbai, Maharashtra, India
e-mail: ade@actrec.gov.in

G. Das
Bioinformatics and Computational Biology Facility, Advanced Centre for Treatment, Research and Education in Cancer (ACTREC), Tata Memorial Centre, Navi Mumbai, Maharashtra, India



M. K. Ghosh, M. Basu (eds.), *Personalized Cancer Therapy*,
https://doi.org/10.1007/978-981-95-6845-1_10

**Keywords**

Breast Cancer · Personalized therapy · Biomarker · Antibody · drug conjugates · Neoantigen vaccine · Drug resistance

## 1 Introduction

Breast cancer (BC) is the most prevalent form of cancer and is still recognized as the leading cause of death among women worldwide. Extensive research has been done showing the heterogeneous nature of this disease. According to the current clinical practices, BC is classified to have four primary subtypes based on the presence or absence of a few well-established cell surface biomarkers. These are classically recognized as luminal A (ER+/PR+/HER2−), luminal B (ER+/PR+/HER2+), HER2 (ER-/PR-/HER2+), and triple negative breast cancer (ER-/PR-/HER2−) [1]. Out of these subtypes, TNBC is the most aggressive form that shows high metastatic potential in critical organs like the lung, liver, and brain [2]. In the treatment landscape, surgery, chemotherapy, radiation therapy, and various refined combinatorial administration processes are widely practiced based on pathological characterization, which shows some degree of improvement in outcome measures. However, these orthodox treatment options remain futile for many patients, predominantly for cases that are diagnosed at a late stage [3]. Due to delayed diagnosis, the overall treatment efficacy reduces mainly for two reasons, i.e., increased manifestation of metastatic disease, and intensive treatment requirement for advanced disease, which itself may cause treatment intolerance [4].

In the post-genome era, rapid advancements in molecular understanding brought in a new molecular classification subdividing BC into ten interclasts, each having unique risk factors for incidence, predicting disease progression, therapeutic response, and favored organ sites of metastatic spread [5]. As far as the global clinical practices go, though interclasts are not integrated in practice, the enriched molecular understanding has majorly impacted the practicing guidelines and treatment landscape for BC. Over the past few decades, various targeted treatment strategies have been developed based on patient-specific genomic alterations and brought in fresh hope for BC treatment. To pave the path of personalized cancer treatment approach, rigorous efforts were made to build a robust computational framework and in silico design tools by the computational biologists [6]. Eventually, a majority of the toolsets and pipelines are robust, specific, and focused cancer-related in silico application design tools for cancer immunotherapy application [7], or systems biology-based protocols for precision medicine applications are now present [8]. A series of targeted therapies against the overexpressed or hyperactivated proteins like steroid hormone receptors, receptor tyrosine kinases, downstream signaling molecules, or other regulatory candidates were developed and strategic implementation of many of these medicines is integrated in clinical practices by now. More recently, personalized vaccine development efforts are ongoing to target neoantigens, which can be tailored based on individual patient-specific molecular aberrations present in the cancer cells [9]. Neoantigen vaccines are to activate patients' immune cells and thereby eradicate the cancer cells by breaking the anti-tumor immunity barriers that the BC tumors present. In this chapter, we will provide a brief overview of the current and future targeted medicines developed for use in breast cancer. The chapter will draw a special reference to neoantigen vaccines under development and the involved processes of their development.

# 2 Evolving Paradigm of Breast Cancer Therapeutics

## 2.1 Current Treatment Landscape

Treatment decisions for breast cancer patients solely rely on disease stage as well as subtypes. To distinguish the diseased stage, tumor size (T), lymph node metastasis (N), and distant organ metastasis (M) are examined, a process commonly known as TNM staging (Fig. 1). Patients with only one or two detectable primary tumors are categorized into stage-IA or IIA, respectively and patients with multiple lymph node metastases are of stage-IIIA. However, patients with one to three primary

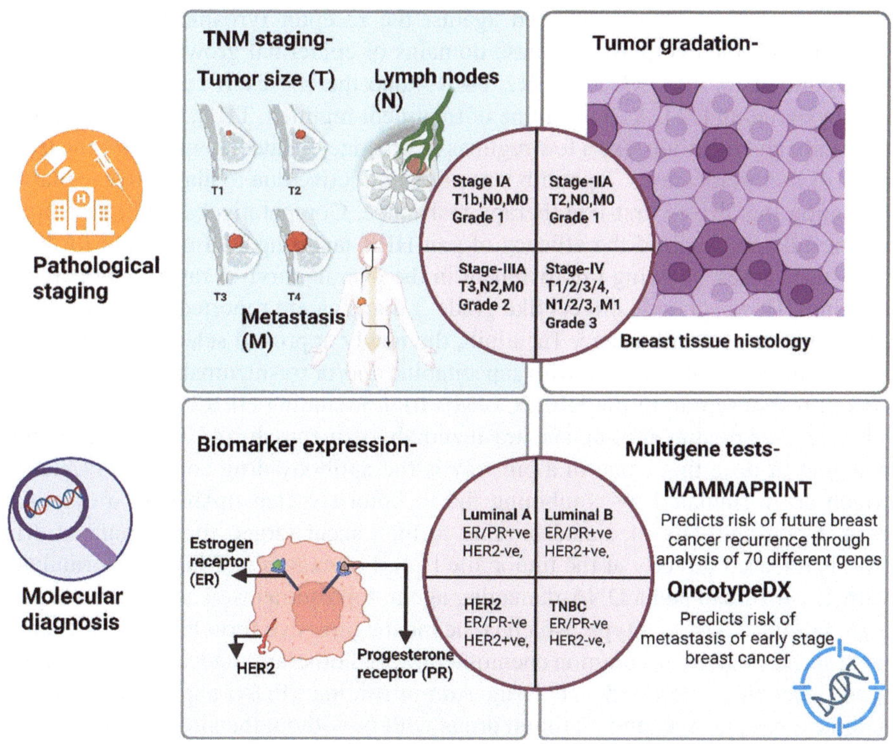


**Fig. 1** Schematic explaining breast cancer disease diagnosis and staging, considering tumor size, tumor burden, its appearance in histological analysis, and also breast cancer subtyping, considering the presence of different biomarkers in the cancer cell

tumors, both lymph node and distant organ metastasis, are categorized as stage-IVA with the worst prognosis. Further, tumor biopsies are widely used to depict poorly differentiated cancer tissues with worse prognosis vs. highly differentiated breast tissues with better prognosis. In addition, assessment of BC biomarker expression is essential to determine disease subtype for planning the treatment regimen. Alongside these key steps of disease staging and subtype determination, molecular profiling using OncotypeDX or Mammaprint is often recommended to offer patients with available tailored personalized medicines for treatment.

Coming to the management of breast cancer treatment, luminal A subtype tumors are the least aggressive subtypes which responds well to hormonal interventions like Tamoxifen. On the other hand, luminal B subtype typically represents poorer prognosis, but responds better to the standard chemotherapy regimen. Since this subgroup of patients often shows *HER2* overexpression, targeted therapies for *HER2* might also be employed in some cases. The HER2+ BC subtype typically represents amplification or overexpression of the HER2/ERBB2 oncogene. One of the first sensational targeted medicines, the monoclonal antibody (mAb) drug trastuzumab for BC was developed against this target. Trastuzumab binds to over-expressing HER2 on the cancer cell surface and inactivates the downstream signaling and stalls cancer cell growth [10]. Another class of targeted therapy comprises small molecule inhibitors selected against the receptor tyrosine kinases (TKIs), which inhibit the cytoplasmic kinase domains of epidermal growth factor receptor family members like EGFR, HER2, HER3, and thereby exert control over downstream signaling [11]. As a third line of treatment regimen, TKIs, such as lapatinib, neratinib, and tucatinib, are a few regimes in practice to date. However, for the treatment of metastatic cases, lapatinib was not so effective due to alternative signaling activation that drives first-line therapy resistance. Comparatively, multiple clinical trial results emphasized the efficacy of pan-HER targeting TKI neratinib for brain metastasis cases, showing improvement in the overall survival rate as compared to lapatinib. However, side effects like grade-3 diarrhea are reported as the main disadvantage of neratinib usage. Tucatinib, the newly approved selective TKI against HER2 kinase in combination with capecitabine and/or trastuzumab, showed a better overall response rate in the HER2CLIMB trial, including HER2 + ve BC patients who received treatment of taxane, trastuzumab, pertuzumab, and T-DM1. The other new line of drug that came in a big way is the antibody-drug conjugates (ADCs), which are formulated by combining the receptor-targeted mAb with a cytotoxic drug. Upon antibody-mediated binding to the cancer target, the conjugated drug load is released directly at the tumor site [12]. For example, T-DXD is formulated with Trastuzumab and a DNA-damaging agent, topoisomerase I as a cytotoxic payload. For the TNBC subtype, which is the most aggressive form but lacks the targetable surface antigens, common chemotherapy, radiotherapy, and surgery are mainly used. The only US Food and Drug Administration (FDA)-approved therapy is anthracycline, taxane, and platinum drugs with or without the anti-angiogenic drug bevacizumab [13]. However, considering the suboptimal treatment outcome with standard therapeutic agents in use, identification of immune targets for developing a new therapy strategy has been undertaken. To engage the patient's own immune

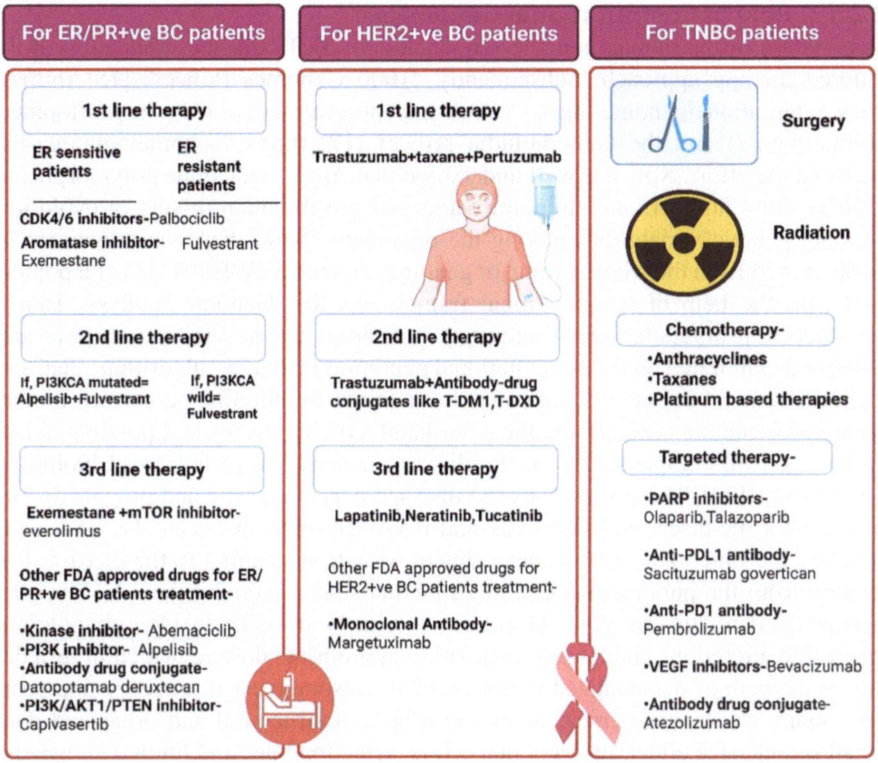


**Fig. 2** Current treatment regimen for breast cancer

system against the tumor burden, another mAb drug, Pembrolizumab, developed against cancer cell immunity marker programmed cell death-1 receptor (PD-1) was introduced [14]. The BC subtype-specific common treatment plan is summarized in Fig. 2.

## 2.2 Advanced Computing in Designing Tailored Therapy Protocol

Advancements in the field of computational science and technology have aided the paradigm of personalized therapy for breast cancer in several ways. Starting from analytical software development, clinical data mining, interpretation, in silico model building, to the development of integration tools for multi-omics data and their association with clinical outcomes, big data warehousing, etc., were done. The impact of advanced computing in personalized therapy can be categorized into the following segments:

### 2.2.1 Role of Computational Genomics

Completion of the human genome sequencing in 2001 built the foundation of the tailored therapy approach. Subsequently "1000 Genomes Project" [15] (https://www.internationalgenome.org/), "TOPMed programme" [16] (https://topmed.nhlbi.nih.gov/) and the GenomeIndia project [17] (https://genomeindia.in/) has enriched the effort. Now it is well understood that single-nucleotide polymorphisms (SNPs), copy number, and structural variations are the main drivers of evolution, creating genetic variabilities among the organisms. Now, these variations can be easily mined from the whole exome or genome sequences (WES/WGS) of a population with the help of computational frameworks like Genome Analysis Toolkit (GATK) (https://gatk.broadinstitute.org/). Numerous recent publications have also listed and elaborated on the computational genomics resources, algorithms, national and international initiatives, and consortia required for building personalized treatment infrastructure [18–22]. On the other hand, GWAS has made it possible to link variations with complex traits of the disease, hence discovering novel biological mechanisms [23]. These variations can also serve as diagnostic and prognostic biomarkers for the diseases. Microarray and RNA-seq techniques can also be used to measure the functional role of these genetic variations leading to the disease. It is evident from the publications that these methods are becoming the workhorse for personalized treatments [24]. Moreover, it is now possible to identify potential druggable alterations and targets, estimating appropriate dosage, which in turn can enrich the path of personalized treatment [25]. Despite these discoveries and progress, many challenges persist, including a lack of structured and organized data, small sample size effect, instrumental errors, selection bias, and limited knowledge to build hypotheses to reveal complex disease mechanisms, which are the current pitfalls of computational genomic research in tailored therapy.

### 2.2.2 Role of Structural Biology

After identification of druggable targets, studying the macromolecular structure of the targets is recommended, which includes finding or predicting three-dimensional (3D) structures, screening the suitable drug molecules, examining the drug-target interactions, checking the stability of the drug-target complex, toxicity measurements, and quantifying their activity relationships. Often, comparative structural studies of the targets carrying driver mutations are beneficial to deduce the individualized drug response and therapeutic interventions [26]. Databases like Protein Data Bank (https://www.rcsb.org/), The Nucleic Acid Knowledgebase (NAKB) (https://nakb.org/), and AlphaFold (https://alphafold.ebi.ac.uk/) are the prime resources for the 3D structural analysis, which can be the initiation point for personalized structural biology. After scrutiny of the target structure quality, many small-molecule synthetic drugs and vaccines can be docked to check energetically favorable complex formation, followed by inspection of its stability through molecular dynamics simulations. However, the study will remain incomplete without assessing the patient's response to the drug/vaccine through quantitative structure–activity relationship (QSAR) models [27, 28]. It is highly expected that a particular drug may not respond equally to all BC patients having identical clinical conditions. Hence, patient-centric structural analysis of targets along with their identification, design, interaction study, and activity prediction can also be performed systematically with the use of high-performance computers in a limited time.

### 2.2.3 Role of Systems Biology

Drug resistance is another major problem in breast cancer research. To deal with this, system-level studies are often required, which follow a holistic approach to integrate information from multi-omics experiments to reveal not only resistance mechanisms but also disease risk estimation, summarizing probable preventive measurements, molecular profile monitoring, etc. [29]. Complex resistance mechanisms can be hypothesized by investigating the pharmacokinetics and pharmacodynamics properties of drug/vaccine response, elucidating affected pathways, activation-deactivation of the key enzymes, and perturbations in the metabolic networks [30]. With the help of software like COPASI (https://copasi.org/), JWS (https://jjj.biochem.sun.ac.za/), and MATLAB (https://www.mathworks.com/), system-level metabolic or stoichiometric model building per patient has become very user-friendly, time and memory-efficient. However, the lack of experimentally determined enzyme kinetic parameters for breast cancer metabolic reactions and population-level model validation are the major bottlenecks of system biology-powered personalized treatments [31].

### 2.2.4 Role of High-Performance Computing and Computational Intelligence

High-performance computing using computationally intelligent algorithms has opened up a new avenue where big data can be analyzed economically in terms of time and computer memory. Parallel and CUDA programming have sped up many biological and biomedical tools and programs that are essential for big data analysis [32]. It is not very far when multi-omics data analysis, integration, validation, and extracting key information to build a patient-centric knowledge base will be achievable with the help of high-end computing solutions. The true power of personalized medicine can be unraveled by the integration of computational intelligence (CI) with multi-omics, text, and image data [33]. CI can handle a vast amount of data to identify patient-specific patterns, responses, and associations leading to the discovery of novel biomarkers and therapeutic targets. CI has great potential to enhance preventive care by estimating risk factors, early detection by learning time-dependent discriminatory features, and improving treatment plans for the betterment of the overall health of the patients [34]. Nevertheless, applications of CI in tailored therapy are not flawless. Though CI holds a huge promise, still data privacy, algorithmic bias, sufficient sample size for efficient learning, class imbalance, and the curse of dimensionality are some major issues to be looked at prior [35].

## 2.3 Future Landscape of Personalized Medicine

It is now evident from recent publications that with the robust multi-dimensional next-generation sequencing data available, the traditional "one-size-fits-all" treatment approach can shift more towards tailored or targeted therapy [36]. Beyond the current FDA-approved therapy regimens, active clinical trials are exploring the scope of other targeted therapeutics for BC (Fig. 3). In this regard, for Luminal A subtype, selective estrogen receptor modulators (SERM) and selective estrogen receptor degraders (SERD) class of drugs developed are now under trial investigation. SERM drugs like lasofoxifene competitively inhibit the binding of estrogen receptor (ER) and thereby stall the downstream signaling cascade. Lasofoxifene showed promising results by reducing both primary and metastatic tumors in ESR1-mutated Luminal A subtype patients in the ELAINE I trial [37]. Recently, the ELAINE III trial (NCT05696626) is investigating the superiority of the lasofoxifene plus abemaciclib combination over the fulvestrant plus abemaciclib combination strategy for the same patient cohort. SERD class of drugs aids in proteasome-mediated degradation of ER, thereby even combating against aromatase inhibitor resistance in Luminal A patients. For example, Giredestrant, Camizestrant are such kinds of drugs under clinical investigation as monotherapy

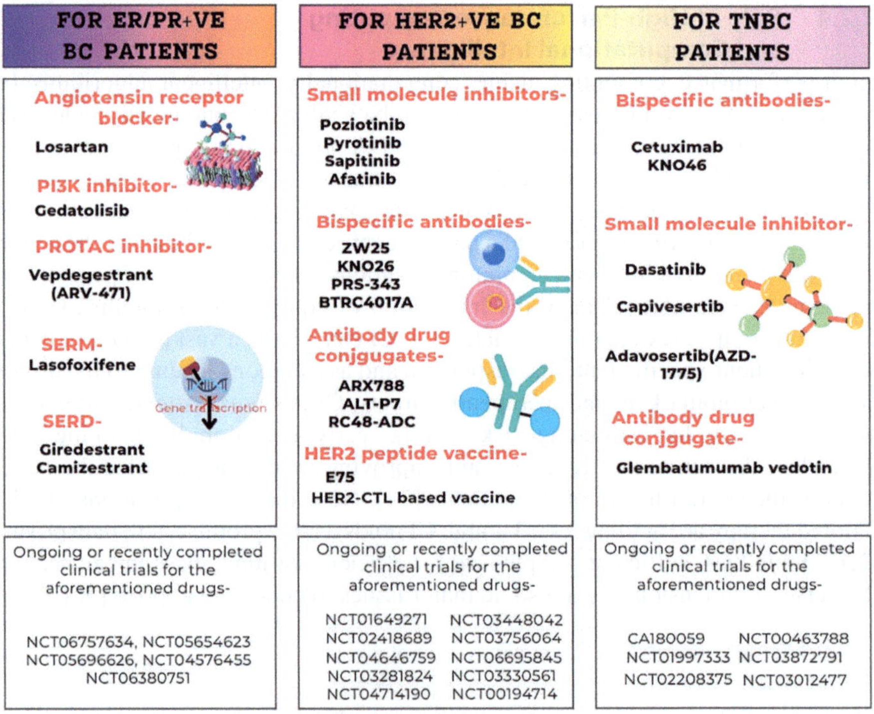


**Fig. 3** Future targeted medicines for use against each subtype of BC patients

or combination therapy with other FDA-approved targeted drugs [37]. Research advancements have also introduced bispecific antibodies like ZW25, KNO26 for the treatment of HER2 + ve BC, which is known to target cancer cells and recruit immune cells simultaneously, showing more effectiveness over the tumor burden [38]. Recently, HER2 vaccine (NeuVax) using the extracellular domain derived E75 peptide has drawn significant attention, showing stimulated immune response *in vivo* [39]. Later, a phase-II clinical trial (NCT02636582) also found increased immune boosting by using NeuVax plus GM-CSF administration as compared to the GM-CSF only arm, proving HER2 vaccine as a promising therapeutic approach [40]. Newer antibody-drug conjugate small molecule inhibitors have also entered the pipeline of clinical trials. Among these, an under-trial small molecule inhibitor, pyrotinib, as an adjuvant in the NCT05880927 trial for trastuzumab-treated patient cohort with grade-3 tumors and lymph node micro-metastasis is notable. ARX788 is an emerging ADC under clinical trial (NCT06578286) for efficacy and safety profile in HER2 + ve BC patients who were pre-treated with a second-line therapy regimen T-DXD. For the TNBC subtype, targeted therapy regimens including bispecific antibody cetuximab, small molecule inhibitor- dasatinib, ADC- Glembatumumab vedotin, etc., are undergoing clinical investigations. Cetuximab, being an EGFR targeting antibody, the NCT01097642 trial highlighted a higher complete response rate (32%) and recurrence-free survival rate (68%) of the cetuximab plus Ixabepilone-treated TNBC patient cohort compared to the Ixabepilone alone arm. Glembatumumab vedotin, being a formulation of targeting antibody for glycoprotein NMB (nonmetastatic melanoma protein B), overexpressed on cancer cell surface, conjugated with monomethyl auristatin E (MMAE) drug, has been studied in several clinical trials NCT01997333 to gain an insight into overall survival or progression-free survival. Another rapidly developing strategy is via patient-specific tumor antigens or neoantigens. The neoantigens arise from patient-specific mutations in tumor cells, which is a common phenomenon in BC. Here, T-cells are sensitized against a specific neoantigen, and the activated T-cell in turn eliminates the cancer cells. Therefore, personalized neoantigen vaccines are tailored for individual patients with antigen-presenting cells (APC).

## 3 Progress on Neoantigen Vaccine Development

Neoantigens distinctly differ from tumor-associated antigens (TAA), since TAAs are not unique to patient-specific tumor tissue. Non-synonymous mutations in the tumor cell genome give rise to neoantigens, which are recognized through comparative sequence analysis of tumor cells and normal cells using high-throughput sequencing techniques. Studies have highlighted the role of neoantigens as an immune modulator, since neoantigens are known to trigger the body's anti-tumor immune response through MHC-mediated TCR activation [40]. Research advancements have introduced different neoantigen vaccine platforms, e.g., SLP vaccine, DNA vaccine, RNA vaccine, and DC vaccine.

The design and optimization of a synthetic long peptide (SLP) vaccine depends on in-silico prediction of immunopeptidomics and in vitro assays to identify optimal epitopes. SLP vaccine platform relies on the immunoproteasome-mediated cleavage and antigen processing followed by binding of short and long peptides to Major histocompatibility complex (MHC) molecules and CD4 T-cell-mediated immune response [41]. SLP-based neoantigen vaccine is already in Phase-II clinical trial (NCT03606967) for TNBC patients. Two treatment arms, with or without the SLP neoantigen vaccine, are under investigation to measure the overall survival benefit due to the vaccine administration. In another ongoing clinical trial (NCT04197687) for HER2 + ve BC cases with residual disease after neoadjuvant treatment, efficacy evaluation is underway for one HER2 peptide-derived neoantigen vaccine in activating an immune response against HER2 tumor growth. On the other hand, DNA vaccine development is much easier and less costly when compared to other neoantigen strategies [42]. DNA vaccine development includes whole-exon sequencing to identify somatic mutations in a patient's genome, followed by neoantigen prediction using an in silico tool. A Phase-I clinical trial (NCT03199040) for grade I-III TNBC patients compared a DNA vaccine alone or in combination with Durvalumab to gain insights about the DNA vaccine's safety profile. Another phase-II clinical trial (NCT04329065) is investigating the immune response and side effects of using WOKVAC DNA vaccine along with chemotherapy and targeted therapy in HER2+ve BC patients. The mRNA vaccine is also a very promising approach in terms of engaging the body's immune system against tumors. Development of RNA-based vaccine involves in vitro transcription (IVT) from a template DNA, in addition with 5'cap and a 3′-poly(A) tail for stability purpose [43]. For RNA neoantigen vaccine delivery, encapsulation of the vaccine within lipid nanoparticles is well practiced, since it ensures stability of the RNA vaccine by mitigating the chance of degradation. RNA vaccines, once administered, translation of antigens and presentation along with MHC molecules over the APCs like DC takes place. Further, this gets recognized by T cells, and subsequent immune response through cytotoxic T-cell activation comes into action by recognizing the same antigens present on the tumor cells [44, 45]. There are also reports of dendritic cell (DC) vaccines, which are primarily ex vivo generated and cost-effective. Neoantigen vaccines can be loaded into DC, or even DC can be transfected with RNA-encoded neoantigen through electroporation [41]. DC-loaded neoantigens, once administered in patients, DC trafficking happens towards the lymph node, where Th1 activation-mediated immune response comes into action [46]. To date, the HER2-sensitized DC vaccine is being investigated in a clinical trial (NCT03387553) to gain more insights about an appropriate treatment strategy for HER2+ve invasive BC, considering the vaccine can be used as a neoadjuvant treatment regimen. This vaccine was developed from patients' blood cells through leukapheresis and has been tested in two arms- single vaccination vs. double vaccination per week, followed by chemotherapy and surgery. Moreover, another clinical trial is also investigating the efficacy of the DC vaccine in shrinking tumors of HER2+ve BC and TNBC patients with brain metastasis.

## 4 Unified Computational Pipeline for Neoantigen Vaccine Development

The design of any vaccine through conventional methods requires a significant amount of human effort, money, and time. To reduce the burdens, bioinformatic strategies have evolved to modernize the vaccine design processes, including reverse vaccinology, immune-informatics, structural vaccinology, etc. These vaccine-designing methods are robust and can be used to design any kind of nucleic acid, peptide-based vaccine, e.g., Neoantigen vaccine (Fig. 4).

Utilizing the cancer cell genome information, with the help of computational tools, plausible candidate antigen targets can be detected, followed by testing and validation for its effectiveness. NERVE, GLIMMER, ORF-FINDER, and GS-Finder

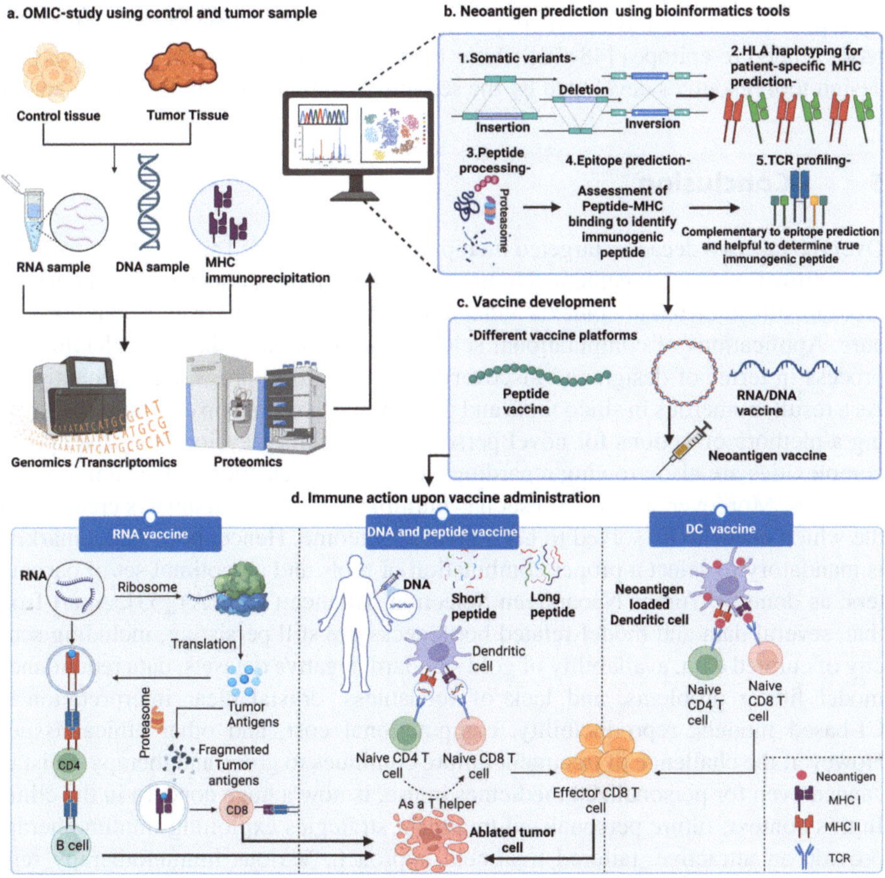


**Fig. 4** Schematic showing neoantigen development pipeline and mechanisms of action. (**a**) Tumor sample and control tissue collection from the patient for neoantigen analysis. (**b**) Identification of cancer-specific mutation mapping by sequencing. (**c**) Neoantigen production process. (**d**) Immune-action mechanisms upon administration of various types of neoantigens

are the main toolsets that provide screening power of potential ORFs from the genome data. This is followed by vaccine candidate identification using another set of in silico tools, namely ProDom, Pfam, and Prosite. Immuno-informatics and structural biology-based vaccine design performs the following steps starting with a screening of candidate tumor antigens based on mutations, gene expression, and gene fusions identified from NGS data, then analysis of protein structures in terms of their stability, physicochemical properties, and antigenicity, followed by T and B-cell epitope prediction, characterization and evaluation and finally Protein-epitope interaction study through docking molecular dynamics simulations [47]. Several standard tools are already in use to execute the aforesaid steps, including NGS data analysis, mutation mining, protein structure prediction, calculation of properties, and interaction studies. Many new tools are coming up, like MixMHCpred, NetMHCpan, MHCflurry, CEDAR, ElliPro, DiscoTope, and Epitope-Evaluator for potential epitope prediction and evaluation. In this context, MHCBN, Immune Epitope Database (IEDB), BciPep, and BepiPred v2.0 are some of the useful databases related to epitopes [48–52]. Table 1 provides information on various vaccine design tool resources developed by the scientific communities to deal with cancer.

## 5 Conclusion

Over the last two decades, targeted therapies have contributed significantly to breast cancer treatment management. Numerous studies have dissected the underlying mechanisms in different contexts and attempted to find an alternative solution for a cure. Applications of computational science and technology have accelerated this process in terms of design and discovery of personalized drugs, e.g., neoantigens. As a result, numerous in silico tools and pipelines are coming up continuously, making a plethora of options for novel personalized vaccine development. Eventually, complexities are also growing regarding suitable tool selection or building boosted pipelines. Moreover, as each toolset has multiple parameters, it always create a riddle which needs to be solved to get a precise outcome. Hence, prior benchmarking is mandatory to select a proper combination of tools and an optimal set of parameters, as done by Tumor Neoantigen Selection Alliance (TESLA) [53]. Apart from that, several data and model-related bottlenecks are still persisting, including scarcity of curated data, availability of gold standard negative datasets, data redundancy, model fitting problems, and lack of robustness, crystal clear interpretation of CI-based models, reproducibility, computational cost, and other ethical issues. However, the challenge of treatment failure continues to grow, and therapy-resistant cancer, even for personalized medicines in use, is now a huge concern in the clinic. In this context, future personalized treatment strategies exploiting immunotherapy provide an attractive, tailored treatment approach. Various immunotherapy regimens like immune checkpoint inhibitors, anti-PD1, and anti-PD-L1 as immunotherapeutic agents for BC patients have entered in practice [13]. However, considering the altered expression and activation status of different biomarkers in BC, the mutational aspect was highlighted, which eventually aided the emergence

**Table 1** List of methods, tools, and pipelines for neoantigen vaccine development[a]

| No. | Tool | Short description | Weblink | PMID/DOI |
|---|---|---|---|---|
| 1 | NeoDisc | A hybrid algorithm of rule-based methods and machine learning integrated with genomics, transcriptomics, and immunopeptidomics data and multiple in silico tools | https://neodisc.unil.ch/ | 39394480 |
| 2 | NeoDesign | A tool identifying optimized protein sequences for tumor polyvalent neoantigen mRNA vaccine design | https://github.com/HuangLab-Fudan/neoDesign | 39331572 |
| 3 | Neo-intline | A pipeline utilizing SNP and Indel data from WGS to invigorate the peptide presentation process and vaccine design | https://github.com/zoolie/neoantigenPre | 37848417 |
| 4 | NeoAgDT | An integrated method of multi-omics data with a special focus on optimal vaccine composition selection by integer linear programming | https://github.com/nec-research/neoagdt | 38614133 |
| 5 | PNVAC | A deep learning model based on peptide and HLA datasets | https://github.com/roubaokai/NUCC.git | 38214336 |
| 6 | Customized pipeline | A pipeline that uses machine learning methods and harmonized datasets for the accurate prediction of neoantigens | https://doi.org/10.1016/j.immuni.2023.09.002 | 37816353 |
| 7 | Hybrid pipeline | A systematic approach combining in silico tools and wet lab methods for efficient detection and validation of immunogenic neoantigens | https://doi.org/10.1158/1541-7786.MCR-19-1118 | 33257508 |
| 8 | Splicing neo antigen Finder (SNAF) | A highly accurate, user-friendly, open-source ep learning based tool for immunogenicity prediction, tumor specificity ranking, and mis-splicing regulator mining | https://github.com/frankligy/SNAF | 38232136 |
| 9 | ScanNeo2 | An automated workflow for neoantigen detection and immunogenicity prediction using various sequencing data | https://github.com/ylab-hi/ScanNeo2/ | 37882750 |
| 10 | NeoHunter | A flexible tool for neoantigen identification and validation from sequencing data | https://github.com/XuegongLab/NeoHunter | https://doi.org/10.1002/qub2.28 |
| 11 | NeoSplice | A program for the detection of splice variant neoantigen utilizing genomics and mass spectrometry data | https://github.com/Benjamin-Vincent-Lab/NeoSplice | 35669345 |
| 12 | ImmuneMirror | A web server for neoantigen discovery and selection built on machine learning models trained using features extracted from WES and peptide data | https://github.com/weidai2/ImmuneMirror, http://immunemirror.hku.hk/App/ | 38343325 |

[a]Apart from the tools/pipelines/methods listed in https://doi.org/10.1016/j.iotech.2021.100052, https://doi.org/10.1515/jib-2023-0043, https://doi.org/10.3389/fimmu.2024.1394003

of tumor-specific antigen targeting vaccines or neoantigen vaccines. Multiple variants of cancer-targeted vaccines are ongoing, some of which have also progressed in clinical trial phases for efficacy and safety profile in HER2+ve and TNBC subtype patients. However, considering the tumor heterogeneity of BC patients and the aspects of drug resistance they represent, a more integral treatment approach utilizing the genetic, immunogenic, as well as molecular landscape of each patient is required to further strengthen the personalized treatment strategy for BC.

**Author's Contributions** Conceived the idea and review structure: AD; Wrote the manuscript: AD, GD, and SG; Revised and edited the manuscript: AD, GD, and SG. All authors have read and approved the final manuscript before submission.

**Conflicts of Interests** There is no competing and conflict of interests.

**Funding** This work is supported by the available research framework at ACTREC, supported by the Department of Atomic Energy (DAE), Govt. of India, funding. DAE DPR [Ref. DAE-ACTREC Grant No: 1/3 (7)/2020/TMC/R&D-II/8823 Dt.30.07.2021] funding gratefully acknowledged.

**Data Availability** The software Biorender (https://www.biorender.com/) and Canva (canva.com) are used for making figures.

## References

1. Siegel RL, Miller KD, Jemal A (2017) Cancer statistics, 2017. CA Cancer J Clin 67:7–30. https://doi.org/10.3322/caac.21387
2. Ensenyat-Mendez M et al (2021) Current triple-negative breast cancer subtypes: dissecting the most aggressive form of breast cancer. Front Oncol 11:681476. https://doi.org/10.3389/fonc.2021.681476
3. Ho PJ et al (2020) Impact of delayed treatment in women diagnosed with breast cancer: a population-based study. Cancer Med 9:2435–2444. https://doi.org/10.1002/cam4.2830
4. Dey P, Rathod M, De A (2019) Targeting stem cells in the realm of drug-resistant breast cancer. Breast Cancer Targets Ther 11:115–135. https://doi.org/10.2147/BCTT.S189224
5. Curtis C et al (2012) The genomic and transcriptomic architecture of 2,000 breast tumours reveals novel subgroups. Nature 486:346–352. https://doi.org/10.1038/nature10983
6. Suwinski P et al (2019) Advancing personalized medicine through the application of whole exome sequencing and big data analytics. Front Genet 10:49. https://doi.org/10.3389/fgene.2019.00049
7. Kashyap D, Garg VK, Goel N (2021) Intrinsic and extrinsic pathways of apoptosis: role in cancer development and prognosis. Adv Protein Chem Struct Biol 125:73–120. https://doi.org/10.1016/bs.apcsb.2021.01.003
8. Filipp FV (2017) Precision medicine driven by cancer systems biology. Cancer Metastasis Rev 36:91–108. https://doi.org/10.1007/s10555-017-9662-4
9. Xu P et al (2020) Cancer neoantigen: boosting immunotherapy. Biomed Pharmacother 131:110640. https://doi.org/10.1016/j.biopha.2020.110640
10. Hudis CA (2007) Trastuzumab-mechanism of action and use in clinical practice. N Engl J Med 357:39–51. https://doi.org/10.1056/NEJMra043186
11. Xuhong JC, Qi XW, Zhang Y, Jiang J (2019) Mechanism, safety and efficacy of three tyrosine kinase inhibitors lapatinib, neratinib and pyrotinib in HER2-positive breast cancer. Am J Cancer Res 9(10):2103–2119. PMID: 31720077; PMCID: PMC6834479

12. Najjar MK, Manore SG, Regua AT, Lo HW (2022) Antibody-drug conjugates for the treatment of HER2-positive breast cancer. Genes 13(11):2065. https://doi.org/10.3390/genes13112065
13. Ye F et al (2023) Advancements in clinical aspects of targeted therapy and immunotherapy in breast cancer. Mol Cancer 22(1):105. https://doi.org/10.1186/s12943-023-01805-y
14. Angelis V, Okines AFC (2024) Systemic therapies for HER2-positive advanced breast cancer. Cancer 16(1):23. https://doi.org/10.3390/cancers16010023
15. Gustafson JA et al (2024) High-coverage nanopore sequencing of samples from the 1000 genomes project to build a comprehensive catalog of human genetic variation. Genome Res 34:2061–2073. https://doi.org/10.1101/gr.279273.124
16. Taliun D et al (2021) Sequencing of 53,831 diverse genomes from the NHLBI TOPMed program. Nature 590:290–299. https://doi.org/10.1038/s41586-021-03205-y
17. Bhattacharyya C et al (2025) Mapping genetic diversity with the GenomeIndia project. Nat Genet 57(4):767–773. https://doi.org/10.1038/s41588-025-02206-1
18. Naithani N, Sinha S, Misra P, Vasudevan B, Sahu R (2021) Precision medicine: concept and tools. Med J Armed Forces India 77(3):249–257. https://doi.org/10.1016/j.mjafi.2021.06.021
19. Stimson NF (2009) Personalized medicine: selected web resources. Dialogues Clin Neurosci 11(4):464–469. https://doi.org/10.31887/DCNS.2009.11.4/nfstimson
20. Taherdoost H, Ghofrani A (2024) AI's role in revolutionizing personalized medicine by reshaping pharmacogenomics and drug therapy. Intell Pharm 2:643–650. https://doi.org/10.1016/j.ipha.2024.08.005
21. Gomez-Lopez G, Dopazo J, Cigudosa JC, Valencia A, Al-Shahrour F (2017) Precision medicine needs pioneering clinical bioinformaticians. Brief Bioinform 20(3):752–766. https://doi.org/10.1093/bib/bbx144
22. Fernald GH, Capriotti E, Daneshjou R, Karczewski KJ, Altman RB (2011) Bioinformatics challenges for personalized medicine. Bioinformatics 27:1741–1748. https://doi.org/10.1093/bioinformatics/btr295
23. Tam V et al (2019) Benefits and limitations of genome-wide association studies. Nat Rev Genet 20(8):467–484. https://doi.org/10.1038/s41576-019-0127-1
24. Zhang W et al (2015) Comparison of RNA-seq and microarray-based models for clinical endpoint prediction. Genome Biol 16(1):133. https://doi.org/10.1186/s13059-015-0694-1
25. Su J, Yang L, Sun Z, Zhan X (2024) Personalized drug therapy: innovative concept guided with proteoformics. Mol Cell Proteomics 23(3):100737. https://doi.org/10.1016/j.mcpro.2024.100737
26. Mukherjee S et al (2022) Personalized structural biology reveals the molecular mechanisms underlying heterogeneous epileptic phenotypes caused by de novo KCNC2 variants. HGG Adv 3(4):100131. https://doi.org/10.1016/j.xhgg.2022.100131
27. Vasilev B, Atanasova M (2025) A (Comprehensive) review of the application of Quantitative Structure–Activity Relationship (QSAR) in the prediction of new compounds with anti-breast cancer activity. Appl Sci 15:1206. https://doi.org/10.3390/app15031206
28. Tropsha A, Isayev O, Varnek A, Schneider G, Cherkasov A (2024) Integrating QSAR modelling and deep learning in drug discovery: the emergence of deep QSAR. Nat Rev Drug Discov 23:141–155. https://doi.org/10.1038/s41573-023-00832-0
29. Bhattacharyya S, Ehsan SF, Karacosta LG (2023) Phenotypic maps for precision medicine: a promising systems biology tool for assessing therapy response and resistance at a personalized level. Front Netw Physiol 3:1256104. https://doi.org/10.3389/fnetp.2023.1256104
30. Hegde S, Andrian SM, Sajal H, Mallikarjunappa AKK, Benny A, Elayaperumal S (2024) Systems biology approaches for precision medicine. In: Systems biology approaches: prevention, diagnosis, and understanding mechanisms of complex diseases. Springer Nature Singapore, pp 271–297. https://doi.org/10.1007/978-981-99-9462-5_11
31. Chen R, Snyder M (2012) Systems biology: personalized medicine for the future? Curr Opin Pharmacol 12(5):623–628. https://doi.org/10.1016/j.coph.2012.07.011
32. Lightbody G et al (2019) Review of applications of high-throughput sequencing in personalized medicine: barriers and facilitators of future progress in research and clinical application. Brief Bioinform 20(5):1795–1811. https://doi.org/10.1093/bib/bby051

33. Cirillo D, Valencia A (2019) Big data analytics for personalized medicine. Curr Opin Biotechnol 58:161–167. https://doi.org/10.1016/j.copbio.2019.03.004
34. Alum EU, Ugwu OPC (2025) Artificial intelligence in personalized medicine: transforming diagnosis and treatment. Discov Appl Sci 7:193. https://doi.org/10.1007/s42452-025-06625-x
35. Kumar A, Dixit S, Srinivasan K, M D, Vincent PMDR (2024) Personalized cancer vaccine design using AI-powered technologies. Front Immunol 15:1357217. https://doi.org/10.3389/fimmu.2024.1357217
36. Molla G, Bitew M (2024) Revolutionizing personalized medicine: synergy with multi-omics data generation, main hurdles, and future perspectives. Biomedicine 12(12):2750. https://doi.org/10.3390/biomedicines12122750
37. Patel R, Klein P, Tiersten A, Sparano JA (2023) An emerging generation of endocrine therapies in breast cancer: a clinical perspective. NPJ Breast Cancer 9(1):20. https://doi.org/10.1038/s41523-023-00523-4
38. Meric-Bernstam F et al (2022) Zanidatamab, a novel bispecific antibody, for the treatment of locally advanced or metastatic HER2-expressing or HER2-amplified cancers: a phase 1, dose-escalation and expansion study. Lancet Oncol 23(12):1558–1570. https://doi.org/10.1016/S1470-2045(22)00621-0
39. Clifton GT, Peoples GE, Mittendorf EA (2016) The development and use of the E75 (HER2 369-377) peptide vaccine. Future Oncol 12:1321–1329. https://doi.org/10.2217/fon-2015-0054
40. Xiong X et al (2025) Breast cancer: pathogenesis and treatments. Signal Transduct Target Ther 10(1):49. https://doi.org/10.1038/s41392-024-02108-4
41. Wu DW et al (2024) Personalized neoantigen cancer vaccines: current progression, challenges and a bright future. Clin Exp Med 24(1):229. https://doi.org/10.1007/s10238-024-01436-7
42. Pandya A et al (2023) The future of cancer immunotherapy: DNA vaccines leading the way. Med Oncol 40(7):200. https://doi.org/10.1007/s12032-023-02060-3
43. Sahin U, Karikó K, Türeci Ö (2014) MRNA-based therapeutics-developing a new class of drugs. Nat Rev Drug Discov 13(10):759–780. https://doi.org/10.1038/nrd4278
44. Zhang L, Zhou X, Sha H, Xie L, Liu B (2022) Recent Progress on therapeutic vaccines for breast cancer. Front Oncol 12:905832. https://doi.org/10.3389/fonc.2022.905832
45. Li JY, Jiang RY, Wang J, Wang XJ (2025) Advances in mRNA vaccine therapy for breast cancer research. Discov Oncol 16(1):673. https://doi.org/10.1007/s12672-025-02542-y
46. Peng M et al (2019) Neoantigen vaccine: an emerging tumor immunotherapy. Mol Cancer 18(1):128. https://doi.org/10.1186/s12943-019-1055-6
47. Richters MM et al (2019) Best practices for bioinformatic characterization of neoantigens for clinical utility. Genome Med 11(1):56. https://doi.org/10.1186/s13073-019-0666-2
48. Huber F et al (2024) A comprehensive proteogenomic pipeline for neoantigen discovery to advance personalized cancer immunotherapy. Nat Biotechnol. https://doi.org/10.1038/s41587-024-02420-y
49. Ismail S, Barakat K (2025) Designing a multi-neoantigen vaccine for melanoma: integrating immunoinformatics and biophysics methods. Comput Biol Med 190:110081. https://doi.org/10.1016/j.compbiomed
50. Aggeletopoulou I, Pantzios S, Triantos C (2025) Personalized immunity: Neoantigen-based vaccines revolutionizing hepatocellular carcinoma treatment. Cancer 17(3):376. https://doi.org/10.3390/cancers17030376
51. Braun DA et al (2025) A neoantigen vaccine generates antitumour immunity in renal cell carcinoma. Nature 639(8054):474–482. https://doi.org/10.1038/s41586-024-08507-5
52. Katsikis PD, Ishii KJ, Schliehe C (2024) Challenges in developing personalized neoantigen cancer vaccines. Nat Rev Immunol 24(3):213–227. https://doi.org/10.1038/s41577-023-00937-y
53. Wells DK et al (2020) Key parameters of tumor epitope immunogenicity revealed through a consortium approach improve neoantigen prediction. Cell 183(3):818–834.e13. https://doi.org/10.1016/j.cell.2020.09.015

# Personalized Therapy in Non-Small-Cell Lung Cancer

Chandreyee Datta, Arpita Mukherjee, Sourav Paul, and Ashish Bhattacharjee

**Abstract**

The most advanced and effective approach for early prediction, prevention, and treatment of cancers is predictive, preventive, and personalized medicine, widely known as PPPM or 3P medicines. This is one of the most important and new directions for the treatment of cancer. Lung cancer is a principal cause of cancer mortality globally, of which non-small-cell lung cancer (NSCLC) accounts for more than 85% deaths. Variations in genetic mutations, histological differences, and heterogeneity in gene expression are also linked to lung cancer patients of the same subtype. These variations closely correlate with diverse therapeutic outcomes, often accompanied by resistance and discrepancies in drug response. Therefore, to prevail in the battle against lung cancer, it is essential to identify the precise biomarkers that will unequivocally indicate a patient's likelihood of responding to chemotherapy. For example, various molecular profiling techniques, including polymerase chain reaction (PCR)-based tests and next-generation sequencing (NGS), help oncologists to identify the key mutation and changes in various important genes such as *EGFR, ALK,* and *ROS1*, and can also help to make individualized treatment regime choices. On the plus side, liquid biopsy offers a minimally invasive method of analysing ctDNA for ongoing treatment response monitoring and resistance mutation detection, which further personalizes treatment. In this respect, it may enable health care professionals to formulate and execute a variety of personalized treatment approaches based on the molecular characteristics of a patient's tumour, which could result in better treatment outcomes with fewer or minor adverse effects compared to regular chemotherapy. So, for the management of NSCLC (i.e. those who are more vul-

C. Datta · A. Mukherjee · S. Paul · A. Bhattacharjee (✉)
Department of Biotechnology, National Institute of Technology, Durgapur, Burdwan, West Bengal, India
e-mail: abhattacharjee.bt@nitdgp.ac.in

 
M. K. Ghosh, M. Basu (eds.), *Personalized Cancer Therapy*,
https://doi.org/10.1007/978-981-95-6845-1_11

nerable to have significant genetic abnormalities, particularly the patients with adenocarcinoma or non-smokers), the incorporation of molecular diagnostics across clinical practice has grown into a very essential thing. Personalized approaches are given for individualized treatment strategies, considering not only the molecular profile of the tumour but also the patient's overall health and performance status. Personalized plans for the treatment of individual patients are best when developed by considering not only molecular features of the tumour, but also other implications that involves the patient's health and functional status. So, in this chapter we would like to discuss the molecular profiling of tumours and their associated different techniques that help us to properly diagnose the type and stage of NSCLC and help in developing targeted therapy and personalized medicine.

**Keywords**

NSCLC · Biomarkers · Precision medicine · Immunotherapy · Liquid biopsy · CTCs

## 1 Introduction

The idea of personalized medicine (PM) has evolved as a significant study domain to enhance conventional treatment methods. It aids in identifying disease patterns based on an individual's medical history, genetics, and lifestyle, and this allows us to better understand disease patterns to identify the most suitable treatment for each patient [1]. This paradigm change is expected to make breakthroughs in disease prediction, followed by preventative measures such as pharmaceuticals, diagnostic tests, and treatments, including the administration of targeted therapies for specific genetic profiles. Diagnosis and appropriate treatment of pathological disorders is laborious because physicians need to consider the patient's medical history, disease stage, and population susceptibility. Conventional techniques established by physicians may not guarantee a cure for the patient and can be detrimental to apply in certain conditions [1]. Personalized medicine (PM) offers markedly improved treatment through its innovative method, which involves comprehending the patient's genetics and forecasting the most effective treatment options for various ailments.

Lung cancer represents one of the most common forms of malignancies worldwide and represents a significant global health concern. According to the American Cancer Society report in 2024, there were approximately 1.7 million individuals who succumbed to lung cancer [2]. Based on the histopathological classification, around 80–85% of cases have been identified as non-small-cell lung cancer (NSCLC) globally. The three common subtypes of NSCLC are lung adenocarcinoma (LUAD), large cell carcinoma (LCC), and lung squamous cell carcinoma (LUSC). Among these three variants, LUAD is the most prevalent, accounting for around 40–45% of all NSCLC, distinguished by larger cells and a glandular appearance [3]. NSCLC is associated with numerous genetic abnormalities, some of which are targetable mutations that represent prospective therapeutic targets [4]. Identification of the NSCLC's histological and molecular subtypes is crucial for finding the appropriate treatment options, formulating personalized therapy, and enhancing clinical prognosis [5]. Consequently, the existing understanding of the NSCLC landscape and its genetic alterations aids in the implementation of tailored and precision medicine applications [5, 6]. The classification of NSCLC based on histological characteristics and/or molecular abnormalities is critical for selecting relevant medicines, devising personalized treatment regimens, and ultimately improving patient outcomes [5]. Consequently, knowledge of the molecular characteristics and genetic alterations in NSCLC has opened the door for the implementation of precision medicine, enabling treatments to be tailored to the specific profiles of individual patients.

Determining the appropriate therapy for each lung tumour relies on various parameters, including histological subtype and molecular features [7]. Following the identification of the histological subtype, NSCLC cases undergo molecular testing utilizing a variety of techniques in order to detect genetic changes that can be addressed by currently available therapies (Fig. 1). At present, all currently approved NSCLC treatments are delineated in the standard guidelines provided by the National Comprehensive Cancer Network (NCCN), the American Society of Clinical Oncology (ASCO), and the European Society for Medical Oncology (ESMO) [8].

The tumour heterogeneity often complicates therapy selection, as diverse molecular profiles greatly affect treatment options. Patients with the same histological classification have unique molecular traits of each tumour subtype, resulting in varying responses to treatment. In spite of the difficulties, the molecular variety of NSCLC offers a promising approach customized for every individual and enhancing the treatment effectiveness [7, 8]. Consequently, in light of the challenges linked to NSCLC therapy and ongoing advancements, we will examine the existing therapeutic options and highlight the issues concerning drug resistance in this book chapter. Recent insights and research in this domain will be examined, emphasizing

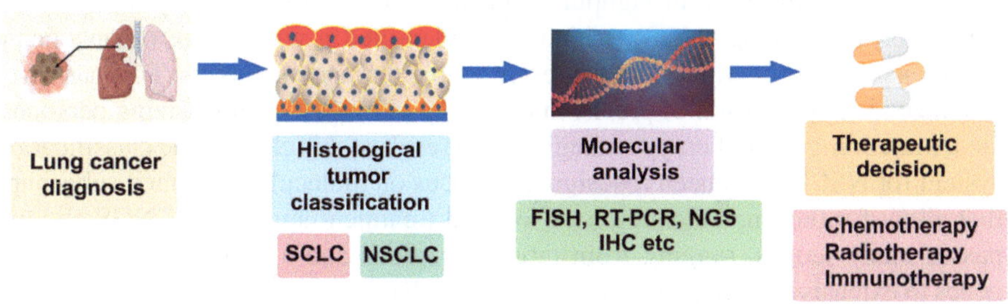


**Fig. 1** Steps involved for a proper therapeutic decision of lung cancer. The figure describes different steps, like histological analysis and different molecular analyses like FISH, RT-PCR, NGS etc., for identifying the appropriate therapy (Chemotherapy/radiotherapy/immunotherapy) required for lung cancer treatment

prospective novel pharmaceuticals and underscoring the significance of personalized medicine in NSCLC treatment.

## 2 Molecular Profiles and Biomarkers

Recently, there has been a surge in prognostic biomarkers and corresponding targeted medicines that modify treatment trajectories, resulting in more substantial changes in outcomes for NSCLC compared to most other malignancies [9]. In accordance with the latest guidelines from the NCCN, individuals diagnosed with NSCLC are recommended to undergo comprehensive molecular testing to identify various genetic alterations. They include rearrangement during transfection (RET) mutations, *ALK* gene rearrangements, *ROS1* fusions, *EGFR* & *KRAS* mutations—more especially, the G12C variant. Other suggested tests are the V600E mutation in the *BRAF* gene, HER2(ERBB2) activating mutations, *NTRK* gene fusions, MET exon 14 skipping mutations, and evaluation of PD-L1 expression levels [10]. Within the EMSO Scale of Clinical Actionability for molecular target (ESCAT) paradigm, EGFR mutations and *ALK* gene fusions are classified as Level I-A, highlighting their proven function as clinically actionable indicators that have a direct impact on therapy decisions [11].

The presence of epidermal growth factor receptor mutation (EGFR) was first identified in non-smoker Asian descent with lung adenocarcinoma (LUAD), and this mutation was responsive to EGFR TKIs [12]. Furthermore, the epidemiology of lung cancer is being gradually shaped by changing smoking behaviour trends, including an increase in the incidence among younger female populations and non-smokers [13]. Current guidelines follow the recommended checking for certain genetic changes, including EGFR mutations, ALK and ROS1 rearrangements, and BRAF V600E mutations in individuals with newly diagnosed advanced lung adenocarcinoma (LUAD). Beyond this, however, the approved targeted therapies have been incorporated, including KRAS G12C variants, alterations within exon 14 of MET, as well as rearrangements in NTRK and RET [14].

The molecular understanding of NSCLC has been completely transformed with the emergence of high-throughput sequencing technologies, particularly next-generation sequencing (NGS), which have identified a wide range of oncogenic mutations, several of which now represent viable therapeutic targets. This has revolutionized the application of precision oncology in NSCLC, emphasizing personalized, mutation-guided therapy strategies. As a result of this change, the classification system for NSCLC has been redesigned to focus on molecularly defined subgroups that allow for the implementation of focused treatment approaches. NGS encompasses many scopes, ranging from targeted regions and a focussed panel of key genes to extensive coverage of numerous genes that drive cancer growth, facilitating comprehensive genomic profiling (CGP). This method is becoming more and more utilized as a crucial part of modern clinical management since it provides direct molecular insights into tumour behaviour. Apart from that, it identifies all known indicators for NSCLC, as well as additional abnormalities in the tumour that may be treated. Additionally, it offers insightful information on the microsatellite status and tumour mutational burden, which can aid in predicting how well a patient will respond to precision therapeutics [8]. CGP performed on tumour tissue samples in NSCLC is frequently constrained by insufficient biopsy samples, decreasing its therapeutic significance. The growing number of biomarkers poses a challenge for single-biomarker testing. As a result, liquid CGP of plasma-derived circulating free DNA has found comparable accuracy to tissue analysis and may serve as an optimal diagnostic tool for certain patients [15, 16]. Nevertheless, a negative liquid biopsy result does not exclude the presence of actionable mutations within the tumour [17].

## 2.1 Proteomic Biomarkers

Mass spectrometry (MS), combined with powerful statistical software, allows for measuring a wide variety of peptides and proteins with high precision and sensitivity. Over the last decade, mass spectrometry-based proteomics has made significant progress and is increasingly merging with other techniques to improve reproducibility and accuracy [18]. In addition to label-free approaches, isotope-coded labelling techniques have emerged as a widely utilized technique for proteomics research. Recent advances in proteomic research have identified ultra-stable isotope labelling techniques (SILAC), which offer a more effective approach for reliably quantifying protein expression in human tissue samples [19].

A quantitative proteomic approach utilizing SILAC and combined with immunoprecipitation has been employed to examine differences in Karyopherin alpha 2(KPNA2)-interacting protein complexes among lung cancer cell lines with different invasive potentials. The abundance of complexes formed between KPNA2-Vimentin-phosphorylated ERK (pERK) was considerably increased in advanced-phase lung cancer tissues compared to early-stage tissues. The overexpression of KPNA2 correlates with cancer invasiveness and cell migration; therefore, evaluating KPNA2 may function as a possible marker of prognosis for LUAD [20]. Notably, elevated serum levels of molecules regulating inflammation - Serum Amyloid A (SAA) and C-reactive protein (CRP) have been identified as predictive markers for lung cancer [21, 22]. Through the application of differential proteomics, complexes of CRP–SAA (C-reactive protein associated with serum amyloid A) have been observed in the serum of individuals diagnosed with lung cancer, indicating their possible function as biomarkers associated with the disease. Significantly higher levels of CRP and SAA were found in individuals with lower survival rates than in healthy controls, suggesting a possible correlation between these parameters and the prognosis of the disease. Moreover, both univariate and multivariate studies based on Cox regression indicated that CRP–SAA may serve as an independent predictive marker for patients' diagnosis [23]. Techniques such as shotgun Mass spectrometry-based, together with two-dimensional imaging, were applied to characterize the variations in phospholipid profile in NSCLC and corresponding non-malignant lung tissue samples. The most significant alterations were a reduction in sphingomyelins and an elevation of phosphatidylinositol. These lipid indicators may signify the existence and subtype of NSCLC, presenting promise for the development of diagnostic tools for NSCLC [24]. Utilizing various methods, MALDI-TOF, and reverse-phase protein assay analysis (RPPA) involving lung cancer cohort versus normal cohort, Calnexin has been identified as a novel serodiagnostic biomarker, which has the potential for the early identification of lung cancer as well as in the clinical management of the disease. The expression level of Calnexin was markedly increased in patients with lung cancer and detected in the early (stage I) phase of the disease. The concentration of this protein in blood may be utilized to identify early lung cancer [25]. Comprehensive label-free quantitative proteomic analysis of pleural effusion samples has demonstrated significantly elevated protein levels of Mesenchymal epithelial transition protein (MET), together with dipeptidyl peptidase-4 (DPP4), and protein tyrosine phosphatase receptor type F (PTPRF) in NSCLC compared to benign disease. This technique allows for the development of a new proteomic library of differential pleural effusion, enabling the biomarker identification that distinguishes NSCLC from benign lung cancers [26].

## 2.2 Metabolomic Biomarkers

Metabolomics is a multidisciplinary domain that has arisen in the post-genomic approach. Consequently, employing metabolomics techniques to investigate associated biomarkers holds significant clinical relevance for lung cancer detection. Numerous prospective biomarkers for NSCLC in plasma or serum have been established [27, 28]. The metabolomics approach not only helps to identify novel biomarkers but also gives a novel insight into the pathogenic mechanisms of NSCLC. For example, Qi et al. demonstrated by examining all plasma metabolites using non-targeted metabolomics to identify biomarkers associated with lung adenocarcinoma and assess prognosis depending on tumour subtypes [29]. Metabolomic analyses were performed using advanced ultra-performance liquid chromatography methods, and multivariate statistical approaches and machine learning algorithms were applied to analyse and validate the biomarkers associated with different kinds and stages of lung cancer. Five metabolic products—palmitic acid, heptadecanoic acid, 4-oxoproline, tridecanoic acid, and ornithine—were identified as the most potential biomarkers in early-stage lung cancer. This outcome supports the adaptation of the diagnostic method into a rapid, point-of-care system for both diagnosis and prognosis of lung cancer [29].

Zheng et al. carried out a study that included 69 NSCLC patients and 76 healthy individuals for an initial untargeted analysis in a discovery cohort to quantify alterations, followed by the recruitment of 26 NSCLC patients and 30 HCs for the validation cohort [28]. The researchers utilized various cutting-edge techniques, allowing for the accurate detection of complex metabolites. To evaluate the accuracy of a diagnostic panel for detecting NSCLC, this study identified two metabolites that were significantly dysregulated: arachidonic acid and 13(S)-HODE. By integrating metabolomic and transcriptomic analyses, two prospective biomarkers associated with Akt pathways were found, potentially offering a novel perspective for investigating the pathophysiology of NSCLC [28]. Interestingly, this observation was in harmony with our recent study, which demonstrated that IL-13 and the hydroperoxy fatty acid 13-(S)-HpODE control A549 non-small-cell lung carcinoma cell aggressiveness by inducing apoptosis via MAO-A/ROS/p53/p21 signalling cascade, indicating a potential therapeutic role of 13-(S)-HpODE in NSCLC [30]. In another study, Kowalczyk et al. carried out a very interesting observation, which, in addition to previously established data, revealed additional metabolites that differentiate NSCLC subtypes, particularly in the early stages of cancer. Tissue and/or plasma specimens from 137 NSCLC patient samples underwent comprehensive metabolomics studies utilizing the liquid chromatography–mass spectrometry (LC–MS) technology. Analysis of early-stage tissue discovered over 200 metabolites that differentiate adenocarcinoma (ADC) and squamous cell lung carcinoma (SCC) subtypes from normal tissue. Most predominantly observed metabolic intermediates comprised amino acids, fatty acids, carnitine derivatives, and related compounds. Furthermore, it has been found that some metabolites that are involved in the biosynthesis of N-acyl ethanolamine (NAE), specifically glycerophospho (N-acyl) ethanolamine (GP-NAE), can effectively differentiate between adenocarcinoma (ADC) and early-stage squamous cell carcinoma (SCC). Conversely, the plasma analysis in chronic obstructive pulmonary disease (COPD) and NSCLC cohorts facilitated the removal of inflammation-associated metabolic products and enabled the detection of compounds (lysoglycerophospholipids, glycerophospholipids, and sphingomyelins) that represent distinct characteristics of cancer [31].

## 2.3 Immune System-Related Biomarkers

Tumour-induced immunosuppression facilitates the evasion of immune surveillance by tumours, while immunotherapy can stimulate the immune system and elicit a prolonged antitumour immune response. Anti-tumour immunity depends on both innate and adaptive immunity. Adaptive immunity begins with the interplay of dendritic cells and T-cells during the priming stage, followed by the effector stage, in which T-cells directly attack and eliminate cancer cells [32]. Both stimulatory signals and inhibitory signals are involved during these two stages. During the priming phase, T-cells exhibit cytotoxic T lymphocyte antigen 4, which plays a role in modulating immune activation by transmitting inhibitory signals. In the case of the effector phase, inhibitory signals are predominantly mediated by PD-1, expressed by T lymphocytes, and its ligand PD-L1, commonly expressed on malignant cells. Lung cancer cells can evade anti-tumour immunity through various ways, including the activation of inhibitory signals [32, 33]. A comprehensive understanding of antitumour immunobiology and the strategies employed by tumour cells to evade the immune surveillance has resulted in the advancement of immunotherapeutic agent like PD-L1.

Important immune regulatory factors, including PD-L1, PD-1, PD-L2, and T-cell immunoglobulin and mucin-domain-containing-3 (TIM-3), have been identified as targets for immune checkpoint inhibitors (ICIs). In this book chapter, to explore the predictive importance of immune-related variables in resected NSCLC, we integrated them. Patients exhibiting PD-L1 overexpression demonstrated markedly reduced survival rates in stage I–III resectable LUAD, particularly in cases of acinar/papillary-dominant LUAD [32]. Patients with the initial stage of LUAD exhibiting positive PD-L1 expression were associated with unfavourable disease-free survival (DFS) and overall survival (OS) at the 1% cut-off threshold. In patients with stage I–III LUSC ($n = 205$), PD-L1 served as a significant marker for deteriorating overall survival at the 1% cut-off level. In a similar manner, PD-L1 at 1% was associated with decreased DFS and OS in stage I–IIIA EGFR-mutated NSCLC [33]. In the context of resected NSCLC, the prognostic implications of CD8 (cut-off values: 543 per $mm^2$) and PD-L1 (cut-off values: 1%) indicate that CD8 positivity combined with PD-L1 negativity is associated with improved recurrence-free survival (RFS) and OS [32]. Handa et al. confirmed that NSCLC patients (Stage I) exhibiting elevated PD-L1 levels (>5%) experienced a poorer prognosis compared to those with lower PD-L1 levels. Intratumoural CD8 level affected the prognostic and predictive value of PD-L1 in relation to surrounding immune cells. The study found that PD-L1 did not predict survival in high-level cohorts of intratumoural CD8. However, PD-L1 overexpression was associated with significantly poorer RFS in cohorts with low-level CD8, with a five-year RFS of 63.4% for PD-L1 (>5%) compared to 81% for PD-L1 (≤5%), $p = 0.034$ [33]. So, altogether, it can be concluded that elevated PD-L1 levels are associated with reduced survival rates in resected NSCLC when compared to lower levels.

In partial-onset of LUAD patients, PD-L2 aberrant expression (>1%) significantly correlated with deteriorating DFS and OS [32]. TIM-3 positivity in tumour cells (24%), as well as tumour-infiltrating lymphocytes, was associated with poor RFS and OS in early-stage LUAD (TILs, 11%). The TIM-3/PD-1/CD8 expression levels for LUAD prognosis were investigated in subgroup analysis. A subset with TIM-3 positivity/PD-1 positivity/CD8 negativity had the poorest prognosis, whereas a subgroup with TIM-3 negativity/PD-1 negativity/CD8 positivity had the best outlook [32]. Tumour mutation burden (TMB) seemed to be a negative predictive factor for OS in early-stage nonsmoking patients [32]. In contrast, TMB was not shown to have any predictive relevance in stage I–II resected LUAD, according to another research [34]. In stage I NSCLC, elevated OX40 (CD134) expression on TILs was linked to poor RFS and OS [35]. In stage I–III NSCLC, on the other hand, OX40, which is found on tumour-immune infiltrating cell membranes, was associated with a good OS [35].

## 3 Blood-Based Biomarkers and Their Role in Prognosis of NSCLC

Although the most preferred therapy for early-phase NSCLC is extensive resection, the incidence of recurrence for the lung carcinoma remains significantly elevated, suggesting the presence of micro-metastases throughout the preoperative phase. So, consideration of other biomarkers in blood, including circulating tumour cells

(CTCs), standard blood tests, circulating DNA/RNA and tumour-specific markers, is very crucial to investigate their predictive significance in early-stage NSCLC.

CTCs are generally discharged from the primary neoplastic lesion and circulate in the bloodstream, which might lead to tumour recurrence or distant metastasis [36]. Numerous prospective studies have demonstrated that preoperative peripheral venous circulating tumour cells (CTCs) correlate with poor DFS and OS in NSCLC [34]. Regarding peripheral folate receptor-positive CTCs in NSCLC, elevated preoperative levels were associated with reduced RFS and OS [37]. A comprehensive analysis by Wankhede et al. revealed that both baseline and postoperative circulating tumour cells (CTCs) correlated with poorer DFS and OS in early-stage NSCLC [38]. Hence, it is quite evident from those reports that circulating tumour cells in preoperative and postoperative peripheral blood may have a role in cancer spread and relapse, ultimately resulting in an unfavourable prognosis.

Apart from CTCs, during hospitalization, different routine blood tests are often performed on cancer patients, and the results could be helpful to indicate the probable prognosis and OS of NSCLC patients. Among different blood tests, level of Haemoglobin (Hb), Platelets (PLT), White Blood Cell count (WBC) (specifically neutrophil and lymphocyte count), CRP (C-reactive protein), Fibrinogen, D dimer, Albumin, Osteopontin (OPN) are reported to be crucial for determination of RFS and OS of NSCLC patients in both preoperative and post operative stages [34].

## 3.1 Circulating Nucleic Acids

Circulating nucleic acids (CNAs) are a wide range of DNA and RNA molecules that are usually released into the blood as a consequence of apoptosis (programmed cell death) and necrosis (uncontrolled cell death) [39]. Multiple investigations have validated that circulating DNA and RNA may serve as prognostic indicators in NSCLC [39].

### 3.1.1 Circulating Tumour DNA (ctDNA)

Through apoptosis, necrosis, and secretion, the cancerous cells release ctDNA, circular RNA (circRNA), microRNA (miRNA), and exosomes. In the circulatory system, the circulating tumour DNA (ctDNA) can exist alone or in combination with extracellular vesicles, such as exosomes [39]. CtDNA has a half-life that ranges from 16 to 150 min. It appears that both the liver and kidneys are the primary organs in charge of removing ctDNA. It is anticipated that ctDNA would have exactly the same biological traits as the tumour cells from where it starts, such as mutation and methylation, as a component of cell-free tumour-specific DNA fragments in the circulation [40]. From sample extraction to sequencing, a number of variables might have a substantial impact on the outcomes.

Plasma cell-free DNA (cfDNA) fragments, on the other hand, originate primarily from tissues undergoing programmed cell death or necrosis, where they are released into the bloodstream via enzymatic chromatin degradation. However, cfDNA levels in 50% of cancer patients were observed to fall into the normal limits (0–50 ng/ml), which further suggests difficulties with the study's credibility. Subsequently, ctDNA, or tumour-derived cfDNA, was identified to harbour numerous mutations associated with tumorigenesis, mutations frequently observed in genes such as *TP53, KRAS,* and *EGFR* recognized as hallmarks of lung cancer [10, 14, 39, 40]. Circulating tumour DNA (ctDNA) is becoming recognized as a reliable, independent biomarker that enhances diagnostic accuracy and offers clinical direction and prognostic information. The Food and Drug Administration (FDA) subsequently approved the first circulating tumour DNA (ctDNA)-based plasma assay for the detection of EGFR mutations, representing an important milestone in clinical practice [41].

Attempts were made to understand the prognostic significance of ctDNA in early-stage NSCLC. A multicentre, randomized phase II trial (DYNAMIC) was carried out, and results indicated that identifiable ctDNA at 3 days along with one month following surgery was correlated with reduced RFS and OS in comparison to undetected ctDNA [42]. Plasma ctDNA detected after surgery was linked to poorer predictions for progression-free survival (PFS) and OS [40]. Another study classified stage II–III NSCLC patients based on their receipt of adjuvant chemotherapy (ACT) and the presence of detectable postoperative ctDNA. The findings indicated that patients exhibiting significant amounts of ctDNA experienced substantially lowered RFS in comparison to those with minimal ctDNA levels, regardless of ACT administration [40, 41]. A multicentre cohort study, LUNGCA-1, confirmed that elevated levels of ctDNA prior to surgery are predictive of unfavourable RFS. Using subsequent procedures and ctDNA-based molecular residual disease (MRD) identification, the scientists came to the conclusion that MRD suggested the reappearance of the tumour [43]. Additionally, multiple studies indicated that stage I–III NSCLC patients exhibiting positive ctDNA both presurgically and postsurgically experienced reduced survival compared to those with negative ctDNA [39–41]. In early-stage NSCLC, Detectable ctDNA during the perioperative phase has emerged as a significant predictive marker for poor survival outcomes.

Apart from ctDNA and cfDNA, some recent reports demonstrated that circulating miRNAs have some prognostic values, and they often serve as non-invasive biomarkers. A study by Hu et al. established a microRNA profile comprising four serum miRNAs, including miR-486, miR-30d, miR-1, and miR-499, demonstrating that elevated miR-486 and miR-30d levels, along with diminished miR-1 and miR-499 levels, were substantially associated with poor OS in early to advanced (stage I-IIIA) NSCLC [44]. Another study was carried out to investigate the prognostic implication of matched plasma and serum miRNAs in stage I–II NSCLC and only the elevated plasma level of miR-let-7b was shown to be associated with enhanced cancer-specific survival, while no serum miRNAs correlated with patient prognosis [32]. Details regarding liquid biopsy have been depicted in Fig. 2.

### 3.1.2 Tumour Markers

There are several interesting reports that have identified several interesting tumour markers that may be very helpful for the prediction of overall survival and prognosis of NSCLC patients. Notably, carcinoembryonic antigen (CEA), neuron-specific

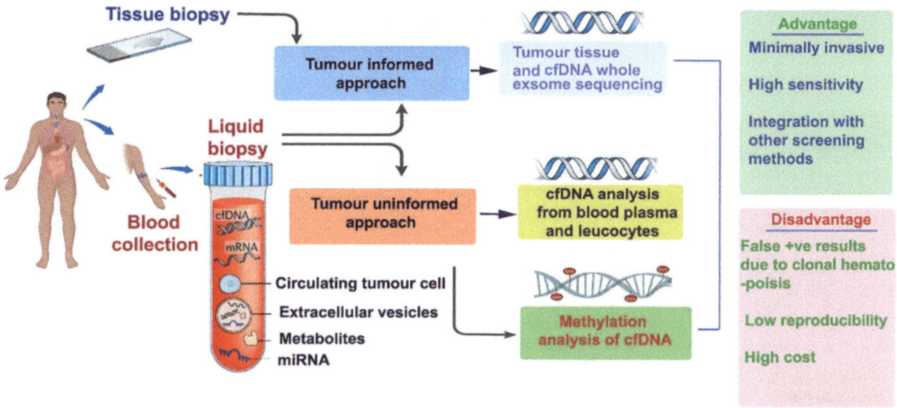


**Fig. 2** Liquid biopsy. This figure depicts about detailed method as well as advantages and disadvantages for liquid biopsy

enolase (NSE), and CYFRA 21-1 were reported to contribute significantly to survival and prognosis of NSCLC patients. For example, a study reported that an elevated level of CEA at preoperative stages of grade I-II NSCLC patients showed abrogated five-year survival rates compared to patients with a CEA level in a normal range. So, naturally, the preoperative level of CEA could serve as a very effective predictor for the prognosis of NSCLC patients [32, 33]. Conversely, growing evidence suggests that heightened preoperative blood CYFRA21-1 levels serve as a distinct prognostic indicator linked to adverse outcomes in individuals with preliminary-stage NSCLC [33, 44]. Specifically, the research conducted by Muley et al. integrated preoperative CYFRA21-1 (cut-off value: 3.3 ng/mL) and CEA (cut-off value: 9.8 ng/mL) to establish a novel metric termed tumour marker index (TMI), revealing that an elevated TMI (>0.54) adversely impacted overall survival in patients with stage I–II NSCLC [45].

## 4 Precision Oncology and Therapeutic Strategy for NSCLC

The therapeutic strategy for NSCLC is intricate and necessitates more testing to choose the best appropriate treatment. In stage I NSCLC, surgery is the primary treatment option; afterward, a subsequent chemotherapy or radiation treatment is employed in order to minimize the chance of the disease's reappearance. In case of Stage II and III NSCLC, surgical intervention may be complemented with chemotherapy, radiation, or immunotherapy. When carcinoma has progressed to phase IV, it has spread across the organs, and surgical resection is not feasible; it is essential to identify specific mutations to implement targeted therapy. In the absence of actionable oncogenic drivers, it is recommended that patients receive treatment with platinum doublets [8, 46]. These treatment strategies have been depicted in Fig. 3.

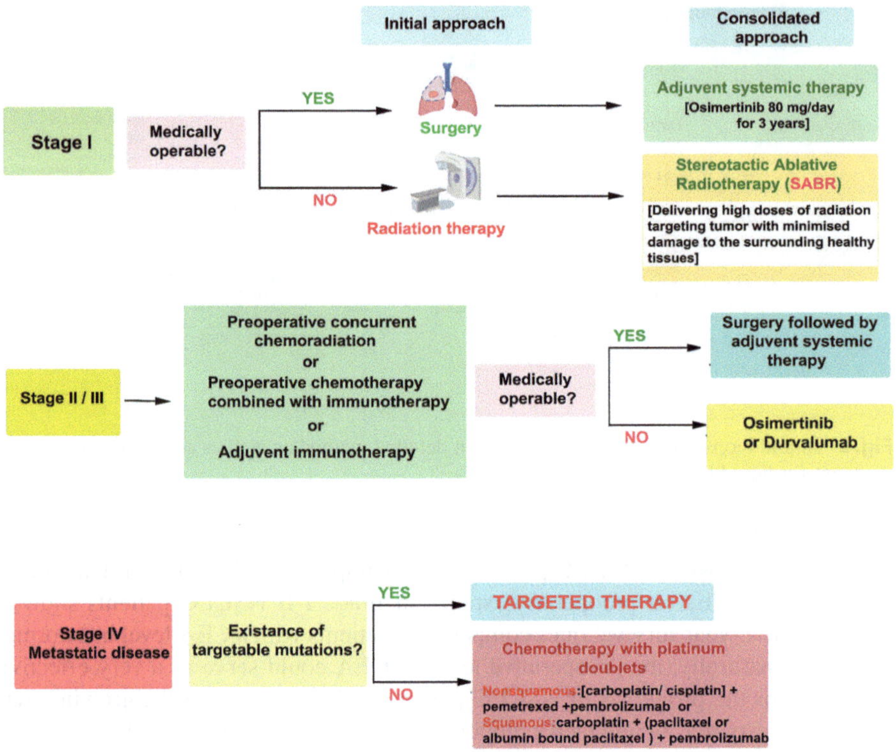


**Fig. 3** Schematic diagram of various therapeutic approaches for different stages of NSCLC. This figure indicates different initial recommended as well as consolidated treatment approaches for different stages of NSCLC

Recent developments in the diagnosis and treatment of NSCLC have established the framework for precision oncology. Significant laboratory-based efforts and extensive epidemiologic studies have identified numerous critical genetic alterations that lead to cellular growth, division, and oncogenesis. Patients require close monitoring to assess disease progression and modify the therapeutic strategy in response to loss of efficacy and the emergence of resistance resulting from molecular changes. The identification of genetic alterations employs laboratory techniques, specifically next-generation sequencing (NGS). Additionally, various techniques can be employed to identify specific alterations. For instance, EGFR, BRAF, and KRAS mutations can be identified through quantitative PCR, alongside IHC or FISH are employed for the identification of ALK and ROS rearrangements as well as amplification of the *MET* gene [8]. Genetic screening typically necessitates a new biopsy; however, a liquid biopsy presents an effective substitute in cases when this process is not possible. A positive result in liquid biopsy for certain mutations, such as T790M in the case of EGFR genetic mutation, can eliminate the necessity for a re-biopsy [8, 40].

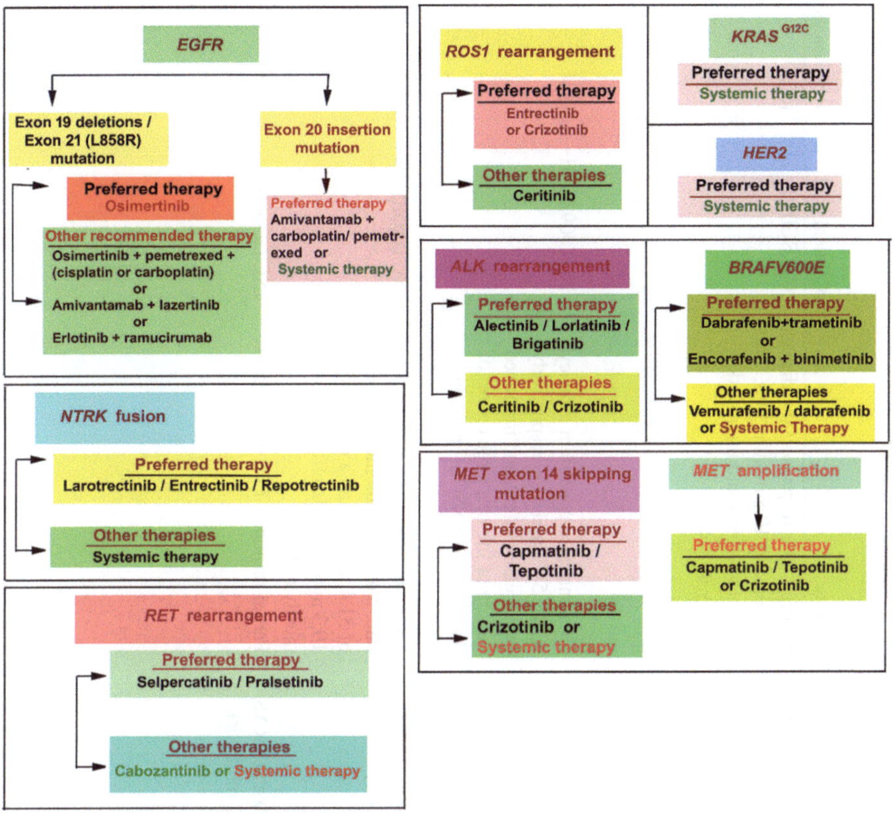


**Fig. 4** Recommended preferred, and alternative targeted therapy for each biomarker in NSCLC. This figure describes the FDA-approved drugs used for the targeted therapy of lung cancer

The latest NCCN guidelines recommend systemic therapy as an alternative to targeted drugs in cases of tumour progression. Given the advanced stage of NSCLC and the recommended therapeutic approach, Fig. 4 illustrates the proposed treatment strategy according to each genetic alteration. Additionally, all the available clinical trial data for the genetic mutation and its outcome have been described in Table 1.

Immunotherapy has been extensively explored, but only a small number of patients benefit from treatment, owing to therapeutic resistance resulting from the complexity and variety of the tumour microenvironment. As a result, novel methodologies to enhance this therapy are under investigation. An instance is tumour-derived exosomes (TEXs), which are capable of transporting a range of molecular cargos and have the ability to influence immunological responses, demonstrating both immunosuppressive and immunostimulatory effects in tumours like lung cancer [47]. TEXs may be able to transport tumour antigens to dendritic cells (DCs), making them especially suitable for use in LC immunotherapy, acting as transporters of self-antigens for driving immune activation and improving anti-tumour

**Table 1** List of clinical trials performed till date and their outcome based on the genetic mutations in NSCLC

| SL No | Mutation | Patient characteristics | Drug of choice | Results | Clinical trial | References |
|---|---|---|---|---|---|---|
| 1. | EGFR Mutation (Exon 19 Deletion) | Twenty patients (≥70 years), all with adenocarcinoma (NSCLC), stage IIIB, IV, or recurrent [eight individuals had an L858R point mutation, while twelve of them had an exon 19 deletion.] | Gefitinib (EGFR inhibitor) | Showed 70% response, 90% disease control, and 10.8 months PFS, with mild side effects and improved quality of life. | CJLSG0901 | [53] |
| 2. | ALK rearrangement | 1069 patients with the majority (66%) non-smoking Asian (46%), all had ALK-positive NSCLC with locally advanced or metastatic disease | Crizotinib (ALK inhibitor) | Demonstrated consistent efficacy (ORR 54% and 41% respectively), manageable safety, and improved quality of life in a large cohort of ALK-positive NSCLC patients | NCT00932451 | [54] |
| 3. | ROS1 fusion | 53 elderly male and female patients with ROS1-rearranged advanced NSCLC and 87% had received prior treatments | Crizotinib (ROS1 inhibitor) | The study reported a 72% response rate, having a combined survival rate of 51.4 months and a median PFS of 19.3 months. Side effects were mostly mild | NCT00585195 | [55] |
| 4. | KRAS mutation (G12C) | Male patient, heavy smoker, stage IV NSCLC | Sotorasib (KRAS G12C inhibitor) | Partial response, PFS of 6 months | NCT03600883 | [56] |

| | | | | | | |
|---|---|---|---|---|---|---|
| 5. | MET amplification | 84 individuals without EGFR or ALK alterations who had more sophisticated, MET-amplified NSCLC (stage IIIB or IV NSCLC) | Capmatinib (MET inhibitor) | Demonstrated clinical activity with higher response rates in treatment-naïve patients (cohort 5a), which includes manageable and consistent safety profile | NCT02414139 | [57] |
| 6. | BRAF V600E mutation | The study included 93 adult patients with metastatic BRAF V600E-mutant NSCLC, regardless of smoking status or histological subtype | Dabrafenib + Trametinib (BRAF + MEK inhibitors) | Independent of prior treatment history, shown significant and long-lasting therapeutic improvement with a reasonable safety profile | NCT01336634 | [58] |
| 7. | PD-L1 expression (immunotherapy) | 305 advanced (stage IV) NSCLC treatment-naïve patients with no EGFR mutations or ALK rearrangements | Pembrolizumab (PD-1 inhibitor) | Significantly increased overall survival relative to chemotherapy in individuals with elevated PD-L1 expression and treatment-naive advanced NSCLC, and was also well tolerated | NCT02142738 | [59] |
| 8. | NTRK fusion | Twenty heavily pretreated adults (primarily of younger age) with advanced-stage lung cancers, including some with brain metastases | Larotrectinib (TRK inhibitor) | Showed significant and long-lasting response rates, CNS activity, extended life, and acceptable tolerability | NCT02576431 and NCT02122913 | [60] |
| 9. | EGFR mutation (T790M) | Four hundred and nineteen patients with EGFR T790M resistance mutation and advanced NSCLC (approx. 34% had baseline CNS metastases) | Osimertinib (EGFR inhibitor) | When compared to platinum-pemetrexed chemotherapy, osimertinib was more effective (longer PFS, better ORR) and more tolerable | NCT02151981 | [61] |

(continued)

**Table 1** (continued)

| SL No | Mutation | Patient characteristics | Drug of choice | Results | Clinical trial | References |
|---|---|---|---|---|---|---|
| 10. | HER2 mutation | Forty-five advanced NSCLC patients with HER2 mutations patients with majority female (72%), including smokers (35%) and brain metastases (30%) | Trastuzumab + Pertuzumab + docetaxel | Demonstrated promising effectiveness and controllable toxicity. [ORR: 29% and PFS: 6.8 months (95% CI: 4.0 to 8.5)] | NCT03845270 | [62] |
| 11. | EGFR mutation (L858R) | Three hundred and forty five patients with IIIB/IV lung stage adenocarcinoma and EGFR mutation (exon 19 deletions, L858R point mutations) | Afatinib (second-generation EGFR inhibitor) | Afatinib showed superior efficacy over standard chemotherapy (PFS in Afatinib: 13.6 months and chemotherapy: 6.9 months) | NCT00949650 | [63] |
| 12. | MET Exon 14 Skipping Mutations | MET exon 14 skipping mutation in 364 individuals with progressed or aggressive NSCLC | Capmatinib (MET inhibitor) | Demonstrated remarkable response rates in MET exon 14-mutated NSCLC (68% in naïve and 41% in pretreated) with minor adverse effects like nausea and oedema | NCT02414139 | [64] |
| 13. | RET fusion | Two hundred and forty seven RET fusion-positive NSCLC patients, most of whom were older, who had received prior therapy, and 69 patients who had never received treatment. Of these, 26 patients had CNS metastases at baseline, and the median follow-up period was between 38 and 43 months | Selpercatinib (RET inhibitor) | Showed extended PFS (22–26.2 months), good response rates (62–83%), significant CNS activity (85% response), and controllable safety | NCT03157128 | [65] |

efficacy [47]. The application of TEXs for generating DC vaccines offers a potential therapeutic strategy that is presently being explored.

## 5 Drug Repurposing in NSCLC

NSCLC frequently exhibits therapy resistance due to its diverse nature and the flexibility of cancer stem cells. By repurposing established drugs provides a more expedient and cost-effective method for developing novel antineoplastic applications for existing medications, aimed at addressing therapeutic resistance [48]. Different types of NSAIDs, such as celecoxib, indomethacin, anti-hypertensive drugs, such as telmisartan, different beta blockers, anti-hyperlipidaemic drugs, majorly statins like atorvastatin, lovastatin, anti-diabetic drugs such as metformin (biguanide) and thiazolidinediones (TZDs) etc., are often reported to play a crucial role in managing cancer aggressiveness either alone or in combination with any other conventional chemotherapeutic drugs, both at in vitro as well as in vivo level [48]. For example, our recent publication has demonstrated that metformin, a biguanide antidiabetic drug, reduces A549 lung cancer cell aggressiveness by activation of AMPK and AMPK/p38MAPK/ROS-mediated signalling cascade. It was also demonstrated that SRT1720, as a non-biguanide antidiabetic small molecule, has been shown to directly activate AMPK alone or in combination with Metformin in LKB1 mutant A549 cells, regulating aggressiveness in cancer cells [49].

Utilizing the aforementioned numerous repurposing strategies, researchers successfully identified possible candidates for drug repurposing based on their mode of action. Nevertheless, only a small fraction of these medications progressed to the clinical trial phase. Glucocorticoids constitute the predominant type of pharmaceuticals in the ongoing trials assessing repurposed treatments for NSCLC. A phase II experiment seeks to evaluate the therapeutic potential of combining prednisone with Afatinib in advanced NSCLC. The study's results encompass PFS, OS, and overall response rate (ClinicalTrials.gov; NCT04497584). A Phase I/II trial investigates the impact of dexamethasone on diminishing the FLT-PET imaging in NSCLC (ClinicalTrials.gov; NCT04037462). Likewise, another phase II/III trial investigates the impact of dexamethasone on the overall survival of advanced NSCLC (ClinicalTrials.gov; NCT05096663). Metformin, utilized for the management of type 2 diabetes, is investigated in phase III clinical research. The trial evaluates PFS, OS, and overall RR in patients with EGFR mutation in NSCLC (ClinicalTrials.gov; NCT05445791) [48]. For the purpose of determining whether the combination of statins and PD-1/PD-L1 inhibitors shows safety and tolerability in NSCLC, a clinical trial in Phase I has been launched (ClinicalTrials.gov; NCT05636592). Additionally, an initial Phase I trial is examining the Th1-directed immune effects induced by the combination of Vancomycin and Stereotactic Body Radiation Therapy (SBRT) (ClinicalTrials.gov; NCT03546829). Hydroxychloroquine, in conjunction with Binimetinib for advanced KRAS-driven NSCLC, is under evaluation in a mid-stage clinical study assessing the objective response rate (ClinicalTrials.gov; NCT04735068) [48].

## 6 Role of Multidisciplinary Teams in Personalized Care

Successful implementation of personalized therapy in NSCLC necessitates a cooperative strategy that includes a multidisciplinary team of healthcare experts in the clinical sector. Many clinical centres and departments have adopted an interdisciplinary management mode. The multidisciplinary team includes radiation oncologists, surgeons, pulmonologists, pathologists, and other staff [50]. These experts work together to extract intricate information about the patient's genomic and proteomic data, develop potential therapeutic drugs, and plan individualized treatment procedures unique to each patient according to their tumour profile [51]. Together, this team plays a key role in monitoring treatment response and educating patients and their families to make future decisions about their care [52]. The European Society for Medical Oncology has recently revised its lung cancer guidelines to emphasize the vital role of multidisciplinary teams (MDTs) [52]. Members of a multi-disciplinary team engage in the patient's treatment decisions (including surgical radical resection, radical combined with minimally invasive therapy, medical systemic therapy, palliative treatment, lung transplant surgery, etc.), rehabilitation strategy, and observation plan to develop a comprehensive and tailored treatment plan, while ensuring timely adherence to treatment milestones [50–52]. All of these investigations proved effective, increasing the number of cases handled by the team and improving overall patient satisfaction. The development of multidisciplinary teams has demonstrated an enhancement in compliance with clinical recommendations and a reduction in the interval between diagnosis and treatment, leading to better survival rates and quality of life [52].

## 7 Conclusion and Future Perspective

In the past decade, advancements in molecular pathology have enhanced our understanding of the pathophysiology and heterogeneity of NSCLC, leading to substantial improvements in the treatment of patients using improved and highly effective methodologies. Targeted therapy in conjunction with immunotherapy has become a promising strategy for lung cancer treatment, showing notable improvements in patient outcomes. Several networks of signalling and specific carcinogenic driver mutations were recently proven to be causal causes of malignant transformations. Apart from the epidermal growth factor receptor (EGFR), novel targets are constantly being discovered, comprising microRNAs, VEGF, ALK, MET, HER2, immune checkpoint suppressors (PD-1 and CTLA4), and major intracellular signalling networks like PI3K/AKT/mTOR, RAS/MAPK, and NTRK/ROS1. This drives towards the invention of new treatments. These components and routes have been the focus of many medications that have shown clinical effectiveness. Presently, several pharmaceuticals, such as EGFR inhibitors (gefitinib and erlotinib), PI3K/AKT/mTOR inhibitors (everolimus), and NTRK/ROS1 inhibitors (entrectinib), are employed as first-line treatments in lieu of chemotherapy. Many NSCLC patients exhibit a favourable response to checkpoint inhibitors, including anti-PD1 antibodies such as nivolumab and pembrolizumab.

Furthermore, evidence suggested that specific selective medicines in conjunction with immunotherapeutic agents yield favourable outcomes in NSCLC patients. Pharmacological resistance in the tumours is unavoidable despite the efficacy of targeted therapy for NSCLC in managing the disease. Enhancing treatment efficacy necessitates the development of combination therapies and a comprehensive understanding of resistance mechanisms.

Numerous clinical trials assessing targeted treatment and immunotherapy drugs are under consideration, and the preliminary results are promising. The results of studies will elucidate the function of immune-based monotherapies, in conjunction with immunotherapies, and targeted therapy-plus-immunotherapy combination, as well as the role of personalized therapy in the therapeutic management of NSCLC. Precision treatment could finally transform the treatment of lung cancer for patients with limited beneficial alternatives. Notwithstanding the existing therapy pipeline endorsed by NCCN international recommendations, continuous research to enhance treatment options for lung cancer patients is essential. Pharmaceutical companies and research organizations are collaborating. Dedicated to the discovery of novel pharmaceuticals, consistently emphasizing personalized medicine to deliver tailored treatments for each patient and enhance their survival rates. Nevertheless, it is imperative for regulators in each country to expedite the drug approval process while considering the hazards and acceptable limits of individual patients to different therapies. Confronting these hurdles while acknowledging the significance of integrating personalized medicine into clinical practice is crucial for attaining substantial advancements and enhanced results for lung cancer patients.

Clinical research is ongoing to identify the predictive characteristics associated with targeted therapy. Detailed understanding of molecular biomarkers and the development of combination medicines may yield the most efficacious treatment. For proper personalized therapy in NSCLC, appropriate patient-specific biomarker selection will be very crucial for better survival and prognosis of lung cancer patients. Future improvement of patient survival can be expected if the side effects of different conventional chemotherapeutic drugs can be minimized by using appropriate doses of inhibitors alone or in combination with existing chemotherapeutic agents.

## References

1. Masucci M, Karlsson C, Blomqvist L, Ernberg I (2024) Bridging the divide: a review on the implementation of personalized cancer medicine. J Pers Med 14:6. https://doi.org/10.3390/jpm14060561
2. Siegel RL, Giaquinto AN, Ahmedin Jemal DVM (2024) Cancer statistics, 2024. CA Cancer J Clin 74:12–49. https://doi.org/10.3322/caac.21820

3. Nicholson AG, Tsao MS, Beasley MB, Borczuk AC, Brambilla E, Cooper WA, Dacic S, Jain D, Kerr KM, Lantuejoul S et al (2022) The 2021 WHO classification of lung tumors: impact of advances since 2015. J Thorac Oncol 17:362–387. https://doi.org/10.1016/j.jtho.2021.11.003
4. Alduais Y, Zhang H, Fan F, Chen J, Chen B (2023) Non-small cell lung cancer (NSCLC): a review of risk factors, diagnosis, and treatment. Medicine (Baltimore) 102:e32899. https://doi.org/10.1097/MD.0000000000032899
5. Rocco G, Morabito A, Leone A, Muto P, Fiore F, Budillon A (2016) Management of non-small cell lung cancer in the era of personalized medicine. Int J Biochem Cell Biol 78:173–179. https://doi.org/10.1016/j.biocel.2016.07.011
6. Yang SR, Schultheis AM, Yu H, Mandelker D, Ladanyi M, Büttner R (2022) Precision medicine in non-small cell lung cancer: current applications and future directions. Semin Cancer Biol 84:184–198. https://doi.org/10.1016/j.semcancer.2020.07.009
7. Marino FZ, Bianco R, Accardo M, Ronchi A, Cozzolino I, Morgillo F, Rossi G, Franco R (2019) Molecular heterogeneity in lung cancer: from mechanisms of origin to clinical implications. Int J Med Sci 16:981–989. https://doi.org/10.1200/JCO.21.01626
8. Ramos R, Moura CS, Costa M, Lamas NJ, Correia R, Garcez D, Pereira JM, Lindahl T, Sousa C, Vale N (2025) Lung cancer therapy: the role of personalized medicine. Cancer 17:725. https://doi.org/10.3390/cancers17050725
9. Tan AC, Tan DSW (2022) Targeted therapies for lung cancer patients with oncogenic driver molecular alterations. J Clin Oncol 40:611–625. https://doi.org/10.1200/JCO.21.01626
10. NCCN. NCCN Clinical practice guidelines in oncology, non-small cell lung cancer. Version 2.2023; NCCN: Plymouth meeting, PA, USA, 2023
11. Hendriks L, Kerr K, Menis J, Mok T, Nestle U, Passaro A, Peters S, Planchard D, Smit E, Solomon B et al (2023) Oncogene addicted metastatic non-small-cell lung cancer: ESMO clinical practice guideline for diagnosis, treatment and follow-up. Ann Oncol 34:339–357. https://doi.org/10.1016/j.annonc.2022.12.009
12. Lynch TJ, Bell DW, Sordella R et al (2004) Activating mutations in the epidermal growth factor receptor underlying responsiveness of non small-cell lung cancer to gefitinib. N Engl J Med 350:2129–2139. https://doi.org/10.1056/NEJMoa040938
13. Jemal A, Miller KD, Ma J et al (2018) Higher lung cancer incidence in young women than young men in the United States. N Engl J Med 378:1999–2009. https://doi.org/10.1056/NEJMoa1715907
14. Lindeman NI, Cagle PT, Aisner DL et al (2018) Updated molecular testing guideline for the selection of lung cancer patients for treatment with targeted tyrosine kinase inhibitors. J Thorac Oncol 13:323–358. https://doi.org/10.5858/arpa.2017-0388-CP
15. Aisner DL, Marshall CB (2012) Molecular pathology of non–small cell lung cancer: a practical guide. Am J Clin Pathol 138:332–346. https://doi.org/10.1309/AJCPFR12WJKCEEZZ
16. Woodhouse R, Li M, Hughes J, Delfosse D, Skoletsky J, Ma P, Meng W, Dewal N, Milbury C, Clark T et al (2020) Clinical and analytical validation of FoundationOne Liquid CDx, a novel 324-gene cfDNA-based comprehensive genomic profiling assay for cancers of solid tumor origin. PLoS One 15:e0237802. https://doi.org/10.1371/journal.pone.0237802
17. Li B, Janku F, Jung B, Hou C, Madwani K, Alden R, Razavi P, Reis-Filho J, Shen R, Isbell J et al (2019) Ultra-deep next generation sequencing of plasma cell-free DNA in patients with advanced lung cancers: results from the actionable genome consortium. Ann Oncol 30:597–603. https://doi.org/10.1093/annonc/mdz046
18. Altelaar AF, Munoz J, Heck AJ (2013) Next generation proteomics: towards an integrative view of proteome dynamics. Nat Rev Genet 14:35–48. https://doi.org/10.1038/nrg3356
19. Shenoy A, Geiger T (2015) Super-SILAC: current trends and future perspectives. Expert Rev Proteomics 12:13–19. https://doi.org/10.1586/14789450.2015.982538
20. Wang CI, Wang CL, Wu YC et al (2015) Quantitative proteomics reveals a novel role of karyopherin alpha 2 in cell migration through the regulation of vimentin-pErk protein complex levels in lung cancer. J Proteome Res 14:1739–1751. https://doi.org/10.1021/pr501097a

21. Jin Y, Sun Y, Shi X et al (2014) Prognostic value of circulating C-reactive protein levels in patients with non-small cell lung cancer: a systematic review with meta-analysis. J Cancer Res Ther 10(Suppl):C160–C166. https://doi.org/10.4103/0973-1482.145854
22. Cho WC, Yip TT, Cheng WW et al (2010) Serum amyloid A is elevated in the serum of lung cancer patients with poor prognosis. Br J Cancer 102:1731–1735. https://doi.org/10.1038/sj.bjc.6605700
23. Zhang XY, Zhang G, Jiang Y et al (2015) The prognostic value of serum C-reactive protein-bound serum amyloid A in early-stage lung cancer. Chin J Cancer 34:39. https://doi.org/10.1186/s40880-015-0039-1
24. Marien E, Meister M, Muley T et al (2015) Non-small cell lung cancer is characterized by dramatic changes in phospholipid profiles. Int J Cancer 137:1539–1548. https://doi.org/10.1002/ijc.29517
25. Kobayashi M, Nagashio R, Jiang SX et al (2015) Calnexin is a novel sero-diagnostic marker for lung cancer. Lung Cancer 90:342–345. https://doi.org/10.1016/j.lungcan.2015.08.015
26. Liu PJ, Chen CD, Wang CL et al (2015) In depth proteomic analysis of six types of exudative pleural effusions for non-small cell lung cancer biomarker discovery. Mol Cell Proteomics 14:917–932. https://doi.org/10.1074/mcp.M114.045914
27. Sun Q, Zhao W, Wang L, Guo F, Song D, Zhang Q, Zhang Da, Fan Y, Wang J (2019) Integration of metabolomic and transcriptomic profiles to identify biomarkers in serum of lung cancer. J Cell Biochem 120(7):11981–11989. https://doi.org/10.1002/jcb.28482
28. Zheng AJ, Zheng Y, Li W, Zhi J, Huang X, Zhu W, Liu Z, Gong L (2022) Combined metabolomics with transcriptomics reveals potential plasma biomarkers correlated with non-small-cell lung cancer proliferation through the Akt pathway. Clin Chim Acta 530:66–73. https://doi.org/10.1016/j.cca.2022.02.018
29. Qi S, Wu Q, Chen Z, Zhang W, Zhou Y, Mao K, Li J, Li Y, Chen J, Huang Y, Huang Y (2021) High-resolution metabolomic biomarkers for lung cancer diagnosis and prognosis. Sci Rep 11:11805. https://doi.org/10.1038/s41598-021-91276-2
30. Biswas P, Swaroop S, Dutta N, Arya A, Ghosh S et al (2022) IL-13 and the hydroperoxy fatty acid 13(S) HpODE play crucial role in inducing an apoptotic pathway in cancer cells involving MAO-A/ROS/p53/p21 signaling axis. Free Radic Biol Med 195:309–328. https://doi.org/10.1016/j.freeradbiomed.2022
31. Kowalczyk T, Kisluk J, Pietrowska K, Godzien J, Kozlowski M, Resze´c J, Sierko E, Naumnik W, Mróz R, Moniuszko M et al (2021) The ability of metabolomics to discriminate non-small-cell lung cancer subtypes depends on the stage of the disease and the type of material studied. Cancer 13:3314. https://doi.org/10.3390/cancers13133314
32. Aldarouish M, Wang C (2016) Trends and advances in tumor immunology and lung cancer immunotherapy. J Exp Clin Cancer Res 35:157. https://doi.org/10.1186/s13046-016-0439-3
33. Handa Y, Tsutani Y, Shiroma N, Kai Y, Mimae T, Miyata Y, Takeshima Y, Arihiro K, Okada M (2020) Prognostic impact of programmed death-ligand 1 and surrounding immune status on stage I lung cancer. Clin Lung Cancer 21:e302–e314
34. Bubendorf L, Zoche M, Dafni U, Rüschoff JH, Prince SS, Marti N, Stavrou A, Kammler R, Finn SP, Moch H et al (2022) Prognostic impact of tumour mutational burden in resected stage I and II lung adenocarcinomas from a European thoracic oncology platform lung scape cohort. Lung Cancer 174:27–35. https://doi.org/10.1016/j.lungcan.2022.09.014
35. He Y, Zhang X, Jia K, Dziadziuszko R, Zhao S, Deng J, Wang H, Hirsch FR, Zhou C (2019) OX40 and OX40L protein expression of tumor infiltrating lymphocytes in non-small cell lung cancer and its role in clinical outcome and relationships with other immune biomarkers. Transl Lung Cancer Res 8:352–366. https://doi.org/10.21037/tlcr.2019.08.15
36. Lin D, Shen L, Luo M, Zhang K, Li J, Yang Q, Zhu F, Zhou D, Zheng S, Chen Y et al (2021) Circulating tumor cells: biology and clinical significance. Signal Transduct Target Ther 6:404. https://doi.org/10.1038/s41392-021-00817-8
37. Li H, Li B, Pan Y, Zhang Y, Xiang J, Zhang Y, Sun Y, Yu X, He W, Hu H (2020) Preoperative folate receptor-positive circulating tumor cell level is a prognostic factor of long-term out-

come in non-small cell lung cancer patients. Front Oncol 10:621435. https://doi.org/10.3389/fonc.2020.621435
38. Wankhede D, Grover S, Hofman P (2022) Circulating tumor cells as a predictive biomarker in resectable lung cancer: a systematic review and meta-analysis. Cancer 14:6112. https://doi.org/10.3390/cancers14246112
39. Szilágyi M, Pös O, Márton É, Buglyó G, Soltész B, Keserű J, Penyige A, Szemes T, Nagy B (2020) Circulating cell-free nucleic acids: main characteristics and clinical application. Int J Mol Sci 21:6827. https://doi.org/10.3390/ijms21186827
40. Li C, Shao J, Li P, Feng J, Li J, Wang C (2023) Circulating tumor DNA as liquid biopsy in lung cancer: biological characteristics and clinical integration. Cancer Lett 577:216365. https://doi.org/10.1016/j.canlet.2023.216365
41. Cobas EGFR Mutation test V2, FDA summary of safety and effectiveness data, 2016. https://www.accessdata.fda.gov/cdrh_docs/pdf15/P150044B.pdf
42. Chen K, Zhao H, Shi Y, Yang F, Wang LT, Kang G, Nie Y, Wang J (2019) Perioperative dynamic changes in circulating tumor DNA in patients with lung cancer (DYNAMIC). Clin Cancer Res 25:7058–7067. https://doi.org/10.1158/1078-0432.CCR-19-1213
43. Xia L, Mei J, Kang R, Deng S, Chen Y, Yang Y, Feng G, Deng Y, Gan F, Lin Y et al (2022) Perioperative ctDNA-based molecular residual disease detection for non-small cell lung cancer: a prospective multicenter cohort study (LUNGCA-1). Clin Cancer Res 28:3308–3317. https://doi.org/10.1158/1078-0432.CCR-21-3044
44. Hu Z, Chen X, Zhao Y, Tian T, Jin G, Shu Y, Chen Y, Xu L, Zen K, Zhang C et al (2010) Serum microRNA signatures identified in a genome-wide serum microRNA expression profiling predict survival of non-small-cell lung cancer. J Clin Oncol 28:1721–1726. https://doi.org/10.1200/JCO.2009.24.9342
45. Muley T, Fetz T-H, Dienemann H, Hoffmann H, Herth FJF, Meister M, Ebert W (2008) Tumor volume and tumor marker index based on CYFRA 21-1 and CEA are strong prognostic factors in operated early-stage NSCLC. Lung Cancer 60:408–415. https://doi.org/10.1016/j.lungcan.2007.10.026
46. Fox AH, Nishino M, Osarogiagbon RU, Rivera MP, Rosenthal LS, Smith RA, Farjah F, Sholl LM, Silvestri GA, Johnson BE (2023) Acquiring tissue for advanced lung cancer diagnosis and comprehensive biomarker testing: a National Lung Cancer Roundtable best-practice guide. CA Cancer J Clin 73:358–375. https://doi.org/10.3322/caac.21774
47. Wu J, Li S, Zhang P (2022) Tumor-derived exosomes: immune properties and clinical application in lung cancer. Cancer Drug Resist 5:102–113. https://doi.org/10.20517/cdr.2021.99
48. Doumat G, Daher D, Zerdan MB, Nasra N, Bahmad HF, Recine M, Poppiti R (2023) Drug repurposing in non-small cell lung carcinoma: old solutions for new problems. Curr Oncol 30:704–719. https://doi.org/10.3390/curroncol30010055
49. Datta C, Das P, Dutta S, Prasad T, Banerjee A et al (2024) AMPK activation reduces cancer cell aggressiveness via inhibition of monoamine oxidase A (MAO-A) expression/activity. Life Sci 352:122857. https://doi.org/10.1016/j.lfs.2024.122857
50. Hardavella G, Frille A, Theochari C, Keramida K, Bellou E, Fotineas A et al (2021) Multidisciplinary care models for patients with lung cancer. Breathe (Sheff) 16(4):200076
51. Popat S, Navani N, M Kerr K, F Smit E, Batchelor TJP, Van Schil P et al (2021) Navigating diagnostic and treatment decisions in non-small cell lung cancer: expert commentary on the multidisciplinary team approach. Oncologist 26(2):e306–e315. https://doi.org/10.1002/onco.13586
52. Batra U, Munshi A, Kabra V, Momi G (2022) Relevance of multi-disciplinary team approach in diagnosis and management of Stage III NSCLC. Indian J Cancer 59(Suppl 1):S46–S55. https://doi.org/10.4103/ijc.IJC_51_21
53. Takahashi K et al (2012) Ann Oncol 23:ix434. https://doi.org/10.1016/S0923-7534(20)33924-7
54. Blackhall F et al (2017) ESMO Open 2(3):e000219. https://doi.org/10.1136/esmoopen-2017-000219
55. Shaw AT et al (2019) Ann Oncol 30:1121–1126. https://doi.org/10.1093/annonc/mdz131

56. Skoulidis F et al (2021) The New England J Med 384(25):2371–2381. https://doi.org/10.1056/NEJMoa2103695
57. Wolf J et al (2020) JCO 38:9509–9519. https://doi.org/10.1200/JCO.2020.38.15_suppl.9509
58. Planchard D et al (2022) J Thoracic Oncol 17(1):103–115. https://doi.org/10.1016/j.jtho.2021.08.011
59. Martin R et al (2019) J Clin Oncol 37(7):537–546. https://doi.org/10.1200/JCO.18.00149
60. Lin JJ et al (2021) JCO 39:9109–9119. https://doi.org/10.1200/JCO.2021.39.15_suppl.9109
61. S, M. K et al (2017) New England J Med 376:629–640. https://doi.org/10.1056/NEJMoa1612674
62. Mazieres J et al (2022) JCO 40:719–728. https://doi.org/10.1200/JCO.21.01455
63. Lecia VS et al (2013) J Clin Oncol 27:3327–3334. https://doi.org/10.1200/JCO.2012.44.2806
64. Wolf J et al (2020) New England J Med 383:944–957. https://doi.org/10.1056/NEJMoa2002787
65. Gautschi O et al. JCO 0, JCO-24-02076. https://doi.org/10.1200/JCO-24-02076

# Personalized Medicines and Prospects in Prostate Cancer

Yogesh Vaishnav, Shikha Agarwal, Joohee Pradhan, Mayank Garhewal, and Ashish Baldi

**Abstract**

One of the most common cancers in males worldwide is prostate cancer, which calls for creative ways to improve patient outcomes and treatment effectiveness. Prostate cancer treatment has great promise; owing to personalized medicine, which is defined by the customization of medical care to each patient's unique characteristics. This approach explores proteomic, genomic, and epigenomic profiling to identify distinct tumor features, allowing physicians to group patients according to their response levels to particular treatments. Targeted treatments, such as androgen receptor inhibitors and PARP inhibitors, have been made possible by recent advancements in biomarker discovery, such as the identification of genetic mutations and abnormalities linked to the advancement of prostate cancer. Additionally, further improvement of individualized treatment regimens is anticipated through the use of machine learning and artificial intelligence tools in the analysis of large datasets. This chapter comprises the application of personalized medicine in prostate cancer, with particular emphasis on recent developments, ongoing clinical studies, and the challenges associated with delivering customized treatments to a range of patient demographics. Finally, the avenues

Y. Vaishnav · M. Garhewal
Department of Pharmacy, Guru Ghasidas Vishwavidyalaya (A Central University), Bilaspur, India

S. Agarwal
Department of Chemistry, Mohanlal Sukhadia University, Udaipur, India

J. Pradhan (✉)
Department of Pharmaceutical Sciences, Mohanlal Sukhadia University, Udaipur, India

A. Baldi (✉)
Department of Pharmaceutical Sciences and Technology, Maharaja Ranjit Singh Punjab Technical University, Bathinda, India



M. K. Ghosh, M. Basu (eds.), *Personalized Cancer Therapy*,
https://doi.org/10.1007/978-981-95-6845-1_12

and possibilities of switching from a one-size-fits-all strategy to tailored ones by having the potential to momentously augment the quality of life and therapeutic benefits for patients with prostate cancer will be presented.

**Keywords**

Prostate cancer · Targeted therapy · Genomic profiling · Biomarker identification · Molecular genetics · PARP inhibitors

## 1 Introduction

Prostate cancer is one of the most frequently diagnosed cancers in males globally. It has seen an increased incidence with longer life in general, greater awareness, and improved diagnostic methods. It is a significant health burden [1, 2]. That said, the challenge of detecting prostate cancer at an early, treatable stage remains, despite advances in diagnostic technologies [2]. While useful, PSA—prostate specific antigen—testing has limitations, including limited specificity and the potential for overdiagnosis, which often leads to unnecessary biopsies and over-treatment [1–3].

High-throughput technologies such as mass spectrometry, RNA sequencing, and next-gen sequencing (NGS) have changed the field of oncology [4]. These tools provide massive amounts of genomic and proteomic data that, with appropriate analytics, can potentially provide new biomarkers that suggest the earliest stages of cancer [5].

Deep learning (DL), a subset of artificial intelligence (AI), has proven to be a disruptive technology in medical research; deep learning models are highly effective at analyzing high volumes of unstructured data, as is commonly seen in proteomic and genomic sensing where they work on higher levels of abstraction and can identify complex patterns in data, unlike standard machine learning algorithms [6]. The recent advent of deep learning models has altered how individual tumors are diagnosed, prognosis determined, and treatment predicted, through reliable computational models that drive the precision of identifying and classifying biomarkers [7]. In addition, the derived targeted therapies that leverage the documented genetic mutations and aberrations, such as androgen receptor inhibitors and PARP inhibitors, are a direct consequence of the evolution of biomarker discovery [8]. This chapter examined the application of deep learning models to proteomic and genomic data analysis for the early detection of prostate cancer, with a great deal of detail being provided on the types of data utilized, preprocessing strategies, model-building strategies, and validation strategies. The case studies and the described real-life examples provide insight into the transformational potential of AI in personalized cancer care (Fig. 1).

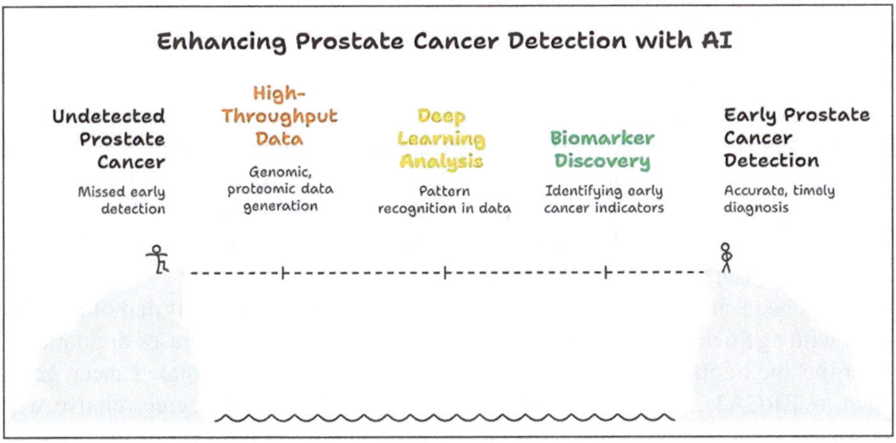


**Fig. 1** Enhancing prostate cancer detection with AI/ML. The figure shows how artificial intelligence and machine learning tools assist in improving the detection accuracy of prostate cancer

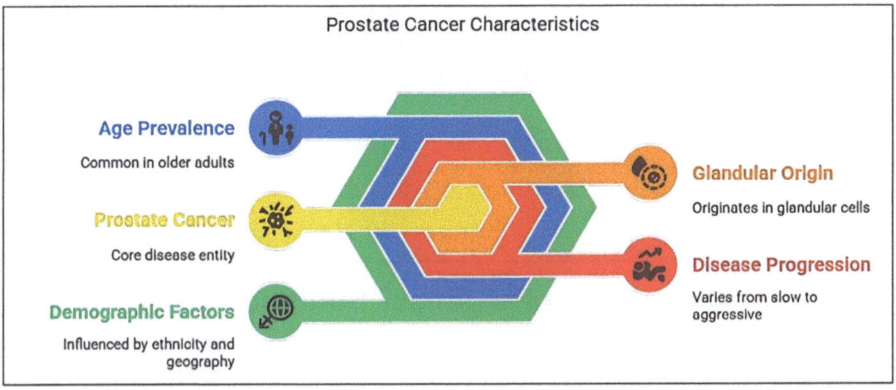


**Fig. 2** Characteristics of prostate cancer. The figure depicts key features such as tumor type, stage, and progression patterns

## 2 Overview of Prostate Cancer and Its Clinical Background

Prostate cancer (PCa) is also heterogeneous histologically and genetically, and almost always begins in glandular cells of the prostate. It can develop as a localized, non-aggressive disease to aggressive types that metastasize. PCa is more common in older adults and incidence varies by race and geographic region (Fig. 2) [1, 9–14].

## 2.1 Epidemiology

Worldwide, prostate cancer is fifth on the list of most common cause of cancer death in men and the second most commonly diagnosed cancer. The Global Cancer Observatory (GLOBOCAN) 2020 estimates there were approximately 375,000 cancer deaths and 1.4 million new cases in 2020 (Fig. 3). As a result of widespread prostate specific antigen (PSA) screening, the incidence varies significantly based on geographic region, with the highest incidence evident in North America, Northern and Western Europe, and Australia [15–17].

It has been shown in Fig. 4 that prostate cancer is linked to a myriad of risk factors, with age being the primary risk factor, because the risk increases dramatically after the age of 50 and above [18]. Genetic predisposition to prostate cancer, genes such as BRCA1, BRCA2, and HOXB13, specifically, a first-degree relative with prostate cancer increases one's risk. Race and ethnicity also play a factor; African-American men had a higher incidence and mortality rate than other groups. Dietary and lifestyle factors that have been associated with increased risk include eating red meat, high-fat dairy, being obese, and being physically inactive [2]. Also, other hormones play an important role in the development and progression of prostate cancer; androgens (high levels of testosterone and dihydrotestosterone) have the most considerable influence on the development and progression of prostate cancer [19, 20].

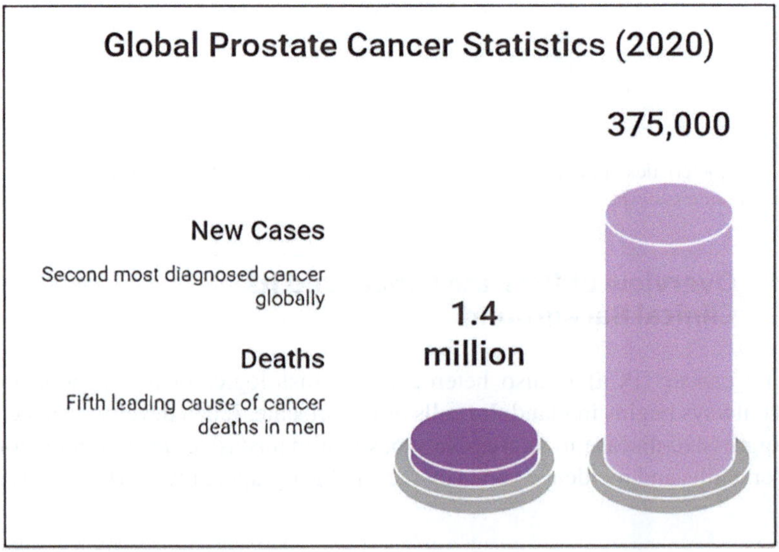


**Fig. 3** Global prostate cancer statistics. The figure illustrates worldwide incidence and mortality rates, highlighting geographic variations

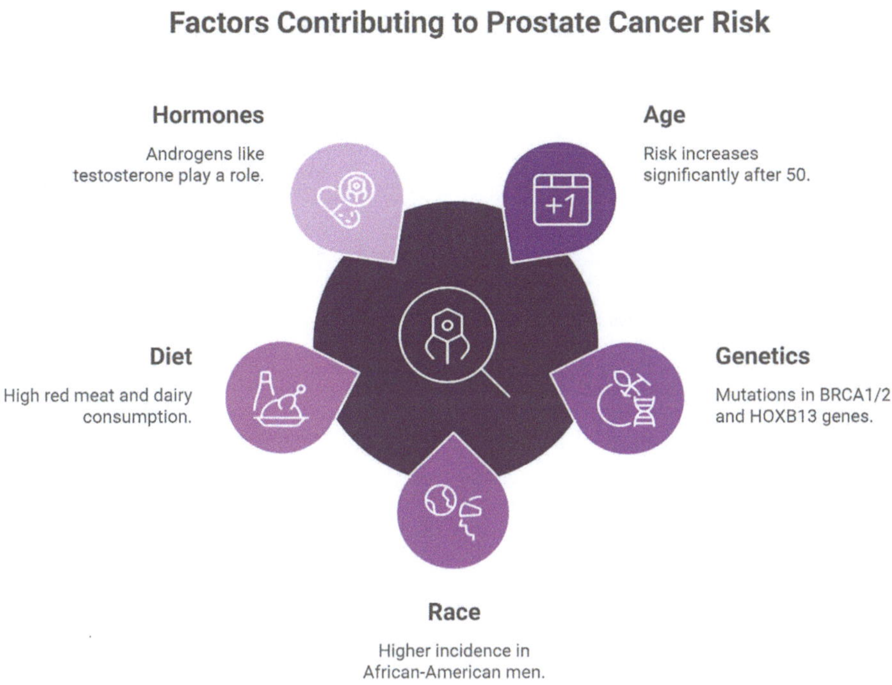


**Fig. 4** Factors affecting prostate cancer. The figure presents major risk factors, including genetics, diet, lifestyle, and environment

## 2.2 Pathophysiology and Molecular Features

In most cases, prostate cancer arises from prostatic acinar cells, progressing from prostatic intraepithelial neoplasia (PIN) to invasive adenocarcinoma. The most prevalent mechanism in the pathogenesis of prostate cancer is the deregulation of androgen receptor (AR) signaling, which drives tumor growth and survival. Pathogenesis, oncogenesis, and progression of the disease, at least in part, result from genomic events that include the deletion of the PTEN tumor suppressor gene, a fusion of the TMPRSS2-ERG gene, and mutations of SPOP. Additionally, epigenetic mechanisms such as aberrant DNA methylation and histone modifications impact malignancy and altered gene expression. The tumor microenvironment, which is typified by pro-inflammatory cytokines, immune cell infiltrates, and the complex interactions of various stromal components, contributes support of angiogenesis, tumor growth, and immune evasion, all of which lead to prostate cancer development and resistance to therapeutic treatment, as can be seen in Fig. 5 [21, 22].

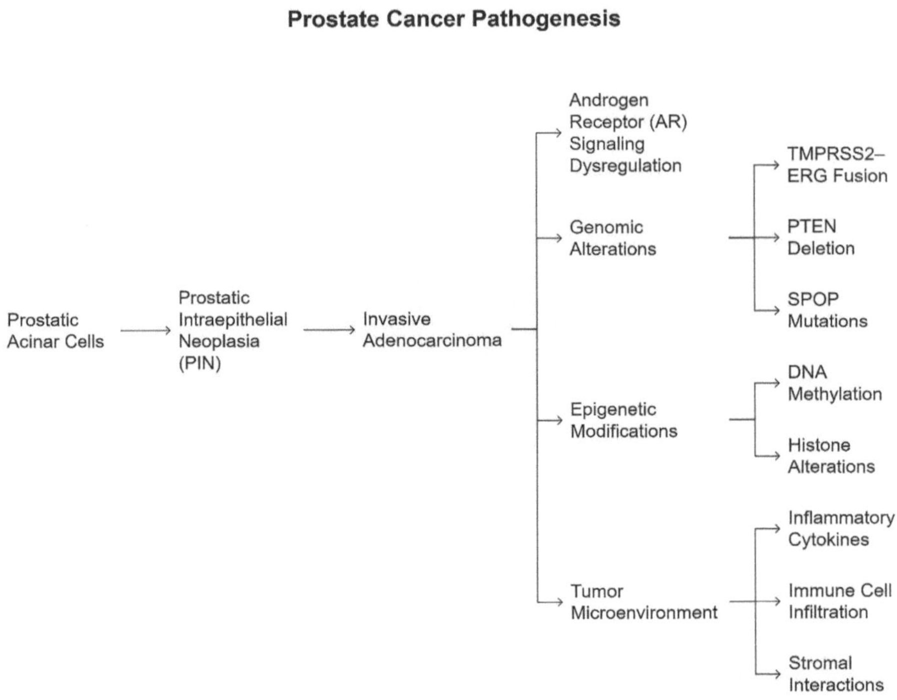


**Fig. 5** Pathway of prostate cancer pathogenesis. The figure outlines the biological and molecular steps from normal cells to cancerous transformation

## 2.3 The Drawbacks of Conventional Diagnostic Methods

Despite the fact that PSA testing has been the foundation for PCa screening, there is a mountain of data available outlining some of its limitations. Elevated PSA levels can have a variety of female or male pathways besides cancer, including benign prostatic hyperplasia (BPH) and prostatitis. This leads to an abundance of false positives, resulting in overtreatment and unnecessary biopsies. Additionally, PSA testing may not be able to detect aggressive tumors with undetectable PSA secretion [23].

## 2.4 Molecular Biomarkers Are Needed

Diagnostic procedures for benign diseases and early-stage cancers reveal a lack of specific, reliable molecular biomarkers, which could aid in distinguishing between them. The application of many omics datasets: genomic, transcriptomic, proteomics, epigenomics, and metabolomics, to discover biomarkers also provides the opportunity for prognosis, risk stratification, targeted therapy, and early detection [24].

## 3 Fundamentals of Biomarker Discovery

Biological indicators known as biomarkers identify disease or physiological state. Biomarkers for prostate cancer include differentially expressed genes, proteins or metabolites that distinguish malignant tissue from normal tissue. Historically statistical analyses have been the primary methods of biomarker discovery, but AI enables subtle patterns and nonlinear interactions to be identified from large omics datasets [25].

### 3.1 Types of Biomarker

***Diagnostic Biomarkers*** Biological signs called diagnostic biomarkers are employed by the medical personnel to diagnose diseases or other health states. These markers are crucial to identify the early signs of a problem, well before symptoms worsen or persist. The diagnostic biomarkers that confirm what it is and usually direct the clinician to order additional tests by tallying with biomarkers that are associated with an illness expressed in elevation of specific substances, cells, or genomic changes [26]. For example, the well-known Prostate-specific Antigen (PSA) test. Assesses the level of PSA protein in the blood corresponding to prostate cancer, or any other possible problems connected with the prostate. These diagnostic biomarkers have to be gold-standard and reliable in order for us to hop in and provide patients with the best opportunity [27].

***Prognostic Biomarkers*** Prognostic biomarkers, of whatever treatment zielorientiert, provide crucial information about the likely trajectory or outcome for a disease. They support predicting relapse-free survival, overall survival and the rate of disease progression given biomarkers The prognosis of the condition can be used to stratify patients and determine what monitoring, strategy medical intervention should be used Clinicians can stratify patients by risk to inform both surveillance and management strategies based on the natural history of disease One can use prognostic biomarkers such as specific genetic alterations or expression levels of proteins in tumors to predict how aggressive (or indolent) cancer is going to be. Therefore, prognostic biomarkers support therapeutic decisions as they inform patient management based on disease activity [28].

***Predictive Biomarkers*** By integrating predictive biomarkers, treatment regimens can be made more patient individualized, what makes patients respond to a given drug. This is indicative of who would likely respond well to the medication/intervention and someone who may not, which might have serious side effects. This approach maximizes drug efficacy while minimizing waste of treatment. HER2 overexpression in breast cancer, for example, is a predictive biomarker that would expect patients to respond well to therapies targeting HER2 (like trastuzumab), such as HER2 overexpressing patients. Precision medicine is largely based on predictive biomarkers for safer, more efficient, and accurate tailor-made treatment plans [29].

# 4 Proteomic and Genomic Indicators of Early Prostate Cancer

## 4.1 Changes to the Genome

Numerous mutations and structural variants have been linked to prostate cancer by genomic studies, including changes in genes like TP53, PTEN, BRCA1/2, and SPOP. These genetic anomalies, as depicted in Fig. 6, are essential for the development, spread, and treatment resistance of tumors. For instance, mutations in the BRCA2 gene offer a therapeutic target for PARP inhibitors and are associated with more severe illness and worse prognosis [30].

## 4.2 Markers of Transcriptomics

RNA sequencing (RNA-seq)-based transcriptomic profiling makes it possible to identify genes that are expressed differently in malignant and healthy tissues (Fig. 7). In the early identification of prostate cancer, biomarkers such as PCA3, TMPRSS2-ERG fusion transcripts, and SCHLAP1 have become important [30].

## 4.3 Proteomic Indications

The identification of proteins that are differently expressed in prostate cancer has been made easier by proteomics, especially mass spectrometry-based methods.

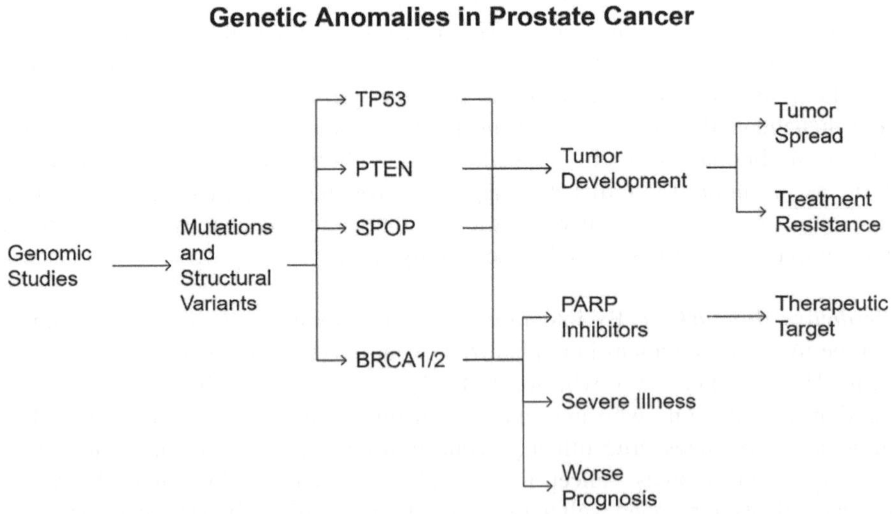


**Fig. 6** Genetic anomalies in prostate cancer. The figure displays common mutations, chromosomal alterations, and their implications

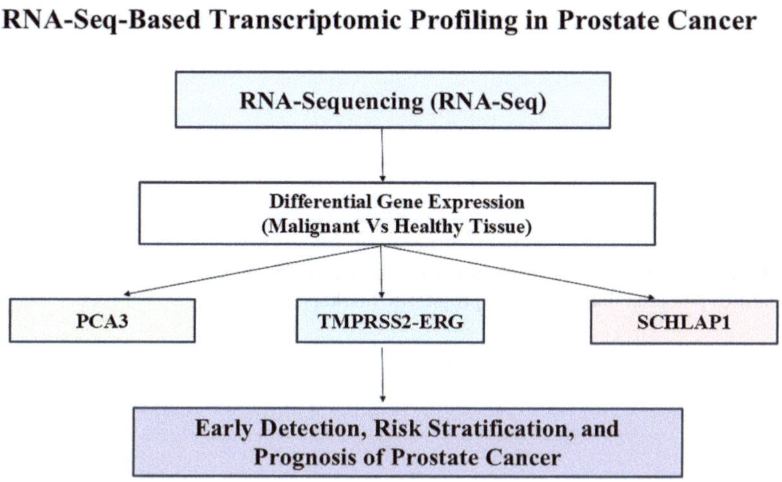


**Fig. 7** RNA-Seq based transcription profiling in prostate cancer. The figure shows differential gene expression patterns obtained from sequencing

Figure 8 illustrates alpha-methylacyl-CoA racemase (AMACR), annexin A3, and KLK3 (PSA) are proteins that have demonstrated potential as predictive or diagnostic biomarkers [31].

## 5 Deep Learning Models for the Identification of Biomarkers

Neural networks with several layers make up deep learning models (Fig. 9), a subclass of machine learning that can learn intricate data representations. For biomedical applications, they are especially well-suited due to their capacity to analyze high-dimensional datasets and simulate non-linear interactions [9, 10].

### 5.1 Common Deep Learning Architectures

In image-based diagnostics, convolutional neural networks, or CNNs, are the most commonly employed model for spatial data processing based on pixel inputs. Moreover, CNNs are adept at modelling organized omics databases. Recurrent neural networks, or RNNs, are capable of modelling sequential data due to their ability to understand what happened preceding that moment, particularly genomic sequences or epigenetic markers such as DNA methylation. Autoencoders, as unsupervised models, can help with feature extraction and dimensionality reduction with high-dimensional biological datasets. Because of their ability to model long-range relationships, transformers are increasingly utilized to model complicated sequence data, such as DNA and RNA sequences, while graph neural networks (GNNs) are

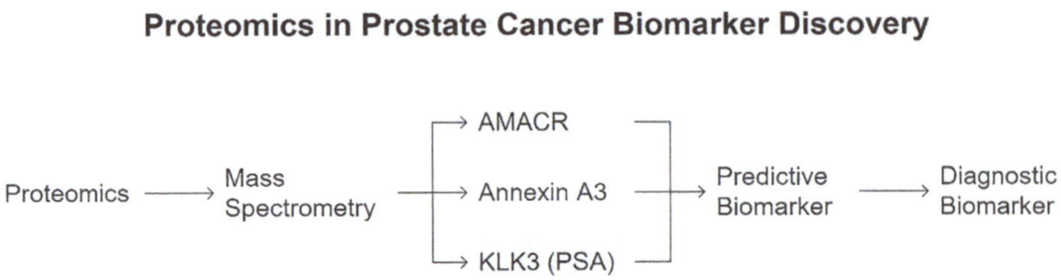


**Fig. 8** Proteomics in prostate cancer biomarker discovery. The figure illustrates how protein analysis identifies potential diagnostic and prognostic markers

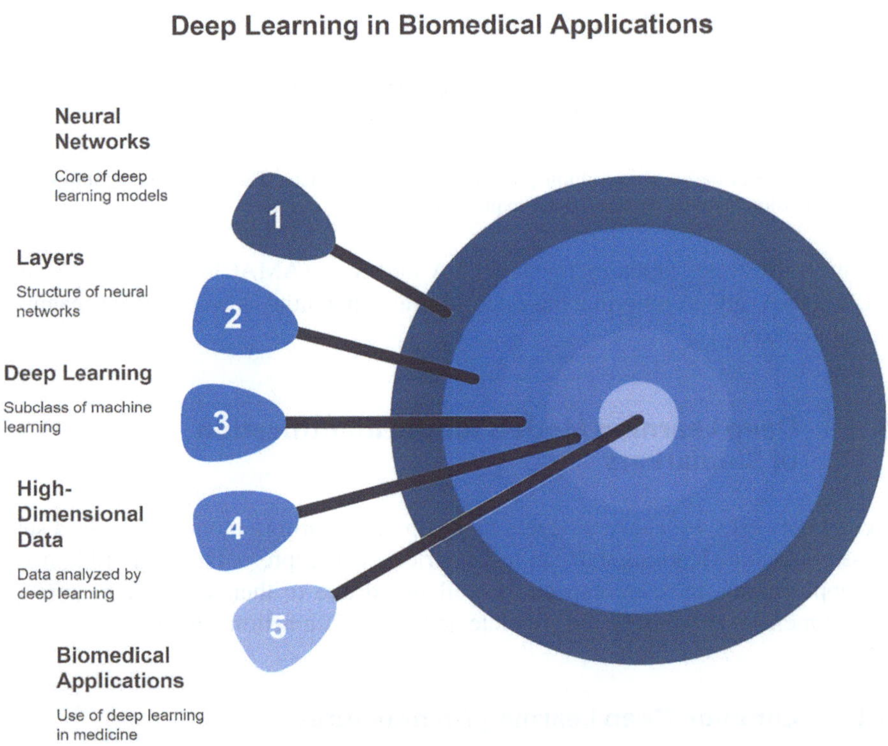


**Fig. 9** Deep learning in biomedical application. The figure demonstrates neural network applications in analyzing medical data for prostate cancer

specifically created to model graph structure data, including molecular graphs or protein–protein interaction networks, to help elucidate structural roles and biological relationships [32].

### 5.2 Developing and Validating Models

Common steps in a deep learning pipeline for biomarker discovery include data preparation, model training, model validation, and model result interpretation. Independent test sets and cross-validation ensure robustness, while explainability techniques such as SHAP or LIME can clarify model results.

## 6 Targeted Therapies Made Possible by Biomarker Identification

The advancement of precision medicine in prostate cancer has made tremendous strides to uncover molecular abnormalities and genetic variations. The androgen receptor (AR) inhibitors, targeting the androgen signaling system, fundamental to the development of prostate cancer, are an important step forward. Inhibiting AR activity, the most effective medications include enzalutamide and apalutamide, specifically in castration-resistant prostate cancer. Another significant class of drugs includes PARP inhibitors, which exploit defects in DNA repair systems; agents such as olaparib and rucaparib are particularly effective against tumors with BRCA1/2 mutations and other homologous recombination defects and can provide focused treatment for patients with these genetic alterations. Together, these classes of treatment illustrate how biomarker-driven approaches are changing the clinical landscape and allowing for individualized treatment plans to improve patient outcomes and efficacy [33] as demonstrated in Fig. 10.

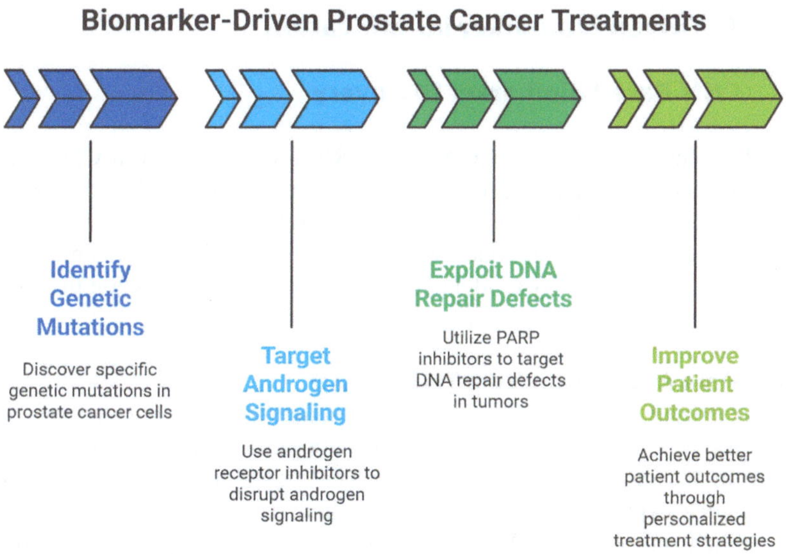


**Fig. 10** Biomarker-driven prostate cancer treatment. Figure highlights treatment strategies tailored to specific molecular markers

**Table 1** Examples of targeted therapies enabled by biomarker discovery

| Therapy type | Target biomarker | Drug | Clinical relevance |
|---|---|---|---|
| Androgen receptor inhibitor | Androgen receptor | Enzalutamide | Treats advanced PCa |
| PARP inhibitor | BRCA1/2 mutation | Olaparib | Treats metastatic PCa with DNA repair defects |

**Table 2** Deep learning models and applications in PCa biomarker discovery

| Model type | Data type | Biomarkers identified | Key outcome | Performance |
|---|---|---|---|---|
| CNN | RNA-seq | AMACR, PCA3 | Early detection | AUC 0.93 |
| Autoencoder | Proteomics | KLK3, Annexin A3 | Feature extraction | Accuracy 89% |
| RNN (LSTM) | Methylation | GSTP1, RASSF1A | Epigenetic marker detection | Sensitivity 91% |
| Transformer | Multi-omics | PTEN, MYC | Integrated biomarker discovery | AUC 0.95 |
| GNN | PPI + expression | BCL2, FGFR1, MYC | Network analysis | F1 score 0.88 |

Tables 1 and 2 contain some examples of targeted therapies enabled by biomarker discovery and deep learning models and applications in PCa biomarker discovery, respectively.

## 7 AI-Driven Biomarker Identification

### 7.1 CNN-Based Biomarker Discovery

Researchers investigated the TCGA-PRAD RNA-seq data with a CNN analysis. The CNN was given gene expression profiles that had been altered to heatmaps demonstrated in Fig. 11. Notably, AMACR, PCA3, and HOXC6 were significant biomarkers according to the model, which had an AUC of 0.93. This method demonstrated the model's ability to identify spatial patterns of expression that are associated with cancer [34].

#### 7.1.1 Data Acquisition from TCGA-PRAD

Researchers used the RNA-Seq gene expression data from the TCGA-PRAD cohort to begin the process of collecting data. This data provided a comprehensive understanding of the molecular differences between malignant and healthy tissue and included gene expression profiling of tumor and normal prostate tissue from the subject. Each sample for each individual is represented by gene-level expression counts, which can be presented in various ways, including raw counts, transcripts per million, and fragments per kilobase of transcript per million mapped reads.

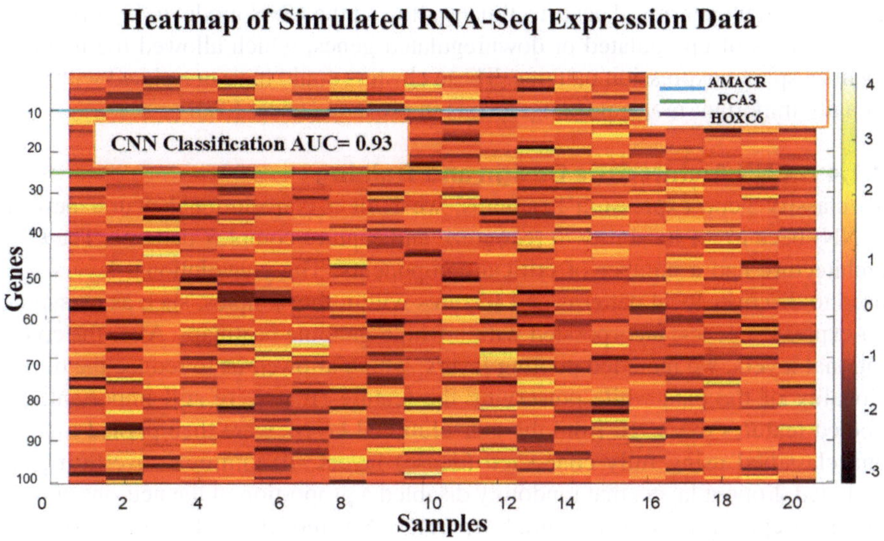


**Fig. 11** Heatmap of simulated RNA-seq expression data. Figure visualizes clustering patterns in gene expression for comparative analysis

These expression levels will be examined in a follow-up analysis for biomarker discovery and other cancer-related research and form a crucial foundation for understanding the underlying molecular mechanisms of prostate cancer [35].

### 7.1.2 Preprocessing and Normalization

Preprocessing is a range of strategies applied to the raw RNA-Seq data to reduce noise and address batch effects, and to establish a solid foundation for subsequent analyses. Normalizing methods, etc., were used to normalize expression values across all samples, such as quantile normalization and log2 transformation, to minimize technical variation so that the expression values could be statistically compared. With normalization procedures performed, genes with low levels of expression across the entire sample were eliminated. The elimination of genes with consistently low expression levels was important to emphasize biologically meaningful gene expression patterns related to prostate cancer that minimized background noise and the effects of computational complexity [36].

### 7.1.3 Feature Transformation—Heatmap Generation

The one-dimensional gene expression vectors were converted into two-dimensional representations (heatmaps) to reproduce spatial patterns since CNNs are designed to work with images. The transformed heatmap was composed of pixels that indicated the expression level of a certain gene, with the gene expression values arranged on a 2D grid. This 2D grid format could have been arranged in a number of different ways: genes could have been arranged alphabetically, randomized, or, more importantly, arranged according to biological importance (e.g., chromosomal positions or co-expression patterns). These heat maps allowed the CNN model to identify localized clusters of upregulated or downregulated genes, which allowed the model to identify spatial expression patterns. These clusters could have correlated with functionally important molecular fingerprints in prostate cancer [37].

### 7.1.4 CNN Architecture Design and Training

To classify gene expression heatmaps, we built a convolutional neural network (CNN) because of its ability to recognize and capture complex patterns. The CNN architecture had a number of critical components. The convolutional layers were used to discover local patterns and features within the samples, such as clusters of co-expressed genes, incorporated within whole-genome heatmaps. After the convolutional layers, we included pooling layers, which allowed us to reduce the dimensionality of the data while preserving the features that were still important. The pooled features extracted from the convolutional layers were then pooled together and classified using fully connected (dense) layers. To minimize overfitting, we included dropout layers that randomly disabled a proportion of the neurons at training each epoch. We created initial separate CNN models to distinguish between normal and malignant prostate tissue as a binary classification problem, using either cross-validation or a train/test split of data to assess its performance and generalizability, and ensure that the classification of tissue would hold when predictable data were being used [38].

### 7.1.5 Model Evaluation

Methods for evaluating the performance of a CNN model to distinguish between normal and cancerous tissue samples from the prostate were assessed using classical classification metrics. One of the parameters that warrants evaluation is the area under the curve (AUC) in the receiver operating characteristic (ROC). The ROC AUC provides a measure of how well the model can categorize positive and negative instances across all thresholds. The CNN model's AUC of 0.93 indicates an ability to discriminate really well—the model underwent exceptional discrimination between samples with cancer and normal samples based on gene expression profiles [39].

### 7.1.6 Biomarker Identification

After the training phase was complete, we performed feature importance analysis in order to identify the most important genes for the CNN model classification. Subsequently, we examined activation maps and weight visualization to ascertain the genes that were frequently identified by the CNN model during classification. The most prominent biomarker genes that were featured assessment were AMACR, PCA3, and HOXC6, in that order. All three significantly implicated a conspicuously well-known diagnostic marker for prostate cancer, subsequently alpha-methylacyl-CoA racemase (AMACR). PCA3 (prostate cancer antigen 3) is also a well-established non-coding RNA genetic marker that has been widely utilized in diagnostic tests and was significantly overexpressed in prostate cancer. The transcription factor, HOXC6 which modulates gene expression associated with cellular proliferation and differentiation, has been linked to prostate cancer development. The consistent prominence of these genes in the interpretability outputs, furthermore, to confirm the relevance of distinguishing prostate cancer from normal prostate tissue, affirmatively enhances the prognostic and diagnostic potentiality as biological markers [40].

### 7.1.7 Biological Validation

The identified biomarkers often undergo additional validation in follow-up analyses, which may occur in the same study or in subsequent research, to assess their biological and clinical relevance. In the validation phase, typically independent patient cohorts are used to validate that the biomarkers perform consistently within diverse populations. Reaction methods like immunohistochemistry and quantitative PCR (qPCR) are used to verify the expression levels of the genes at the mRNA or protein levels, respectively. Subsequently, pathway enrichment analysis is conducted to advance the understanding of processes and mechanisms relevant to the identified biomarkers, while also providing a context for their roles in important cancer processes such as metastasis, apoptosis, and cell proliferation, which suggest their potential to be used as targets for the treatment of prostate cancer or as diagnostic markers [41].

### 7.1.8 Significance and Interpretation

In this research, convolutional neural networks (CNNs) have been demonstrated to analyze high-dimensional RNA-Seq data by innovatively converting gene expression profiles into deep learning-ready image-like heatmaps. Using this means of converting high-dimensional data into a recognizable image, the CNN was able to identify spatial linkages in data, and local expression patterns, these types of local patterns are nuanced which advanced statistical or traditional machine learning methods do not frequently recognize. The CNN was able to identify clinically important biomarkers and classify normal versus malignant prostate tissue, with strong diagnostic accuracy, employing an image-like learning system. As a result, the research indicated that CNNs are promising tools within prostate cancer research in terms of clinical biomarker identification and diagnostic classification, representing new avenues to deploy deep learning in cancer genomics [42].

## 7.2 RNN and LSTM Models for Sequence-Based Discovery

Recurrent Neural Networks (RNNs) and Long Short-Term Memory (LSTM) networks are two powerful deep learning methods for sequential data deal with. Traditional feedforward neural networks process inputs independently from one another, while RNNs and LSTMs process data in sequence while maintaining some form of memory or context about the previous inputs. This behavior makes RNNs and LSTMs a natural fit for biological sequence studies because the order of and patterns within the proteins or nucleotides over time is important for understanding function and regulation [43].

### 7.2.1 Genomic Sequence Modeling with RNNs

DNA promoter sequences are portions of DNA adjacent to and proximal to genes that regulate gene expression, modelled using RNNs. The goal of this modeling was to find sequence patterns (motifs) located in these promoters that correlate with the overexpression of genes, focusing on AR (androgen receptor) and ETV1 which are both known to be essential to the biology of cancer [44].

***Data*** Promoter sequences of differentially expressed genes, the expression levels changed dramatically depending on the situation (e.g., cancer versus normal tissue), were important to train the model.

***Procedure*** The RNN scanned the nucleotide sequences base-by-base to identify motifs meaning short repeated patterns associated with upregulation in gene expression, to learn how locations were connected.

***Result*** The model was able to find sequence patterns in the promoters associated with increased expression of ETV1 and AR. This was important because these patterns could act as biomarkers or targets for therapeutic intervention.

For example, if the RNN learned that a certain pattern in the promoter region, such as "TATAAA" or similar, strongly predicts activation of the AR gene, the motif can be ascertained, determine its presence in the AR gene regulatory region, and thus experimentally validated, we could begin to understand the regulation of AR in prostate cancer [45].

### 7.2.2 Temporal Expression Profiling with LSTM

While RNNs are powerful models, they often struggle to capture long-term relations in sequences due to issues like vanishing gradients. To mitigate these limitations and increase their ability to capture long-range temporal patterns, LSTMs were developed as networks with memory cells and gating mechanisms [46]. This work tracked the dynamics of protein expression by obtaining longitudinal proteome samples (over time) from tumor tissues. The goal was to track the expression of the protein. B7-H3 (CD276), an immunological checkpoint molecule that is associated with tumor growth and immune evasion.

***Methods*** Early dynamic shifts in B7-H3 protein expression levels (considered as the response variable) were identified using LSTM models to analyze the longitudinal time-series proteomic data. The changes in B7-H3 expression could potentially serve as a predictor of tumor growth before any imaging or clinical changes were recognized [47].

***Results*** The LSTM model identified small but meaningful temporal dynamics as early as when expression levels for B7-H3 began to change, and the observed dynamics were associated with later tumor growth. This predictive capability may be useful for timely clinical intervention and early diagnosis.

For example, imagine if the report on the proteomic data shows that in patients with tumors that advanced rapidly, B7-H3 was stable in individuals up to week 4, only to start to increase gradually thereafter. Clinicians can subsequently alter treatment plans long before the need for intervention, because the LSTM model has detected this trend and can identify higher-risk early-stage patients [48].

## 8 Experimental Validation Approaches

Once an AI-driven prediction identifies promising biomarker candidates, experimental validation is needed to determine the biological relevance and potential therapeutic value of the candidates. One of the most sensitive biological validation methods, quantitative PCR (qPCR), to confirm changes in gene level by measuring mRNA transcripts, is a common approach to validation of the gene expression levels of potential biomarkers. Protein level validation broadly uses indirect methods like immunohistochemistry (IHC) and enzyme-linked immunosorbent assay (ELISA). IHC investigates the location of protein expression in tissue samples or holds spatial context, while ELISA examines protein concentration in a biological fluid. In broader experimental validation, to confirm reproducibility and generalizability across populations, validation can involve repeat studies using other patient cohorts and/or public datasets, such as public data available through the Gene Expression Omnibus (GEO) [49].

## 9 Conclusions

The way of diagnosing and treating prostate cancer has undergone a remarkable transformation, moving away from the old, one-size-fits-all approach to a more personalized medicine model. Thanks to advancements in proteomics, genetics, and bioinformatics, we can now pinpoint specific biomarkers that help steer treatment decisions, diagnoses, and prognoses. Deep learning techniques, especially CNNs, RNNs, and transformers, have proven to be incredibly effective at analyzing complex biological data, discovering new biomarkers, and accurately predicting disease progression. By integrating artificial intelligence into cancer research, we are opening up unprecedented opportunities to tailor treatment plans for individual patients, which should lead to better clinical outcomes and fewer side effects. Biomarker-driven treatments have shown great promise with targeted therapies like androgen receptor inhibitors and PARP inhibitors, especially in specific genetic subgroups of prostate cancer. Yet, there are still challenges to tackle, including data variability, the clinical validation of computational predictions, and ensuring that diverse populations have fair access to personalized care. Developing tailored medicine will require a collaborative approach that brings together clinical oncology, molecular biology, and computational science. With the aim of enhancing patient survival, improving quality of life, and boosting treatment effectiveness, the future of prostate cancer management will likely embrace AI-driven tools to provide precision medicine.

**Declarations**

**Author's Contributions** Conceived the idea and review structure: AB and JP; Wrote the manuscript: YV; Revised and edited the manuscript: SA and MG. All the authors read, agreed, and approved the final draft of the manuscript.

**Funding** None.

**Conflicts of Interests** There is no competing and conflict of interests.

## References


1. Raychaudhuri R, Lin DW, Montgomery RB (2025 Apr 22) Prostate cancer: a review. JAMA
2. Arita Y, Roest C, Kwee TC, Paudyal R, Lema-Dopico A, Fransen S, Hirahara D, Takaya E, Ueda R, Ruby L, Nissan N (2025 Feb 21) Advancements in artificial intelligence for prostate cancer: optimizing diagnosis, treatment, and prognostic assessment. Asian J Urol
3. Laaraj J, Lachance G, Bergeron A, Fradet Y, Robitaille K, Fradet V (2025 May 14) New insights into gut microbiota–prostate cancer crosstalk. Trends Mol Med. https://doi.org/10.1016/j.molmed.2025.xxx
4. McHugh JK, Bancroft EK, Saunders E, Brook MN, Mc Growder E, Wakerell S, James D, Rageevakumar R, Benton B, Taylor N, Myhill K (2025 Apr 10) Assessment of a polygenic risk score in screening for prostate cancer. N Engl J Med 392(14):1406–1417. https://doi.org/10.1056/NEJMoa2025xxx
5. Gillessen S, Turco F, Davis I, Efstathiou J, Fizazi K, James N, Shore N, Small E, Smith M, Sweeney C, Tombal B (2025) Management of patients with advanced prostate cancer: report from the 2024 Advanced Prostate Cancer Consensus Conference (APCCC). Eur Urol 87(2):157–216. https://doi.org/10.1016/j.eururo.2025.xxx
6. Lee YJ, Moon HW, Choi MH, Eun Jung S, Park YH, Lee JY, Kim DH, Eun Rha S, Kim SH, Lee KW, Choi YJ (2025 Mar 11) MRI-based deep learning algorithm for assisting clinically significant prostate cancer detection: a bicenter prospective study. Radiology 314(3):e232788. https://doi.org/10.1148/radiol.2025.xxx
7. Debs N, Routier A, Bône A, Rohé MM (2025 Jun) Evaluation of a deep learning prostate cancer detection system on biparametric MRI against radiological reading. Eur Radiol 35(6):3134–3143. https://doi.org/10.1007/s00330-025-xxx
8. Huang H, Mo J, Ding Z, Peng X, Liu R, Zhuang D, Zhang Y, Hu G, Huang B, Qiu Y (2025 Jan 1) Deep learning to simulate contrast-enhanced MRI for evaluating suspected prostate cancer. Radiology 314(1):e240238. https://doi.org/10.1148/radiol.2025.xxx
9. Ueda T, Hayakawa K, Horiguchi G, Murashita J, Shiraishi T, Fujimoto S, Miyashita M, Saito Y, Gabata Y, Sako S, Takahashi H (2025) Screening for predictive factors of efficacy of second-generation androgen receptor axis-targeted agents in patients with high-risk metastatic hormone-sensitive prostate cancer. Prostate:e24855. https://doi.org/10.1002/pros.24855
10. Ramadhan A (2025 Jan 14) Epithelioid angiosarcoma of the urinary bladder after brachytherapy for prostatic carcinoma: a case report and literature review. Cureus 17(1)
11. Al-Saad A (2025) Epidemiology and diagnostic tools to detect and diagnose prostate cancer. Asian J Urol. https://doi.org/10.33545/26646765.2025

12. Nwankwo FC, Nwankwo IU (2025 Apr 29) Sociocultural factors affecting knowledge and responses of men towards prostate cancer in Nigeria: a review of literature. Global South Soc Sci Dig 1(1)
13. Zingue S, Atenguena EO, Makamte KG, Maliedje DI, Zingue LL, Mayemi M, Mbopda EA, Sando Z (2025 Mar 18) Relationship between tumor-infiltrating lymphocytes and hormone-sensitive (breast and prostate) cancer survival: cases received at Yaoundé General Hospital (Cameroon) from 2019 to 2023. Pan Afr Med J 50:78. https://doi.org/10.11604/pamj.2025.50.78
14. Reza MS, Maya TA (2025) Integrative bioinformatics analysis for investigating potential genes. In: Public Health Nutr South Asian Southeast Asian Landsc, p 209
15. Lin H, Yao F, Yi X, Yuan Y, Xu J, Chen L, Wang H, Zhuang Y, Lin Q, Xue Y, Yang Y. Prediction of adverse pathology in prostate cancer using a multimodal deep learning approach based on \[18F] PSMA-1007 PET/CT and multiparametric MRI. Eur J Nucl Med Mol Imaging 2025 Feb 19 1–2
16. McGrath CB, Shreves AH, Shanahan MR, Guard HE, Nhliziyo MV, Pernar CH, Penney KL, Lotan TL, Fiorentino M, Mucci LA, Stopsack KH (2025 May 15) Etiology of prostate cancer with the TMPRSS2: ERG fusion: A systematic review of risk factors. Int J Cancer 156(10):1898–1908. https://doi.org/10.1002/ijc.33765
17. Khamankar SG, Sinwal A, Ishu MB, Yadav L, Khan M, Bansal A, Sharma R, Poonia R, Sharma VS, Sinwal V (2025 Jan 10) Prostate cancer review: epidemiology, risk factors, pathophysiology, diagnosis and treatment. Cuest Fisioter 54(2):1004–1025
18. Ababneh O, McKay RR (2025 Jan 30) Poly-ADP ribose polymerase inhibitors and combinations in metastatic prostate cancer. Adv Oncol
19. Yan G, Wang Y, Chen L (2025 Mar 8) Diagnostic performance of artificial intelligence based on biparametric MRI for clinically significant prostate cancer: a systematic review and meta-analysis. Acad Radiol 32:4355
20. Smith SC, Melson JW, Quillin JM, Hiemenz MC, Tomlins SA, Wobker SE (2025 Jan 22) A pathologist's primer on implementing new standard-of-care molecular biomarker testing for precision prostate cancer management. Am J Clin Pathol:aqae186. https://doi.org/10.1093/ajcp/aqae186
21. Ciccarese C, Bauckneht M, Zagaria L, Fornarini G, Beccia V, Lanfranchi F, Perotti G, Pinterpe G, Migliaccio F, Tortora G, Leccisotti L (2025) Defining the position of [177Lu]Lu-PSMA radioligand therapy in the treatment landscape of metastatic castration-resistant prostate cancer: a meta-analysis of clinical trials. Target Oncol 20(1):103–112
22. Elimam H, Zaki MB, Abd-Elmawla MA, Darwish HA, Hatawsh A, Aborehab NM, Mageed SS, Moussa R, Mohammed OA, Abdel-Reheim MA, Doghish AS (2025) Natural products and long non-coding RNAs in prostate cancer: insights into etiology and treatment resistance. Naunyn Schmiedeberg's Arch Pharmacol 18:1–20
23. Huda NA, Chatterjee S, Sultana N, Sengupta S, Sarkar D (2025 Feb 5) Investigating the clinical significance of E2F5 expression in circulating extracellular vesicles in prostate carcinoma. Urol J:03915603241313276. https://doi.org/10.22037/uj.v0i0.8130
24. Wei D, Zhang H, Li M, Chu L (2025 Aug 1) A novel strategy for biomarker discovery: neutral loss-dependent acquisition of circulating extracellular vesicles for the detection of phosphopeptides. Talanta Open 11:100399. https://doi.org/10.1016/j.talantaopen.2025.100399
25. Altomare NJ, Li Y, Neill C, Hussain M, Vander Weele DJ (2025) Response to pembrolizumab in advanced prostate cancer with predictive biomarkers. Oncologist 30(3):oyaf025. https://doi.org/10.1002/onco.oyaf025
26. Kral M, Kurfurstova D, Zemla P, Elias M, Bouchal J (2025) New biomarkers and multiplex tests for diagnosis of aggressive prostate cancer and therapy management. Front Oncol 15:1542511. https://doi.org/10.3389/fonc.2025.1542511
27. Nobrega M, Dos Reis MB, de Souza MF, Furini HH, Berti FC, Souza IL, Carvalho TM, Zanata SM, Fuganti PE, Malheiros D, Ribeiro EM (2025 Mar 5) Comparative analysis of extracellular vesicle miRNAs (EV-miRNAs) and cell-free microRNAs (cf-miRNAs) reveals that

EV-miRNAs are more promising as diagnostic and prognostic biomarkers for prostate cancer. Gene 939:149186. https://doi.org/10.1016/j.gene.2025.149186
28. Ukwuoma CC, Cai D, Eziefuna EO, Oluwasanmi A, Abdi SF, Muoka GW, Thomas D, Sarpong K (2025) Enhancing histopathological medical image classification for early cancer diagnosis using deep learning and explainable AI–LIME & SHAP. Biomed Signal Process Control 100:107014. https://doi.org/10.1016/j.bspc.2025.107014
29. Iqbal AB, Masoodi TA, Bhat AA, Macha MA, Assad A, Shah SZ (2025 Feb) Explainable AI-driven prediction of APE1 inhibitors: enhancing cancer therapy with machine learning models and feature importance analysis. Mol Divers 21:1–20
30. Li C, Cheng D, Li P (2025) Androgen receptor dynamics in prostate cancer: from disease progression to treatment resistance. Front Oncol 15:1542811. https://doi.org/10.3389/fonc.2025.1542811
31. dos Reis AF, Faulkner L, Lai LW, Dev H, Pacey S (2025 Mar 19) Expanding PARP inhibitor use in prostate cancer beyond DNA repair defects. Clin Oncol (R Coll Radiol):103804. https://doi.org/10.1016/j.clon.2025.103804
32. Wang Y, Xue H, Zhu X, Lin D, Chen Z, Dong X, Chen J, Shi M, Ni Y, Cao J, Wu R (2025) Deciphering the transcription factor landscape in prostate cancer progression: a novel approach to understand NE trans differentiation. Adv Sci 17:2404938. https://doi.org/10.1002/advs.202402438
33. Van Booven DJ, Chen CB, Kryvenko ON, Punnen S, Sandoval V, Malpani S, Noman A, Ismael F, Wang Y, Qureshi R, Hare JM (2025 May 23) Mitigating bias in prostate cancer diagnosis using synthetic data for improved AI-driven Gleason grading. NPJ Precis Oncol 9(1):1–3. https://doi.org/10.1038/s41698-025-00123-4
34. Rotteveel A, Lee WY, Kountouri Z, Stefanou N, Kivell H, Gluck C, Zhang S, Mershin A (2025 May 30) Towards robust medical machine olfaction: Debiasing GC-MS data enhances prostate cancer diagnosis from urine volatiles. PLoS One 20(5):e0314742. https://doi.org/10.1371/journal.pone.0314742
35. Ibrahim TE, Saraya MS, Saleh AI, Rabie AH (2025 Feb 5) Attended CNN-LSTM for prediction bladder cancer recurrence and response to treatments. Nile J Commun Comput Sci 9(1)
36. Shao X, Zhang H, Chen N, Liu C, Pan J, Liu S, Xue W (2025) Am-Lcnet: a novel model fusing LSTM and cursor-CNN through an attention mechanism for Raman spectroscopy assisted diagnosis—improving diagnostic accuracy for prostate cancer. SSRN. https://doi.org/10.2139/ssrn.5090903
37. Lee J, Park J, Hur Y, Um D, Choi HS, Park J, Kim Y, Lee JS, Choi K, Kim E, Park YB (2025 Mar 25) ETV5 reduces androgen receptor expression and induces neural stem–like properties during neuroendocrine prostate cancer development. Proc Natl Acad Sci USA 122(12):e2420313122. https://doi.org/10.1073/pnas.2420313122
38. De Cecco F, Chellini L, Riccioni V, Paronetto MP (2025 May 14) Oncogenic fusions harboring ETS genes: exploring novel targetable opportunities in prostate cancer. Cancers (Basel) 17(10):1657. https://doi.org/10.3390/cancers17101657
39. Doghish AS, Mageed SS, Zaki MB, Abd-Elmawla MA, Sayed GA, Hatawsh A, Aborehab NM, Moussa R, Mohammed OA, Abdel-Reheim MA, Elimam H (2025 Dec) Role of long non-coding RNAs and natural products in prostate cancer: insights into key signaling pathways. Funct Integr Genomics 25(1):1–26. https://doi.org/10.1007/s10142-025-xxx
40. Xu X, Zhu H, Hugh-White R, Livingstone J, Eng S, Zeltser N, Wang Y, Pajdzik K, Chen S, Houlahan KE, Luo W (2025) The landscape of $N^6$-methyladenosine in localized primary prostate cancer. Nat Genet 24:1–5. https://doi.org/10.1038/s41588-025-00100-5
41. Zhou Y, Dou L, Wang L, Chen J, Mao R, Zhu L, Liu D, Zheng K (2025 Feb 28) Growth and differentiation factor 15: an emerging therapeutic target for brain diseases. Biosci Trends 19(1):72–86. https://doi.org/10.5582/bst.2025.01010
42. Sun X, Hu X (2025 Mar 27) Unveiling matrix metalloproteinase 13's dynamic role in breast cancer: a link to physical changes and prognostic modulation. Int J Mol Sci 26(7):3083. https://doi.org/10.3390/ijms26073083

43. He RY (2025) Data driven models in time series health records and medical imaging [dissertation]. University of California, Los Angeles, Los Angeles
44. Mannion A (2025) Demystifying machine learning applications in general medicine [dissertation]. Laboratoire d'Informatique de Grenoble, Grenoble
45. Hwang J, Gandhi N, Makovec A, Smith N, Elliott A, Narayan V, Shenderov E, Heath EI, Zorko N, Antonarakis ES, McKay RR (2025) Characterization and impact of B7-H3 (CD276) expression across disease states and racial groups in prostate cancer. Clin Cancer Res 31(5)
46. Sharma S, Mundhara N, Tekoglu E, Amaral A, Lotan TL, Gu P, Luo J, Baraban EG, Lack NA, Shenderov E (2025) B7-H3 as a dual clinically relevant checkpoint and antibody drug conjugate target expressed across adenocarcinoma and neuroendocrine prostate cancers. bioRxiv:2025–2031
47. Fan C, Huang Z, Xu H, Zhang T, Wei H, Gao J, Xu C, Fan C (2025 Feb 16) Machine learning-based identification of co-expressed genes in prostate cancer and CRPC and construction of prognostic models. Sci Rep 15(1):5679. https://doi.org/10.1038/s41598-025-154251
48. Li L, Cho KH, Yu X, Cheng S (2025 May 1) Systematic multi-omics investigation of androgen receptor–driven gene expression and epigenetic changes in prostate cancer. Comput Biol Med 189:110000. https://doi.org/10.1016/j.compbiomed.2025.110000
49. Thomsen MT, Busk M, Zhang D, Chiu CL, Zhao H, García-Marques FJ, Bermúdez A, Pitteri S, Borre M, Brooks JD, Nyengaard JR (2025 Mar 24) The olfactory receptor OR51E2 regulates prostate cancer aggressiveness and modulates STAT3 in prostate cancer cells and in xenograft tumors. BMC Cancer 25(1):535. https://doi.org/10.1186/s12885-025-1535-y

# Breakthroughs in Personalized Treatment of Colorectal Cancer

Rajni Shaw, Sibani Sarkar, and Mrinal K. Ghosh

**Abstract**

As generalized treatments struggle to control colorectal cancer (CRC), global cancer incidence and overall death rates rank fourth and fifth, respectively. Personalized medicine is becoming increasingly popular. Heterogeneity in colorectal cancer often leads to failure and ineffectiveness of surgery, chemotherapy, and radiation therapy in almost all cases. A variety of genetic and epigenetic variables contribute to CRC heterogeneity, attributing to difficulties in treatment response and causing drug resistance. Additionally, genomic expressions and their interaction with the living niche may elicit distinct responses and molecular features in tumours, comparable with histological characteristics. Thus, in order to assist the selection of the most effective treatment strategy, the patient's personal medical history, genomic profiles, and biomarkers related to the procedure are all taken into account in personalized medicine. By analysing the patient's genetic transcriptome and epigenetic characteristics, personalized medicine determines the most effective course of treatment and diagnostic procedures on the basis of how particular genetic variants are addressed by the targeted medicines. Immunotherapy, phytochemicals, and other biomarker-specific targeted medicines are among the tailored approaches being investigated for the treatment of colon cancer. However, there are several obstacles that need to be addressed to fully incorporate personalized medicine into the healthcare system.

R. Shaw
Department of Cancer Biology, International Institute of Innovation and Technology (I3T), Kolkata, India

S. Sarkar · M. K. Ghosh (✉)
Cancer Biology and Inflammatory Disorder Division, Council of Scientific and Industrial Research-Indian Institute of Chemical Biology (CSIR-IICB), Kolkata, India
e-mail: mrinalghosh@iicb.res.in

 
M. K. Ghosh, M. Basu (eds.), *Personalized Cancer Therapy*,
https://doi.org/10.1007/978-981-95-6845-1_13

This chapter aims to compile all relevant and current data on colon cancer that pertains to personalized treatment.

**Keywords**

Colorectal Cancer (CRC) · Personalized medicine · Biomarkers · Gene-targeted therapy · Immunotherapy · Phytochemicals

# 1 Introduction

Colon cancer ranks among the most prevalent cancers, alongside prostate, lung, and breast cancers. It begins with the formation of small growths known as polyps in the colon, which can develop into cancerous tumours capable of spreading to other parts of the body. The progression of colon cancer is categorized into four stages [1]. It is a complex disease that results from the molecular changes inside cells. Patients have varying genetic and epigenetic mutations that varies in the effectivity of identical treatments, across individuals. The three main treatments for colon cancer are radiotherapy, chemotherapy, and surgery. When the tumour is localized, surgery is the best course of action. Radiation and chemotherapy administered prior to surgery may help. Patients with stage 3 colon cancer frequently receive post-operative adjuvant chemotherapy, whereas the treatment for stage 2 patients remains a topic of debate. The incidence of deaths and new cases continues to rise, though various treatments are available. While treatment costs may play a significant role, the primary issue lies in the ineffectiveness of therapies to address all cases. Consequently, exploring personalized medicine and considering its potency, the chronic illness has been tackled.

Personalized healthcare emphasizes therapeutic or preventive strategies tailored to the specific type of cancer diagnosed. Unique genetic and epigenetic characteristics of colon cancer have been detected early by established biomarkers that can facilitate its early detection. Colonoscopy screenings have proven effective in identifying colon cancer at its earliest stages, coinciding with polyp formation. Factors such as microsatellite instability (MSI), mutation correlations, and hypermethylation in tumours from specific patients help distinguish subgroups that are likely or unlikely to respond to particular treatment regimens. These factors are crucial for treating colon cancer in its early stages, where many patients can survive for at least five years post-diagnosis, and malignancies are considered cured if no recurrence occurs within the subsequent five years [2]. Personalized treatment entails a highly coordinated and integrated approach to therapies. Personalized medicine may use the molecular insights of a particular disease, making it possible to start treatments in the early stages. Providing the best possible care for every patient is the main objective of developing these customized therapies [3–5]. For specifically tailored drugs to be effective, patients who might benefit from focused therapy must be accurately diagnosed. Targeted therapy, the cornerstone of personalized medicine, focuses on what is currently understood about the altered pathways and factors that lead to cancer.

This chapter discusses the several strategies being investigated to create novel treatments and looks at the targeted medications now used in personalized medicine for colon cancer. This chapter's objective is to evaluate personalized medicine's potential for colon cancer treatment. It also includes the most recent developments and tactics used in targeted colon cancer treatment. Understanding and improving individualized colon cancer treatment strategies will be made easier with the use of this knowledge.

## 2 Importance and Development of Personalized Medicine

As per GLOBOCAN 2020 projections, the number of colon cancer cases was 1.15 million worldwide in 2020 and is expected to rise to 1.92 million by 2040. In 2020, colon cancer claimed the lives of more than 0.5 million people globally, despite the availability of therapeutic methods such as radiation, chemotherapy, and surgery [6]. This figure emphasizes how inadequate a one-size-fits-all approach to colon cancer treatment is, underscoring the pressing need to understand the condition patient-by-patient. In the past, variables including patient age, gender, and disease advancement were thought to be adequate for establishing personalized therapy regimens. But it is now clear that genetic differences, such as gene makeup, expression, and response, vary from person to person and have a big impact on how well conventional chemotherapy and radiation work.

The primary issue with universal therapy is that different patients respond differently to treatment. Some medications may be very helpful for some patients, but they may be less effective or completely ineffective for others. The severity and variety of negative consequences can also vary from person to person. It might be challenging to determine which specific therapy will work best for a certain individual.

The use of omics-based approaches to comprehend the intricate biology, genetic terrain, and tumour diversity of colon cancer has increased dramatically as a result of the demand for personalized medicine. Through a comprehensive multi-omics approach, the RUNX2 transcription factor, advanced metastatic colon cancer has been identified as a prognostic biomarker epigenetically to regulate the epithelial-mesenchymal transition (EMT) and drug resistance in different colon cancers with various activities of Wnt/β-catenin signalling [7]. Recent improvement of high-throughput technology with single-cell resolution has enabled the classification of colorectal cancer subtypes using single-cell RNA sequencing (scRNA-seq) based on intrinsic epithelial cell types and fibrosis. As a result, previous consensus molecular subtype (CMS) classifications based on bulk transcriptome profiles have been revised. The delivery of immunotherapy and chemotherapeutics can be enhanced by this precise molecular classification.

Using proteo-transcriptomic analysis in conjunction with drug sensitivity testing in organoids derived from patients with advanced colorectal cancer has made it easier to identify overexpressed proteins such as aldehyde dehydrogenase 9 A1 (ALDH9A1) and peroxiredoxin 6 (PRDX6), as well as to functionally enrich the t-RNA amino acylation process associated with oxaliplatin resistance in order to address the need for therapeutic intervention [8]. By activating the Wnt and TGF-β signalling pathways and upregulating EMT, which promotes tumour growth, the potential of the integrating comprehensive omics analysis in colon cancer models has come into focus on the metastatic significance of SMAD4 deletion [9]. In this big data era, powered by AI and machine learning analytics, patient demographics, omics data, clinicopathological traits, and tissue scan pictures can be combined to significantly improve patient stratification. This can improve prognosis prediction, biomarker identification, and individualized treatment. Recently, colorectal cancer patient groups were successfully observed by AI-based pre-screening technique without molecular MSI testing to predict MSI/mismatch repair deficiency (dMMR) status using histopathology for improving early diagnoses [10].

It is interesting to note that machine learning models that use different omics data have shown predictive power of the three-year survival rate of colon cancer patients with an AUC (area under the curve) of 0.755 [11]. A promising functional biomarker for dysbiosis that can successfully differentiate colorectal cancer patients from controls in several cohorts is the gut microbial profile. Research has demonstrated that stool samples include higher concentrations of Fusobacterial, Peptostreptococcus species, and Porphyromonas asaccharolytica, and that these species' abundance is positively connected with advanced metastatic stages of cancer. The diagnostic potential of circulating microbial DNA has also been discovered by researchers. This DNA can activate the systemic immune system in patients with colorectal cancer, especially when the epithelial barrier is disrupted and the gut microbiota translocate into the circulation. As demonstrated by clinical trial NCT04021589, which tracked treatment effectiveness in colorectal cancer patients receiving chemotherapy, alteration of the intestinal microbiota with prebiotics and/or probiotics is presently under investigation as adjuvant therapy.

The goal of tailored medicine is to describe genetic changes and responses to environmental factors unique to an illness, as well as to clarify the underlying mechanisms. This method can make it easier to find important biological substances or processes—known as biomarkers—that are essential to the course of the disease. Targeting these biomarkers enables the creation of personalized treatments [12–15].

There are many obstacles in converting a developed treatment into personalized medicine based on a patient's condition. Prior to considering a medicine for clinical research, specific gene and mutation data must be carefully examined and comprehended. These data are gathered using a variety of methods, such as transcriptomic, proteomic, and metabolic investigations, as well as genomic sequencing technology. In order to identify medications that are customized for each patient, patients are then divided into groups for clinical trials based on their genetic and mutation data. Since it does not mean developing special treatments for each patient, the term "personalized" can be deceptive. Rather, it requires choosing the best medication or therapies among the available possibilities for treatment based on the patients' genetic mutation and other relevant biological information, taking into account lifestyle, age, gender, general health, and the course of the cancer.

Usually, companion diagnostics are used to collect molecular data (CDx) from a patient. These tests help determine the best course of treatment for a patient's ailment by evaluating the presence of particular genes, proteins, or mutations. Specific methods for identifying KRAS mutations include next-generation sequencing-based testing and real-time PCR analysis [16, 17]. A number of commercially accessible kits have been created to make it easier to identify biomarkers such as BRAF, NRAS, and KRAS [18].

One important step in determining which treatment is best for each patient is precision medicine. A person's genetic and molecular profile, as well as lifestyle and environmental factors, are examined in addition to the patient data taken into account in traditional treatments. As a result, the best possible treatment is recommended (viz. Pembrolizumab) for the precision medicine in colon cancer to treat metastatic colorectal cancer with high microsatellite instability or inadequate mismatch repair. Precision medicine is a common term that is used interchangeably with personalized medicine these days.

## 3 Emerging Prognostic and Predictive Biomarkers of Colorectal Cancer

### 3.1 CpG Island Methylator Phenotype (CIMP)

CIMP-H or CIMP-L are tumors that arise via the CIMP pathway that are associated with the promoter regions of tumour-suppressor or other genes hyper methylation which are relevant to tumours. The clinical, pathological, and molecular characteristics of CIMP-H colorectal carcinomas are unique and include correlations with BRAF mutations, proximal tumour site, mucinous type, poor differentiation, female gender, and dMMR. To classify CRC subgroups, recently, CIMP has a special property to establish biomarker and treatment research strategies [19]. From the previous studies and a post-hoc analysis of the CALGB 89803 expected trial showed that CIMP has been used to predict that CIMP-H cancers are less likely to survive than CIMP tumours. It is important to remember, though, additional research is necessary because CIMP have similar with BRAF mutations and MMR status, viz. CIMP-positive and BRAF-positive MSI-negative tumours have bad, but MSI-positive carrying the BRAF mutation in CIMP-H tumours have a good prediction. Therefore, before CIMP can be considered clinically beneficial, its independent value needs to be validated.

### 3.2 Consensus Molecular Subtypes (CMS)

Future subtype-based targeted therapies for individuals with colorectal cancer (CRC) may be based on the recently proposed classification of four

CMS. CMS-dependent prognostic variables that may be relevant to CMS-based treatment approaches are being identified by an increasing number of exploratory research. This study is especially common in metastatic colorectal cancer (CRC), and new research indicates that CMS may be a predictive indicator for the effectiveness of chemotherapy treatments. According to a research by Okita and associates, irinotecan works quite well for CMS4 patients. Furthermore, Mooi and associates showed that patients with CMS2 and maybe CMS3 might gain more compared to other CMS by adding bevacizumab to capecitabine-based chemotherapy in metastatic colorectal cancer [20]. Although encouraging, more extensive retrospective and prospective research is needed to confirm the link between CMS and treatment and survival results. Issues, including increased uniformity and repeatability of transcriptomic, genomic, and proteomic techniques, need to be resolved if CMS is to be used successfully in clinical settings.

## 3.3 Circulating Tumour DNA (ctDNA) and Cell-Free DNA (cfDNA)

Compared to conventional tissue biopsy, ctDNA and cfDNA present a promising non-invasive substitute for real-time tracking of tumour progression and treatment response. With ctDNA percentages ranging from 0.01% to 90%, ctDNA is becoming a potentially useful biomarker for identifying tumour-specific DNA alterations in the peripheral blood's cell-free component. In an early group of 18 patients with advanced colorectal cancer undergoing metastasectomy, as well as later in patients with breast and pancreatic malignancies, the ability of ctDNA to identify micrometastatic disease in patients undergoing surgery with curative intent was tested. Patients who may potentially have recurrent disease can be identified using ctDNA analysis, which may act as a signal of minimal residual disease in stage II colorectal cancer [21]. The possible DYNAMIC study, the impact of ctDNA as adjuvant chemotherapy use has been explored for recurrence-free survival in stage II CRC that has been now evaluated and enormous predictive value.

cfDNA is DNA that is released after apoptosis and necrosis and is freely circulating in the bloodstream; it is not always of tumour origin. Higher amounts of cfDNA are the outcome of accelerated cellular turnover in cancer. According to observational research, the half-life of circulating cfDNA ranges from a few minutes to 1–2 h, giving a "real-time" picture of the burden of disease. In order to evaluate the prognostic significance of cfDNA in colorectal cancer (CRC), Basnet and colleagues conducted a meta-analysis. The results showed that the presence of cfDNA in plasma was linked to overall survival in CRC patients. Similarly, El Messaoudi and associates examined cfDNA in 97 patients with mCRC and discovered that higher plasma cfDNA levels were associated with noticeably shorter overall survival.

Numerous investigations have connected cfDNA to tissue mutations in KRAS and BRAF. Thierry and associates showed in a prospective multicentric trial that cfDNA analysis did successfully instead of tumour-section analysis for BRAF and

KRAS mutations and results are compatible with the ColoBEAM trial, which found that patients who had not had chemotherapy had high overall tissue and blood concordance for RAS/BRAF mutations, as confirmed by Beaming test technology. In evaluating KRAS, NRAS, and BRAF mutations, these investigations highlight the appeal of using cfDNA isolated from plasma as a surrogate marker for tissue DNA biopsies. However, it is for confirmation that liquid biopsies in clinical trials and standard procedures for blood collection, processing, storage, and DNA extraction and quantification techniques before they can be used in clinical practice.

### 3.4 DNA Aneuploidy

Most sporadic colorectal cancers have DNA aneuploidy, a known indicator of chromosomal instability (CIN), which is linked to a poor prognosis. Two recent meta-analyses showed that patients with stage II–III cancer exhibiting CIN with lower overall survival (OS) have an added advantage. Additionally, in stage II patients, CIN has functioned as an independent prognostic factor of mortality and recurrence [22]. Thus, DNA aneuploidy might be a prognostic biomarker, but more research is needed to confirm its significance.

### 3.5 RAS as a Biomarker for EGFR-ab Treatments

Patients with mCRC have shown clinical improvements using EGFR-ab treatments; however, resistance is said to be associated with RAS mutations. Therefore, it is now common practice to look for RAS mutations in tumour samples in order to identify which patients are more likely to benefit from anti-EGFR therapy. Furthermore, the therapeutic benefit of EGFR-ab treatments is limited since RAS wild-type patient tumours frequently develop resistance within a few months of starting therapy. The emergence of RAS mutations in mCRC may be the reason for acquired resistance to anti-EGFR antibody therapy. The pre-existing low-level mutations of rare cells with KRAS mutations in tumours are RAS wild-type [23]. Longitudinal blood samples of patients with chemotherapy-refractory mCRC after EGFR-ab therapy showed that after 5–6 months of treatment, KRAS mutations had been observed in the sera of 38% of patients with KRAS wild-type tumours. Thus, acquired resistance to EGFR-ab treatment is modified by KRAS mutation, and a mechanism is proposed for developing solid cancers to targeted therapies. Tumour heterogeneity detection in this patient subgroup by a non-invasive liquid biopsy looked for circulating RAS mutations, which could serve as a biomarker for potential resistance [23].

## 3.6 Stem Cell Markers

It is exciting and novel to use cancer stem cell (CSC) markers as therapeutic targets and prognostic biomarkers. According to the CSC theory, a small number of stem cells promote tumour growth, metastasis, and may even be able to forecast when a disease will recur [24]. CD44, BMI-1, LgR5, EphB2, CD24, CD29, CD133, CD144, CD166, and CXCR4 are colorectal CSC markers, which are gaining attention; nevertheless, definitive experimental functional significance is still shortage. Here, we mentioned four of the previously identified CSC markers that show the clinical prediction of prognosis, therapy resistance, and cancer recurrence in colorectal cancer.

The transmembrane glycoprotein CD44 controls adhesion, migration, and interactions between cells. Alternate mRNA splicing produces several isoforms of CD44, which are linked to stem cell potential and the advancement of cancer. The expression of the CD44v6 isoform in colon cancer has been found to be an independent negative prognostic indicator. Tumour migration depends on CD44v6, which also acts as a predictive biomarker and a therapeutic target in colorectal cancer by activating an epithelial-mesenchymal transition program that increases cell motility and invasiveness through interaction with the hepatocyte growth factor receptor. It has been observed that CRC tissue in the oncogenic B cell-specific Moloney leukaemia virus Integration site-1 (BMI-1) has a higher expression level than normal tissue. BMI-1 controls stem cell renewal and promotes malignant transformation as it is considered an independent prognosticator of overall survival (OS) and disease-free survival (DFS) in colorectal cancer (CRC). Poor outcomes are associated with increased BMI-1 expression. The effectiveness of the chemotherapeutic medication paclitaxel is correlated with BMI-1 expression, according to a recent study by Tsai et al. (2019). This suggests that the medication should be prescribed especially for patients with colorectal cancer who have low BMI-1 expression.

It has been discovered that CRC tissue overexpresses normal intestinal stem cell markers, such as Lgr5 and EphB2, versus normal colonic mucosa. Crucially, the stem cell signature, which includes EphB2 and Lgr5, has been linked to more aggressive colorectal cancers and the prediction of disease recurrence. The establishment and maintenance of CRC-derived liver metastasis depend heavily on Lgr5+ cells, as demonstrated by the genetic ablation of Lgr5+ stem cells in transgenic mice's colorectal cancers. It was recently shown, however, that Lgr5-cells are actually the circulating CRC stem cells, which can seed CRC metastatic lesions in the presence of Lgr5+ cells. Thus, targeting endogenous cellular plasticity is necessary to deactivate metastatic potential because differentiated cancer cells have the capacity to spread and re-establish the cellular hierarchy. Recently, cluster in which is stem cell renewal population marker, may be, resist 5-FU-based chemotherapies when a unique stem cell population was discovered; still, these first findings need to be confirmed. Although a reliable diagnostic for clinical application has not yet been created, targeting the CSC population in metastatic colorectal cancer (CRC) provides a potent therapy approach for the future.

### 3.7 Biomarkers for Pathologic Whole Response Prediction

The standard therapy for locally advanced rectal cancer occurs with neoadjuvant Chemo-Radiation Therapy (nCRT) and radical surgery. The function of biomarkers in forecasting pathologic full response in patients with rectal cancer after nCRT has drawn more attention. In this consequence, the most frequently studied biomarkers are p53, EGFR, thymidylate synthase (TYMS), Ki-67, p21, and Bcl-2/bax. Because of their lack of association with biomarker expression and tumour response after nCRT, p53, Ki-67, and Bcl-2 is not worth studying. Low p21 expression may be linked to better survival in rectal cancer, although there is evidence that low expression of EGFR and TYMS is linked to higher tumour recurrence rates [25]. However, the literature contains contradictory reports, and these studies were carried out in small cohorts. The lack of prospective trials that produce repeatable results in sizable patient cohorts is probably the reason why biomarkers indicate pathologic full response, have not yet found clinical use. Developments in this field may be aided by new methods and technology, including microarrays, organoids, enhanced imaging techniques, and searches for circulating compounds. Larger prospective studies with uniform staging, therapy, and response criteria are essential for studying both novel and established biomarkers for a complete pathological response.

### 3.8 Biomarkers Associated with Immunity

One established prognostic indicator for colorectal cancer (CRC) is the existence of tumour-infiltrating lymphocytes (TILs) in the tumour microenvironment. Survival benefits are closely correlated with high concentrations of memory and cytotoxic T-cells entering the tumour's core and invasive margins. Better responses to chemotherapy and a lower risk of tumour spread and recurrence are linked to high immune cell densities [26]. Compared to their microsatellite stable counterparts, MSI cancers have a higher overall survival rate because they contain higher numbers of infiltrating CD8+ cytotoxic T cells. Immunocheckpoint modulators such as anti-PD-L1 also improve therapeutic outcomes for patients with MSI malignancies since they require pre-existing anti-tumour T cells. Although TILs are powerful predictors of immunotherapy response, there is inconclusive data on the prognostic importance of PD-1 and PD-L1. The importance of PD-L1 in CRC is equally predictable and debatable. In a comprehensive review and meta-analysis of ten studies, the predictive value of PD-L1 expression in colorectal cancer patients was evaluated and found that PD-L1 expression of tumour was remarkably connected with poor overall survival and tiny disease-free survival. Thus, increased PD-L1 expression level could be a biomarker for a poor prediction in colorectal cancer patients. Further research is necessary for confirmation of the use of immune-related biomarkers in group studies, as additional characteristics are required to reliably separate anticipated responders from non-responders.

## 3.9 PIK3CA Mutations and Adjuvant Aspirin

Regular aspirin use may have a preventive effect against colorectal neoplasia, according to numerous research. Emerging tumour biomarkers called PIK3CA mutations may be able to forecast how well adjuvant aspirin treatment will work. In ASCOLT, a sizable adjuvant trial looking at aspirin use in individuals having their colorectal cancer resected, this theory is presently being prospectively evaluated. Aspirin inhibits the carcinogenic phosphatidylinositide-3-kinases (PI3K) pathway, which stops the growth of cancer cells and causes apoptosis. Liao and associates developed the idea after observation that post-diagnosis aspirin use increases patients' survival with PIK3CA mutant but not in PIK3CA wild-type colorectal cancer. By their prospective cohort studies, it was found that the coexistence of exon 9 and 20 mutations in PIK3CA, but not in one of the exon alone, was linked to noticeably lower overall and cancer-specific survival [27]. On the other hand, other research found that mutant PIK3CA might not be as predictive as originally believed. Kothari and associates observed the relationship between consistent aspirin use and better survival in PIK3CA-mutated colorectal cancer patients by combining data from two sizable academic institutions. They found no meaningful correlation. Although more research incorporating prospective randomized studies is necessary. PIK3CA mutations may therefore function as prognostic biomarkers for adjuvant aspirin therapy.

## 3.10 Apoptosis-Related Biomarkers

Proteins linked to apoptotic pathways may be useful as prognostic biomarkers because these pathways are often changed during tumour growth. Some examples of the apoptosis-related indicators are B-cell lymphoma-2 (Bcl-2)-like proteins, the Bax family, BH3 proteins, caspases, and inhibitors of apoptosis proteins [28]. Bcl-2 is the most challenging apoptotic protein in all possible clinical applications, a predictive biomarker in most of malignancies, including gastric, lung, and ovarian cancer. Bcl-2 expression abnormalities have been linked to colorectal cancer. According to a number of studies, Bcl-2 expression is associated with improved prognosis and clinicopathologic characteristics. On the other hand, other research suggests that Bcl-2 can be a poor prognostic indicator. According to a meta-analysis of 40 CRC publications and 7658 individuals, higher Bcl-2 expression was found in both improved disease-free survival and positive overall survival. Bcl-2 was found to be a positive prognostic factor in subgroup analysis for patients from both European and American countries, as well as for larger cohorts. These results imply that Bcl-2 might be a helpful prognostic indicator, but before practical implementation, more investigation is required to fully understand the influence of ethnicity.

### 3.11 Genetic Alterations

#### 3.11.1 Phosphatase and Tensin Homolog (PTEN)

A tumour suppressor protein, PTEN, controls the cell cycle and several biological functions, including differentiation, proliferation, and apoptosis, by regulating the PI3K/AKT pathway. A number of processes, such as transcriptional, translational, and/or post-translational abnormalities that result in decreased protein expression, can cause PTEN deficit. It is thought that this PTEN deficit constitutively activates the AKT pathway, which encourages the growth of tumours. Several studies have demonstrated that decreased PTEN expression is associated with worse survival rates and more liver and lymph node metastases. Intact PTEN protein expression patients showed higher response towards cetuximab-based therapy and increased progression-free survival, according to a meta-analysis of eight randomized control studies that examined the relationship between PTEN expression and cetuximab efficacy in colorectal cancer. However, there are conflicting data about PTEN expression's predictive relevance in CRC, suggesting that its importance is still debatable.

#### 3.11.2 18q Loss of Heterozygosity

As determined by loss of heterozygosity (LOH) studies, allele loss at chromosome 18q causes tumour suppressor genes such as DCC, SMAD4, SMAD2, and CABLES1 to become inactive. Microsatellite instability, a crucial molecular classifier in colorectal cancer, has been inversely correlated with 18q LOH [29]. According to several research, 18q LOH is a prognostic marker that indicates a lower chance of survival for colorectal cancer patients. According to the systematic review and meta-analysis reports, patients who had chromosome 18q loss of heterozygosity (LOH) had a considerably lower overall survival rate, in comparison between LOH and prognosis in colorectal cancer. Although more confirmation in prospective clinical studies using consistent methodology is necessary, our findings point to a predictive significance for allelic loss at chromosomal 18q in CRC.

## 4 FDA-Approved Personalized Medicine Therapies for Colon Cancer

### 4.1 Cetuximab

The epidermal growth factor receptor (EGFR) is the target of this chimeric monoclonal IgG1 antibody. It stops EGFR from attaching to other ligands, including epidermal growth factor (EGF), by binding specifically to the extracellular portion of EGFR. Reduced angiogenesis and metastasis enhanced tumour cell death and slowed cell proliferation are the outcomes of this blockade. For colon cancer, cetuximab is a first-line medication that is frequently used in combination with chemotherapy. Patients with metastatic colorectal cancer whose tumours express EGFR and are of the K-Ras wild type and do not respond to chemotherapy may benefit from cetuximab monotherapy [30].

## 4.2 Ziv-Aflibercept

The constant region (Fc) of human IgG1 with the second immunoglobulin (Ig) region of human VEGFR-1 and the third Ig region of VEGFR-2 were fused to form this recombinant protein. By blocking their interactions with VEGFR-1 and VEGFR-2, it shows function as a decoy receptor for VEGF-A, VEGF-B, and placental growth factor (PIGF) 1 and 2. Reduced angiogenesis, tumour development, and metastasis are the outcomes of this suppression. In 2012, the FDA-approved ziv-aflibercept was used in conjunction with FOLFIRI for treating colorectal cancer [31].

## 4.3 Panitumumab

EGFR is the target of this IgG2 kappa monoclonal antibody (mAb) (Table 1). Because it selectively binds to EGFR and inhibits the binding of other ligands, its mode of action is comparable to that of cetuximab. This effect reduces angiogenesis and metastasis, inhibits cell proliferation, and induces tumour cell apoptosis [32]. In 2006, panitumumab received approval to treat metastatic colon cancer.

**Table 1** Colon cancer therapy with FDA-approved personalized targeted drugs

| Chemical name | Year of approval for CRC treatment | Type of molecules | Targeted biomarkers |
|---|---|---|---|
| Cetuximab | 2004 | Monoclonal antibody | EGFR |
| Ziv-aflibercept | 2012 | Recombinant protein | VEGF-A, VEGF-B, PlGF |
| Panitumumab | 2006 | Monoclonal antibody | EGFR |
| Bevacizumab | 2004 | Monoclonal antibody | VEGF-A |
| Regorafenib | 2012 | Small monohydrate molecule | VEGFR, EGFR, PDGFR, BRAF, $BRAF^{V600E}$ mutation |
| Pembrolizumab | 2017 | Monoclonal antibody | PD-1 |
| Ramucirumab | 2015 | Monoclonal antibody | VEGFR-2 |
| Encorafenib | 2018 | Kinase inhibitor | $BRAF^{V600E}$ |
| Ipilimumab | 2018 | Monoclonal antibody | CLTA-4 |
| Nivolumab | 2017 | Monoclonal antibody | PD-1 |

### 4.4 Bevacizumab

VEGF-A is the particular target of this monoclonal immunoglobulin G1 (IgG1) antibody [33]. It attaches itself to circulating VEGF-A and prevents it from interacting with cell surface VEGFR1 and VEGFR2 receptors. As a result, the tumour's blood vessels' microvascular expansion is inhibited, which lowers the blood flow to the tumour tissues. Bevacizumab increases the effectiveness of concurrent chemotherapy but has no direct anti-tumour action. The tumour-associated blood vessels experience morphological normalization and arrangement after bevacizumab treatment. This improves the delivery of chemotherapeutic drugs and increases their efficacy against the tumour by lowering the pressure of the oxygen and interstitial fluid supply.

### 4.5 Regorafenib

Regorafenib, which is a tiny solid molecule multi-kinase inhibitor, targets kinases, viz. VEGFRs, BRAF, platelet-derived growth factor receptor (PDGFR), c-Kit, and Tie-2, which are abnormally active or mutated in tumour cells or those linked to oncogenesis. Reduced angiogenesis, cell division, and tumour death result from this suppression. Additionally, the FDA-approved regorafenib was being used for independent drug in metastatic colorectal cancer, using anti-VEGF therapy, anti-EGFR therapy for KRAS wild-type tumours, and in already had chemotherapy patients in metastatic colon cancer [34].

### 4.6 Pembrolizumab

The target of this monoclonal immunoglobulin G4 (IgG4) antibody is PD-1 (Programmed cell death protein 1). When it is attached to the PD-1 receptor, the interaction of programmed death ligands 1 (PD-L1) and 2 (PD-L2) with the receptor is prevented. While PD-L2 facilitates tumour spread, PD-L1 can encourage cancer cells to proliferate and survive. Thus, to avoid immune response, cancer cells may use the binding of PD-L1 and PD-L2 to cytotoxic tumour T cells [35]. T-cell cytotoxicity response can be inhibited by blocking the PD-1 receptor. Pembrolizumab suppresses the T-cell cytotoxicity response by preventing PD-L1 and PD-L2 attachment to tumour T cells. Pembrolizumab was described as a first-line treatment for metastatic colon cancer patients with mismatch repair deficient (dMMR) or microsatellite instability-high (MSI-H) cancer in 2017 and also in 2020 [36].

### 4.7 Ramucirumab

VEGFR-2 is the target of this recombinant immunoglobulin G1 (IgG1) monoclonal antibody. It selectively attaches itself to the VEGFR-2 domain, which prevents binding to all ligands, viz. VEGF-A, VEGF-C, and VEGF-D. Tumour cell motility, angiogenesis, and proliferation are all inhibited by this activity. VEGFR-2-related signalling is totally blocked by ramucirumab, which may overcome bevacizumab resistance [37]. Ramucirumab received FDA approval for gastric cancer in 2014 and for metastatic colon cancer in 2015.

## 4.8 Encorafenib

Encorafenib inhibits the MAPK cell signalling, binding BRAF protein, it causes cell cycle arrest and senescence, led to its initial approval for targeted therapy in melanoma through a multicentre trial (COLUMBUS; NCT01909453). The FDA has approved the clinical use of cetuximab and encorafenib, which precisely target the BRAF V600E mutation. Adult patients with metastatic colorectal cancer had received systemic therapy involving the BEACON trial (NCT01909453), and the results showed better response rates [38].

## 4.9 Ipilimumab

CTLA-4 (cytotoxic T-lymphocyte antigen 4) is the target of this monoclonal IgG1 antibody. Ipilimumab selectively binds to CTLA-4, preventing ligands such as CD80 and CD86 from interacting with CTLA-4. By increasing T cell and T-effector cell activation and proliferation, this blocking mechanism strengthens the immune response to tumour activity. In order to treat metastatic colorectal cancer with MSI-H or dMMR, the FDA authorized ipilimumab in 2018 for use in combination therapy with nivolumab. After fluoropyrimidine, irinotecan, and oxaliplatin treatment for MSI-H or dMMR metastatic colorectal cancer in patients aged 12 and older, it is administered in combination with nivolumab [39].

## 4.10 Nivolumab

The PD-1 receptor is the target of this monoclonal IgG4 antibody. It binds selectively to PD-1 and prevents PD-L1 and PD-L2 from interacting with the PD-1 receptor, such as pembrolizumab. In 2017, after receiving the approval for treating metastatic colorectal cancer nivolumab was done by either mismatch repair deficient (dMMR) or microsatellite instability-high (MSI-H). Fluoropyrimidine, irinotecan, and oxaliplatin can be taken either alone or with ipilimumab for the treatment of MSI-H or dMMR metastatic colorectal cancer [40].

## 5 Potential Targeted Therapies and Molecular Pathways Involved in Normal and Colon Cancer Cells

### 5.1 Microsatellite Instability Targeted Therapy and Immunotherapy

MSI is present in about 16% of instances of stage 2 and stage 3 colon cancer. Fluor pyrimidines have been used to treat MSI colon cancers, but in MSI stage 2 instances, they do not produce good results; in fact, some studies have suggested that they may even be harmful to overall survival. As a result, they are not advised for MSI stage 2 patients at this time. Fluor pyrimidines and oxaliplatin give good outcomes in disease-free survival (DFS) and overall survival (OS) of stage 3 patients (MSS) as compared to stable microsatellite tumours. Despite mixed findings, some research has suggested that irinotecan may be susceptible to MSI cancer [41]. Recently, bevacizumab has been investigated in conjunction with conventional therapies. The accepted immunotherapy is deficient mismatch repair (dMMR) and high levels of microsatellite instability (MSI-H). Immune checkpoint inhibitors (ICI), including anti-PD-1, anti-PD-L1, and anti-CTLA-4, are now used in immunotherapy [42]. In stage 3 MSI colon cancer, FOLFOX and the anti-PD-L1 drug atezolizumab are investigated. Another anti-PD-L1 medicine, Avelumab, has been assessed for stage 3 cancer. In a phase II clinical study, ipilimumab, an anti-CTLA-4 IgG1 monoclonal antibody, and nivolumab, an anti-PD-1 IgG4 antibody, demonstrated encouraging activity together [43]. It has been observed that pembrolizumab, another anti-PD-1 IgG4 antibody, gave longer progression-free survival (PFS) than chemotherapy in a phase III clinical trial, when used as a first-line treatment.

### 5.2 BRAF Mutation Targeted Therapy

FOLFOXIRI has been used in conjunction with bevacizumab for the initial treatment of BRAFV600E mutations. In second-line BRAFV600E mutation studies, [44, 45] anti-angiogenic drugs such as ramucirumab and aflibercept, when used in conjunction with FOLFIRI, have demonstrated strong activity. BRAFV600E inhibitors have been employed in treatment recently, although trials using the BRAFV600E inhibitor vemurafenib have shown that they are less effective when administered alone. These inhibitors are currently being used in combination with other drugs, including MEK inhibitors and anti-EGFR medicines. The FDA-approved BRAFV600E inhibitor encorafenib has generated a lot of interest because research suggests that it works better than standard FOLFIRI therapy, combining with the cetuximab (anti-EGFR inhibitor) and binimetinib (MEK inhibitor) in a triple therapy or with two cetuximabs (doublet therapy).

## 5.3 Targeted Therapy of RAS Mutation

Although bevacizumab combined chemotherapy is the approved drugs for KRAS mutations, not much information is currently available. Targeting KRASG12C and permanently blocking it by targeted therapy employing micro molecules, adagrasib, and sotorasib, has recently demonstrated encouraging activity. When used with cetuximab, adagrasib monotherapy has shown promise. Phase III trials are currently being conducted to assess combinations of sotorasib and panitumumab, as well as adagrasib and cetuximab. Furthermore, there has been evidence of synergistic tumour regression when MEK inhibitors and anti-PD-L1 are combined. Phase Ib trials are being conducted to examine atezolizumab and cobimetinib (an MEK inhibitor) together (NCT03785249). The addition of atezolizumab to the FOLFOXIRI, bevacizumab, and fluorouracil regimen produced noteworthy outcomes in a phase II trial [46]. Immunotherapy has been tried in patients with RAS mutations. In patients with the KRASG12D mutation, tumour-infiltrating lymphocyte (TIL) treatment has shown a good effect.

## 5.4 Phytochemical-Based Targeted Therapy

Significant resources for cancer treatment have continuously been made available by phytochemicals. Well-known natural compounds with therapeutic promise include camptothecin, which inhibits DNA topoisomerase I, and vinca alkaloids, which inhibit microtubules [47]. About 69% of cancer medications used between 1980 and 2002 were either developed based on or derived from natural ingredients, according to a survey done by the National Cancer Institute in the United States. Curcumin is capable of inhibiting cell growth in HCT-116 and HT29 colon cancer and has been thoroughly investigated for a variety of cancer treatments. It interacts with a variety of targets, affecting important signalling pathways linked to colon cancer, including p53, Wnt, Notch, and receptor tyrosine kinase (RTK) pathways. A naturally occurring phenol found in grapes, raspberries, and mulberries, known as resveratrol has been shown to have anti-colon cancer potential and to have strong anti-breast cancer properties similar to curcumin and affects multiple pathways, viz. RTK, notch, and Wnt signalling. In SW480 and HL60 colorectal cell lines, it has demonstrated inhibitory and antigenotoxic effects, respectively. Preventing VEGF to attach its receptor and interference with the Wnt signalling cascade, luteolin (a flavonoid) derived from Reseda luteola L., prevent colon cancer by targeted therapy. Another phytochemical that affects T-cell signalling receptors and Wnt signalling pathways is quercetin, which may be used as a targeted therapeutic medication. The T-cell signalling receptor pathway is modulated by gallic acid, which also inhibits EGFR, whereas apigenin inhibits HER2. With different phytochemicals and their interactions with colon cancer pathways (Fig. 1), the incorporation of phytochemicals in targeted therapy may help solve drug resistance difficulties.

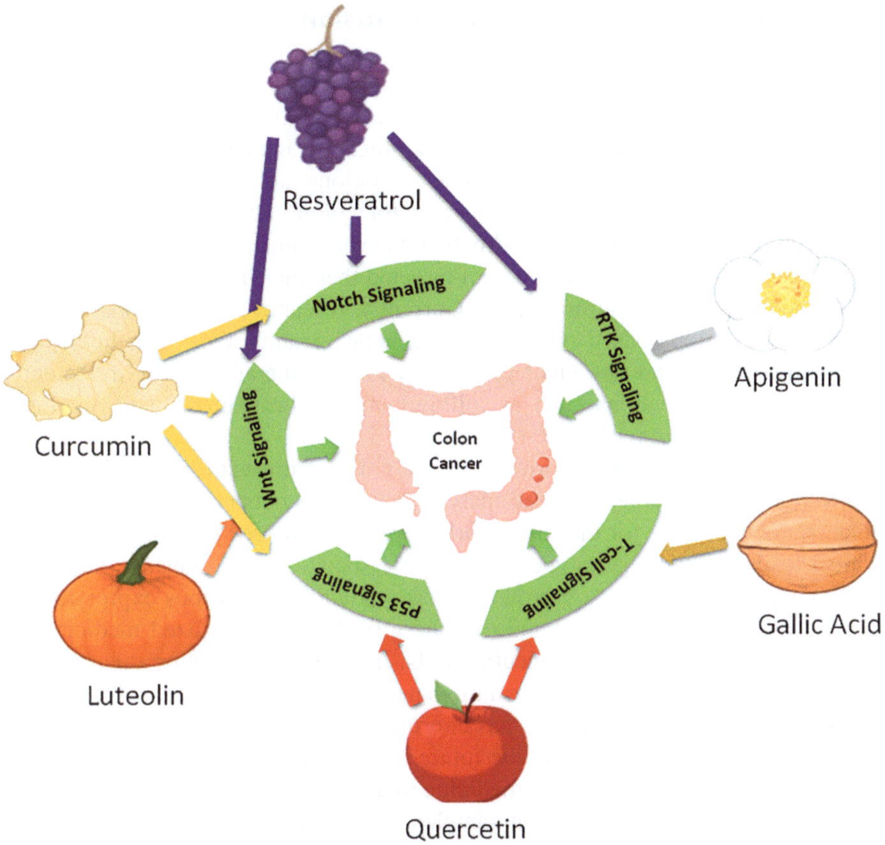


**Fig. 1** Phytochemical-based targeted therapy in colon cancer

## 5.5 NTRK Gene Fusion Targeted Therapy

Cancer driver, i.e. neurotrophic tyrosine receptor kinase (NTRK) genes, is more prevalent in MSI-H/dMMR and metastatic patients with RAS/BRAF wild-type encoding defective tropomyosin receptor kinase (TRK) proteins in NTRK fusion tumour types. These tumour types occur at frequencies ranging from 5% to 15%. In the NAVIGATE basket trial (NCT02576431), larotrectinib (Vitrakvi) treated a subgroup of adult and paediatric patients with different gastrointestinal malignancies and had targeted TRK mutations. For NTRK fusions, FDA-approved drugs such as larotrectinib and entrctinib are currently advised therapy options. Larotrectinib has a longer life expectancy than entrectinib due to its comparative long-term efficacy [48].

## 6 Future Perspectives and Conclusion

The way colon cancer is treated has changed significantly in recent years. A shift to a more customized approach to treatment has been spurred by the disease's unpredictability. Overall, patient survival has significantly increased as a result of the identification of KRAS, BRAF, and other mutations and the development of targeted biomarker therapy. Targeted therapy for colon cancer became possible in 2004 with the introduction of cetuximab and bevacizumab. By reducing the necessity for some surgeries, the accuracy of targeted therapies has produced favourable results and decreased treatment costs. Currently, research is being done on a new class of medications based on particular genetic variants. One such field that is being investigated and has great promise is immunotherapy. In the treatment of colon cancer, medications such as ipilimumab, pembrolizumab, and nivolumab have produced encouraging outcomes.

There are currently few real-world examples of customized therapy. Infrastructure and technology for molecular test analysis, as well as cooperation from all stakeholders, are necessary for the successful adoption of tailored pharmaceuticals. It is critical to improve the genotype–phenotype associations and advance individualized treatment by finding key proteins, gene variants, and various expression patterns in the disease. It is expected that the growth of personalized medicine will benefit the diagnostic and therapeutic fields. In addition to core products and services, personalized medications are anticipated to help numerous consumer-focused industries. Personalized medicine has enormous potential to enhance and progress cancer therapy in the present and the future as the discipline steadily develops.

The development of precision medicine treatments for cancer has several obstacles that include procuring patient samples, developing reliable testing models to label the presence of inherent genetic and non-genetic diversity in ever-changing cancer. Although there is a fair amount of hope, the truth is that there is a dearth of supporting research. Oncologists and physicians have few options for customized treatments, despite the benefits of personalized therapy. Numerous studies are being conducted to detect biomarkers, but the results have not always been translated into effective treatments. The biomarkers that have already been discovered must be used efficiently. To find biomarkers, new molecular techniques must be created, and those that already exist should be examined and improved. To improve the precision and quality of data derived from human tissues, substantial technological advancements are required. Another significant obstacle for customized medicine is creating more precise assays and integrating them into clinical trials for validation. Nonetheless, there is optimism that these investigations may soon lead to treatments that can save lives.

## References

1. Hossain MS et al (2022) Colorectal cancer: a review of carcinogenesis, global epidemiology, current challenges, risk factors, preventive and treatment strategies. Cancers (Basel) 14:1732

2. Verma M (2012) Personalized medicine and cancer. J Pers Med 2:1–14
3. Das A et al (2023) An updated review on the role of nanoformulated phytochemicals in colorectal cancer. Medicina (B Aires) 59:685
4. Huyghe N, Benidovskaya E, Stevens P, Van den Eynde M (2022) Biomarkers of response and resistance to immunotherapy in microsatellite stable colorectal cancer: toward a new personalized medicine. Cancers (Basel) 14:2241
5. Wu CWK, Reid M, Leedham S, Lui RN (2022) The emerging era of personalized medicine in advanced colorectal cancer. J Gastroenterol Hepatol 37:1411–1425
6. Sung H et al (2021) Global cancer statistics 2020: GLOBOCAN estimates of incidence and mortality worldwide for 36 cancers in 185 countries. CA Cancer J Clin 71:209–249
7. Yi H et al (2020) Integrative multi-omics analysis of a colon cancer cell line with heterogeneous Wnt activity revealed RUNX2 as an epigenetic regulator of EMT. Oncogene 39:5152–5164
8. Papaccio F et al (2023) Proteotranscriptomic analysis of advanced colorectal cancer patient derived organoids for drug sensitivity prediction. J Exp Clin Cancer Res 42:8
9. Dijkstra JJ et al (2023) Multiomics of colorectal cancer organoids reveals putative mediators of cancer progression resulting from SMAD4 inactivation. J Proteome Res 22:138–151
10. Echle A et al (2022) Artificial intelligence for detection of microsatellite instability in colorectal cancer—a multicentric analysis of a pre-screening tool for clinical application. ESMO Open 7:100400
11. Yang M et al (2022) A multi-omics machine learning framework in predicting the survival of colorectal cancer patients. Comput Biol Med 146:105516
12. Zhang X, Sun X-F, Shen B, Zhang H (2019) Potential applications of DNA, RNA and protein biomarkers in diagnosis, therapy and prognosis for colorectal cancer: a study from databases to AI-assisted verification. Cancers (Basel) 11:172
13. Zhang X, Zhang H, Shen B, Sun X-F (2019) Chromogranin-a expression as a novel biomarker for early diagnosis of colon cancer patients. Int J Mol Sci 20:2919
14. Jiang X et al (2021) ITGB4 as a novel serum diagnosis biomarker and potential therapeutic target for colorectal cancer. Cancer Med 10:6823–6834
15. Zhang X et al (2022) Loss of CHGA protein as a potential biomarker for colon cancer diagnosis: a study on biomarker discovery by machine learning and confirmation by immunohistochemistry in colorectal cancer tissue microarrays. Cancers (Basel) 14:2664
16. Sekita-Hatakeyama Y et al (2022) Evaluation and diagnostic value of next-generation sequencing analysis of residual liquid-based cytology specimens of pancreatic masses. Cancer Cytopathol 130:202–214
17. Matsunaga M et al (2016) A comparison of four methods for detecting KRAS mutations in formalin-fixed specimens from metastatic colorectal cancer patients. Oncol Lett 12:150–156
18. Buckley AF, Kakar S (2007) Comparison of the Dako EGFR pharmDx kit and Zymed EGFR antibody for assessment of EGFR status in colorectal adenocarcinoma. Appl Immunohistochem Mol Morphol 15:305–309
19. Ogino S et al (2007) CpG Island methylation, response to combination chemotherapy, and patient survival in advanced microsatellite stable colorectal carcinoma. Virchows Arch 450:529–537
20. Mooi JK et al (2018) The prognostic impact of consensus molecular subtypes (CMS) and its predictive effects for bevacizumab benefit in metastatic colorectal cancer: molecular analysis of the AGITG MAX clinical trial. Ann Oncol 29:2240–2246
21. Kannarkatt J, Joseph J, Kurniali PC, Al-Janadi A, Hrinczenko B (2017) Adjuvant chemotherapy for stage II colon cancer: a clinical dilemma. J Oncol Pract 13:233–241
22. Hveem TS et al (2014) Prognostic impact of genomic instability in colorectal cancer. Br J Cancer 110:2159–2164
23. Diaz LA Jr et al (2012) The molecular evolution of acquired resistance to targeted EGFR blockade in colorectal cancers. Nature 486:537–540
24. Batlle E, Clevers H (2017) Cancer stem cells revisited. Nat Med 23:1124–1134

25. Bertolini F et al (2007) Prognostic and predictive value of baseline and posttreatment molecular marker expression in locally advanced rectal cancer treated with neoadjuvant chemoradiotherapy. Int J Radiat Oncol Biol Phys 68:1455–1461
26. Halama N et al (2011) Localization and density of immune cells in the invasive margin of human colorectal cancer liver metastases are prognostic for response to chemotherapy. Cancer Res 71:5670–5677
27. Liao X et al (2012) Prognostic role of PIK3CA mutation in colorectal cancer: cohort study and literature review. Clin Cancer Res 18:2257–2268
28. Hector S, Prehn JHM (2009) Apoptosis signaling proteins as prognostic biomarkers in colorectal cancer: a review. Biochimica et Biophysica Acta (BBA) 1795:117–129
29. Watanabe T et al (2001) Molecular predictors of survival after adjuvant chemotherapy for colon cancer. N Engl J Med 344:1196–1206
30. Fornasier G, Francescon S, Baldo P (2018) An update of efficacy and safety of Cetuximab in metastatic colorectal cancer: a narrative review. Adv Ther 35:1497–1509
31. El-Deiry WS et al (2019) Disease control with FOLFIRI plus Ziv-aflibercept (zFOLFIRI) beyond FOLFIRI plus bevacizumab: case series in metastatic colorectal cancer (mCRC). Front Oncol 9
32. Taha AM, Aboulwafa MM, Zedan H, Helmy OM (2022) Ramucirumab combination with sorafenib enhances the inhibitory effect of sorafenib on HepG2 cancer cells. Sci Rep 12:17889
33. Gürel B et al (2022) Optimized methods for analytical and functional comparison of biosimilar mAb drugs: a case study for Avastin, Mvasi, and Zirabev. Sci Pharm 90:36
34. Ma C-J et al (2019) Oncologic outcomes in metastatic colorectal cancer with Regorafenib with FOLFIRI as a third- or fourth-line setting. Transl Oncol 12:502–512
35. Xu Y et al (2019) A tumor-specific super-enhancer drives immune evasion by guiding synchronous expression of PD-L1 and PD-L2. Cell Rep 29:3435–3447.e4
36. André T et al (2020) Pembrolizumab in microsatellite-instability–high advanced colorectal cancer. N Engl J Med 383:2207–2218
37. Clarke JM, Hurwitz HI (2013) Targeted inhibition of VEGF receptor 2: an update on ramucirumab. Expert Opin Biol Ther 13:1187–1196
38. Ascierto PA et al (2020) Update on tolerability and overall survival in COLUMBUS: landmark analysis of a randomised phase 3 trial of encorafenib plus binimetinib vs vemurafenib or encorafenib in patients with BRAF V600–mutant melanoma. Eur J Cancer 126:33–44
39. Korman AJ, Garrett-Thomson SC, Lonberg N (2022) The foundations of immune checkpoint blockade and the ipilimumab approval decennial. Nat Rev Drug Discov 21:509–528
40. Rajan A, Kim C, Heery CR, Guha U, Gulley JL (2016) Nivolumab, anti-programmed death-1 (PD-1) monoclonal antibody immunotherapy: role in advanced cancers. Hum Vaccin Immunother 12:2219–2231
41. Vilar E, Tabernero J (2013) Molecular dissection of microsatellite instable colorectal cancer. Cancer Discov 3:502–511
42. Huang Z, Yang M (2022) Molecular network of colorectal cancer and current therapeutic options. Front Oncol 12
43. Overman MJ et al (2018) Durable clinical benefit with Nivolumab Plus Ipilimumab in DNA mismatch repair–deficient/microsatellite instability–high metastatic colorectal cancer. J Clin Oncol 36:773–779
44. Yoshino T et al (2018) Pan-Asian adapted ESMO consensus guidelines for the management of patients with metastatic colorectal cancer: a JSMO–ESMO initiative endorsed by CSCO, KACO, MOS, SSO and TOS. Ann Oncol 29:44–70
45. Van Cutsem E et al (2016) ESMO consensus guidelines for the management of patients with metastatic colorectal cancer. Ann Oncol 27:1386–1422
46. Ebert PJR et al (2016) MAP kinase inhibition promotes T cell and anti-tumor activity in combination with PD-L1 checkpoint blockade. Immunity 44:609–621
47. Efferth T et al (2017) Integration of phytochemicals and phytotherapy into cancer precision medicine. Oncotarget 8:50284–50304
48. Yu IS, Kopetz S (2021) The emergence of targetable pathways in colorectal cancer. Clin Adv Hematol Oncol:774–783

# Innovative Approaches in Pancreatic Cancer Treatment

Arisha Arora, Basab Ghosh, Pijush Kanti Khanra, Sailayee Choudhury, and Siddhartha Sankar Ghosh

**Abstract**

Pancreatic cancer is characterized by its aggressive behavior and limited treatment options, with tumors typically diagnosed at advanced stages that show resistance to standard therapies. The need for innovative and effective treatments has become more critical than ever, especially considering that survival rates have consistently remained discouragingly low, around 10% on average. The recent advancements in pancreatic cancer treatment highlight the integration of molecular profiling, advanced diagnostics, and personalized therapies to enhance patient outcomes. This chapter discusses how molecular profiling, using tools like next-generation sequencing (NGS) and liquid biopsies (including circulating tumor DNA), enables the analysis of tumors at a molecular level. It highlights how these techniques help identify crucial mutations and biomarkers, allowing for personalized treatment strategies tailored to each patient's unique profile. Furthermore, we address the role of integrating molecular diagnostics into clinical practice, which helps identify biomarkers that contribute to early detection and better cancer prognosis. In recent years, liquid biopsies have emerged as a

Arisha Arora, Basab Ghosh, Pijush Kanti Khanra and Sailayee Choudhury contributed equally for this book chapter.

A. Arora · B. Ghosh · P. K. Khanra · S. Choudhury
Department of Biosciences and Bioengineering, Indian Institute of Technology Guwahati, Guwahati, Assam, India

S. S. Ghosh (✉)
Department of Biosciences and Bioengineering, Indian Institute of Technology Guwahati, Guwahati, Assam, India

Centre for Nanotechnology, Indian Institute of Technology Guwahati, Guwahati, Assam, India
e-mail: sghosh@iitg.ac.in

 
M. K. Ghosh, M. Basu (eds.), *Personalized Cancer Therapy*,
https://doi.org/10.1007/978-981-95-6845-1_14

non-invasive tool for collecting vital information about tumor behavior, though their full clinical potential is still being investigated through ongoing research. Additionally, the advancements in personalized treatment strategies through targeted therapies focused on individual molecular profiles have been discussed. It includes case studies that showcase the successful implementation of these therapies while emphasizing the role of multidisciplinary teams in creating personalized treatment plans. Overall, the chapter offers a comprehensive understanding of ongoing research into innovative approaches, from molecular subtyping to personalized treatment, in pancreatic cancer, with the goal of improving outcomes and enhancing the quality of life for patients with pancreatic cancer.

**Keywords**

Biomarkers · Liquid biopsies · Molecular profiling · Pancreatic cancer · Precision medicine

## 1 Introduction

Pancreatic cancer (PC) is one of the most lethal and aggressive forms of solid malignancy. According to GLOBOCAN Global Cancer Statistics, there were 511,000 new cases of pancreatic cancer and 467,000 deaths in the year 2022 alone. Pancreatic ductal adenocarcinoma (PDAC), the most common form, is particularly lethal due to its biological complexity and aggressive progression, leading to the poorest prognosis among all cancers, with a 5-year survival rate of just 10–12% [1, 2]. The poor prognosis is further compounded by the lack of routine screening protocols, non-specific symptoms, and the absence of reliable early-stage biomarkers, often leading to diagnosis at advanced stages. This makes the treatment of pancreatic cancer more challenging, with a majority of cases already metastatic at the time of initial diagnosis and only 9.7% identified at a localized stage.

Over the past two decades, significant advancements have been made in understanding pancreatic cancer, revealing that both inherited and somatic genetic mutations contribute to its development. Key genes implicated in the disease include *BRCA1*, *BRCA2*, *PALB2*, *STK11*, *KRAS*, *CDKN2A*, *TP53*, and *SMAD4/DPC4* [3]. Furthermore, PDAC exhibits significant variability both between patients and within individual tumors, affecting the tumor composition and treatment responses. This variability is partly due to homologous recombination deficiency (HRD), linked to mutations in DNA repair genes. Surgical resection is the only curative option for pancreatic cancer, but only 15–20% of patients qualify due to late diagnosis, and despite resection, the prognosis remains poor.

Although challenges persist, advancements in clinical management and a deeper understanding of the disease have resulted in a twofold increase in survival rates over the past decade [1, 4]. Therefore, to further mitigate the rising burden of tumor incidence and heterogeneity, it is crucial to develop highly sensitive and specific molecular profiling markers and diagnostic techniques, which would enable more personalized and targeted treatment approaches [3]. As a result, ongoing research is focused on uncovering the genetic and molecular mechanisms driving pancreatic cancer heterogeneity. Additionally, the clinical trial landscape for pancreatic cancer continues to evolve, emphasizing personalized and biologically driven strategies, including targeted therapies directed at specific molecular pathways and immunotherapies.

Therefore, this chapter provides a comprehensive overview of the recent advancements in analyzing molecular markers along with developments in diagnostic markers and techniques. It also describes the evolution and findings of the clinical trial landscape of pancreatic cancer over the past decade, all aimed at facilitating personalized treatment strategies for better pancreatic cancer management.

## 2 Molecular Profiling of Pancreatic Cancer

Molecular profiling of pancreatic tumors involves a comprehensive analysis of genetic, transcriptomic, and proteomic alterations to improve diagnosis and predict disease progression and patient outcomes. Studies emphasize that both germline and somatic molecular testing are essential for patients with pancreatic ductal adenocarcinoma (PDA), aiding in personalized treatment approaches [3].

Traditional tissue enrichment methods, such as manual microdissection, have been employed to isolate tumor cells for molecular analysis. However, advancements in techniques like fine-needle aspiration (FNA) and fine-needle biopsy (FNB) have improved the quality of genetic material obtained from pancreatic tumors, facilitating thorough molecular studies [5]. Furthermore, comprehensive molecular characterization, including whole-exome and RNA sequencing and the integration of genomic and transcriptomic data from other sensitive molecular techniques, enables the identification of biomarkers and potential therapeutic targets along with the classification of PDAC into molecular subtypes with distinct prognostic and therapeutic implications. This enables informed therapeutic decisions and potentially improves outcomes by tailoring interventions based on tumor biology and characterization.

### 2.1 Comprehensive Molecular Profiling Techniques for Pancreatic Cancer

Molecular profiling has emerged as a transformative approach in the management of pancreatic cancer, particularly pancreatic ductal adenocarcinoma (PDAC), by enabling precision medicine strategies. Several advanced techniques are utilized to analyze tumor biology and guide personalized treatment plans. Techniques such as next-generation sequencing (NGS), fluorescence in situ hybridization (FISH), immunohistochemistry (IHC), liquid biopsies, and polymerase chain reaction (PCR) are crucial for analyzing tumor biology and profiling. NGS has transformed genomic research with various advantages such as high-throughput analysis, the ability to analyze multiple genetic loci simultaneously, and comprehensive data output [6]. It is especially valuable for identifying actionable mutations in genes like *KRAS, TP53, SMAD4*, and *BRCA2*, providing insights into molecular subtypes of PDAC with distinct biological traits and targetable features. Furthermore, another technique, FISH, has significantly progressed the realm of guiding targeted therapies by precisely detecting chromosomal abnormalities and gene rearrangements and alterations, such as amplifications or translocations. Furthermore, PCR-based methods mainly analyze genetic alterations, including specific mutations or gene fusions with high sensitivity and cost-effectiveness. In pancreatic cancer profiling, the IHC technique is another valuable tool, mainly employed to evaluate protein expression levels in tumor tissues, which can indicate the presence of actionable proteomic alterations, in cell cycle regulation genes (such as *CCND1/2/3*, *CDK4/6*) or DNA repair pathway genes (such as *BRCA1/2*, *ATM*), which may predict responsiveness to PARP inhibitors or CDK inhibitors [7]. Additionally, IHC, along with molecular assays such as polymerase chain reaction (PCR), plays a crucial role in analyzing microsatellite instability (MSI) and mismatch repair (MMR) status. These analyses detect mutations or epigenetic changes in genes responsible for DNA replication and recombination, providing critical insights into patients responsive to immunotherapy treatments.

Currently, liquid biopsies hold promise for pancreatic cancer profiling by enabling minimally invasive analysis of circulating tumor cells, DNA, and other biomarkers in blood and body fluids, aiding in early detection, treatment selection, and rapid disease monitoring [8]. Cumulatively, these approaches have significantly improved patient outcomes by enabling focused therapies and clinical trials. Also, programs such as the Know Your Tumor (KYT) initiative, developed by the Pancreatic Cancer Action Network (PanCAN), have proven instrumental in advancing molecular profiling for pancreatic cancer patients. By utilizing multi-omics profiling, KYT identifies actionable molecular alterations in tumor biology and matches patients with therapies tailored to their specific genetic and proteomic profiles. Approximately 50% of patients tested through the program were found to have actionable alterations, with 27% being highly actionable. Patients receiving matched therapies based on these findings demonstrated significantly improved progression-free survival rates compared to those on unmatched treatments [6]. Overall, integrating molecular profiling techniques such as NGS, FISH, IHC, PCR, and liquid biopsy into clinical practice is transforming pancreatic cancer care by outlining and refining treatment selections that can improve disease outcomes.

## 2.2 Emerging Role of Liquid Biopsies and Circulating Tumor DNA in Pancreatic Cancer Profiling

Liquid biopsy is a minimally invasive molecular profiling method that analyzes body fluids, such as blood, saliva, or urine, tumor-derived components such as circulating tumor DNA (ctDNA), circulating tumor cells (CTCs), extracellular vesicles, cell-free DNA, microRNA, and exosomes. This provides a comprehensive determination of tumor heterogeneity and genetic mutations in PDAC. The technique bypasses the challenges of obtaining tissue samples in pancreatic cancer, which is predominantly diagnosed at advanced stages when surgical sampling is difficult or impossible. Additionally, in PDAC patient treatments, tissue-based genomic sequencing often delays the first-line treatment decisions due to prolonged processing times (3–6 weeks). Liquid biopsies, on the other hand, have revolutionized the spectrum of molecular profiling in PDAC patients, offering rapid molecular profiling with a shorter turnaround time along with the ease for repeated sampling, detection of therapy-resistant clones, dynamic monitoring of the disease, and easier access to tumor information (Fig. 1). These features overcome the limitations of traditional tissue sequencing and accelerate patient enrolment in targeted therapies and clinical trials in pancreatic cancer, where early intervention and dynamic monitoring are critical for optimizing outcomes [9].

Liquid biopsies offer a minimally invasive approach, requiring just a small blood sample, which improves patient compliance and enables frequent testing. Furthermore, the various components of the technique provide a more comprehensive representation of tumor heterogeneity than a single tissue biopsy. Circulating tumor DNA, a key component of liquid biopsies, is DNA released by cancer cells into the bloodstream. It can be used to identify the genetic makeup of the tumor, including tumor-specific mutations such as *KRAS* and epigenetic changes, monitor disease progression, predict recurrence, and detect minimal residual disease or

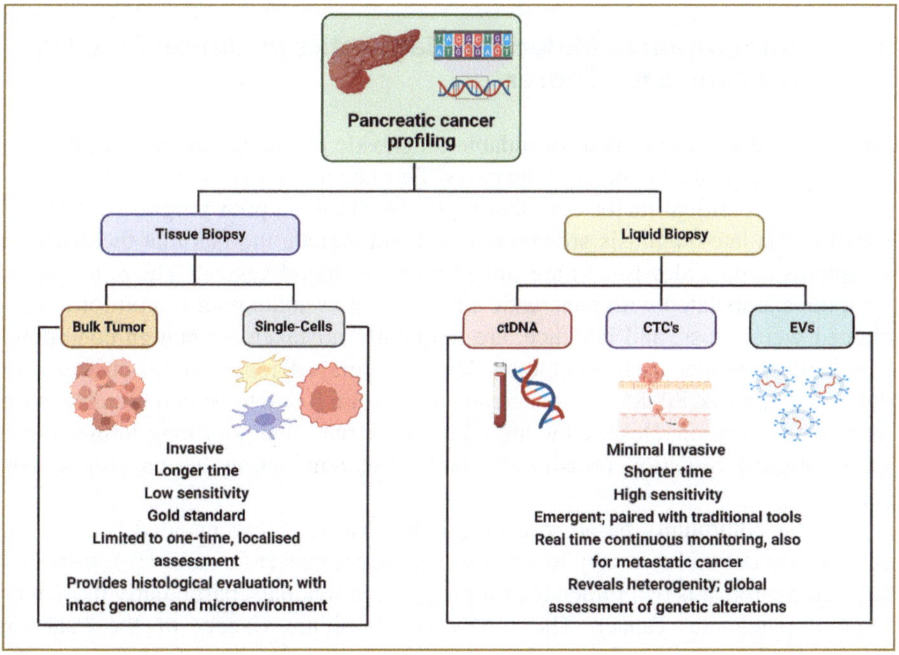


**Fig. 1** Molecular Profiling tools in pancreatic cancer with distinct features and clinical applications

treatment resistance. More recently, next-generation sequencing-based methods have enabled ctDNA profiling as a superior alternative to routine tumor tissue sequencing. Other components of liquid biopsies, such as CTCs, which are intact cancer cells, ctRNA, which includes RNA, including messenger RNA (mRNA), microRNA (miRNA), and other types of RNA that circulate in the bloodstream, serve as sensitive biomarkers for information on cancer progression and metastasis. Additionally, EVs are small membrane-bound vesicles released by the tumor cells, which can carry proteins, lipids, and nucleic acids. EVs can serve as biomarkers for cancer diagnosis and monitoring. They are thought to provide a more stable source of tumor-derived genetic material, potentially increasing sensitivity for molecular profiling [10, 11].

Recent studies have underscored the role of liquid biopsies in facilitating patient enrolment for genome-based precision medicine. For instance, a study by Yoshino and colleagues found that circulating tumor DNA (ctDNA) was detectable in 91.4% of patients, and it detected multiple biomarkers relevant to the selection of treatment, including *KRAS*, *NRAS*, *BRAF*, and *PIK3CA* mutations; *ERBB2*, *FGFR1/2*, and *MET* amplifications; *FGFR2/3*, *ALK*, *NTRK1*, and *RET* fusions; and *MSI* [12]. Therefore, with subsequent successful case studies, liquid biopsies are poised to drive innovation in precision medicine, particularly in cancers such as pancreatic cancer, where they complement traditional diagnostic methods and enhance early detection through advancements in multi-marker approaches and new technologies.

## 3 Integration of Molecular Diagnostics in Clinical Practice for Pancreatic Cancer

Pancreatic cancer presents a formidable diagnostic challenge, as evidenced by its late-stage appearance in most of the cases. This late diagnosis often results in metastatic form, a critical factor contributing to the disease's poor prognosis. The reasons for this late diagnosis are complicated and include the fact that the disease's symptoms appear slowly and are linked to major blood vessels. The non-specific symptoms associated with pancreatic cancer, such as abdominal discomfort, unexplained weight loss, and jaundice, are frequently mistaken for benign conditions. The lack of proper early symptoms leads to delayed report and, consequently, delayed diagnosis. 80–85% of pancreatic tumors are found to be unresectable at the time of presentation. Despite the high recurrence rates and poor long-term survival rates, surgical resection remains the only treatment option for pancreatic cancer [13].

Regular community-based screening is not usually recommended because the general population has a very low risk of pancreatic cancer (around 1%). Instead, a targeted approach is recommended for high-risk individuals, particularly those with familial pancreatic cancer. The CAPS (International Cancer of the Pancreas Screening Consortium) plays an important role in defining these high-risk populations and establishing screening guidelines. However, even within this high-risk group, there has been debate regarding the optimal age for initiating screening. While a consensus has emerged advocating for screening to begin at 50 years of age, the nuances of individual risk assessment and the timing of interventions continue to be refined. Non-suspicious cysts require regular monitoring, typically every 6–12 months, to track any changes that may indicate malignant transformation. Conversely, solid lesions that are not immediately resectable and main pancreatic duct structures necessitate closer scrutiny, with repeat imaging recommended after 3 months. These protocols are essential for ensuring timely intervention and optimizing patient outcomes [14].

Despite the effective identification of high-risk groups for pancreatic cancer screening, the definitive selection of appropriate diagnostic imaging techniques and the characterization of target lesions are still under research. Secretin-enhanced magnetic resonance imaging (MRI) and magnetic resonance cholangiopancreatography (MRCP) have demonstrated similarities with endoscopic ultrasound (EUS) results when employed as a single screening procedure, notably mitigating exposure to ionizing radiation. Nevertheless, EUS demonstrates greater sensitivity in the detection of small solid pancreatic lesions, especially those measuring less than 2 cm in diameter, when compared to computed tomography (CT) and MRI. Furthermore, EUS facilitates fine-needle aspiration cytology, thereby enabling precise characterization of suspect lesions and yielding critical diagnostic data [14].

Historically, the identification of pancreatic intraepithelial neoplasia (PanIN) through imaging was considered unreliable. Nevertheless, recent investigations and evidence suggest a potential correlation between pancreatic intraepithelial neoplasia (PanIN) and lobulocentric atrophy, resulting in EUS imaging patterns that mimic those observed in chronic pancreatitis. This development expands the diagnostic capabilities of EUS and enhances its potential for early detection. A combination of EUS and MRI/MRCP approach is currently being used as the screening protocol for high-risk individuals, as it effectively integrates the unique diagnostic capabilities of both technologies. This combined approach aims to improve the sensitivity and specificity of early detection, ultimately enhancing patient outcomes [15].

## 3.1 Diagnostic Biomarkers for Pancreatic Cancer

The early diagnosis of pancreatic cancer is evolving with the exploration of various biomarkers, each with distinct advantages and limitations. Serum cancer antigen 19-9 (CA19-9), the most commonly employed biomarker, while invaluable for monitoring disease progression, recurrence, and therapy response, falls short in early diagnosis due to its insufficient sensitivity and specificity. Its utility is thus confined to tracking established disease rather than detecting it at its nascent stages. Similarly, circulating tumor cells (CTCs), while potentially diagnostic, are primarily detectable in patients with metastatic disease, limiting their application in early-stage detection [16].

In contrast, circulating tumor DNA (ctDNA) analysis offers a more promising avenue. The identification of mutant *KRAS* ctDNA in a substantial percentage of patients with localized disease indicates that an extensive ctDNA panel, encompassing genes such as *KRAS* and *TP53*, may function as a non-invasive early diagnostic tool. This approach offers the potential for detecting pancreatic cancer before the onset of overt symptoms or advanced disease [17]. Moreover, identifying systemic metabolic changes, including elevated levels of circulating branched-chain amino acids, offers an additional possible pathway for early diagnosis. These metabolic alterations, signifying tissue protein degradation, have been identified in patients several years before clinical diagnosis. This implies that metabolic profiling holds potential as a predictive method, facilitating the identification of individuals exhibiting increased susceptibility to pancreatic cancer. The combination of these biomarkers, alongside advanced imaging techniques, offers a promising avenue for substantially enhancing the early diagnosis and management of pancreatic cancer [18].

## 3.2 Imaging Techniques in Pancreatic Cancer Diagnosis

When pancreatic cancer is suspected, imaging becomes an indispensable technique for diagnosis, staging, and treatment process. The selection of imaging modality depends on the clinical context, the availability of resources, and the specific diagnostic question being addressed (Fig. 2).

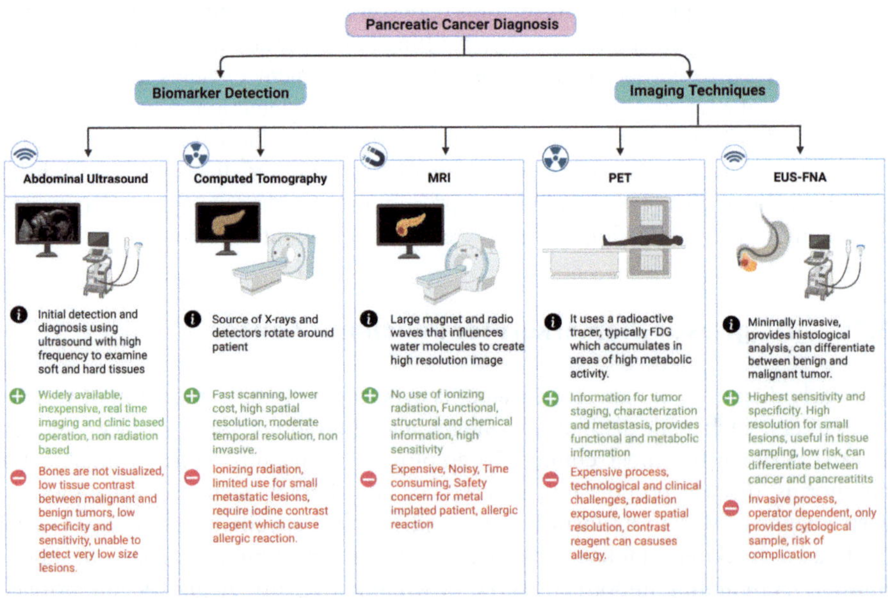


**Fig. 2** Diagnostic imaging techniques of pancreatic cancer with their advantages and disadvantages

### 3.2.1 US

Abdominal ultrasonography (US) is often considered a primary imaging modality for evaluating abdominal pain or jaundice with a relatively low accuracy of 50–70%. US can identify hypoechoic masses, pancreatic duct dilatation, and common bile duct dilatation. Contrast-enhanced Doppler US can differentiate between pancreatic adenocarcinoma, endocrine cell tumors, and pancreatitis-associated masses based on their vascularity. Specifically, pancreatic adenocarcinoma typically exhibits reduced blood flow (hypovascularity), while endocrine tumors demonstrate increased blood flow (hypervascularity), and masses associated with pancreatitis show similar blood flow to surrounding tissue (isovascularity). Further advancements in ultrasound technology, such as contrast-enhanced coded phase inversion harmonic imaging, have been investigated for their ability to differentiate pancreatic tumors [19].

### 3.2.2 CT and MDCT

CT is the major imaging technique used in pancreatic cancer diagnosis. On contrast-enhanced CT, pancreatic cancer usually appears as a hypoattenuating region due to its hypovascularity and dense fibrous stroma. Nonetheless, certain tumors may present as iso-attenuating, requiring the identification of indirect indications such as mass effect, interrupted duct sign, and atrophic distal parenchyma. Dynamic CT, involving scanning during pancreatic and hepatic phases, enhances the detectability of pancreatic adenocarcinoma. Helical CT has demonstrated high sensitivity in revealing pancreatic carcinoma and depicting local extension and vascular invasion [20]. However, limitations include challenges in detecting liver and lymph node metastasis [21].

MDCT (Multidetector Computed Tomography) is the gold standard for staging and resectability assessment. Its ability to provide high-resolution cross-sectional 3D images of the pancreas and surrounding vasculature is paramount for evaluating tumor resectability and staging. MDCT protocols are meticulously designed to optimize visualization of the pancreas, incorporating multiphase contrast enhancement. These protocols generally encompass early arterial phase imaging, which evaluates the involvement of the superior mesenteric artery, coeliac trunk, and other arteries. The pancreatic phase, obtained during maximal pancreatic parenchymal enhancement, elucidates pancreatic lesions and their attributes. MDCT identifies solid hypodense lesions, the hallmark appearance of pancreatic adenocarcinoma. It also plays a crucial role in detecting distant metastases, particularly in the liver, significantly impacting treatment decisions. Lastly, the portal venous phase assesses the involvement of the superior mesenteric vein, portal vein, and other venous structures [22].

### 3.2.3 MRI

MRI offers distinct advantages in pancreatic cancer imaging, particularly in its superior soft tissue contrast and non-invasive visualization of the pancreatobiliary system. MRI enhances soft tissue contrast and provides detailed visualization of the ductal system. It is particularly valuable for characterizing cystic lesions and detecting liver metastases, where it may be more sensitive than MDCT. MRCP utilizes extended echo times on T2-weighted magnetic resonance imaging and facilitates the detailed visualization of the biliary and pancreatic ducts. In terms of signal intensity characteristics, pancreatic carcinoma commonly exhibits hypo-intensity on T1-weighted imaging and either hyper-intensity or iso-intensity on T2-weighted imaging [23]. Dynamic MRI with contrast agents like gadolinium can further characterize tumor vascularity. Mangafodipir trisodium, another contrast agent, shows uptake in the pancreatic parenchyma while exhibiting no uptake in tumors, enhancing tumor delineation [24].

### 3.2.4 EUS

EUS demonstrates a high degree of sensitivity in the detection of small pancreatic tumors and in assessing vascular invasion. EUS provides high-resolution images of the pancreas from within the gastrointestinal tract, enabling the identification of hypoechoic masses with irregular borders, characteristic of pancreatic carcinoma. EUS-guided fine-needle aspiration biopsy (EUS-FNA) facilitates the acquisition of tissue samples, enabling definitive histopathological diagnosis and subsequent confirmation of pancreatic cancer. EUS is superior to MRI, PET (Positron Emission Tomography), and CT in detecting small tumors and assessing vascular invasion and lymph node metastasis [25]. However, EUS is operator-dependent and has a limited field of view for detecting distant metastases.

### 3.2.5 PET

PET plays a valuable, albeit complementary, role in the diagnostic workup of pancreatic cancer. Its utility originates from its ability to visualize metabolic activity and provide functional information that complements the anatomical details obtained from other imaging modalities like CT and MRI. $^{18}$F-fluorodeoxyglucose or FDG-PET uses a radiolabelled glucose analog that can differentiate between benign and malignant types based on glucose uptake. Malignant tumors often exhibit elevated glucose uptake, a phenomenon attributed to the overexpression of glucose transporter proteins and increased activity of major glycolytic enzymes, hexokinase and phosphofructokinase. Previous studies showed that FDG-PET's ability to identify malignant pancreatic tumors varies, with sensitivity ranging from 71% to 100% and specificity from 64% to 90%. However, PET has limitations in pancreatitis and hyperglycemic patients, which can lead to false-negative or false-positive results. This facilitates its use as a complementary imaging modality in conjunction with MRI or CT [26].

## 4 Personalized Treatment for Pancreatic Cancer

A malignancy with poor prognosis and limited patient tolerance to traditional therapies. Personalized targeted therapies improve treatment tolerability, reduce systemic side effects, and maintain efficacy at lower drug doses, enhancing patient quality of life. Precision-based strategies are shaping a new paradigm in cancer care, offering a more tailored and effective approach for challenging malignancies such as PDAC.

## 4.1 Development and Application of Targeted Therapies

The clinical application of targeted therapies has significantly improved treatment outcomes across diverse patient populations. These approaches demonstrate superior efficacy over single-agent therapies by integrating novel targeted agents with conventional chemotherapy or employing multitargeted strategies to address complex disease mechanisms. This combinatorial approach enhances therapeutic efficacy and helps to overcome treatment challenges. These advancements are particularly effective in managing pancreatic ductal adenocarcinoma (PDAC) [27].

## 4.2 Successful Targeted Therapies

Advances in precision medicine have paved the way for ground-breaking discoveries, offering researchers new avenues for exploration while significantly improving patient outcomes. However, overcoming the formidable challenges posed by diseases such as pancreatic ductal adenocarcinoma (PDAC) requires urgent and transformative breakthroughs. Addressing these barriers is critical to developing effective therapeutic strategies and ultimately improving survival rates in patients facing this aggressive malignancy [3, 27]. The emergence of targeted therapies for pancreatic cancer is becoming more viable with tumor-agnostic approaches, characterized by sequential identification of oncogenic biomarkers and functionalization of individualized therapeutic strategies (as detailed in Table 1). PARP inhibitors offer therapeutic benefit for patients with *gBRCA 1/2* mutation, while respective immunotherapies (Pembrolizumab against PD-L1) and small molecule inhibitors (BRAF inhibitor, first-generation *KRAS* G12 inhibitor) have shown efficacy in cases with high microsatellite instability and elevated mutational burden, specifically against $BRAF^{V600E}$ and *KRAS* G12 mutations. Furthermore, novel therapeutic agents have been developed to target specific genomic alteration, genetic aberration (*FGFR*/*FGF* aberration), and abnormal fusion proteins (*NRG1* and *NTRK* fusions).

### 4.2.1 Case Study_1: An Individualized Neoantigen Vaccine (Autogene Cevumeran) Based on mRNA—Lipoplex Nanoparticles Containing up to 20 Distinct Patient Neoantigens: Phase I Clinical Trial Study

*Patient profile:*

The trial enrolled 16 patients with surgically resectable pancreatic ductal adenocarcinoma (PDAC). Patients had Eastern Cooperative Oncology Group performance status of 0–1 with surgically resectable, radiographically suspicious and single PDAC tumor, no distant metastases, and ≥5 identified neoantigens. All patients underwent surgical resection with R0/R1 margins, and the median primary tumor size was 2.6 cm (range 1.0–4.2 cm).

**Table 1** Key biomarkers in PDAC and the personalized therapies: evidence from human studies

| Key biomarker(s) | Prevalence in PDAC | Associated targeted therapy | Clinical response rate (RR) | Key human studies |
|---|---|---|---|---|
| *BRCA1/2* (*gBRCA1/2* or *gPALB2*) mutations | 8% [33] | PARP inhibitors (olaparib, rucaparib, cisplatin-gemcitabine, veliparib) | 21.7–41.7% [3] | POLO-phase III trial (Olaparib), O'Reilly-phase II trial (cisplatin-gemcitabine) [3] |
| MSI-H/ dMMR | 4.8% [34] | Pembrolizumab | Objective RR of 34.3% [35] | KEYNOTE-158 phase II trial [35] |
| *KRAS G12C* | 90% [3] | Sotorasib, Adagrasib, Diravasib | 21%, 33%, and 36%, respectively | CODEBREAK-100 phase Ib/II, KRYSTAL-1 phase II & Sacher phase I trial [3] |
| *NRG1* fusions | 0.5% [3] | Zenocutuzumab, Seribantumab | 40% and 30%, respectively | eNRGy trial (NCT02912949) [30] and CRESTONE study [31] |
| *NTRK* fusions | 0.3% [3] | Larotrectinib, Entrectinib | 75% and 61%, respectively | NAVIGATE phase III trial [36], STARTRK-2 phase II basket study [29] |
| *FGF/FGFR* aberration | 4.7% [3] | Pemigatinib, Futibatinib | Objective RR of 37% and 25.4%, respectively | FIGHT-101 phase I/II trial (NCT02393248) [37], FOENIX-CCA2 phase I expansion study [38] |
| *BFAF*$^{V600E}$ mutation | 3% [3] | BRAF inhibitor | Case reports only | Nil [39] |

- Initially, 34 patients were enrolled from December 2019 to August 2021. 28 out of the 34 patients underwent surgical resection of the PDAC tumor.
- 19 out of the 28 patients received anti-PD-L1 immunotherapy (Atezolizumab intravenously administered during the post-surgery (PS) period of 4.3–7.9 weeks). During the incubation period of Atezolizumab immunotherapy, mRNA vaccines targeting up to 20 neoantigens per patient were manufactured (although one patient had insufficient neoantigens for vaccine manufacturing).
- 16 out of the 19 patients went ahead to receive the personalized vaccine (Autogene Cevumeran, administered during the range of 7.4–11.0 weeks PS). 15 out of the 16 vaccinated patients received chemotherapy (mFOLFIRINOX, administered at PS >12 weeks).
- Three out of 16 patients did not complete all 9 vaccine doses due to the seriousness of disease progression, death or mFOLFIRINOX toxicity [16].

*Treatment course:*

- A single dosage of 1200 mg of Atezolizumab was administered intravenously.

- Seven subsequent weekly dosage of 25 μg Autogene cevumeran, followed by an eighth dose at Week 17, and finally a ninth booster dose at Week 46 were administered intravenously.
- Chemotherapeutic administration of mFOLFIRINOX up to 12th cycles [16].

*Outcome:*

- *Safety assessment profile*:
  - In the safety-evaluation cohort analysis, none of the 19 patients treated with Atezolizumab had Grade 3 or higher adverse events (AEs).
  - Only one out of 16 (6%) patients treated with Autogene cevumeran had Grade 3 AEs (fever and hypertension).
  - The most common grade 1–2 AEs were reported in Autogene cevumeran-treated patients [16].

- *Immunogenicity response*:
  - 8 out of 16 patients generated T cell responses (detected by ex vivo IFNγ ELISpot assay) after receiving autogene cevumeran dosage.
  - Single-target level induced a T cell response, which was detected in the case of 25/300 patients after neoantigen vaccine administration.
  - Half patient population of the neoantigen receivers mounted substantial neoantigen-specific T cell response against at least a single dosage.
  - Furthermore, half of the responders were polytopic targeted by multiple neoantigen vaccines.
- Single-cell RNA sequencing data revealed autogene-cevumeran-expanded T cells are primarily CD8+ effector cells expressing lytic markers (perforin 1 and granzyme B) and cytokines (IFNγ), therefore, phenotypically resembling effector T cells [16].

*Clinical implication:*

- Responders with survival efficiency: The significant recurrence-free survival (RFS) and overall survival (OS) had better results in case of Autogene cevumeran responders, as per safety-evaluable and biomarker-evaluable cohort analysis (median RFS not reached in case of 8 autogene cevumeran responders, and for 8 non-responders median RFS was 13.4 months).
- Landmark analysis: Patients who completed 8 priming doses of autogene cevumeran resulted in no median RFS values compared to the 11.0 months of RFS in case of non-responders.
- Correlation between tumor clonality and T cell responses: On a regular basis, autogene cevumeran responders resulted in generating more clonal tumors and showed vaccine-induced high-magnitude T-cell responses after neoantigen administration. Hence, it describes the correlation between tumor clonality and T-cell responses.
- CA19-9 levels: In comparison to the non-responders, responders persistently had a lower level of serum CA19-9, which indicates the vaccinated patients with neoantigens had null capacity to disseminate micro metastasis. But, exceptionally, in

Patient 29, the serum level of CA19-9 was detected more with a dense lymphoid infiltrate and no liver lesions, indicating autogene cevumeran had the capacity to eradicate micro metastasis by maximizing expanded blood T cell population [16].

### 4.2.2 Case Study_2: *NRG1* Fusion-Positive Pancreatic Cancer—Novel Targeted Therapy

- Larotrectinib (a selective inhibitor of TRKA, B, C encoded by genes *NTRK1,2,3*) trial:
  *Patient profile*: 55 patients with *NTRK* fusion across multiple tumor types were enrolled in a single-arms 3 phase I/II clinical trial. Among the 55 enrolled patients, 14 were assessed with gastrointestinal malignancy (GM), including PDAC.
  *Outcome:* The obtained response rate (RR) is 75% with median DoR (duration of response) and PFS (progression-free survival) not reached. Specific to the patients with GM (14 patients), global RR was 43% with long-term efficacy [28].
- Entrectinib (Oral tyrosine kinase inhibitor targeting TRAK A, B, C) trial:
  *Patient profile:* 121 patients harboring *NRTK* alteration underwent single-arms 3 phase I/II clinical trial. 16, among 121 patients exhibiting GM and more specifically 4 among 16 were diagnosed with PDAC.
  *Outcome:* The patients with GM demonstrated 61% RR (16% with complete response) with DoR of 20 months. In 4 patients with PDAC, the RR was 75% [29].
- Zenocutuzumab (a bispecific anti-HER2/3 monoclonal antibody) trial:
  *Patient profile:* 33 patients were evaluated with several tumors.
  *Outcome:* 8 patients with RR of 27% and 40% RR in 4 out of 10 patients with PDAC and 90% disease control rate (DCR) in 9 out of 10 patients [30].
- Seribantumab (selective inhibitor of *NRG1* fusions) trial:
  *Patient profile*: 55 patients were selected in a multi-center single-arm phase II trial with locally advanced or metastatic tumors (majority with lung cancer).
  *Outcome:* RR of 30% and DCR of 90% were observed [31].

Clinical implications of Larotrectinib, Entrectinib, Zenocutuzumab, and Seribantumab trials:

- The demonstrated targeted therapies against rare genomic alterations (0.3% for NRTK and 0.5% for NRG1 fusions) can be effectively treated with a specific tumor-agnostic approach, ultimately yielding higher RR than conventional chemotherapy [3].
- The ongoing eNRGy trial for Zenocutuzumab and CRESTONE study for seribantumab works on further validating these approaches and expanding treatment options (the action of Zenocutuzumab & seribantumab is displayed in Fig. 3) alongside previously limited therapeutic alternatives [30, 31].
- These therapeutic agents emerge as the transformative treatment paradigm towards precision medicine [3].

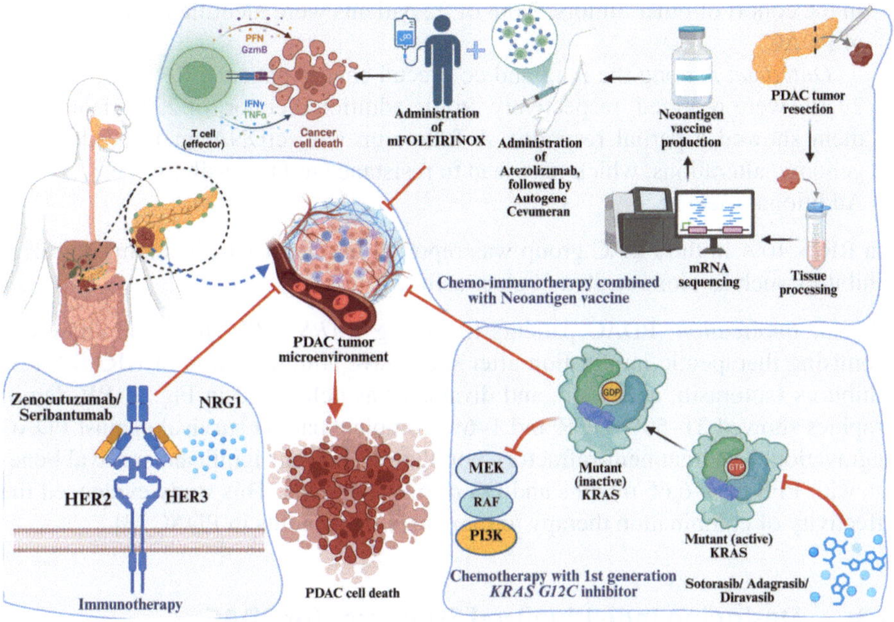


**Fig. 3** Personalized treatment strategies against PDAC aggravation: Three distinct approaches of PDAC personalized treatment strategies are outlined—sequential administration of chemo-immunotherapy in combination with Neoantigen vaccine, conventional chemotherapy in combination with small molecule inhibitor, and solo administration of immunotherapeutic agent

### 4.2.3 Case Study_3: *KRAS G12C*-Mutated Pancreatic Cancer—Targeted Inhibition

- CODEBREAK-100 trial:
  *Patient profile:* Enrolled PDAC patient numbers are 38, who had progressed after at least one line of standard chemotherapy in phase Ib/II trial testing Sotorasib.
  *Outcome:* RR of 21% and 4 months of median PFS and 6.9 months of median OS (overall survival) [3].
- KRYSTAL-1 trial:
  *Patient profile:* 12 patients, orally administered with Adagrasib for treating advanced PDAC. Follow-up studies were also conducted with 21 patients.
  *Outcome:* The above basket studies for 12 patients revealed the RR and DCR of 50% and 100%, respectively, with 6.6 months of median PFS. For 21 patients, the observed RR and DCR were 33.3% and 81% with a median PFS and OS of 5.4 months and 8 months, respectively [3].
- GO42144 phase I trial:
  *Patient profile:* The study was conducted with the *KRAS* G12C inhibitor, Divarasib. The treatment study demonstrated promising therapeutic efficacy in 58 patients across multiple solid tumors (including lung and colorectal cancer).

In the cohort of other tumors, 3 out of 22 patients were identified in the category of PDAC.

*Outcome*: Among the lung and colorectal cancer patients, RR of 53.4% and 29.1% were reported, respectively. In the additional cohort of 22 patients, 8 of them showed a partial response. A follow-up with ctDNA analysis identified genomic alterations, which may lead to resistance to Divarasib [32].

- Additional trial:

An RR > 40% in the PDAC group was reported in other trials, including selective inhibitors such as olomorasib and glecirasib [3].

*Clinical implication*: PDAC patients harboring *KARS G12C* mutation showed a promising therapeutic implication after successive administration of KRAS G12C inhibitors (sotorasib, adagrasib, and divarasib) as delineated in Fig. 3. RR demographics showed 21–50% DCR and 1–6% complete patient survival against PDAC aggravation. The treatment-refractory population showed significant survival benefits with PFS of 4–6.66 months and OS of 6.9–8 months. This study evidenced the effectivity of combination therapy to maximize the benefits in PDAC [3].

## 4.3 Designing Individualized Strategies for PDAC

Conventional therapies targeted to treat solid tumors like pancreatic cancer, have predominantly been structured based on the general anatomical criteria of the tumor. The heterogeneity characterized in different individuals poses a critical challenge for therapeutic design. The paradigm tilt towards precision oncology leverages a high-throughput genomic and transcriptomic analysis to predict biomarkers and individualize this field of therapeutics for Pancreatic Ductal Adenocarcinoma (PDAC).

### 4.3.1 Molecular Profiling

Molecular profiling has become the cornerstone of tailored pancreatic cancer therapy. There are specific molecular targets, viz., HER2 amplification, *KRAS* wild type status, and DNA repair mutation, which have been identified via the IMPaCT trial. Additionally, 22 distinct patient candidates have been identified from 76 patients within 21.5 days. Based on this concept, comprehensive biomarker and genetic profiling have been extensively tested, providing wider aspects of personalized treatment recommendations [27]. Here are the particularly encouraging findings that the tumors with altered modification of DNA repair pathways (including *BRCA* mutations) have shown better response to platinum-based chemotherapy regimens such as mFOLFIRINOX [40]. Furthermore, chemotherapeutic failure against unidentified *KRAS* mutations has been overcome via technological advancements by using combination drug therapies, where the conventional chemotherapeutic drug has been functionalized with first-generation *KRAS* G12C inhibitors, which showed better outcome and better patient survival in PDAC [3]. In addition to this, NGS has been used to identify actionable mutations in genes, *TP53, KRAS, CDKN2A,* and

*SMAD4* [41]. Recent advancement of multi-omics analysis integrates transcriptomics, proteomics, and metabolomics to facilitate the detection of the abnormal cancer-causing genetic flux. The liquid biopsy technique has been used for detecting non-invasive ctDNA monitoring, which facilitates real-time assessment of tumor dynamics [42].

### 4.3.2 Immunotherapeutic Strategies

The tumor microenvironment is enriched with immunosuppressive macrophages and surrounded by dense stroma and limited T-cell infiltration. Thereby, scientific advancement has been broadly strategized based on the specific molecular subtypes that demonstrate potential for immunotherapeutic approaches.

- Microsatellite instability: Approximately 1–2% cases of PDAC showed responsiveness to immune checkpoint inhibition due to the presence of the microsatellite instability biomarker, MSI-H/dMMR. Revealed by KEYNOTE-158 study, pembrolizumab rendered to exhibit 18.2% RR in MSI-H PDAC with some notable cases of durable responses after successive treatments [3].
- Abundance of tumor mutation: A range of mutational burden often confers chemotherapeutic failure. Therefore, identifying robust predictive biomarkers helps in functionalizing the immunotherapeutic agents (Zenocutuzumab against *NRG1* fusion, atezolizumab against PD-L1) against it, ultimately benefiting from the mutation-mediated drug resistance [30, 43].
- Combination therapy: Combining the immunotherapeutic agent with chemotherapeutic drug (PARP inhibitors olaparib and rucaparib, FGFR inhibitor Pemigatinib, Futibatinib) potentiates in converting the immunologically inactive tumor to immunologically active tumor by increasing the neoantigen load and triggering the T-cell infiltration [3, 37, 38].

## 4.4 Role of Multidisciplinary Teams in Personalized PDAC Treatment

Multidisciplinary teams (MDT) acquire collaborative workers, healthcare professionals, and specialists from different fields and plan personalized treatment strategies. These collaborative approaches ensure seamless delivery of patient care by addressing the unique needs, challenges, and circumstances of each patient. For some special cases, they incorporate clinical and translational research teams by integrating cutting-edge research findings for optimal clinical practice [44]. In this regard, the crucial role of bioinformaticians, pharmacologists, ethicists, and finally Molecular Tumor Boards (MTBs) [45] comes into action for collaborative information of genomic data for high-throughput molecular profiling to curate the personalized treatment. The key role and contributions of MDT specialists are detailed in Table 2.

**Table 2** Roles and contribution of multidisciplinary teams (MDTs) for personalized PDAC care

| Team specialist | Key role in personalized care | Contribution to individualized treatment |
|---|---|---|
| Medical oncologist | Treat cancer with conventional and advanced drug therapies such as chemotherapy, immunotherapy, targeted therapy | Curating the personalized treatment strategies based on critical molecular findings with clinical factors, viz., specific mutation (Olaparib maintenance against *gBRCA* mutation as guidelined in clinical trial NCT02184195 [3]), along with genetic and molecular level complication [46] |
| Radiation oncologist | Acquires data from radiologist and systemic designs the radiation therapy based on post-analysis | Integrating the molecular information from the radiologists to guide the radiation therapy by selecting the radio-sensitizing agents for treating broadline resectable disease [46, 47] |
| Pancreatic surgeon | Detection and segregation of the tumor tissues from the pancreas and adjacent organs via surgical resection process, ultimately procuring the tumor tissue | Hands-on resection of the tumor from the body sites (via Whipple procedure) followed by sample procurement from the resection site for clinical assessment and functionalizing neoadjuvant therapies to evaluate patient's benefit [46, 47] |
| Clinical pharmacist | Optimization of the medication regimens by managing the dosage control and assessing the treatment-mediated side effects | Assessing the interaction profile between drug and biological molecules to guide the dosage limit and modification, considering organ dysfunction |
| Gastroenterologist | Diagnosing and resolving the digestive system-related disorders both beyond and after surgery | Performs endoscopic procedures for visualizing and detecting the organ-level abnormalities, delivers the targeted therapies directly to the PDAC sites, and dissects the tissues to perform biopsies [47] |
| Palliative care specialist | Focuses on patient Well-being by relieving symptoms by improving the patient's quality of life | Integration of the qualitative guidance in consideration of critical treatment decisions, focuses on symptom management by improving nutritional status of the patient [46, 48] |
| Medical counsellor | Fostering personal moral values by bridging the gap between patient understanding and clinical-translational information | Providing emotional support throughout the challenging course of the treatment and helping the patients to articulate their personal choices and preferences in the key decision-making steps [46, 48] |
| Nutritionist | Offers the nutritional support by aligning with patient care guidelines | Develops and optimizes the personalized nutrition strategies for managing the pancreatic metabolic alteration both before and after surgery [47] |

## 5 Conclusion

In summary, integrating various molecular profiling techniques, from traditional microdissection to advanced tools, has significantly enhanced our understanding of pancreatic tumor biology. Integrated analyses of the genome, epigenome, and transcriptome are yielding biological insights with potential therapeutic relevance in PDAC. Despite notable progress, challenges in pancreatic cancer diagnosis and treatment remain. The diagnostic challenges of PC arise from late-stage presentation and non-specific symptoms, necessitating a multifaceted approach. Targeted screening for high-risk individuals, particularly those with familial pancreatic cancer, is crucial, utilizing a combination of EUS and MRI/MRCP for improved sensitivity. While EUS excels in detecting small lesions and enabling tissue diagnosis, MDCT is the gold standard for staging cancer. Liquid biopsies, especially ctDNA, have shown considerable promise for early detection, complementing imaging techniques. Therefore, future research requires a focus on refining screening protocols, optimizing imaging modalities, and developing more sensitive biomarkers for earlier diagnosis and improved patient outcomes.

In addition, the shift towards personalized treatment has transformed pancreatic cancer management. Personalized treatment in pancreatic cancer has evolved from aspirational to executable through accumulated approaches, including molecular profiling to adaptive treatment strategies. Extensive genomic and transcriptomic assessment enables precision in addressing the specific alterations and mutations, including KRAS mutations or NRG1 fusions. This framework also integrates novel therapeutic combinations, such as traditional chemotherapy with agents such as mRNA vaccines, viz., autogenous cerumen, to overcome the barriers exclusive to pancreatic cancer. Ultimately, personalized therapy, tailored to the distinct and evolving molecular characteristics of each tumor, is paving the way for transforming pancreatic cancer from a fatal disease into a manageable condition.

Overall, while significant strides have been made in understanding and managing pancreatic cancer, continued research and innovation are essential. The integration of advanced molecular profiling and diagnostics into clinical practice will be instrumental in designing personalized therapies, ultimately improving survival rates and enhancing the quality of life for patients with pancreatic cancer.

**Declarations**

**Author's Contributions** AA, BG, PKK, SC, and SSG conceived the idea and review structure. AA, BG, PKK, SC & SSG wrote the manuscript. AA, BG, PKK, SC, and SSG revised and edited the manuscript. All the authors have read, agreed, and approved the final draft of the manuscript.

**Funding** This work was supported by the Department of Biotechnology, Government of India, for Translation Program in the DBT Program Support Facility (BT/PR44695/NER/95/1880/2021), Indian Institute of Technology, Guwahati, India. We also acknowledge the Center for Nanotechnology (ICMR Center for Excellence, Grant no. 5/3/8/20/2019-ITR, INUP, MeitY Grant no. 5(1)/2021-NANO) at IIT Guwahati.

**Data Availability** The software Biorender has been used for illustrations in the book chapter.

**Conflicts of Interest** There is no conflict of interest.

## References

1. Siegel RL, Miller KD, Wagle NS, Jemal A (2023) Cancer statistics, 2023. CA Cancer J Clin 73(1):17–48. https://doi.org/10.3322/caac.21763
2. Bray F et al (2024) Global cancer statistics 2022: GLOBOCAN estimates of incidence and mortality worldwide for 36 cancers in 185 countries. CA Cancer J Clin 74(3):229–263. https://doi.org/10.3322/caac.21834
3. Brozos-Vázquez E et al (2024) Pancreatic cancer biomarkers: a pathway to advance in personalized treatment selection. Cancer Treat Rev 125:102719. https://doi.org/10.1016/j.ctrv.2024.102719
4. Jemal A, Siegel R, Xu J, Ward E (2010) Cancer statistics, 2010. CA Cancer J Clin 60(5):277–300. https://doi.org/10.3322/caac.20073
5. Razzano D et al (2022) Comprehensive molecular profiling of pancreatic ductal adenocarcinoma in <scp>FNA</scp>, biopsy, and resection specimens. Cancer Cytopathol 130(9):726–734. https://doi.org/10.1002/cncy.22589
6. Pishvaian MJ et al (2018) Molecular profiling of patients with pancreatic cancer: initial results from the know your tumor initiative. Clin Cancer Res 24(20):5018–5027. https://doi.org/10.1158/1078-0432.CCR-18-0531
7. Chrzanowska NM, Kowalewski J, Lewandowska MA (2020) Use of fluorescence in situ hybridization (FISH) in diagnosis and tailored therapies in solid tumors. Molecules 25(8):1864. https://doi.org/10.3390/molecules25081864
8. Wang K, Wang X, Pan Q, Zhao B (2023) Liquid biopsy techniques and pancreatic cancer: diagnosis, monitoring, and evaluation. Mol Cancer 22(1):167. https://doi.org/10.1186/s12943-023-01870-3
9. Zhen DB, Safyan RA, Konick EQ, Nguyen R, Prichard CC, Chiorean EG (2023) The role of molecular testing in pancreatic cancer. Ther Adv Gastroenterol 16:17562848231171456. https://doi.org/10.1177/17562848231171456
10. Wang S et al (2021) The molecular biology of pancreatic adenocarcinoma: translational challenges and clinical perspectives. Signal Transduct Target Ther 6(1):249. https://doi.org/10.1038/s41392-021-00659-4
11. Halbrook CJ, Lyssiotis CA, Pasca di Magliano M, Maitra A (2023) Pancreatic cancer: advances and challenges. Cell 186(8):1729–1754. https://doi.org/10.1016/j.cell.2023.02.014
12. Miyabayashi K, Nakagawa H, Koike K (2021) Molecular and phenotypic profiling for precision medicine in pancreatic cancer: current advances and future perspectives. Front Oncol 11:682872. https://doi.org/10.3389/fonc.2021.682872
13. Vincent A, Herman J, Schulick R, Hruban RH, Goggins M (2011) Pancreatic cancer. Lancet 378(9791):607–620. https://doi.org/10.1016/S0140-6736(10)62307-0
14. Unger K et al (2018) Metabolomics based predictive classifier for early detection of pancreatic ductal adenocarcinoma. Oncotarget 9(33):23078–23090. https://doi.org/10.18632/oncotarget.25212
15. Ottenhof NA, de Wilde RF, Maitra A, Hruban RH, Offerhaus GJA (2011) Molecular characteristics of pancreatic ductal adenocarcinoma. Pathol Res Int 2011:1–16. https://doi.org/10.4061/2011/620601
16. Bettegowda C et al (2014) Detection of circulating tumor DNA in early- and late-stage human malignancies. Sci Transl Med 6(224):224ra24. https://doi.org/10.1126/scitranslmed.3007094

17. Mayers JR et al (2014) Elevation of circulating branched-chain amino acids is an early event in human pancreatic adenocarcinoma development. Nat Med 20(10):1193–1198. https://doi.org/10.1038/nm.3686
18. Karlson B-M, Ekbom A, Lindgren PG, Källskog V, Rastad J (1999) Abdominal US for diagnosis of pancreatic tumor: prospective cohort analysis. Radiology 213(1):107–111. https://doi.org/10.1148/radiology.213.1.r99oc25107
19. Schulte SJ, Baron RL, Freeny PC, Patten RM, Gorell HA, Maclin ML (1991) Root of the superior mesenteric artery in pancreatitis and pancreatic carcinoma: evaluation with CT. Radiology 180(3):659–662. https://doi.org/10.1148/radiology.180.3.1871275
20. Andersson R, Vagianos CE, Williamson RCN (2004) Preoperative staging and evaluation of resectability in pancreatic ductal adenocarcinoma. HPB 6(1):5–12. https://doi.org/10.1080/13651820310017093
21. Catalano C et al (2003) Pancreatic carcinoma: the role of high-resolution multislice spiral CT in the diagnosis and assessment of resectability. Eur Radiol 13(1):149–156. https://doi.org/10.1007/s00330-002-1473-4
22. Miura F, Takada T, Amano H, Yoshida M, Furui S, Takeshita K (2006) Diagnosis of pancreatic cancer. HPB 8(5):337–342. https://doi.org/10.1080/13651820500540949
23. Gabata T et al (1994) Small pancreatic adenocarcinomas: efficacy of MR imaging with fat suppression and gadolinium enhancement. Radiology 193(3):683–688. https://doi.org/10.1148/radiology.193.3.7972808
24. Vander Noot MR et al (2004) Diagnosis of gastrointestinal tract lesions by endoscopic ultrasound-guided fine-needle aspiration biopsy. Cancer 102(3):157–163. https://doi.org/10.1002/cncr.20360
25. Delbeke D, Pinson CW (2004) Pancreatic tumors: role of imaging in the diagnosis, staging, and treatment. J Hepato-Biliary-Pancreat Surg 11(1):4–10. https://doi.org/10.1007/s00534-002-0775-x
26. Chantrill LA et al (2015) Precision medicine for advanced pancreas cancer: the individualized molecular pancreatic cancer therapy (IMPaCT) trial. Clin Cancer Res 21(9):2029–2037. https://doi.org/10.1158/1078-0432.CCR-15-0426
27. Rojas LA et al (2023) Personalized RNA neoantigen vaccines stimulate T cells in pancreatic cancer. Nature 618(7963):144–150. https://doi.org/10.1038/s41586-023-06063-y
28. Drilon A et al (2018) Efficacy of larotrectinib in *TRK* fusion–positive cancers in adults and children. N Engl J Med 378(8):731–739. https://doi.org/10.1056/NEJMoa1714448
29. Demetri GD et al (2022) Updated integrated analysis of the efficacy and safety of entrectinib in patients with *NTRK* fusion-positive solid tumors. Clin Cancer Res 28(7):1302–1312. https://doi.org/10.1158/1078-0432.CCR-21-3597
30. Schram AM et al (2022) Zenocutuzumab, a HER2xHER3 bispecific antibody, is effective therapy for tumors driven by *NRG1* gene rearrangements. Cancer Discov 12(5):1233–1247. https://doi.org/10.1158/2159-8290.CD-21-1119
31. Carrizosa DR et al (2022) CRESTONE: initial efficacy and safety of seribantumab in solid tumors harboring *NRG1* fusions. J Clin Oncol 40(16_suppl):3006–3006. https://doi.org/10.1200/JCO.2022.40.16_suppl.3006
32. Sacher A et al (2023) Single-agent Divarasib (GDC-6036) in solid tumors with a *KRAS* G12C mutation. N Engl J Med 389(8):710–721. https://doi.org/10.1056/NEJMoa2303810
33. Ricci V et al (2022) Pancreatic cancer: beyond Brca mutations. J Pers Med 12(12):2076. https://doi.org/10.3390/jpm12122076
34. Lai E et al (2021) BRCA-mutant pancreatic ductal adenocarcinoma. Br J Cancer 125(10):1321–1332. https://doi.org/10.1038/s41416-021-01469-9
35. Marabelle A et al (2020) Efficacy of pembrolizumab in patients with noncolorectal high microsatellite instability/mismatch repair–deficient cancer: results from the Phase II KEYNOTE-158 study. J Clin Oncol 38(1):1–10. https://doi.org/10.1200/JCO.19.02105
36. Shen L et al (2024) Updated efficacy and safety of larotrectinib (laro) in patients (pts) with TRK fusion gastrointestinal (GI) cancer. J Clin Oncol 42(3_suppl):109–109. https://doi.org/10.1200/JCO.2024.42.3_suppl.109

37. Subbiah V et al (2022) FIGHT-101, a first-in-human study of potent and selective FGFR 1-3 inhibitor pemigatinib in pan-cancer patients with FGF/FGFR alterations and advanced malignancies. Ann Oncol 33(5):522–533. https://doi.org/10.1016/j.annonc.2022.02.001
38. Meric-Bernstam F et al (2022) Futibatinib, an irreversible FGFR1–4 inhibitor, in patients with advanced solid tumors harboring *FGF/FGFR* aberrations: a phase I dose-expansion study. Cancer Discov 12(2):402–415. https://doi.org/10.1158/2159-8290.CD-21-0697
39. Li H-S, Yang K, Wang Y (2022) Remarkable response of *BRAF* V600E-mutated metastatic pancreatic cancer to BRAF/MEK inhibition: a case report. Gastroenterol Rep (Oxf) 10:goab031. https://doi.org/10.1093/gastro/goab031
40. Wattenberg MM et al (2020) Platinum response characteristics of patients with pancreatic ductal adenocarcinoma and a germline BRCA1, BRCA2 or PALB2 mutation. Br J Cancer 122(3):333–339. https://doi.org/10.1038/s41416-019-0582-7
41. Stefanoudakis D, Frountzas M, Schizas D, Michalopoulos NV, Drakaki A, Toutouzas KG (2024) Significance of TP53, CDKN2A, SMAD4 and KRAS in pancreatic cancer. Curr Issues Mol Biol 46(4):2827–2844. https://doi.org/10.3390/cimb46040177
42. Ma L et al (2024) Liquid biopsy in cancer: current status, challenges and future prospects. Signal Transduct Target Ther 9(1):336. https://doi.org/10.1038/s41392-024-02021-w
43. Herbst RS et al (2020) Atezolizumab for first-line treatment of PD-L1–selected patients with NSCLC. N Engl J Med 383(14):1328–1339. https://doi.org/10.1056/NEJMoa1917346
44. Law NLW, Hong LW, Tan SSN, Foo CJ, Lee D, Voon PJ (2024) Barriers and challenges of multidisciplinary teams in oncology management: a scoping review protocol. BMJ Open 14(2):e079559. https://doi.org/10.1136/bmjopen-2023-079559
45. Boos L, Wicki A (2024) The molecular tumor board—a key element of precision oncology. Mag Eur Med Oncol 17(3):190–193. https://doi.org/10.1007/s12254-024-00977-7
46. Martin-Perez E et al (2020) Multidisciplinary consensus statement on the clinical management of patients with pancreatic cancer. Clin Transl Oncol 22(11):1963–1975. https://doi.org/10.1007/s12094-020-02350-6
47. Rosenzweig A et al (2023) Management of patients with pancreatic cancer using the 'Right Track' model. Oncologist 28(7):584–595. https://doi.org/10.1093/oncolo/oyad080
48. Fogel EL et al (2017) A multidisciplinary approach to pancreas cancer in 2016: a review. Am J Gastroenterol 112(4):537–554. https://doi.org/10.1038/ajg.2016.610

# Advancements in Hematopoietic Cancer Treatment

Subhajit Karmakar, Mouli Chatterjee, and Malini Basu

**Abstract**

Hematopoietic cancers, including leukemia, lymphoma, and multiple myeloma, constitute a heterogeneous group of malignancies characterized by intricate genetic and microenvironmental changes. In recent decades, therapeutic approaches have progressed from traditional chemotherapy and radiotherapy to advanced targeted and immune-based treatments. Hematopoietic stem cell transplantation (HSCT) continues to be a fundamental component of curative therapy enhanced by reduced-intensity regimens and innovative donor sources with increased accessibility. Concurrent advancements in small-molecule inhibitors, such as kinase and proteasome pathway modulators, have enhanced therapeutic precision by selectively targeting oncogenic signaling while minimizing toxicity. The development of monoclonal antibodies, bispecific antibodies, antibody-drug conjugates, and immune checkpoint inhibitors has further expanded therapeutic options by leveraging immune effector mechanisms. Cellular immunotherapies, notably CAR-T cells, have revolutionized outcomes in relapsed and refractory hematologic malignancies, whereas next-generation CAR constructs, CAR-NK cells, and tumor vaccines continue to broaden the scope of immune engineering. Simultaneously, nanotechnology has redefined drug delivery paradigms by improving pharmacokinetics, reducing systemic toxicity, and enabling tumor-

Subhajit Karmakar and Mouli Chatterjee contributed equally with all other contributors.

S. Karmakar · M. Chatterjee
Cancer Biology and Inflammatory Disorder Division, Council of Scientific and Industrial Research-Indian Institute of Chemical Biology (CSIR-IICB), Kolkata, India

M. Basu (✉)
Department of Microbiology, Dhruba Chand Halder (DCH) College, University of Calcutta, Kolkata, West Bengal, India

 
M. K. Ghosh, M. Basu (eds.), *Personalized Cancer Therapy*,
https://doi.org/10.1007/978-981-95-6845-1_15

targeted release of chemotherapeutics, molecular inhibitors, and immunomodulators. Smart nanocarriers, biomimetic systems, and nanodrug delivery platforms integrated with immunotherapies exemplify the shift towards precision and personalized medicine. Collectively, these advancements highlight a paradigm shift in hematopoietic cancer therapy, where the integration of molecular insights, immunological innovation, and nanotechnology is paving the way for durable remission, improved survival, and enhanced quality of life.

**Keywords**

Hematopoietic cancer · Nanotechnology · Personalized medicine · Immunotherapies

## 1 Introduction

Hematopoietic cancers (HCs), including leukemia, lymphoma, and multiple myeloma (MM), originate from the malignant transformation of immune and hematopoietic cells, often driven by specific genetic abnormalities, such as chromosomal translocations [1]. Leukemias typically manifest as "liquid" tumors of myeloid or lymphoid origin, either acute or chronic. Lymphomas form solid masses from transformed lymphocytes, with subtypes including Hodgkin and non-Hodgkin. Myeloma arises from neoplastic plasma cells within bone marrow niches. Contemporary classification systems incorporate architectural features, immunophenotypes, and recurrent genomic alterations to enhance diagnostic precision and tailor therapeutic strategies [2]. These advancements have built upon the success of hematopoietic stem cell transplantation and molecularly targeted therapies, which have significantly improved patient survival rates. Therapeutic advancements have progressed through two complementary trajectories: immune-based strategies and innovations in drug delivery. In the realm of immunotherapy, monoclonal antibodies, initiated by the introduction of anti-CD20 rituximab, have redefined the standards of care. This was succeeded by next-generation agents, such as anti-CD19 and anti-CD38, bispecific T-cell engagers, and antibody-drug conjugates, such as brentuximab vedotin [3]. Checkpoint inhibitors, including PD-1/PD-L1 and CTLA-4, are currently being used in specific contexts. Cellular therapies, most notably CAR-T cells, have achieved durable remission in relapsed/refractory acute lymphoblastic leukemia (ALL), chronic lymphocytic leukemia (CLL), non-Hodgkin lymphoma (NHL), and multiple myeloma, thereby extending their curative potential once predominantly associated with allogeneic transplantation [4]. Concurrently, advancements in nanomedicine have significantly enhanced the therapeutic index and precision of treatment. Liposomal and polymeric platforms have improved pharmacokinetics and tumor targeting, as exemplified by Doxil® and CPX-351 (Vyxeos™), with the latter demonstrating survival benefits in high-risk acute myeloid leukemia (AML) [5]. Additional formulations, such as Marqibo®, and "smart" nanocarriers, engineered for pH/enzyme-responsive release or co-delivery, are effectively reducing toxicity and overcoming resistance in leukemias, lymphomas, and myeloma, supported by an expanding body of clinical evidence.

Recent advancements have delineated an evolving paradigm in HCs, characterized by genomically informed disease classification integrated with immune and nanoscale delivery technologies. These innovations enhance therapeutic efficacy, broaden the scope of curative strategies, and pave the way for personalized combination-driven care. Nanomedicine has transformed the therapeutic landscape of HCs by significantly improving their precision and efficacy. Liposomal and polymeric nanocarriers have emerged as potent tools for optimizing drug pharmacokinetics and tumor targeting. Nanocarrier systems effectively encapsulate and deliver therapeutic agents, safeguarding them from degradation and enhancing their accumulation in tumor tissues. Advanced formulations, such as Marqibo® (liposomal vincristine) and "smart" nanocarriers engineered for stimuli-responsive drug release or co-delivery of multiple agents, address critical challenges in HC treatment, including reducing systemic toxicity and overcoming drug resistance [6]. The integration of nanomedicine with genomic insights has facilitated a more nuanced and personalized approach to HC management. This synergy enables genomically informed disease taxonomy, allowing for more precise patient stratification and treatment selection. The combination of immune-based therapies and nanoscale delivery technologies has expanded the range of treatment options, potentially increasing the number of patients eligible for curative treatments. As the clinical evidence base for these innovative approaches continues to grow, the field is advancing towards a paradigm of personalized, combination-driven care in HCs. This chapter provides a brief explanation of all treatment regimens available for therapeutic perspectives in HC patient profiles and disease characteristics, with the ultimate aim of improving survival rates and quality of life for patients with various forms of leukemia, lymphoma, and myeloma.

## 2 Targeting Oncogenic Signaling with Small-Molecule Inhibitors

Small-molecule inhibitors have transformed the therapeutic landscape of hematopoietic malignancies by providing a targeted strategy to disrupt oncogenic pathways. These inhibitors have been engineered to interfere with specific molecular targets implicated in cancer cell growth, survival, and proliferation. By selectively inhibiting these pathways, small-molecule inhibitors can effectively suppress cancer progression while often minimizing harm to healthy hematopoietic cells. This targeted approach represents a substantial advancement over traditional chemotherapy, which frequently indiscriminately affects both cancerous and normal cells. The development of small-molecule inhibitors has been propelled by an enhanced understanding of the molecular mechanisms underlying hematopoietic malignancies. These inhibitors can target various components of oncogenic signaling cascades, including kinases, transcription factors, and epigenetic regulators. Notable examples include tyrosine kinase inhibitors of CML, Bruton's tyrosine kinase inhibitors of CML, and IDH inhibitors of AML [7]. The specificity of these inhibitors not only enhances treatment efficacy, but also potentially reduces side effects, thereby improving the quality of life of patients. As research in this domain continues to progress, the development of new small-molecule inhibitors and the refinement of existing ones promise to further augment the precision and effectiveness of targeted therapies for hematopoietic malignancies.

### 2.1 Proteasome and Ubiquitin Pathway Inhibitors

Peptide-based molecules, such as melphalan, continue to serve as standard treatments for MM. In contrast, proteasome inhibitors, such as bortezomib, have revolutionized therapy by inhibiting the degradation of pro-apoptotic proteins, thereby reinstating programmed cell death. The emergence of resistance to bortezomib has catalyzed the development of novel ubiquitin-specific protease (USP) inhibitors, including P5091 (USP7 inhibitor) and VLX1570 (USP14 inhibitor), highlighting the therapeutic potential of targeting protein homeostasis [8, 9].

### 2.2 PI3K Pathway Inhibitors

Aberrant PI3K/AKT signaling plays a significant role in leukemogenesis and resistance to therapy. Idelalisib, a PI3Kδ inhibitor, induces apoptosis in CLL by disrupting microenvironmental cytokine support. Duvelisib, a dual PI3Kδ/γ inhibitor, enhances cytotoxicity by impairing T cell and macrophage-mediated survival signals [10]. Next-generation molecules, such as umbralisib, which inhibits both PI3Kδ and CK1ε, demonstrate improved selectivity and tolerability. Additionally, novel candidates, including zandelisib, linperlisib, and acalisib, are currently undergoing clinical evaluation.

### 2.3 Tyrosine Kinase Inhibitors (TKIs)

The introduction of Bruton's tyrosine kinase (BTK) inhibitors, including ibrutinib, acalabrutinib, and zanubrutinib, has significantly transformed the treatment landscape for B-cell malignancies by irreversibly inhibiting B-cell receptor (BCR) signaling and nuclear factor kappa-light-chain-enhancer of activated B cells (NF-κB) activation [11]. In CML, bosutinib demonstrates efficacy against imatinib-resistant BCR-ABL mutations, with the exception of T315I and V299, while maintaining a favorable kinase selectivity profile.

### 2.4 Multikinase and Mutation-Resistant Inhibitors

To address resistance mutations, such as BCR-ABL T315I, FLT3-ITD, and JAK2 V617F, multitargeted inhibitors, including ponatinib, midostaurin, and gilteritinib, have been developed [12]. Ponatinib effectively overcame T315I-mediated resistance in CML and ALL, whereas midostaurin and gilteritinib were effective in FLT3-mutant AML. Additionally, ruxolitinib and fedratinib, which are selective JAK inhibitors, have been approved for the treatment of myelofibrosis and polycythemia vera and contribute to the improvement of splenomegaly and disease-related symptoms [13].

### 2.5 Emerging Inhibitor Classes

In addition to kinase signaling, emerging pathways have been explored. The blockade of CD47, exemplified by Hu5F9-G4, interrupts the "don't eat me" signal to macrophages, thereby facilitating the immune-mediated clearance of AML and lymphoma cells. Neddylation inhibitors, such as pevonedistat (MLN4924) and TAS4464, target the activation of cullin-RING ligases, leading to apoptosis and senescence in AML and MM models [14]. These small-molecule inhibitors collectively underscore a therapeutic transition towards the precise targeting of oncogenic drivers and stress-adaptive pathways in hematopoietic malignancies. Continued efforts to optimize specificity, safety, and resistance management will further enhance their efficacy in achieving sustained disease control.

## 3 Conventional and Advanced Therapies in Hematopoietic Cancers

The management of hematopoietic malignancies has undergone substantial advancements by integrating traditional treatment modalities with novel targeted and immune-based therapies to enhance both survival rates and quality of life. Radiotherapy continues to be a fundamental component in the treatment of hematologic cancers, particularly localized diseases and Hodgkin lymphoma. Its precise application facilitates the effective eradication of tumor masses, alleviation of symptoms, and induction of remission. In addition to its role as a local therapy, radiotherapy is crucial in conditioning regimens for hematopoietic stem cell transplantation (HSCT), where it eliminates residual malignant cells and suppresses immune function, thereby supporting engraftment and minimizing the risk of rejection. HSCT serves as a curative intervention for a broad range of hematologic malignancies, including leukemia, lymphoma, and multiple myeloma [15]. This procedure entails intravenous infusion of healthy hematopoietic stem cells to restore normal hematopoiesis and immune function. Autologous HSCT, which utilizes a patient's own stem cells, is commonly employed in the treatment of multiple myeloma and certain lymphomas. In contrast, allogeneic HSCT, which uses donor-derived stem cells, offers both hematopoietic reconstitution and graft-versus-leukemia effects. Recent advancements, such as reduced-intensity conditioning regimens, have extended HSCT eligibility to older patients and those with comorbidities, whereas peripheral blood stem cells have largely supplanted bone marrow harvests, facilitating faster recovery [16, 17]. Cord blood transplantation, including strategies such as UM171-expanded units, further enhances donor availability and reduces relapse and chronic graft-versus-host disease (GVHD), although the risk of infection remains limited. Monoclonal antibodies (mAbs) have revolutionized treatment by targeting specific antigens in malignant cells. The first-generation anti-CD20 antibody rituximab paved the way for second-generation agents, such as ofatumumab and obinutuzumab, which have improved efficacy in CLL and follicular lymphoma (FL). Additional mAbs, including daratumumab (anti-CD38) and elotuzumab (anti-CS1), have transformed the management of multiple myeloma, particularly in relapsed and refractory settings, and are now being incorporated into frontline therapy. Bispecific antibodies (bsAbs) represent a significant advancement by simultaneously binding tumor antigens and immune effector molecules, thereby redirecting T cells against malignant cells. Bispecific T-cell engagers (BiTEs), such as blinatumomab (anti-CD19/CD3), have demonstrated substantial efficacy in relapsed/refractory ALL. Newer bsAbs target BCMA, CD22, CD33, CD123, and CD30, among others, with several already approved by the FDA for multiple myeloma and lymphoma [18]. Their clinical success has been accompanied by ongoing optimization to overcome immune escape and resistance. Antibody-drug conjugates (ADCs) combine the specificity of mAbs with cytotoxic payloads to selectively eliminate malignant cells while sparing normal tissue. Brentuximab vedotin (anti-CD30) and inotuzumab ozogamicin (anti-CD22) are well established in Hodgkin lymphoma and B-ALL, respectively. Belantamab mafodotin (anti-BCMA) has become an important option in multiple myeloma [19]. Ongoing trials are expanding ADC applications to new targets, such as CD33, CD123, CD74, and GPRC5D. Immune checkpoint inhibitors (ICIs) have expanded therapeutic options by overcoming tumor-induced immune suppression. Agents targeting PD-1, PD-L1, and CTLA-4 have shown efficacy in a subset of hematologic malignancies. Novel checkpoints, including TIM-3, TIGIT, CD70, and CD47, are being actively explored to overcome resistance and broaden applicability [20]. Combinatorial strategies with other immunotherapies and targeted drugs are under investigation to enhance outcomes (Fig. 1).

Collectively, these advancements underscore a therapeutic continuum in hematopoietic malignancies, wherein traditional approaches, such as radiotherapy and HSCT, remain essential, and emerging biologics and immunotherapies are transforming treatment paradigms. The integration of these modalities holds the potential for achieving durable remission, minimizing toxicity, and ultimately, providing curative outcomes for an increasing number of patients.

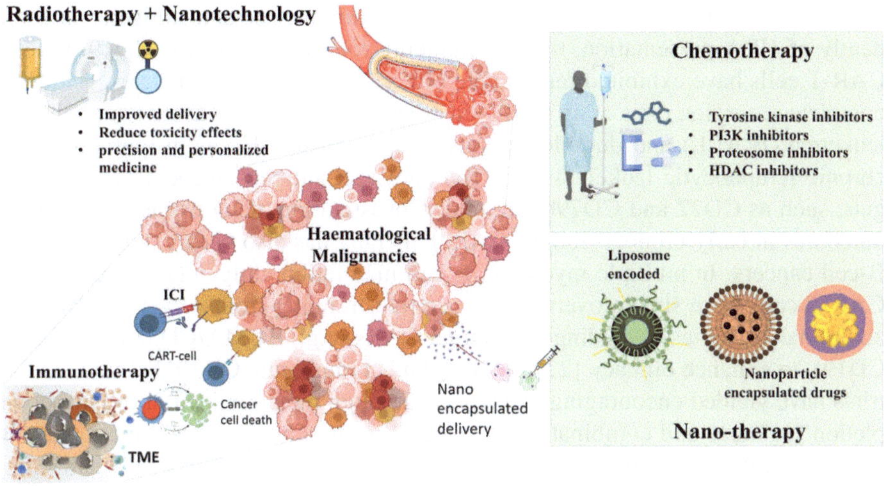


**Fig. 1** Therapeutic strategies and nanotechnology-based interventions in hematological malignancies. This schematic represents outlines of the main therapeutic approaches for hematological malignancies, including chemotherapy, immunotherapy, radiotherapy, and nanotechnology-based delivery systems. Chemotherapy uses tyrosine kinase inhibitors, PI3K inhibitors, proteasome inhibitors, and HDAC inhibitors to disrupt oncogenic pathways. Immunotherapeutic approaches such as immune checkpoint inhibitors (ICI) and CAR-T cell therapy enable targeted immune-mediated cytotoxicity. Radiotherapy combined with nanotechnology improves delivery precision and reduces toxicity. Nanotherapeutic platforms, including liposomes and nanoparticle-encapsulated drugs, enhance the pharmacokinetics and therapeutic efficacy

## 4 Cellular Immunotherapies in Hematopoietic Cancers

Cellular immunotherapy has rapidly emerged as a transformative modality for the treatment of hematopoietic malignancies, particularly for patients with relapsed or refractory conditions, where conventional therapies often prove insufficient. Broadly, cellular therapies are categorized into non-genetically modified and genetically modified strategies, both of which are designed to harness and enhance immune effector cells.

### 4.1 Nongenetic and Genetically Modified Cellular Therapies

Initial strategies in adoptive cell therapy encompassed donor lymphocyte infusions and cytotoxic T lymphocytes, including virus-specific T cells, such as those targeting Epstein-Barr virus (EBV), and tumor-specific T cells directed against tumor-associated or neoantigens. These approaches have established a proof-of-concept that the adoptive transfer of immune cells could elicit durable antitumor responses in hematologic malignancies. The most notable advancements have arisen from genetically engineered products, particularly CAR T-cells.

CAR-T therapy enables T cells to recognize tumor-associated antigens independently of HLA presentation, thereby inducing robust cytotoxicity. CD19-directed CAR-T cells have exhibited remarkable efficacy in B-cell malignancies, achieving complete remission rates of up to 90% in relapsed/refractory B-acute lymphoblastic leukemia (B-ALL) and durable responses in B-cell non-Hodgkin lymphoma and chronic lymphocytic leukemia [21]. To counteract antigen escape, alternative targets, such as CD22 and CD79b, are under investigation, demonstrating promising outcomes in early clinical studies. CAR-T therapy has also been extended beyond B-cell cancers. In multiple myeloma, B-cell maturation antigen (BCMA)-directed CAR-T cells have shown overall response rates ranging from 43% to 100%, with additional strategies, including dual-target CARs (e.g., BCMA/CD38 and BCMA/CD19), to enhance efficacy [22]. In Hodgkin's lymphoma, CD30-directed CAR-T trials have yielded encouraging responses, although the optimization of lymphodepletion regimens and combinatorial strategies is ongoing. For T-cell malignancies, where identifying tumor-specific antigens is more challenging, CD7- and CD5-directed CAR-Ts are being evaluated, whereas in AML, efforts focus on targets such as CLL-1, CD33, and CD123, although balancing specificity and safety remains a significant challenge.

## 4.2 Emerging Cellular Platforms

In addition to chimeric antigen receptor T-cell (CAR-T) therapies, CAR-natural killer (CAR-NK) and CAR-macrophage therapies are being developed to address challenges such as relapse, toxicity, and manufacturing complexity. Natural killer (NK) cell–based therapies, for example, have demonstrated potential in AML, with cytokine support (interleukin-15, interleukin-21) and optimized preconditioning regimens enhancing their expansion and persistence [23]. Clinical trials have indicated that adoptively transferred NK cells can persist in vivo, mediate cytotoxicity, and complement HSCT without increasing graft-related complications.

## 4.3 Tumor Vaccines

Parallel advancements encompass tumor vaccines engineered to prime the immune system against antigens associated with leukemia or lymphoma. Molecular vaccines, such as peptide-based vaccines targeting WT1 or proteinase 3 in AML, along with dendritic cell vaccines, have demonstrated the capacity to elicit specific cytotoxic T cell responses, diminish the leukemic burden, and potentially eradicate minimal residual disease [24]. Although predominantly investigational, tumor vaccines represent a promising strategy for long-term disease management and relapse prevention.

### 4.4 Clinical Impact

Recent clinical trials have highlighted the potential of nanocarrier-based therapies for the management of hematopoietic cancers. Liposomal formulations have demonstrated efficacy in enhancing the therapeutic index of conventional chemotherapeutic agents. For instance, liposomal vincristine (Marqibo®) has exhibited significant clinical activity with an acceptable safety profile in a Phase II trial involving patients with relapsed or refractory ALL. Similarly, liposomal doxorubicin (Doxil®), a PEGylated formulation, has been utilized in various hematologic malignancies, including relapsed/refractory multiple myeloma and non-Hodgkin lymphoma, with a reduction in cardiotoxicity compared to conventional doxorubicin. Another significant advancement was the use of Vyxeos™ (CPX351), a liposomal co-formulation of daunorubicin and cytarabine [25, 26]. In a Phase III trial, Vyxeos™ significantly improved overall survival in elderly patients with newly diagnosed high-risk AML compared to the traditional 7+3 regimen, resulting in its FDA approval in 2017. Treatment regimens in hematological malignancies are represented in Table 1.

This success underscores the capacity of nanotechnology to optimize drug ratios and pharmacokinetics, thereby achieving superior outcomes. Beyond liposomal drugs, antibody-drug conjugates (ADCs) and nanoparticle-antibody conjugates are expanding therapeutic possibilities. Brentuximab vedotin, an ADC targeting CD30, has demonstrated significant efficacy in relapsed Hodgkin lymphoma with reduced systemic toxicity [27]. Additionally, gold nanoparticles conjugated with anti-CD20 antibodies have shown enhanced specificity and improved efficacy for rituximab in B-cell lymphomas. These targeted approaches exemplify the potential of nanotechnology to increase precision while minimizing off-target effects. The nanoparticle albumin-bound paclitaxel (Abraxane®), which was initially developed for solid tumors, is currently being evaluated for hematologic malignancies. Its unique delivery mechanism, which exploits the natural affinity of albumin for tumor cells, offers enhanced drug penetration into cancerous tissues. Furthermore, nanocarrier platforms have been employed to stabilize and improve the pharmacokinetics of established drugs, thereby enabling prolonged circulation and sustained therapeutic efficacy.

In the future, the convergence of nanotechnology with genomics and precision medicine presents significant potential. The development of personalized nanomedicines specifically designed to align with patient-specific molecular profiles could revolutionize the treatment of hematologic cancers. Furthermore, advancements in artificial intelligence and machine learning are being utilized to enhance the design, dosage, and release kinetics of nanomedicines, thereby expediting the transition of next-generation patient-specific nanodrugs from clinical trials to clinical practice. Collectively, these innovations underscore the transformative impact of nanomedicine in hematopoietic cancer therapy, offering safer, more effective, and personalized treatment options for patients with otherwise limited alternatives. The clinical success of FDA-approved CAR-T therapies, such as tisagenlecleucel (Kymriah) and axicabtagene ciloleucel (Yescarta), has firmly established cellular

**Table 1** Therapeutic approaches in hematopoietic malignancies (HM)

| Therapy | Target/Class | Drug(s) | HM type | Status |
|---|---|---|---|---|
| TKIs | BCR-ABL | Imatinib, Nilotinib, Dasatinib, Bosutinib, Ponatinib | CML, ALL | FDA app. |
| BTK | BTK | Ibrutinib, Zanubrutinib | CLL, MCL, MZL | FDA app. |
| Multi-K | FLT3, JAK2 | Midostaurin, Gilteritinib | AML | FDA app. |
| PI3K | PI3Kδ/γ | Idelalisib, Duvelisib, Copanlisib, Zandelisib, Linperlisib | CLL, FL | Mix (FDA/CT) |
| BCL-2 | BCL-2 | Venetoclax | CLL, AML | FDA app. |
| Proteasome | 20S proteasome | Ixazomib citrate | MM | FDA app. |
| HDAC | HDAC | Panobinostat, Belinostat | MM, PTCL | FDA app. |
| mAbs | CD20, CD38 | Rituximab (var.), Obinutuzumab, Daratumumab, Isatuximab | NHL, MM | FDA app. |
| bsAbs | CD19 × CD3, CD20 × CD3, BCMA × CD3 | Blinatumomab, Teclistamab, Elranatamab, Mosunetuzumab | ALL, MM, FL, DLBCL | FDA/CT |
| ADCs | CD22, BCMA, CD30 | Inotuzumab ozogamicin, Belantamab mafodotin, Brentuximab vedotin | ALL, MM, HL | FDA app. |
| ICIs | PD-1, CTLA-4, TIGIT, TIM-3 | Pembrolizumab, Nivolumab, Ipilimumab, Tiragolumab, etc. | HL, NHL, MM | FDA/CT |
| CAR-T | CD19, BCMA, CD30 | Tisagenlecleucel, Brexucabtagene, Axicabtagene, Ide-cel, Allo-715 | B-ALL, NHL, MM, HL | FDA/CT |
| Nanotherapies | Liposomal & NP carriers | Doxil®, AuNPs, Lipid NPs, Polymeric NPs, CNTs, Dendrimers | MM, AML, NHL | FDA/Precl. |

Abbreviations: *App.* Approved, *CT* Clinical trials, *Precl.* Preclinical, *CML* Chronic myeloid leukemia, *ALL* Acute lymphoblastic leukemia, *AML* Acute myeloid leukemia, *CLL* Chronic lymphocytic leukemia, *MCL* Mantle cell lymphoma, *MZL* Marginal zone lymphoma, *FL* Follicular lymphoma, *MM* Multiple myeloma, *PTCL* Peripheral T-cell lymphoma, *DLBCL* Diffuse large B-cell lymphoma, *HL* Hodgkin lymphoma

immunotherapy as the standard of care for certain hematopoietic malignancies [28]. However, challenges such as antigen escape, relapse, resistance, and toxicity, including cytokine release syndrome and neurotoxicity, remain critical areas of ongoing research. Next-generation strategies involve dual- and multi-antigen CARs, universal "off-the-shelf" CAR products, and rational combinations with checkpoint inhibitors or targeted drugs. Collectively, cellular immunotherapies represent a paradigm shift in hematopoietic cancer treatment, moving beyond traditional cytotoxic approaches to precision immune engineering [29]. With continuous refinement, these therapies hold the promise of durable remission and potential cure across a broader spectrum of hematologic cancers.

## 5 Nanotechnology-Based Therapies in Hematopoietic Cancers

Nanotechnology-based therapies have revolutionized the treatment landscape for hematopoietic cancers, offering unprecedented precision and efficacy. These innovative approaches exploit the unique properties of nanomaterials to overcome the traditional limitations of cancer treatment. Nanocarriers, such as liposomes, polymeric nanoparticles, and inorganic nanoparticles, can be engineered to specifically target cancer cells while sparing healthy tissues. This targeted delivery not only enhances the therapeutic index of anticancer drugs, but also significantly reduces systemic toxicity, which is a major concern in conventional chemotherapy. The application of nanotechnology in hematopoietic cancer treatment has been extended beyond drug delivery. In this section, current advancements in nanotechnology-based approaches, along with targeted delivery approaches, are described.

### 5.1 Nanocarriers for Drug Delivery

Liposomes, polymeric nanoparticles, inorganic nanoparticles, and dendrimers have been extensively studied as drug delivery systems for the treatment of leukemia, lymphoma, and multiple myeloma. Liposomal formulations enhance drug solubility, extend circulation time, and facilitate targeted drug accumulation at specific sites. For example, the FDA-approved liposomal formulation Vyxeos® (CPX-351) co-encapsulates daunorubicin and cytarabine at an optimized 1:5 molar ratio, yielding synergistic anti-leukemic effects in AML [30]. Compared with conventional chemotherapy, Vyxeos® has significantly improved survival rates while reducing adverse events, representing a paradigm shift in the management of AML.

### 5.2 Smart and Stimuli-Responsive Nanocarriers

In addition to traditional delivery methods, "smart" nanocarriers have been engineered to release drugs in response to tumor-specific stimuli. pH-sensitive nanocarriers take advantage of the acidic microenvironment characteristic of leukemic niches, whereas enzyme-sensitive systems are designed to respond to elevated levels of proteases such as matrix metalloproteinases, which are found in aggressive leukemias. These strategies facilitate the localized release of therapeutic agents, thereby minimizing systemic exposure. Furthermore, multifunctional carriers capable of co-delivering chemotherapeutics and molecular inhibitors have demonstrated synergistic benefits [31]. For instance, combinations of liposomal doxorubicin and paclitaxel in lymphoma models have been shown to enhance therapeutic efficacy while reducing dose-related toxicity.

### 5.3 Pegylated Systems and Polymeric Nanoparticles

Pegylation enhances the half-life of drugs and diminishes immunogenicity, as demonstrated by pegylated asparaginase and interferons in the treatment of leukemia. Polymeric nanoparticles encapsulating barasertib (AZD2811), an Aurora kinase B inhibitor, are currently undergoing clinical evaluation, providing sustained release and exhibiting significant antileukemic activity in preclinical models [32].

## 6 Nanodrug Delivery Systems in Hematopoietic Cancer Treatment

Nanotechnology has significantly transformed contemporary medicine, particularly in the field of oncology, by facilitating the development of nanodrug delivery systems (NDDS). These systems have emerged as potent tools to overcome existing barriers, enabling targeted and sustained drug release, enhancing therapeutic efficacy, and modulating the tumor microenvironment. NDDS can be engineered to exploit disease-specific molecular markers, aberrant signaling pathways, and niche-associated survival mechanisms. In the context of hematologic cancers, NDDS strategies encompass antibody-decorated carriers, nucleic acid aptamers, peptide- and protein-mediated delivery, cell membrane-camouflaged nanocarriers, and integration with immunotherapy approaches, such as CAR-T cells and checkpoint inhibitors.

### 6.1 Antibody-Mediated Nanodrug Delivery

Antibody-functionalized nanocarriers facilitate the selective recognition of malignant cells by targeting tumor-associated surface antigens. For instance, mesoporous silica nanoparticles (MSNs) loaded with gemcitabine and coated with epidermal growth factor receptor (EGFR) antibodies have demonstrated selective uptake by EGFR-positive leukemia cells while sparing normal hematopoietic cells. Similarly, antibody-mediated delivery targeting FLT3 mutations in acute myeloid leukemia (AML) using poly(amidoamine) (PAMAM) dendrimers conjugated with FLT3 ligands and encapsulating miR-150 has shown enhanced bone marrow accumulation, effective inhibition of AML progression, and minimal hematopoietic toxicity [33]. Liposome-based antibody systems are also being investigated. CD30-modified liposomes carrying cytotoxic payloads have exhibited superior binding and antitumor activity in anaplastic large cell lymphoma (ALCL), outperforming standard liposomal formulations. Similarly, CD19-, CD20-, and CD33-directed nanocarriers have been optimized to enhance drug accumulation in B-cell malignancies and AML, thereby offering a highly selective therapeutic platform.

## 6.2 Nucleic Acid Aptamer-Based Delivery

Aptamers are synthetic single-stranded nucleic acids characterized by their high affinity and specificity for tumor cell surface receptors. Their small size, low immunogenicity, and ease of modification render them highly suitable as targeting ligands for nanodrug delivery systems (NDDS). For example, the Sgc8 aptamer, when conjugated with doxorubicin (DOX), facilitates targeted delivery to protein tyrosine kinase 7 (PTK7)-positive T cell acute lymphoblastic leukemia (ALL) cells, thereby inducing site-specific cytotoxicity with minimal off-target effects. Another example involves a DNA aptamer targeting CD177 conjugated with methotrexate, which selectively inhibits acute myeloid leukemia (AML) cell growth without adversely affecting normal hematopoiesis. Innovative designs, such as ATP-responsive mesoporous silica nanoparticles loaded with DOX and coated with calcium carbonate, enable tumor microenvironment-specific drug release [34]. In lymphoma models, these nanocomposites significantly enhanced apoptosis and improved survival outcomes compared to free drug formulations. Aptamer-drug conjugates are currently being investigated in combination with standard chemotherapeutics (e.g., cytarabine with AS1411 aptamer) to enhance efficacy in refractory AML.

## 6.3 Peptide and Protein-Mediated Nanocarriers

Cell-penetrating peptides (CPPs) present a distinctive mechanism for facilitating the intracellular transport of drugs across cellular membranes. For instance, dendrimers modified with CPP44, which targets the M160 receptor on acute myeloid leukemia (AML) cells, have demonstrated enhanced anti-leukemic efficacy compared to their unmodified counterparts. In addition to CPPs, receptor-targeted peptides, such as AA13, which bind to low-density lipoprotein (LDL) receptors overexpressed in leukemia, have shown promise in improving uptake specificity [35]. Proteins such as transferrin and ferritin have also been extensively studied for their potential in nanodrug delivery. Ferritin-based arsenic nanomedicine targeting CD71 receptors in leukemia models has exhibited significant antitumor efficacy and reduced systemic toxicity compared to free arsenic trioxide. Similarly, transferrin-decorated lipopolyplexes have effectively delivered miR-29b or Bcl-2 antisense oligonucleotides to AML cells, thereby extending survival in preclinical models [36]. These protein-ligand systems highlight the potential of leveraging endogenous molecules for targeted therapeutic applications.

## 6.4 Other Ligand-Targeted Delivery Systems

Hyaluronic acid (HA), which interacts with CD44 receptors, has been integrated into nanocarriers to treat AML [37]. HA-conjugated 6-mercaptopurine prodrugs demonstrated glutathione-responsive drug release, thereby enhancing cytotoxic effects against AML cells while reducing systemic exposure. Similarly, folate-conjugated siRNA lipopolyplexes achieved substantial mRNA knockdown and tumor growth inhibition, demonstrating the adaptability of ligand-mediated strategies. In the context of B-cell lymphoma, pH-sensitive mesoporous silica nanoparticles (MSNs) functionalized with rituximab facilitated selective CD20-targeted uptake and intracellular doxorubicin (DOX) release under acidic conditions, significantly promoting apoptosis in malignant B cells [38]. This approach underscores the integration of molecular targeting with stimuli-responsive drug-release mechanisms.

## 6.5 Cell Membrane-Camouflaged Nanocarriers

A significant advancement in nanodrug delivery systems (NDDS) is the development of cell membrane-coated nanoparticles, which offer biomimetic camouflage to evade immune clearance and extend circulation time. For example, leukemic cell membrane-coated mesoporous silica nanoparticles (MSNs) loaded with daunorubicin and modified with transforming growth factor-beta receptor II (TGFβRII) antibodies effectively accumulated in the bone marrow niche and overcame microenvironment-mediated chemoresistance in AML [39]. Similarly, platelet membrane-coated nanoparticles carrying bortezomib and tissue plasminogen activator (tPA) targeted multiple myeloma, achieving dual antitumor activity and thrombus prevention [40]. Additionally, neutrophil membrane-coated nanocarriers that co-deliver doxorubicin (DOX) and anti-inflammatory agents have demonstrated synergistic tumor apoptosis and immunomodulation in non-Hodgkin's lymphoma models.

## 6.6 Cell-Based Nanodrug Delivery

Engineered living cells constitute a novel frontier in nanodrug delivery systems (NDDS). Hematopoietic stem cells conjugated with anti-PD-1-modified platelets have been developed to facilitate the direct delivery of immune checkpoint inhibitors into the bone marrow. This innovative system has demonstrated enhanced T-cell activation, increased cytokine release, and extended survival in acute myeloid leukemia (AML) mouse models while concurrently minimizing systemic toxicity [41]. Such strategies leverage the inherent homing capabilities of hematopoietic and immune cells to optimize targeted delivery.

### 6.7 Integration with Immunotherapy

Traditional CAR-T cell manufacturing processes are both time-intensive and expensive. The introduction of nanoparticle-assisted CAR delivery has streamlined production, thereby reducing costs and enhancing scalability. For instance, nanoparticles carrying CAR plasmids have successfully reprogrammed T cells in vivo, resulting in the generation of effective CAR-T cells that target leukemia [42, 43]. Nanobody-based selection platforms have further improved anti-AML efficacy while minimizing off-target activation. Although checkpoint blockade therapy has demonstrated potential in the treatment of hematologic malignancies, it is often accompanied by immune-related adverse effects. Nanoparticle formulations conjugated with PD-1/PD-L1 antibodies have shown improved tumor-specific delivery and reduced systemic toxicity. For example, immunoliposomes co-loaded with docetaxel and PD-L1 antibodies exhibit synergistic antitumor effects [44, 45]. Biomimetic nanoplatforms that integrate catalase and DOX can alleviate hypoxia, enhance drug penetration, and potentiate immune responses. The local application of anti-CD47 nanoparticles at surgical sites has reduced the recurrence risk by activating both local and systemic immunity, while magnetic nanostructures activating TLR4 have provided synergistic effects with PD-1/PD-L1 blockade [46]. These strategies exemplify the convergence of nanotechnology and immunotherapy for the treatment of hematopoietic cancers.

## 7 Conclusion and Future Directions

Nanotechnology has a profound impact on leukemia treatment. Liposomal formulations of cytarabine and daunorubicin have improved bioavailability and reduced cardiotoxicity compared to their free drug counterparts. Nanoparticles can be engineered to respond to stimuli within the tumor microenvironment, such as pH fluctuations or enzyme activity, enabling controlled drug release. This capability allows prolonged exposure of cancer cells to therapeutic agents, potentially overcoming drug resistance. Furthermore, nanotechnology facilitates the co-delivery of multiple drugs, addressing cancer cell heterogeneity and targeting diverse molecular pathways. The transition of NDDS from preclinical models to clinical applications faces challenges, including manufacturing, stability, immune compatibility, and regulatory approval. Several antibody- and liposome-based nanocarriers are currently advancing into clinical trials, highlighting their translational potential. Integration with precision medicine, multi-omics profiling, and imaging is expected to refine NDDS strategies for next-generation hematopoietic cancer therapies. Research into optimizing selectivity, safety, and clinical translation is critical for realizing their potential for hematologic malignancies. As research progresses, nanotechnology-based therapies hold promise for personalized treatment strategies tailored to individual patients' hematopoietic malignancies, improving treatment outcomes. Polymeric nanoparticles based on PLGA facilitate the controlled release of agents such as doxorubicin, increasing drug accumulation in leukemic cells while minimizing toxicity. Chitosan-based nanoparticles have improved arsenic trioxide delivery in acute promyelocytic leukemia by addressing solubility challenges. In lymphoma treatment, nanocarriers enhance the delivery of chemotherapeutic agents and mAbs. Liposomal doxorubicin reduces cardiotoxicity in relapsed/refractory lymphoma. Polymeric micelles loaded with paclitaxel offer improved solubility, whereas antibody-conjugated micelles increase specificity for lymphoma cells. AuNPs conjugated with anti-CD20 antibodies target B-cell lymphomas for drug delivery or photothermal therapy, where localized heating induces tumor cell death and enhances the efficacy of chemotherapy. Nanocarriers have been used to treat advanced myeloma. Liposomal bortezomib formulations enhance bone marrow delivery and reduce peripheral neuropathy. Polymeric nanoparticles co-delivering bortezomib and dexamethasone show synergistic anti-myeloma activity, whereas dendrimer-based systems facilitate thalidomide delivery and reduce neurotoxicity. Research is expanding toward multifunctional, tumor-targeted systems that can co-deliver drugs, overcome drug resistance, and synergize with immunotherapy. Nanotechnology continues to transform hematopoietic cancer therapy, enhancing precision and improving patient outcomes.

**Author's Contributions** SK, MC, and MKG conceived the idea and chapter structure. MC contributed to data accumulation. SK and MKG wrote the manuscript. MKG revised and edited the manuscript. All the authors have read, agreed, and approved the final draft of the chapter.

**Conflicts of Interest** There is no competing and conflict of interest.

## References

1. Cardona-benavides IJ, de Ramón C, Gutiérrez NC (2021) Genetic abnormalities in multiple myeloma: prognostic and therapeutic implications. Cells 10. https://doi.org/10.3390/cells10020336
2. Paiva B, Almeida J, Pérez-Andrés M, Mateo G, López A, Rasillo A et al (2010) Utility of flow cytometry immunophenotyping in multiple myeloma and other clonal plasma cell-related disorders. Cytometry B Clin Cytom 78. https://doi.org/10.1002/cyto.b.20512
3. Zheng Y, Si J, Yuan T, Ding S, Tian C (2021) Immune targeted therapy for diffuse large B cell lymphoma. Blood Sci 3. https://doi.org/10.1097/BS9.0000000000000095
4. Cappell KM, Kochenderfer JN (2023) Long-term outcomes following CAR T cell therapy: what we know so far. Nat Rev Clin Oncol 20. https://doi.org/10.1038/s41571-023-00754-1
5. Lemoli RM, Montesinos P, Jain A (2023) Real-world experience with CPX-351 in high-risk acute myeloid leukemia. Crit Rev Oncol Hematol 185. https://doi.org/10.1016/j.critrevonc.2023.103984
6. Ramdass B, Chowdhary A, Koka PS (2013) Hematological malignancies: disease pathophysiology of leukemic stem cells. J Stem Cells 8. https://doi.org/jsc.2014.8.3/4.151
7. Kennedy JA, Hobbs G (2018) Tyrosine kinase inhibitors in the treatment of chronic-phase CML: strategies for frontline decision-making. Curr Hematol Malig Rep 13. https://doi.org/10.1007/s11899-018-0449-7
8. Bin ET, Shahriar A, Mahmud AR, Rahman T, Abir MH, Siddiquee MFR et al (2022) Multidrug resistance in cancer: understanding molecular mechanisms, immunoprevention and therapeutic approaches. Front Oncologia 12. https://doi.org/10.3389/fonc.2022.891652
9. Agathanggelou A, Smith E, Davies NJ, Kwok M, Zlatanou A, Oldreive CE et al (2017) USP7 inhibition alters homologous recombination repair and targets CLL cells independently of ATM/p53 functional status. Blood 130. https://doi.org/10.1182/blood-2016-12-758219

10. Hus I, Puła B, Robak T (2022) PI3K inhibitors for the treatment of chronic lymphocytic leukemia: current status and future perspectives. Cancers (Basel) 14. https://doi.org/10.3390/cancers14061571
11. Rozovski U, Harris DM, Li P, Liu Z, Jain P, Veletic I et al (2017) Activation of the B-cell receptor successively activates NF-κB and STAT3 in chronic lymphocytic leukemia cells. Int J Cancer 141. https://doi.org/10.1002/ijc.30892
12. Ling Y, Xie Q, Zhang Z, Zhang H (2018) Protein kinase inhibitors for acute leukemia. Biomark Res 6. https://doi.org/10.1186/s40364-018-0123-1
13. Loscocco GG, Vannucchi AM (2022) Role of JAK inhibitors in myeloproliferative neoplasms: current point of view and perspectives. Int J Hematol 115. https://doi.org/10.1007/s12185-022-03335-7
14. Fu DJ, Wang T (2023) Targeting NEDD8-activating enzyme for cancer therapy: developments, clinical trials, challenges and future research directions. J Hematol Oncol 16. https://doi.org/10.1186/s13045-023-01485-7
15. Wildes TM, Stirewalt DL, Medeiros B, Hurria A (2014) Hematopoietic stem cell transplantation for hematologic malignancies in older adults: geriatric principles in the transplant clinic. J Natl Compr Canc Netw 12. https://doi.org/10.6004/jnccn.2014.0010
16. Algeri M, Merli P, Locatelli F, Pagliara D (2021) The role of allogeneic hematopoietic stem cell transplantation in pediatric leukemia. J Clin Med 10. https://doi.org/10.3390/jcm10173790
17. Tarlock K, Sulis ML, Chewning JH, Pollard JA, Cooper T, Gamis A et al (2022) Hematopoietic cell transplantation in the treatment of pediatric acute myelogenous leukemia and myelodysplastic syndromes: guidelines from the American Society of Transplantation and Cellular Therapy. Transplant Cell Ther 28. https://doi.org/10.1016/j.jtct.2022.06.005
18. Tian Z, Liu M, Zhang Y, Wang X (2021) Bispecific T cell engagers: an emerging therapy for management of hematologic malignancies. J Hematol Oncol 14. https://doi.org/10.1186/s13045-021-01084-4
19. Barreca M, Lang N, Tarantelli C, Spriano F, Barraja P, Bertoni F (2022) Antibody-drug conjugates for lymphoma patients: preclinical and clinical evidences. Explor Target Antitumor Ther 3. https://doi.org/10.37349/etat.2022.00112
20. Marin-Acevedo JA, Dholaria B, Soyano AE, Knutson KL, Chumsri S, Lou Y (2018) Next generation of immune checkpoint therapy in cancer: new developments and challenges. J Hematol Oncol 11. https://doi.org/10.1186/s13045-018-0582-8
21. Martino M, Alati C, Canale FA, Musuraca G, Martinelli G, Cerchione C (2021) A review of clinical outcomes of car t-cell therapies for b-acute lymphoblastic leukemia. Int J Mol Sci 22. https://doi.org/10.3390/ijms22042150
22. Mishra AK, Gupta A, Dagar G, Das D, Chakraborty A, Haque S et al (2023) CAR-T-cell therapy in multiple myeloma: B-cell maturation antigen (BCMA) and beyond. Vaccines (Basel) 11. https://doi.org/10.3390/vaccines11111721
23. Rafei H, Daher M, Rezvani K (2021) Chimeric antigen receptor (CAR) natural killer (NK)-cell therapy: leveraging the power of innate immunity. Br J Haematol 193. https://doi.org/10.1111/bjh.17186
24. Maslak PG, Dao T, Krug LM, Chanel S, Korontsvit T, Zakhaleva V et al (2010) Vaccination with synthetic analog peptides derived from WT1 oncoprotein induces T-cell responses in patients with complete remission from acute myeloid leukemia. Blood 116. https://doi.org/10.1182/blood-2009-10-250993
25. Cohen A, Spektor TM, Stampleman L, Bessudo A, Rosen PJ, Klein LM et al (2018) Safety and efficacy of pomalidomide, dexamethasone and pegylated liposomal doxorubicin for patients with relapsed or refractory multiple myeloma. Br J Haematol 180. https://doi.org/10.1111/bjh.14992
26. Duggan ST, Keating GM, Ferrandina G, Kesterson JP, Lorusso D, Muggia F (2011) Pegylated liposomal doxorubicin: a review of its use in metastatic breast cancer, ovarian cancer, multiple myeloma and AIDS-related Kaposis sarcoma. Drugs 71. https://doi.org/10.2165/11207510-000000000-00000

27. Vaklavas C, Forero-Torres A (2012) Safety and efficacy of brentuximab vedotin in patients with Hodgkin lymphoma or systemic anaplastic large cell lymphoma. Ther Adv Hematol 3. https://doi.org/10.1177/2040620712443076
28. Pan K, Farrukh H, Chittepu VCSR, Xu H, Pan CX, Zhu Z (2022) CAR race to cancer immunotherapy: from CAR T, CAR NK to CAR macrophage therapy. J Exp Clin Cancer Res 41. https://doi.org/10.1186/s13046-022-02327-z
29. Sanmamed MF, Chen L (2018) A paradigm shift in cancer immunotherapy: from enhancement to normalization. Cell 175. https://doi.org/10.1016/j.cell.2018.09.035
30. Alotaibi S, Niederwieser D, Ahmed SO, Sanz J, Mohty M, Aljurf M (2022) Current status of CPX-351 therapy in acute myeloid leukemia and myelodysplastic syndrome. Clin Lymphoma Myeloma Leuk 22. https://doi.org/10.1016/j.clml.2022.02.008
31. Liu C, Xing J, Akakuru OU, Luo L, Sun S, Zou R et al (2019) Nanozymes-engineered metal-organic frameworks for catalytic cascades-enhanced synergistic cancer therapy. Nano Lett 19. https://doi.org/10.1021/acs.nanolett.9b02253
32. Ashton S, Song YH, Nolan J, Cadogan E, Murray J, Odedra R et al (2016) Aurora kinase inhibitor nanoparticles target tumors with favorable therapeutic index in vivo. Sci Transl Med 8. https://doi.org/10.1126/scitranslmed.aad2355
33. Jiang X, Bugno J, Hu C, Yang Y, Herold T, Qi J et al (2016) Eradication of acute myeloid leukemia with FLT3 ligand-targeted miR-150 nanoparticles. Cancer Res 76. https://doi.org/10.1158/0008-5472.CAN-15-2949
34. Song Y, Li Y, Xu Q, Liu Z (2017) Mesoporous silica nanoparticles for stimuli-responsive controlled drug delivery: advances, challenges, and outlook. Int J Nanomedicine 12. https://doi.org/10.2147/ijn.s117495
35. Vitols S, Angelin B, Ericsson S, Gahrton G, Juliusson G, Masquelier M et al (1990) Uptake of low density lipoproteins by human leukemic cells in vivo: relation to plasma lipoprotein levels and possible relevance for selective chemotherapy. Proc Natl Acad Sci USA 87. https://doi.org/10.1073/pnas.87.7.2598
36. Vidal L, Blagden S, Attard G, De Bono J (2005) Making sense of antisense. Eur J Cancer 41. https://doi.org/10.1016/j.ejca.2005.06.029
37. Misra S, Hascall VC, Markwald RR, Ghatak S (2015) Interactions between hyaluronan and its receptors (CD44, RHAMM) regulate the activities of inflammation and cancer. Front Immunol 6. https://doi.org/10.3389/fimmu.2015.00201
38. Chavda VP, Pandya A, Kumar L, Raval N, Vora LK, Pulakkat S et al (2023) Exosome nanovesicles: a potential carrier for therapeutic delivery. Nano Today 49. https://doi.org/10.1016/j.nantod.2023.101771
39. Mandal T, Beck M, Kirsten N, Lindén M, Buske C (2018) Targeting murine leukemic stem cells by antibody functionalized mesoporous silica nanoparticles. Sci Rep 8. https://doi.org/10.1038/s41598-017-18932-4
40. Raza S, Safyan R, Suzanne L (2017) Immunomodulatory drugs (IMiDs) in multiple myeloma. Curr Cancer Drug Targets 17. https://doi.org/10.2174/1568009617666170214104426
41. Vishwasrao P, Li G, Boucher JC, Smith DL, Hui SK (2022) Emerging CAR T cell strategies for the treatment of AML. Cancers (Basel) 14. https://doi.org/10.3390/cancers14051241
42. Su S, Kang PM (2020) Recent advances in nanocarrier-assisted therapeutics delivery systems. Pharmaceutics 12. https://doi.org/10.3390/pharmaceutics12090837
43. Kong B, Kim Y, Kim EH, Suk JS, Yang Y (2023) mRNA: a promising platform for cancer immunotherapy. Adv Drug Deliv Rev 199. https://doi.org/10.1016/j.addr.2023.114993
44. Fan D, Cao Y, Cao M, Wang Y, Cao Y, Gong T (2023) Nanomedicine in cancer therapy. Signal Transduct Target Ther 8. https://doi.org/10.1038/s41392-023-01536-y
45. Cremolini C, Vitale E, Rastaldo R, Giachino C (2021) Advanced nanotechnology for enhancing immune checkpoint blockade therapy. Nanomaterials 11. https://doi.org/10.3390/nano11030661
46. Kumar CSSR, Mohammad F (2011) Magnetic nanomaterials for hyperthermia-based therapy and controlled drug delivery. Adv Drug Deliv Rev 63. https://doi.org/10.1016/j.addr.2011.03.008

# Nanotherapeutics in Personalized Medicine

# Nanotherapeutics in Cancer Therapy: Targeted Drug-Delivery and Real-Time Monitoring

Sibani Sarkar, Sunny Kumar, and Malini Basu

**Abstract**

Nanotherapeutics has emerged as a promising approach in cancer therapy, combining advanced drug delivery systems with real-time monitoring capabilities to enhance treatment efficacy and minimize side effects. Targeted drug delivery using nanoparticles allows for the precise localization of therapeutic agents at tumor sites. Thus, drug concentration is improved in the tumor by reducing systemic toxicity. Nanoparticles are developed for active targeting, leveraging ligands or antibodies that bind specifically to cancer cell receptors, or for passive targeting through enhanced permeability and retention (EPR) effects. Additionally, nanocarriers can be designed for controlled and sustained drug release, optimizing therapeutic outcomes. For targeted drug delivery, nanotechnology is a approach to monitor distribution of drug, tumor progression, and therapy using imaging techniques such as MRI, fluorescence, and photoacoustic imaging. This dual capability theranostic (therapeutic and diagnostic) offers a unique approach to personalized cancer care, allowing clinicians to track treatment efficacy and adjust regimens based on real-time data. As research continues to address challenges in biocompatibility, scalability, and regulatory approval, nanotherapeutics holds significant potential in revolutionizing cancer treatment paradigms.

S. Sarkar (✉) · S. Kumar
Cancer Biology and Inflammatory Disorder Division, Council of Scientific and Industrial Research-Indian Institute of Chemical Biology (CSIR-IICB), Kolkata, India

M. Basu (✉)
Department of Microbiology, Dhruba Chand Halder (DCH) College, University of Calcutta, Kolkata, West Bengal, India

 
M. K. Ghosh, M. Basu (eds.), *Personalized Cancer Therapy*,
https://doi.org/10.1007/978-981-95-6845-1_16

**Keyword**

Nanotherapeutics · Real-Time Monitoring · Targeted Drug Delivery

## 1 Introduction

Nanotechnology plays an important role in cancer. Though continuous progress has been observed in the medical field and technology, cancer is the primary cause of mortality worldwide due to its dynamic behavior that requires diagnosis for efficient therapy. It starts from the benign tumor that spreads to the body at later stages by certain mechanisms such as apoptosis resistance, inhibition, angiogenesis, tissue invasion, and apoptosis. Generally, cancer diagnosis occurs at later stages due to lack of proper diagnosis at initial stage resulting in the metastasis of oncogenic cells and spreading occurs into different organs in the body. Chemotherapy, radiation, and surgery are the frequently used therapies, but several challenges occur because of acute side effects, especially chemotherapy, leading to toxicity on noncancerous cells along with cancerous ones. Chemotherapy often leads to dose-limited toxicity and thereby inactivates the drugs. Application of the drug by nanotherapy targets cancer cells by minimizing undesired side effects. It has been used to outperform traditional cancer monitoring and for treatment. Individual molecules like peptides, ligands, polymeric materials, and metals are not used for cancer detection and therapy due to their bulk size but when it was covalently linked with nanoparticles, making a notable contribution in medicine, molecular profiling, and biological assays, overcoming their side effects [1, 2].

Nanotherapy is also an important theragnostic tool. Theragnostic uses a radioactive drug in personalized medicine and nuclear medicine, where a single radioactive drug is used for identification and treatment as iodine isotope 131. This is one of the earliest drugs in the theranostic approach, which is used in the identification and treatment of thyroid cancer in the field of nuclear medicine. Nuclear medicine uses various chemicals also for diagnosis and targeting treatment, either individually or in combination. Iodine enters the target by metabolic processes, but others, like receptor, ligand, which are already present on the target sites. Theragnostic processes use image-based localization in diseased tissues and radiation-based targeting that utilizes particles in the nanoscale range and shows important advantages in medicine and treatment. The term "nano-theranostic" approach improves nanomedicine therapy in advanced theragnostic purposes that involve improved nanocarriers such as polymer conjugation, gold-based nanomaterials, silica-based nanomaterials, dendrimers, micelles, carbon nanotubes, and metal and inorganic nanomaterials [1, 3, 4].

## 2 Overview of Nanotherapy

Cancer has long been presented as a tremendous therapeutic obstacle that necessitates a simple treatment process that can overcome the constraints of traditional procedures such as chemotherapy and radiation therapy. This therapy causes severe side effects and shows disappointing outcomes. In this medical critical situation, the development of nanotechnology has become a promising model for cancer detection and treatment. It can modify the compound at the nanoscale range, which is typically between 1 and 100 nm. The various physicochemical characteristics of these small molecules show distinct interactions with cells and tissues, promoting novel targeting strategies that can activate transformational advances for cancer treatment and diagnosis. The present state of cancer therapy emphasizes the urgent need for novel approaches, and the current research paves the way to show the positive outcomes by using nanotechnology, which can fulfill all the vital needs for cancer treatment [1, 4].

Nanocarrier acts as a key player at the intersection of nanotechnology and cancer treatment. These minute carriers encapsulate therapeutic substances such as medicines or DNA that offer abundances benefits such as improved accuracy of targeting cancer cells, reducing damage to healthy neighboring cells, and increased efficacy of therapeutic outload, which are beyond the limitations of treatments. Nanocarrier-based cancer cell targeting has two forms: Passive targeting and Active (aggressive) targeting. Passive targeting has advantages such as permeability of blood arteries which promote the accumulation of nanocarriers into the tumor microenvironment whereas active targeting, on the other hand, comprises surface modifications of nanocarriers with specific targeting ligands that binds to receptors on cancer cell surfaces allowing for more precise and increased cellular engagement [4–6] (Fig. 1).

Nano-vector deposits in the tumor microenvironment by passive targeting due to the specific behavior present in the tumor microenvironment, which is generally not normally present in healthy tissues. Nanoparticle delivery depends on the tumor microvasculature. Additionally, this factor is inherent within the nanoparticle itself, such as shape, size, and surface charge. Targeting strategy enhances the uptake of nanoparticles selectively into the tumor cells. Molecules attached to the nanoparticle's surface, which are overexpressed by the neoplastic cells. As a result of passive targeting, nanoparticles accumulate in the tumor tissues due to leaky vasculature supported by the enhanced permeability and retention effect. Thus, passive targeting allows the localization of nanoparticles within the tumor microenvironment. However, active targeting of nanoparticles occurs when a targeting moiety (antibody, ligand, or peptide) is incorporated into the nanoparticle system to target specific alterations in cancer cell biology that are greatly elevated in comparison to healthy surrounding cells and tissues. Nanoparticles connect to target cells via ligand-receptor interactions and are absorbed before drug release, resulting in less off-target drug release than passively targeted systems. Active targeting to cell membrane receptors, which are overexpressed in cancer cells, is one of the ways to increase the retention of nanoparticles and consequently their therapeutic payload in the tumor. EGFR activation has been linked to the aggressiveness of various

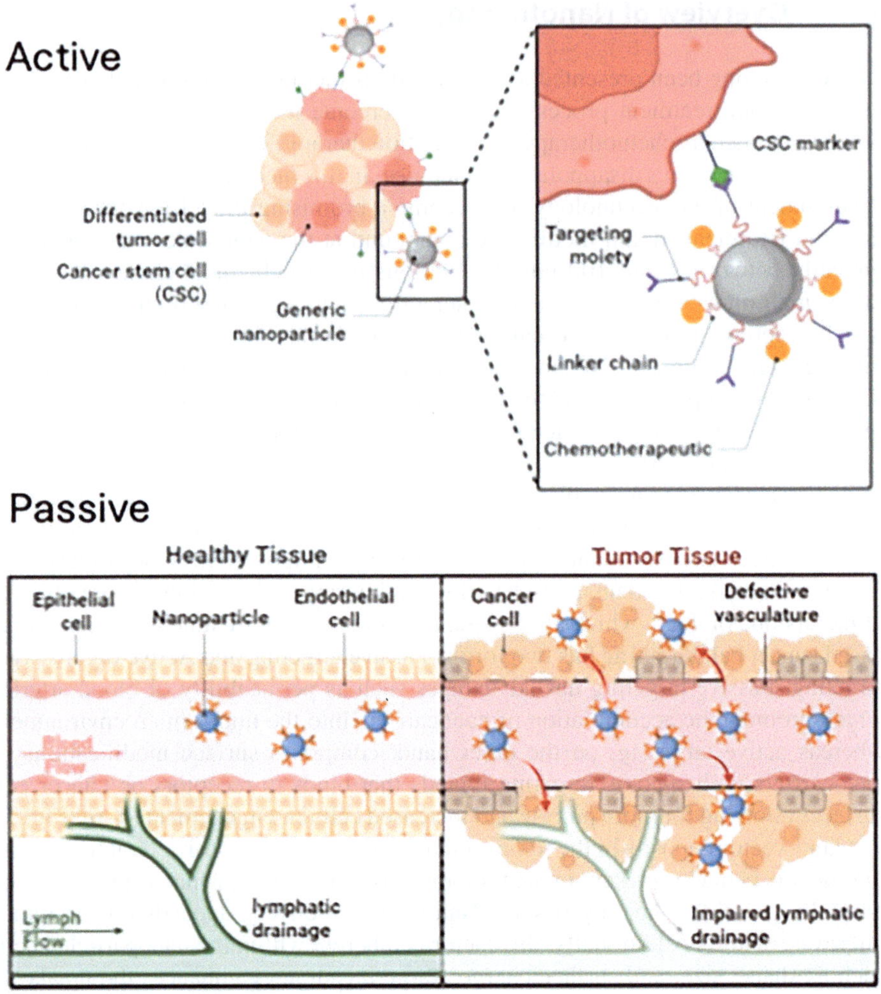


**Fig. 1** Nanoparticle-mediated targeting of cancer cells

malignancies, including breast (generally overexpression is seen in nearly 50% of triple-negative breast cancer), renal, ovarian, colon, and non-small cell lung cancer. Thus, it has been widely accepted that regardless of the targeting technique used in the delivery system, whether active or passive, nanoparticles must be therapeutically relevant, whether intracellular absorption is critical. Furthermore, it is a well-established fact that the size, shape, and surface chemistry of nanoparticles have a significant impact on cellular internalization. However, it is unclear how nanoparticles intracellular trafficking and distribution influence the overall therapeutic effect [6–9].

# 3 Nanoparticles in Drug Delivery

Although a wide range of nanocarriers have been used for preclinical testing that has been applied for human trials, only a few have been used clinically. They have the special moieties which are intended for selecting cancer cell interactions. This characteristic emphasizes the complexities involved in laboratory medicine in effective clinical use, which necessitates more study to improve the nanocarrier's potency in cancer therapy. Nanotechnology has increased therapeutic efficacy in cancer by reducing collateral damage to healthy cells and improving the outcome of patients (Fig. 2).

## 3.1 Lipid-Based Nanoparticles

Lipid formulations of nanoparticles used in oncology have seen significant progress in recent years in terms of preparations and variable composition of nanoformulated compounds. Lipid nanoparticles have been used as drug delivery, viz., liposomes, solid lipid nanoparticles (SLN), and nano lipid carriers that have increased attention in cancer treatment. These types of nanoparticles carry both hydrophobic and

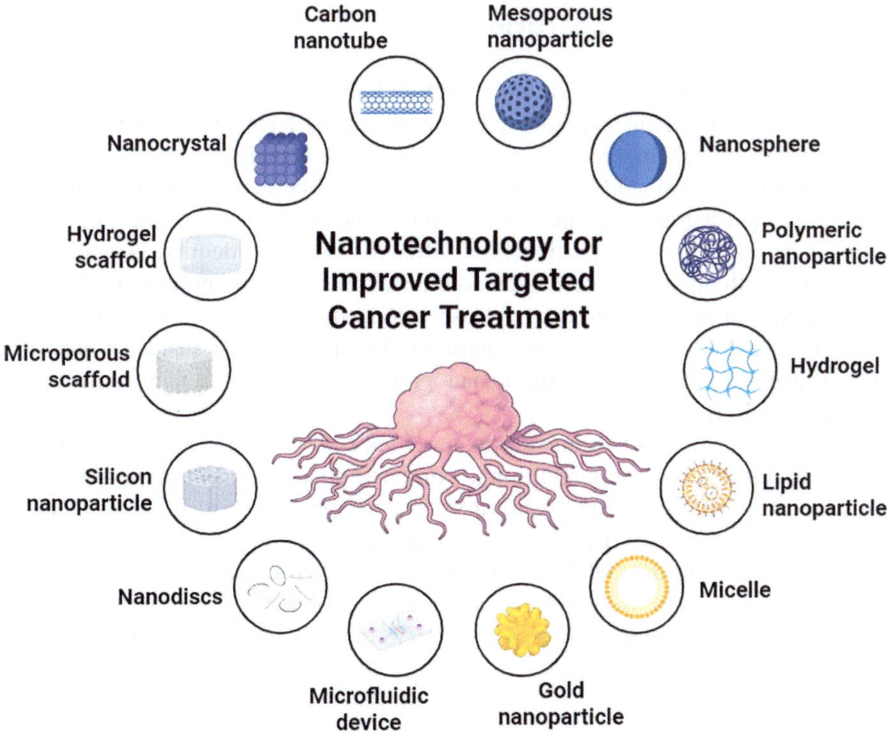


**Fig. 2** Types of nanotechnologies utilized in cancer therapeutics

hydrophilic drugs, which have very low toxicity. Liposome vesiculated molecules have their extended half-life and controlled drug release properties. Thus, it has also improved the time of medication action. Gangliosides or polyethylene glycol (PEG) present in lipid nanoparticles are not detected by the immune system which improve the solubility of molecules. Additionally, they can be prepared in pH-sensitive formulations to increase drug release in the acidic environment or be tagged with an antibody to recognize tumor cells or their receptors (folic acid (FOA)) [10–12].

The most extensively investigated delivery modes are liposomes because of their biocompatibility and biodegradability. Primary constituents of lipid molecules are phospholipids, arranged into the bilayer because of the amphiphilic nature of the compounds. Water vesicles are formed to increase the solubility of anticancer compounds which are encapsulated either hydrophobic or hydrophilic. Along with phospholipids, cholesterol is also added to reduce the volatility of nanoparticles and increase the penetration power of hydrophobic molecules in the membrane. The stability of nanoparticles is increased in circulation. Liposomes are presented in multiple bilayers and sizes are observed from 0.5 to 10 nm, named Multilamellar Vesicles (MLV). Sizes above 100 nm are named as Large Uni-lamellar Vesicles (LUV) whereas between 10 and 100 nm named as Small Uni-lamellar Vesicles (SUV) [13, 14].

## 3.2 Polymeric Nanoparticles

Polymeric nanoparticles have recently captured great attention due to their unique features caused by their small size. Polymeric nanos have several advantages as drug carriers, which include the capacity of controlled drug release, protecting drugs and other compounds with biological activity in their environment, and bioavailability and therapeutic index is also improved. Both nanocapsules and nanospheres are named as nanoparticles but differ in structural identity. Nanocapsules are formed with oily core where the medicine is dissolved, and polymeric shells are surrounded that regulate the drug release profile from the core. Nanospheres are made up of a continuous polymeric network that allows drugs inside or adsorbed into surface. Two forms of polymeric nanoparticles are known as nanocapsules and nanosphere [2, 4].

Polymeric nanoparticles can vary in size, shape, surface characteristics, crystallinity, and dispersion state in addition to composition and concentration. These features are frequently examined using a variety of approaches to fully characterize NPs. Some of the most frequent techniques used include electron microscopy, dynamic light scattering (DLS), electrophoresis, and chromatography. The characterization of polymeric nanoparticles and nanotoxicology exposure in workshop is crucial to assess the safety of health and also in the in manufacturing processes [2].

### 3.3 Inorganic Nanoparticles

Nanoparticles are generally categorized into two types: inorganic and organic. Organic nanoparticles (such as dendrimer, micelles, ferritin, and liposomes) are nontoxic and biodegradable. It is also observed that nanocapsules, viz., micelles and liposomes, have a hollow space. They are sensitive to thermal or electromagnetic radiations. These remarkable characteristics make them excellent molecules for biomedical applications, particularly in drug delivery. Poor stability, short shelf life, and drug encapsulation effectivity may limit use in broader perspectives compared to inorganic nanoparticles.

Inorganic nanoparticles, on the other hand, are very stable and hydrophilic. They have intrinsically superior physicochemical characteristics, viz., magnetic, thermal, optical, and catalytic properties. Nanomaterials show a strong structure into which many substances may be incorporated to provide multiple actions. Organic nanoparticles, in comparison to inorganic ones, have a higher drug loading efficiency, higher stability, and fair breakdown. These inorganic nanoparticles can be divided into two groups: metallic and metallic oxide- nanoparticles. Metallic nanoparticles are prepared by metals and subsequently nanosized using a bottom-up or top-down approach, such as Fe, Cu, Zn, Cd, Au, and Ag. Metallic oxide nanos are prepared by modification of their properties of metallic nanoparticles, viz., ZnO (Zinc oxide), $Fe_2O_3$ (Iron (II) oxide, FeO (Iron (III) oxide), $Al_2O_3$ (aluminium oxide), and so on. These metal oxide-based nanoparticles have the properties to maximize their reactive efficacy. When compared to metal equivalents, these nanoparticles exhibit certain features. The inherent physical and chemical properties of inorganic nanoparticles have these properties (confinement and superparamagnetic behavior) may vary partially or completely the particles interaction with their physical environments. For example, when nanoparticles with their surface modification comes into contact with their physical environments, it Some of its coating may be lost that leads to particle in stabilization and aggregation. As the nanoparticles are absorbed by the cells, the protein coating protects them from enzymatic destruction in the phagosomes or lysosomes. Then the degraded nanoparticles after digestion, is regarded as foreign, which triggers a variety of immunological reactions within the host and may cause changes to the residue's physicochemical properties [15–17].

### 3.4 Stimuli-Responsive Nanocarriers

To improve the therapeutic efficacy of nanoparticles, the researchers have started investigation into environmentally sensitive NPs, where external stimulus create changes to promote release of the drugs at target points, these include (i) physical signals such as temperature, magnetic field, electric field, and ultrasound, and (ii) chemical changes such as pH, ionic strength, redox potential, and enzyme activity. Liposomes, polymeric micelles, multiplexes, and polyplexes are examples of nanoparticle systems that use physical and chemical signals to change the qualities of drug release.

### 3.4.1 pH-Sensitive Nanoparticles

pH gradients are one of the most utilized environmental stimuli to create innovative, sensitive nanoparticles. pH-responsive environmental-based stimuli are acting on three levels: (i) organ, (ii) tissues, and (iii) subcellular compartments by oral drug administration, tumor targeting, and intracellular delivery, highlighting the pH-response nano-architectures. At the organ level, NP based oral delivery systems are involved in variable uptake of medicines via the gastrointestinal (GI) tract. At the tissue level, NP formulations have been developed taking advantage of the pH gradient found in tumor microenvironments to achieve higher drug concentration. Finally, pH-responsive nanos have been made to bypass the endo-lysosomal compartments and release drugs into the cytoplasm. pH gradient sensitive nano-formulations exist in the microenvironments of organs, cell organelles, tissues may be added beneficial in comparison to the other NP-based delivery vesicles. Each GI tract part has its own pH from the stomach lumen (acidic, pH = 1–3) to the duodenum and ileum (alkaline, pH = 6.6–7.5) from digestion to chime neutralization. Oral delivery is a suitable method of medication delivery for its convenience and cost-effectiveness. But orally administered medicines are subjected to enzymatic breakdown, resulting in low systemic exposure. Thus, oral medicines have been proven to be challenging due to their poor bioavailability and direct acceptance to the target sites. NPs made with biodegradable polymers were applied to increase the bioavailability of readily degraded peptide medicines, including insulin, calcitonin, and elcatonin. Recently, novel medicines have been prepared with pH-responsive techniques to boost systemic exposure through increased retention time in the stomach, transepithelial transport, and cellular targeting using functionalized surface ligands [18].

Nanoparticles also prepared with a surface charge reversal following stomach transit with an aim that drugs would be released into the alkaline intestinal tract. Mesoporous silica NPs are prepared with functionalized surface ligands with positively charged trimethyl ammonium (TA) groups (density varied). This positive TA would help to load and trap anionic medicines such as sulfasalazine (an anti-inflammatory prodrug used in bowel disorders) in acidic conditions (pH < 3). The deprotonation of silanol groups occur due to the drug-loaded NPs placing in physiological buffer (pH 7.4), causing the development of a partial negative charge on the surface of NPs, that results in electrostatic repulsion and the prolonged release of the loaded molecules. Thus, nanoparticles are effective vesicles for differential retention and absorption in the GI tract and also increase the effectivity of orally administered medicines [18–20].

### 3.4.2 Redox-Responsive Systems

In drug delivery systems, disulfide bond, di-selenide bond, platin conjugation, thioester bond, and thiol groups are commonly used as redox-responsive agents to target redox-responsive locations. Hydrophobic medicines can be inserted into the NP's internal core, pores, or chemically attached to the surfaces via redox-sensitive connections. The chemical structure of nanocarriers may also contain redox linkages to allow the nanoplatform to have stimulus-responsive, regulated release of drugs at the target sites. Electrostatic interaction between positively charged carriers and negatively charged genes or drugs is an alternative essential technique, especially in gene delivery systems.

Platinum-based anticancer medicines have been utilized for a long time to treat cancers. Prodrugs have been created for Pt (II) and Pt (IV) complexes as square planar and octahedral conformations. Platinum-based anticancer properties are based on the creation of coordinate bonds between reactive platinum complexes and DNA bases. Pt (IV) derivatives were developed as a redox-sensitive smart medication delivery device. Pt (IV) based prodrugs are less toxic than Pt (II) cisplatin. Cisplatin is released from the redox-sensitive drug carriers by the reduction of Pt (IV) complexes. The disulfide bond can be employed as a redox-sensitive linkage that can be reduced to thiol groups in the presence of a reducing agent. When exposed to an oxidizing substance, a thioester link is oxidized to a hydrophilic sulfoxide or sulfone. While thioester moieties are susceptible to oxidation, disulfide linkages can be cleaved by both reducing and oxidizing agents. These redox-sensitive linkers have been used as drug delivery carriers [21–23].

### 3.4.3 Enzyme-Triggered Nanoparticles

The chemical structure and physical properties of stimuli-responsive "smart" nanoparticles are responsible for the enzyme's biologically catalytic activity. They show the crucial role in all the processes, viz., biological and metabolic dysregulation of Expression and activity dysregulation suppressed the many diseases pathological action. In therapeutic treatment, these types of dysregulated enzymes trigger biological systems actively. After triggering of the enzyme, they catalyze chemical reactions in mild temperature, neutral pH and buffered aqueous solution where many conventional reactions do not occur. Additionally, enzymes' action on the substrates selectively permits specific, sophisticated, biologically active chemical reactions. Many nano-range compounds, viz., polymer, phospholipids, and inorganic materials, have been used previously for designing enzyme-active delivery modes. This enzymatic responsive nanomaterial can be formulated bio-specifically and selectively to apply in diverse fields. As, site-specific enzyme-triggered as well as active tumor targeting nano accumulate at the tumor target sites, reducing uptake by the non-targeted tissue [24, 25].

## 4 Real-Time Monitoring with Nanotherapeutics

Real-time monitoring with nanotherapeutics involves the use of nanoparticles or nanomaterials to deliver therapeutic agents while enabling continuous observation of their activity within the body. This approach combines drug delivery with advanced imaging techniques to track the distribution, efficacy, and interactions of nanotherapeutic agents in vivo. It allows for personalized treatment adjustments and better understanding of the treatment dynamics, particularly in cancer therapy and other complex diseases. The integration of real-time monitoring helps optimize therapeutic outcomes and minimize side effects by providing valuable insights into the drug's performance at the target site.

## 4.1 Theranostic Nanoparticles

A combination of diagnostic and therapeutic purposes is fulfilled by theranostic nanoparticles, which are biocompatible and biodegradable. These multifunctional nanosystems are well designed and specific for personalized disease management. Ideally, they must be safe for humans and can (i) target to the specific sites rapidly and selectively, (ii) overview of biochemical and morphological characteristics of the disease, (iii) deliver an adequate amount of drug to the target site efficiently, and (iv) be excreted or biodegraded as nontoxic byproducts within hours. Though many theranostic nanoparticles, both organic and inorganic, have been made nowadays for treating cancers. In case of poorly vascularized metastatic cases ($<100$ $mm^3$), treatment, active targeting can be used where the EPR effect alone may not be effective. Generally, passive targeting can be effective in the case of theranostic nanoparticles [26, 27]. It remains a major challenge in using biocompatible theranostic nanoparticles in the case of in vivo tumor active targeting. Here we will discuss some interesting theranostic nanoparticles as active tumor targeting:

### 4.1.1 Fluorescent Nanoparticles

The analysis of biological, pharmaceutical, and chemical agents uses fluorescent nanosensors in both qualitative and quantitative ways. Many techniques such as gas chromatography, high-performance liquid chromatography, and capillary electrophoresis are used for economical, operational, and sensitive cases. Organic fluorescent probes are all dyes, molecules, nanocarriers, and rare earth metals. All these nanosensors are quantum dots (QDs) and carbon dots (CDs). The typical functions, viz., strong chemical inertia, stable lighting, low toxicity, and biocompatibility, of CDs (a new variety of glowing carbon), make an ideal substance to develop new, sensitive sensors. These CDs can selectively respond to specific ions due to their ionic functions, which are considered intelligent sensors due to their critical roles, such as biodegradable, eco-friendly, and act theranostically against many diseases and cancers.

By fluorescent methods, small amounts of sample can be identified for biochemical analysis and imaging due to their simplicity and high sensitivity. Many other fluorescent nanoparticles have been identified but low cytotoxicity, attractive photostability, easy functionalization and ideal biocompatibility properties makes the CDs most unique and powerful. Mild noble metal nano having limited fluorescence and biocompatible properties target theranostically against many cancers. Similarly, nano-metal organic frameworks (MOFs), which have various functional properties and controlled fluorescent biocompatibility, are used for imaging, biosensing, and therapy of cancer. Triplet annihilation-fluorescent probes help to identify the potassium levels in the body, whereas lysine-sensitive graphene fluorescent QDs detect tumor markers efficiently. CD-based fluorescent nanosensors determine the quality of foods. They also detect the physicochemical properties of DNA and its multifunctional aspects. Besides these properties, they are also used to detect pesticides, herbicide residues, human proteins, and pollutants in pharmaceutical agents [28–30].

### 4.1.2 Magnetic Nanoparticles

Magnetic nanoparticles (MNPs) have emerged as a multifunctional platform in cancer therapy, offering significant advantages for both targeted drug delivery and real-time monitoring. These nanoparticles, typically composed of iron oxide cores coated with biocompatible materials, exhibit superparamagnetic properties that allow them to be guided and concentrated at specific tumor sites using external magnetic fields. This magnetic targeting enhances the accumulation of therapeutic agents within the tumor microenvironment, thereby increasing the local drug concentration while reducing off-target effects and systemic toxicity. To further improve selectivity, MNPs can be surface-functionalized with tumor-sensitive ligands, antibodies, peptides, or aptamers, and are able to target actively in cancer cells based on overexpressed surface markers. Once localized at the tumor site, MNPs can release their drug payloads in a controlled manner. Sometimes they are triggered by stimuli, viz., pH, temperature, or enzymatic activity present in the tumor milieu.

In addition to targeted delivery, MNPs serve for contrasting molecules for magnetic resonance imaging (MRI), allowing for non-invasive, real-time monitoring of biodistribution, tumor accumulation, and therapeutic outcomes. This dual functionality makes them ideal molecules in theranostic applications, where diagnosis and treatment are occur simultaneously into one single cases. Real-time imaging feedback not only facilitates the optimization of treatment regimens but also enables early detection of therapy resistance or disease progression. Furthermore, MNPs can be engineered for hyperthermia therapy, where alternating magnetic fields induce localized heating to damage tumor cells, providing another layer of therapeutic potential. Overall, the integration of targeted delivery, imaging, and therapy in a single nanosystem highlights the immense promise of magnetic nanoparticles in advancing personalized and precision cancer medicine [5, 31, 32].

### 4.1.3 Photoacoustic Imaging Agents

Photoacoustic imaging (PAI) agents have become a transformative tool in cancer therapy and real-time monitoring due to their ability to provide high-resolution, deep-tissue imaging with excellent contrast. PAI combines optical and ultrasound imaging modalities, leveraging the photoacoustic effect: when photo-absorbing agents absorb pulsed laser light, they undergo rapid thermal expansion, emitting ultrasonic waves that are captured to form detailed images. In cancer applications, specially designed photoacoustic imaging agents—such as gold nanorods, carbon nanotubes, semiconducting polymer nanoparticles, and dye-loaded liposomes—can be engineered to preferentially accumulate in tumors through passive (e.g., enhanced permeability and retention effect) or active targeting (e.g., ligand-receptor interactions). These agents enhance the contrast between malignant and normal tissues, enabling precise tumor localization and margin delineation [33].

Beyond diagnosis and tumor mapping, photoacoustic agents are increasingly being used as theranostics—integrated platforms combining therapy and diagnostics. Some photoacoustic agents are loaded with chemotherapeutics or photothermal therapy (PTT) components, allowing for simultaneous treatment and real-time visualization of drug distribution and therapeutic response. The ability to monitor the accumulation of agents in the tumor and assess therapy efficacy non-invasively in real time reduces the need for repeated biopsies and supports adaptive treatment planning. Moreover, PAI can provide valuable information on tumor vasculature, hypoxia, and oxygen saturation, offering insights into the tumor microenvironment that guide therapy decisions. Overall, photoacoustic imaging agents represent a cutting-edge approach in cancer nanomedicine, providing both high-sensitivity imaging and functional assessment to support precision oncology [33, 34].

### 4.1.4 Raman and Surface-Enhanced Raman Spectroscopy (SERS) Nanoparticles

These nanoparticles have emerged as powerful tools for high-sensitivity detection, molecular profiling, and real-time monitoring in cancer diagnostics and therapy. Raman spectroscopy relies on inelastic scattering of light to provide molecular "fingerprints" of biological samples, but its inherently weak signal limits its direct application. SERS overcomes this limitation by using plasmonic nanostructures—typically gold or silver nanoparticles—that dramatically amplify Raman signals of nearby molecules through electromagnetic and chemical enhancement mechanisms. When functionalized with tumor-targeting ligands and Raman reporter molecules, SERS nanoparticles can selectively bind to cancer cells, enabling ultra-sensitive and multiplexed detection of biomarkers with single-molecule resolution.

In cancer therapy, SERS nanoparticles serve both diagnostic and therapeutic purposes. Their high specificity allows for precise tumor imaging, early detection, and intraoperative guidance for tumor resection, especially in anatomically complex or hard-to-reach areas. Additionally, because SERS signals are stable and do not photobleach like traditional fluorophores, they enable long-term tracking of nanoparticle distribution and drug delivery in vivo. Some advanced SERS nanoplatforms are integrated with therapeutic modalities—such as chemotherapy, photothermal therapy, or photodynamic therapy—creating multifunctional agents capable of simultaneous treatment and real-time monitoring. This theranostic capability facilitates personalized therapy, allowing clinicians to adapt treatment based on molecular feedback from the tumor environment. Furthermore, the ability of SERS to perform multiplex detection means multiple cancer biomarkers or therapeutic responses can be monitored concurrently in a single imaging session, providing a comprehensive overview of tumor status and heterogeneity. As such, SERS nanoparticles represent a next-generation nanotechnology platform, combining molecular precision, therapeutic capability, and real-time assessment to significantly enhance the effectiveness of cancer management strategies [35, 36].

## 4.2 Biosensing and In Vivo Tracking

Biosensing and in vivo tracking are essential for early cancer detection, therapy monitoring, and personalized treatment. Nanotechnology has significantly advanced these fields by enabling highly sensitive, specific, and real-time detection of cancer biomarkers using nanomaterials such as gold nanoparticles, quantum dots, and magnetic nanoparticles. These nanosystems facilitate non-invasive monitoring of tumor progression, drug delivery, and microenvironmental changes through imaging and signal-based detection methods. Innovations such as wearable and implantable nanosensors allow continuous physiological monitoring, while microfluidic nanoplatforms support rapid, high-throughput biomarker analysis from small biological samples. Together, these technologies are transforming cancer care by integrating diagnostics with targeted therapy in a personalized and minimally invasive manner.

### 4.2.1 Nanoprobes for Biomarker Detection

Nanoprobes have revolutionized the field of cancer biosensing by enabling highly sensitive, specific, and multiplexed detection of biomarkers at both the cellular and molecular levels. These engineered nanosystems, often composed of materials such as gold, silver, quantum dots, silica, or carbon-based nanoparticles, are functionalized with recognition elements such as antibodies, aptamers, peptides, or nucleic acids that bind selectively to tumor-associated biomarkers. Upon binding, the interaction triggers a measurable signal—such as fluorescence, Raman scattering, or electrochemical change—allowing real-time detection of specific proteins, nucleic acids, or metabolites associated with cancer progression, drug resistance, or metastasis. Compared to conventional methods, nanoprobes offer significantly enhanced sensitivity, enabling detection of biomarkers at femtomolar or even single-molecule levels, which is crucial for early diagnosis and monitoring of minimal residual disease.

Importantly, nanoprobes can be designed for in vivo applications, facilitating non-invasive tracking of dynamic biological processes within living organisms. This capability supports longitudinal studies of tumor growth, therapeutic response, and biomarker fluctuations in real time. Some advanced nanoprobes incorporate activatable or stimuli-responsive elements that enhance signal output only in the presence of specific tumor microenvironmental conditions, such as acidic pH, high glutathione levels, or specific enzymatic activity, further increasing specificity and reducing background noise. Overall, nanoprobe-based biomarker detection offers a transformative approach to cancer diagnostics, enabling precision monitoring and contributing to the advancement of personalized oncology [3, 4, 37, 38].

### 4.2.2 Wearable and Implantable Nanosensors

Wearable and implantable nanosensors represent a cutting-edge frontier in real-time, continuous health monitoring and cancer management. These miniaturized devices integrate nanomaterials—such as graphene, carbon nanotubes, gold nanoparticles, and conductive polymers—with flexible electronics to detect physiological and biochemical signals with exceptional sensitivity and spatial resolution. Wearable nanosensors, typically incorporated into skin patches, textiles, or smart bands, enable non-invasive monitoring of cancer-related biomarkers in sweat, interstitial fluid, or breath, providing insights into patient status, treatment response, and metabolic alterations. Their lightweight and flexible design ensures user comfort and compatibility with daily activity, making them suitable for long-term use in outpatient settings.

In contrast, implantable nanosensors are embedded directly into tissues or near tumor sites to provide continuous, localized detection of biomarkers such as glucose, lactate, oxygen, pH, or specific proteins and cytokines associated with tumor progression or inflammation. These sensors can transmit real-time data wirelessly to external receivers or cloud platforms for analysis, enabling timely therapeutic intervention and dynamic adjustment of treatment strategies. Advanced implantable systems may also be integrated with drug reservoirs or actuators, forming closed-loop platforms that combine sensing with responsive therapy. The biocompatibility, miniaturization, and signal specificity of nanosensors are critical factors that influence their long-term functionality and clinical applicability. As these technologies continue to mature, wearable and implantable nanosensors are poised to transform personalized cancer care by facilitating proactive monitoring, early detection of recurrence, and adaptive therapy, all in real time [39–42].

### 4.2.3 Microfluidic Nanoplatforms

Microfluidic nanoplatforms have emerged as versatile tools for the integration of nanoscale biosensing with high-throughput, miniaturized fluid control, offering powerful capabilities for cancer diagnosis, biomarker profiling, and therapeutic monitoring. These lab-on-a-chip systems combine microfluidic channels with functionalized nanoparticles, such as magnetic beads, gold nanoparticles, or quantum dots, to enable the rapid isolation, enrichment, and detection of cancer-specific biomarkers from minute biological samples such as blood, saliva, or urine. Their ability to manipulate fluids at the microliter or nanoliter scale allows precise control over reagent mixing, reaction timing, and temperature, thereby improving assay sensitivity, reproducibility, and speed. One of the key advantages of microfluidic nanoplatforms lies in their potential for point-of-care diagnostics and liquid biopsy applications. By incorporating nanoparticle-based capture agents, these systems can isolate circulating tumor cells (CTCs), extracellular vesicles (EVs), and cell-free nucleic acids with high specificity and efficiency from complex clinical samples. Furthermore, integration with optical, electrochemical, or fluorescence-based detection methods allows for real-time analysis of tumor-derived markers, enabling continuous disease monitoring and therapy response evaluation. Some advanced microfluidic devices also incorporate programmable elements, such as valves and sensors, to support automated workflows, multi-step assays, or multiplexed detection.

In cancer research and precision medicine, microfluidic nanoplatforms are increasingly used to mimic the tumor microenvironment on-chip, supporting organ-on-chip or tumor-on-chip models for drug screening and functional analysis. These integrated systems offer an unprecedented level of control and analytical power, making them invaluable for translating nanomedicine-based diagnostics and therapeutics into clinically viable tools. As fabrication techniques advance and materials become more biocompatible and cost-effective, microfluidic nanoplatforms are expected to play a central role in next-generation cancer biosensing and personalized therapeutic decision-making [18, 43–45].

## 5 Challenges and Future Directions

Despite the promising advances in nanoparticle-based cancer diagnostics and therapy, several critical challenges remain before their full clinical potential can be realized. Biocompatibility and safety are of foremost concern, as the long-term toxicity, biodistribution, biodegradation, and potential immunogenicity of various nanomaterials must be thoroughly evaluated to ensure patient safety. Many nanoparticles may accumulate in off-target tissues such as the liver, spleen, or kidneys, raising concerns about chronic exposure and systemic effects. Additionally, surface coatings, particle size, and shape play pivotal roles in determining biological interactions and must be carefully optimized. Another significant barrier is scalability and manufacturing, as the reproducible synthesis of nanoparticles with consistent physicochemical properties at industrial scales is technically complex and cost-intensive. Batch-to-batch variability can significantly affect therapeutic efficacy and safety profiles [2, 4, 17].

Equally pressing are the regulatory and clinical translation hurdles. The lack of standardized protocols for nanoparticle characterization, limited long-term clinical data, and the complexity of regulatory approval processes impede the rapid integration of nanotherapeutics into standard oncology care. Collaborative efforts between researchers, industry, and regulatory bodies will be required to develop clear guidelines and streamline approval pathways to face these challenges. Lastly, the future of cancer nanomedicine lies in personalized nanotherapeutics, where treatment regimens are tailored to an individual's genetic, molecular, and phenotypic tumor profile. Advances in genomics, proteomics, and bioinformatics are paving the way for designing smart nanoparticles capable of recognizing patient-specific tumor markers, delivering customized payloads, and adapting to tumor heterogeneity. Integrating these technologies will be crucial for the realization of precision medicine and maximizing the therapeutic benefit of nanomedicine in cancer care [4, 46].

## 6 Conclusion

Nanotherapeutics represents a transformative advancement in cancer treatment, offering enhanced specificity in drug delivery and enabling real-time monitoring of therapeutic outcomes. By utilizing nanoparticles for targeted drug delivery, therapeutic agents can be concentrated precisely at the tumor site, minimizing systemic toxicity and improving treatment efficacy. Additionally, the integration of real-time imaging and biosensing technologies facilitates continuous monitoring of tumor progression, drug distribution, and therapeutic response, allowing for adaptive treatment strategies. These developments have paved the way for more personalized, effective, and less invasive cancer therapies. However, challenges such as biocompatibility, manufacturing scalability, and regulatory hurdles remain, requiring ongoing research and collaboration. As these challenges are addressed, nanotherapeutics are poised to redefine cancer care, offering tailored treatment options that could significantly improve patient outcomes and quality of life.

## References


1. Wang Y, Sun S, Zhang Z, Shi D (2018) Nanomaterials for cancer precision medicine. Adv Mater 30(17):1705660. https://doi.org/10.1002/adma.201705660
2. Sarkar S, Kumar S, Saha G, Basu M, Ghosh MK (2024) Glioma nanotherapy: unleashing the synergy of dual-loaded DIM and TMZ. Int J Pharm 665:124697
3. Ghosh MK, Kumar S, Begam S, Ghosh S, Basu M (2024) GBM immunotherapy: exploring molecular and clinical frontiers. Life Sci:123018. Published online 2024
4. Kumar S, Ansari A, Basu M, Ghosh S, Begam S, Ghosh MK (2025) Carbon nanotubes in cancer diagnosis and treatment: current trends and future perspectives. Adv Ther 8(1):2400283. https://doi.org/10.1002/adtp.202400283
5. Shakeri-Zadeh A, Bulte JW (2025) Imaging-guided precision hyperthermia with magnetic nanoparticles. Nat Rev Bioeng 3(3):245–260
6. Yang X, Kubican SE, Yi Z, Tong S (2025) Advances in magnetic nanoparticles for molecular medicine. Chem Commun. Published online 2025. Accessed 3 May 2025. https://pubs.rsc.org/en/content/articlehtml/2025/cc/d4cc05167j
7. Khan M, Moutaoukil ME, Boucetta A, Spadavecchia J (2025) Bortezomib (BOR)-Pegylated-gold nanovector: synthesis, spectroscopic evaluation and diagnostic tool as Galectin-1 biomarker. Nano Sel 6(3):e202400090. https://doi.org/10.1002/nano.202400090
8. Ghazizadeh Y, Salehi Shadkami H, Madani F, Niknam S, Adabi M (2025) Advances in cancer nanovaccines: a focus on colorectal cancer. Nanomed 20(9):1029–1041. https://doi.org/10.1080/17435889.2025.2486930
9. Qi F, Fu D, Cai H, Zheng Y, Wang N, Xu Z (2025) Metabolic reprogramming of cancer-associated fibroblasts: transforming tumor accomplices into immunotherapeutic allies. Adv Funct Mater 35(13):2418240. https://doi.org/10.1002/adfm.202418240
10. Yao Z, Liu T, Wang J et al (2025) Targeted delivery systems of siRNA based on ionizable lipid nanoparticles and cationic polymer vectors. Biotechnol Adv:108546. Published online 2025
11. Wang C, Xue Y, Markovic T et al (2025) Blood–brain-barrier-crossing lipid nanoparticles for mRNA delivery to the central nervous system. Nat Mater:1–11. Published online 2025
12. Rahdari T, Mahdavimehr M, Ghafouri H, Ramezanpour S, Ehtesham S, Asghari SM (2025) Advancing triple-negative breast cancer treatment through peptide decorated solid lipid nanoparticles for paclitaxel delivery. Sci Rep 15(1):6043
13. Moghaddam SH, Vatankhah A, Oroojalian F, Kesharwani P, Sahebkar A (2025) Liposomes as immunotherapeutic carriers: a game-changer in cancer therapy. J Drug Deliv Sci Technol:106847. Published online 2025
14. Khan IN, Arshad N, Shaheen F, Shakoor R, Hassan A, Waqar MA (2025) Recent composition and applications of liposomes in cancer therapy. Int J Polym Mater Polym Biomater 74(8):720–732. https://doi.org/10.1080/00914037.2024.2368896
15. Razavi ZS, Razavi FS, Alizadeh SS (2025) Inorganic nanoparticles and blood-brain barrier modulation: advancing targeted neurological therapies. Eur J Med Chem:117357. Published online 2025

16. Reddy KTK, Reddy AS (2025) Recent breakthroughs in drug delivery systems for targeted cancer therapy: an overview. Cell Mol Biomed Rep 5(1):13–27
17. Gong Z, Zhou D, Wu D et al (2025) Challenges and material innovations in drug delivery to central nervous system tumors. Biomaterials:123180. Published online 2025
18. Nasirzadeh N, Hajian R, Nadri S (2025) Microfluidic synthesis of pH-sensitive nanoparticles containing curcumin against breast cancer cell. Nanomed J 12(1). Accessed 3 May 2025. https://search.ebscohost.com/login.aspx?direct=true&profile=ehost&scope=site&authtype=crawler&jrnl=23223049&AN=181598479&h=ycjGlULmMLSeSz1RzWSuJGQ9YmZ5U%2FIazxNLbFjT%2FovtaX2mHm6S4uTIxZURRjAQGTPtDY5kvbW7My4w7ibQsQ%3D%3D&crl=c
19. Di Gregorio G, Vallée C, Konate K et al (2025) Enhancing WRAP-based nanoparticles for small interfering ribonucleic acid delivery in pH-sensitive environments. ChemMedChem:2400885. Published online April 10, 2025. https://doi.org/10.1002/cmdc.202400885
20. Yang T, Bian Y, Qu Z et al (2025) Gastrointestinal pH-sensitive pickering emulsions stabilized by glycosylated zein conjugates ferulic acid nanoparticles: improving oral bioaccessibility of coenzyme Q10. Colloids Surf B Biointerfaces 249:114502
21. Jobdeedamrong A, Crespy D (2025) Redox-responsive polyprodrugs: recent innovations in reduction- and oxidation-responsive drug delivery systems. Chem Mater 37(6):2073–2086. https://doi.org/10.1021/acs.chemmater.4c03181
22. Gaddimath S, Payamalle S, Channabasavana Hundi Puttaningaiah KP, Hur J (2024) Recent advances in pH and redox responsive polymer nanocomposites for cancer therapy. J Compos Sci 8(1):28
23. Meng X, Shen Y, Zhao H, Lu X, Wang Z, Zhao Y (2024) Redox-manipulating nanocarriers for anticancer drug delivery: a systematic review. J Nanobiotechnol 22(1):587. https://doi.org/10.1186/s12951-024-02859-w
24. Mathur A, Singhal J, Gupta D, Saxena R (2024) Enzyme-responsive nanoparticles for drug delivery. In: Applications of nanoparticles in drug delivery and therapeutics. Bentham Science Publishers, pp 70–93. Accessed 3 May 2025. https://www.benthamdirect.com/content/books/9789815256505.chapter-4
25. Sahoo PK, Desai DN Nanoparticulate enzymatic mechanism in anti-tumor drug delivery. In: Smart micro-and nanomaterials for pharmaceutical applications. CRC Press, pp 281-292. Accessed 3 May 2025. https://www.taylorfrancis.com/chapters/edit/10.1201/9781003468431-16/nanoparticulate-enzymatic-mechanism-anti-tumor-drug-delivery-sahoo-dhruv-desai
26. Singh V (2024) Theranostics: integrated diagnostics and therapy using nanomedicine. In: Gautam V, Kumar R, Das Manandhar K, Kamble SC (eds) Nanomedicine, Nanotechnology in the life sciences. Springer Nature Switzerland, pp 505–530. https://doi.org/10.1007/978-3-031-72467-1_17
27. Suhag D, Kaushik S, Taxak VB (2024) Theranostics: combining diagnosis and therapy. In: Handbook of biomaterials for medical applications, Volume 1, Biomedical materials for multi-functional applications. Springer Nature Singapore, pp 271–295. https://doi.org/10.1007/978-981-97-4818-1_10
28. Pal J, Kola P, Samanta P, Mandal M, Dhara D (2024) Polymer nanoparticles for preferential delivery of drugs only by exploiting the slightly elevated temperature of cancer cells and real-time monitoring of drug release. Biomacromolecules 25(8):5181–5197. https://doi.org/10.1021/acs.biomac.4c00572
29. El-Sewify IM, Shenashen MA, El-Agamy RF et al (2024) Fluorescent sensor/tracker for biocompatible and real-time monitoring of ultra-trace arsenic toxicants in living cells. J Hazard Mater 478:135429
30. Kalashgrani MY, Mousavi SM, Akmal MH et al (2024) Gold fluorescence nanoparticles for enhanced SERS detection in biomedical sensor applications: current trends and future directions. Chem Rec:e202300303. Published online February 5, 2024. https://doi.org/10.1002/tcr.202300303

31. Rezaei B, Tay Z, Mostufa S et al (2024) Magnetic nanoparticles for magnetic particle imaging (MPI): design and applications. Nanoscale. Published online 2024. Accessed 3 May 2025. https://pubs.rsc.org/en/content/articlehtml/2024/nr/d4nr01195c
32. Bushra R, Ahmad M, Alam K et al (2024) Recent advances in magnetic nanoparticles: key applications, environmental insights, and future strategies. Sustain Mater Technol:e00985. Published online 2024
33. Sun Y, Wang Y, Li W, Li C (2024) Real-time dual-modal photoacoustic and fluorescence small animal imaging. Photoacoustics 36:100593
34. Zhang M, Peng X, Xu H et al (2024) Photoacoustic imaging-guided self-adaptive hyperthermia supramolecular cascade nano-reactor for diabetic periodontal bone regeneration. Adv Sci 11(29):2404143. https://doi.org/10.1002/advs.202404143
35. Choi YS, Son WK, Kwak H et al (2025) Real-time monitoring of volatile organic compound-mediated plant intercommunication using surface-enhanced Raman scattering nanosensor. Adv Sci 12(7):2412732. https://doi.org/10.1002/advs.202412732
36. Khonina SN, Kazanskiy NL (2025) Trends and advances in wearable plasmonic sensors utilizing surface-enhanced Raman spectroscopy (SERS): a comprehensive review. Sensors 25(5):1367
37. Kumar S, Basu M, Ghosh MK (2024) E3 ubiquitin ligases and deubiquitinases in colorectal cancer: emerging molecular insights and therapeutic opportunities. Biochim Biophys Acta BBA-Mol Cell Res:119827
38. Ghosh MK, Kumar S, Ganguly KK et al (2023) Covid-19 and cancer: insights into their association and influence on genetic and epigenetic landscape. Epigenomics 15(4):227–248. https://doi.org/10.2217/epi-2023-0052
39. Krishnan S, Giwa A (2025) Advances in real-time water quality monitoring using triboelectric nanosensors. J Mater Chem A. Published online 2025. Accessed 3 May 2025. https://pubs.rsc.org/en/content/articlepdf/2025/ta/d4ta08871a
40. Xu Y, Zhang J, Ray WZ, MacEwan MR (2024) Implantable and semi-implantable biosensors for minimally invasive disease diagnosis. Processes 12(7):1535
41. Alam F, Ashfaq Ahmed M, Jalal AH et al (2024) Recent progress and challenges of implantable biodegradable biosensors. Micromachines 15(4):475
42. Mahato K (2024) Implantable biosensors for personalized healthcare. In: Mahato K, Chandra P (eds) Biosensors for personalized healthcare. Springer Nature Singapore, pp 375–392. https://doi.org/10.1007/978-981-97-5473-1_14
43. Khondakar KR, Kaushik AK (2024) Next-generation smart biosensing: nano-platforms, Nano-microfluidics interfaces, and emerging applications of quantum sensing. Published online 2024. Accessed 3 May 2025. https://books.google.com/books?hl=en&lr=&id=E3SyEAAAQBAJ&oi=fnd&pg=PP1&dq=Microfluidic+Nanoplatforms+in+real+time+monitoring&ots=jc0F0n-xBJ&sig=Z7fLAtcatuUgTXNiVIu5C%2D%2DtKb0
44. Tian Z, Fu Y, Dang Z, Guo T, Li W, Zhang J (2025) Utilizing nanomaterials in microfluidic devices for disease detection and treatment. Nanomaterials 15(6):434
45. Lou C, Yang H, Hou Y et al (2024) Microfluidic platforms for real-time in situ monitoring of biomarkers for cellular processes. Adv Mater 36(6):2307051. https://doi.org/10.1002/adma.202307051
46. Palakurti NR (2024) Challenges and future directions in anomaly detection. In: Practical applications of data processing, algorithms, and modeling. IGI Global, pp 269–284. Accessed 3 May 2025. https://www.igi-global.com/chapter/challenges-and-future-directions-in-anomaly-detection/345815

# Biomarker Identification Using Nanoparticles

Nahid Rehman and Anjana Pandey

**Abstract**

Identification of biomarkers is essential for prognosis, treatment monitoring, and early disease detection. Because of their special physicochemical characteristics, including high surface-to-volume ratios, adjustable optical and electrical characteristics, and superior biocompatibility, nanoparticles have become effective instruments in biosensing applications. Using a variety of biosensing techniques, such as optical nano-biosensors, electrochemical methods, nanoprobes, and immunoassay sensors, this chapter investigates the function of nanoparticles in biomarker detection. Highly sensitive and selective biomarker detection is made possible by optical nano-biosensors, such as fluorescence-based, surface plasmon resonance (SPR), and surface-enhanced Raman spectroscopy (SERS) sensors. Electrochemical methods, such as electrochemical impedance spectroscopy (EIS) and differential pulse voltammetry (DPV), provide quick, affordable diagnostic platforms with superior detection limits. For the detection of biomarkers, nanoprobes functionalized with biomolecules such as DNA, enzymes, or antibodies offer improved selectivity and signal amplification. Furthermore, point-of-care diagnostics are made more accurate by immunoassay sensors, such as lateral flow assays and nano-enabled ELISA. This chapter compares several approaches in terms of sensitivity, specificity, and practicality to highlight the benefits and difficulties of nanoparticle-based biosensors. Lastly, to demonstrate how nanoparticle-based biosensors have the potential to transform clinical diagnostics, new developments and future directions are explored, including AI integration and multi-modal sensing platforms.

N. Rehman · A. Pandey (✉)
Department of Biotechnology, Motilal Nehru National Institute of Technology Allahabad (MNNITA), Prayagraj, Uttar Pradesh, India
e-mail: anjanap@mnnit.ac.in



M. K. Ghosh, M. Basu (eds.), *Personalized Cancer Therapy*,
https://doi.org/10.1007/978-981-95-6845-1_17

**Keywords**

Nanoparticle-based biosensors · Biomarker detection · Electrochemical and optical sensing · Nano-immunoassays

# 1 Introduction

The early and accurate identification of disease biomarkers is fundamental for timely intervention, improved patient outcomes, and personalized medicine. Biomarkers serve as critical indicators of normal or pathogenic biological processes and responses to therapeutic interventions [1]. Their identification enables the early detection of diseases such as cancer, cardiovascular disorders, and infectious diseases, significantly enhancing treatment efficacy [2]. Nanotechnology plays a transformative role in biomarker detection by enabling high sensitivity, miniaturization, and multiplexing capabilities [3]. Nanomaterials provide unique physicochemical properties such as high surface-to-volume ratios and quantum confinement effects that enhance signal transduction in biosensors. This allows for the detection of ultra-low concentrations of biomarkers, often in complex biological matrices [4]. Detection strategies that use nanoparticles include optical, electrochemical, magnetic methods, and so on [5]. Optical methods leverage fluorescence, plasmon resonance, or Raman scattering, while electrochemical techniques benefit from enhanced electron transfer kinetics [6]. Additionally, magnetic nanoparticles allow for target enrichment and separation. Each strategy offers distinct advantages depending on the intended application and biomarker type [7]. This chapter provides a comprehensive overview of nanoparticle-enabled biosensing technologies for biomarker detection. It discusses various nanoparticles, their functionalization techniques, detection strategies (optical, electrochemical, and immunoassay-based), and future directions in this rapidly evolving field.

# 2 Nanoparticles in Biosensing Applications

Recently, biosensors made of nanoparticles (NPs) have gained very high popularity due to their excellent applications in nearly all fields of science and technology. The use of NPs in the biosensor is usually done to fill the gap between the converter and the bioreceptor, which is at the nanoscale [8]. The following are the types of nanoparticles commonly used in biomarker identification.

## 2.1 Gold Nanoparticles (AuNPs)

AuNPs possess exceptional stability, low resistance, high conductivity, and extensive light interaction, making them well-suited for biological sensing applications [9]. Known for their surface plasmon resonance and ease of functionalization during synthesis, surface functionalization, and tunable optical properties [10], AuNPs are widely used in colorimetric, electrochemical, and SERS-based biosensors [11]. Furthermore, AuNPs are chemically inert, biocompatible, and non-toxic, ensuring their safety and eco-friendliness in medical applications. Different surface modifications and changes in size and shape can also change how sensitive and selective AuNPs are in biological sensing [9]. The diverse array of available shapes for AuNPs comprises nanospheres [12], nanorods, nano-shells, nano-prisms, nano-pyramids, nano-bipyramids, nanocages, nano-rings, nanodisks, nano-stars, nano-rice, nano-bowls, nano-flowers, and nanocrescents [13]. These varied AuNPs shapes showcase distinctive optical characteristics, encompassing the ratio of light absorption and scattering coefficients, local surface plasmon resonance (LSPR), fluorescence, and surface-enhanced Raman scattering (SERS). The optical properties inherent in AuNPs form the cornerstone for their application in biological sensing, specifically as biosensors [14].

## 2.2 Quantum Dots (QDs)

Quantum dots (QDs) are semiconductor nanoparticles with diameters in the nanometer range. QDs are increasingly used in biosensing applications due to their unique optical properties, including high brightness, photostability, and tunable emission spectra. In addition to possessing the advantages of common QDs, carbon quantum dots (CQDs) have a higher solubility, lower toxicity, and easier synthesis, making them highly useful and a promising candidate for biomedical applications. The CQD is a zero-dimensional nanostructure with a size of less than 10 nm in every dimension. The unparalleled features of CQDs, including good biocompatibility and unique optical properties, such as high photoluminescence (PL) and quantum yield (QY), make them attractive candidates for biosensors and bioimaging applications [15]. Semiconductor nanocrystals that offer high photostability and tunable emission properties are suitable for multiplexed fluorescence detection [16].

## 2.3 Carbon-Based Nanomaterials

One of the attractive research areas in which carbon-based nanomaterials have many applications is the science of detection and measurement called biosensors. Some special properties, such as large surface-to-volume ratio, high electrical conductivity, chemical stability, strong mechanical strength, easy functionalization, biocompatibility, and biodegradability, make these nanomaterials suitable for use in biosensors and transducers for improving signal processing. Recently, these biosensors have been used widely in the monitoring of disease progression, environmental pollution detection and measurement, food control, and drug discovery. Considering these substantial properties, using these carbon-based nanomaterials leads to higher sensitivity with a lower limit of detection, wide linear range detection, and reusable sensors compared to the conventional ones. A wide variety of carbon nanomaterials, including graphene, graphene oxide, reduced graphene oxide, carbon nanotubes, carbon dots, diamond, fullerene, carbon nanofibers, carbon nanohorns, nanoporous carbon, carbon black, and their composites, are being explored recently for biosensor development [17].

### 2.4 Magnetic Nanoparticles (MNPs)

Magnetic nanoparticles (MNPs) are widely used in biosensing due to their unique properties and ease of manipulation under magnetic fields. They enhance sensitivity, selectivity, and signal amplification in biosensors, leading to improved detection of various analytes [7]. Synthesized through various methods, especially green synthesis, MNPs, particularly iron oxide nanoparticles (IONPs), enable precise manipulation in biological samples, improving target detection accuracy. Surface functionalization enhances biocompatibility and selectivity, which are crucial for biosensing [18].

## 3 Nano-Based Biosensing for Enhanced Detection

Nanoparticles (NPs) have revolutionized biomarker assays by amplifying weak signals and enabling highly selective binding. NPs amplify assay signals via their large surface area and unique physics. In addition to plasmonic and fluorescence, NPs can carry multiple reporter groups (enzymes, fluorophores) or catalyze colorimetric reactions (e.g., peroxidase-mimicking "nanozymes"). They can also serve as electron mediators on electrodes or a "bridge" between reporter molecules (e.g., in hybridization chain reactions) to amplify currents [5].

NPs are readily functionalized with antibodies, aptamers, or nucleic acid probes. For example, magnetic iron-oxide NPs (MNPs) are often coated with specific antibodies or DNA probes to bind target biomarkers; the resulting "sandwich" complexes are then detected magnetically. Magnetic separation with antibody-coated MNPs can enrich targets from complex samples, concentrating analytes and reducing interference. Likewise, fluorescent QDs or carbon dots conjugated to antibodies provide multiplex detection via different emission colors [19]. Depending on their type, NPs enable diverse readouts: optical (colorimetric SPR shifts, fluorescence, SERS, photothermal/photoacoustic imaging) [20], magnetic (magnetic relaxation or impedance changes using MNPs), electrochemical (NP-modified electrodes yield amplified redox current), or even acoustic (ultrasonic scattering) [21]. In each case, the NP greatly improves the sensor's sensitivity by converting a low-abundance binding event into a strong, measurable signal. Table 1 describes the merits and demerits of various biosensors.

**Table 1** Advantages and limitations of various biosensors

| Biosensing approach | Advantages | Limitations |
|---|---|---|
| Optical biosensors | High sensitivity and specificity | Often require complex optical instrumentation |
| | Multiplexing capability (especially with QDs and FRET) | Background fluorescence or autofluorescence may interfere |
| | Label-free detection is possible with SPR and SERS | Photobleaching and cytotoxicity (e.g., with some QDs) |
| | Rapid detection and real-time monitoring | Surface fouling and nonspecific binding can affect performance |
| Electrochemical biosensors | High sensitivity (pM–fM range) | Surface fouling can reduce reproducibility |
| | Compatible with portable, low-cost devices | Requires calibration and optimization for complex biological fluids |
| | Fast response time (<5 min in optimized systems) | May require electrode surface regeneration |
| | Label-free and label-based options are available | Signal drift under variable environmental conditions |
| Nanoprobe-based sensors | Ultra-sensitive detection (down to attomolar levels) | Complex synthesis and functionalization |
| | High specificity via functionalized recognition elements (aptamers, antibodies) | Risk of nanoparticle aggregation or instability |
| | Enables multiplexed detection in a single assay | Potential toxicity and biocompatibility issues |
| | Multimodal (optical, magnetic, electrochemical) capability | Lack of clinical standardization |
| | Real-time/in situ monitoring in live cells | High production costs |
| Immunoassay-based sensors | High specificity through antigen-antibody binding | Susceptible to cross-reactivity and false results |
| | Widely adaptable formats (ELISA, LFAs, label-free sensors) | Antibodies/enzymes are sensitive to degradation |
| | Nanoparticles enhance signal output and speed | Limited quantification in LFA formats |
| | User-friendly in POC settings (e.g., LFAs) | Variability in antibody quality |
| | Established clinical usage and regulatory pathway | Requires careful optimization and calibration |

## 3.1 Optical Nano-Biosensors for Biomarker Identification

Optical nano-biosensors exploit the interaction of light with nanomaterials and biomolecular targets to enable sensitive, label-based or label-free detection of disease-specific biomarkers. These sensors utilize the unique optical properties of nanoparticles, such as fluorescence emission, localized surface plasmon resonance (LSPR), and enhanced Raman scattering, to generate measurable optical signals upon biomarker recognition.

- *Fluorescence-Based Detection*: Quantum dots (QDs) enable multiplexed detection with high sensitivity due to their narrow emission and broad absorption spectra [22].
- *Forster Resonance Energy Transfer (FRET)-based biosensors***:** FRET enables real-time detection of molecular interactions, where energy transfer between donor and acceptor fluorophores is modulated by biomarker presence.
- *Surface Plasmon Resonance (SPR) and Localized SPR (LSPR):* Gold nanoparticles in SPR sensing enhance the local refractive index change during biomolecular interactions, increasing sensitivity in LSPR-based detection platforms [23].
- *Raman Spectroscopy-Based Detection (SERS):* Nanoparticle-enhanced SERS for ultrasensitive biomarker identification, where nanostructured metals (Au, Ag) amplify Raman signals of biomarkers, achieving single-molecule sensitivity [24].

## 3.2 Electrochemical Techniques for Biomarker Detection

Electrochemical biosensors convert biological interactions into quantifiable electrical signals. These systems are especially attractive for biomarker detection due to their high sensitivity, rapid response, low cost, and compatibility with miniaturization and point-of-care (POC) platforms. Nanomaterials enhance sensitivity, selectivity, and stability by facilitating electron transfer and increasing surface functional groups for biorecognition [25]. The incorporation of nanomaterials such as carbon-based nanostructures, metal nanoparticles, and conducting polymers has significantly improved sensor performance by enhancing electron transfer, increasing surface area, and enabling effective biomolecule immobilization. Electrochemical biosensors typically involve a biorecognition element (e.g., antibody, aptamer, enzyme) immobilized on an electrode surface. Upon specific interaction with a target biomarker, the biochemical event generates an electrical signal (current, voltage, impedance), which is measured by a transducer. These sensors are classified based on the mode of signal generation. These can be *Amperometric:* measures current due to redox reactions, *Potentiometric:* detects changes in potential at zero current, *Impedimetric:* monitors impedance changes at the electrode interface, *Conductometric:* measures conductivity variations in the electrolyte.

- *Electrochemical Impedance Spectroscopy:* EIS monitors changes in charge transfer resistance at the electrode interface upon target binding, enabling label-free detection [26].
- *Differential Pulse Voltammetry (DPV) and Square Wave Voltammetry (SWV):* Incorporating nanomaterials such as AuNPs or graphene increases electroactive surface area and electron transfer, improving detection limits [27].
- *Conductometric and Potentiometric Sensors:* These methods detect conductivity or potential changes upon target interaction. Though less common, they are effective for ion or small-molecule biomarkers.

### 3.3 Nanoprobes in Biomarker Detection

Nanoprobes are nanoscale devices or particles engineered to detect specific biomolecules, often through signal generation or amplification. Nanoprobes are engineered nanoscale structures that integrate recognition elements and signal transducers to detect specific biological targets, such as DNA, RNA, proteins, or small molecules. Their small size, surface tunability, and multifunctionality enable ultra-sensitive and selective detection of disease biomarkers, particularly in complex biological environments such as blood or tissue samples. Nanoprobes are categorized based on their structural composition and functional modality, ranging from simple fluorophore-labeled nanoparticles to sophisticated hybrid systems capable of multimodal detection.

- *DNA and RNA-based nanoprobes:* Molecular beacons (MBs) are single-stranded oligonucleotide probes with a stem-loop structure labeled with a fluorophore and quencher at opposite ends. In the absence of the target, the beacon is in a closed conformation (quenched). Upon hybridization with a complementary nucleic acid sequence (e.g., microRNA, mRNA, or pathogen DNA), the loop opens, separating fluorophore and quencher, resulting in fluorescence emission. Applications include early cancer diagnosis via miRNA detection (e.g., miR-21), viral RNA identification (e.g., SARS-CoV-2, Dengue virus), minimal sample preprocessing, ideal for in situ and real-time analysis [28].
- *Enzyme-Linked Nanoprobes:* Enzyme-nanoparticle conjugate hybrid probes for signal amplification enhance signal through enzymatic reactions, often improving detection limits in colorimetric or electrochemical assays [29].
- Enzyme-nanoparticle conjugates are designed to amplify detection signals through catalytic turnover of substrates. These nanoprobes can be constructed by attaching enzymes (e.g., horseradish peroxidase, alkaline phosphatase) to gold nanoparticles or magnetic nanoparticles. Using the enzyme's activity to convert a substrate into an electroactive or chromogenic product. An example where a single recognition event (e.g., antigen–antibody binding) can lead to multiple enzymatic cycles, generating strong optical or electrochemical signals. This strategy is widely used in nano-enhanced ELISA and electrochemical immunosensors [30].
- *Hybrid Nanoprobes:* Multimodal nanoprobes for enhanced sensitivity and specificity combine multiple detection modalities (e.g., optical and magnetic). Hybrid nanoprobes combine multiple detection modalities to enable simultaneous sensing, imaging, and therapeutic interventions (theranostics). These can integrate: magnetic nanoparticles for separation, fluorescent tags for imaging, and Enzyme or redox probes for signal generation. Examples: Magnetic-fluorescent nanoprobes for dual-mode detection of circulating tumor cells (CTCs), SERS based electrochemical hybrid probes for real-time quantification of low-abundance proteins in serum, Core-shell nanoparticles combining QDs (core) and AuNPs (shell) for both FRET and SERS applications to improve diagnostic accuracy and versatility [31].

## 3.4 Immunoassay Sensors for Biomarker Detection

Immunosensors exploit antigen-antibody interactions for highly specific detection, often enhanced through labeling or nanostructured interfaces. Immunoassay sensors leverage the high specificity of antigen–antibody interactions to detect biomarkers with high confidence and sensitivity. These biosensors have found widespread applications in clinical diagnostics, food safety, and environmental monitoring due to their reliability and adaptability to various detection platforms. Immunosensors operate based on the specific binding between an antibody (or antigen) immobilized on a sensor surface and its target antigen (or antibody) present in the sample. The interaction generates a measurable signal, which may be optical, electrochemical, piezoelectric, or thermal in nature. The magnitude of the signal correlates with the concentration of the target biomarker, allowing qualitative or quantitative analysis.

- Nano-Enabled ELISA (Enzyme-Linked Immunosorbent Assay): Use of nanoparticles as enzyme carriers or signal amplifiers enhances assay sensitivity and reduces detection time [32]. The incorporation of nanomaterials into ELISA protocols enhances signal transduction and enables lower detection limits. Nanoparticles such as gold (AuNPs), magnetic nanoparticles (MNPs), and quantum dots (QDs) can serve as enzyme carriers, signal amplifiers, or detection labels. For example, gold nanoparticles conjugated with secondary antibodies and horseradish peroxidase (HRP) significantly amplify the colorimetric signal in sandwich ELISA assays [33].
- Label-Free Immunosensors: Label-free immunosensors detect biomarker binding events without requiring secondary labels or signal reporters. These sensors rely on techniques such as Surface Plasmon Resonance (SPR), which measures changes in refractive index upon biomolecular interaction. Electrochemical Immunosensors detect changes in current, impedance, or voltage upon antigen–antibody binding.
- Lateral Flow Assays (LFAs) Using Nanoparticles: AuNPs are predominant in LFAs as they provide a visible color change, enabling simple and rapid point-of-care tests. Label-free detection is especially useful in high-throughput screening and for measuring weak or transient interactions. LFAs are portable, paper-based immunoassays widely used for rapid diagnostics, such as in pregnancy tests or COVID-19 antigen detection kits. Nanoparticles, particularly gold nanoparticles, are employed as visible colorimetric labels that indicate the presence or absence of a biomarker. Key features include a simple, equipment-free readout, results in under 15 minutes, and compatibility with whole blood, serum, saliva, and urine. Recent advances incorporate fluorescent or magnetic nanoparticles to enable quantitative LFA readouts with smartphone-based readers or handheld detectors [34].

## 4 Disease-Specific Applications

Nanoparticle-based biosensing platforms have revolutionized disease diagnostics by enabling the early and accurate detection of specific biomarkers across a broad range of pathological conditions. Their high surface-to-volume ratio, tunable surface chemistry, and unique optical/electrical properties allow for sensitive, specific, and multiplexed detection of disease-associated biomolecules. The varying categories of disease, along with the detection approach, have been summarized in Table 2.

**Table 2** Disease-specific applications of nanoparticle-based biomarker detection

| Disease area | Target biomarkers | Nanomaterials used | Detection platform/ mechanism | Key advantages | Ref |
|---|---|---|---|---|---|
| Cancer | PSA, CEA, CA-125, HER2, miRNAs | AuNPs, CNTs, QDs, MNPs | Colorimetric assays, electrochemical sensors, FRET-based probes | High sensitivity; early detection; multiplexing capability | [35] |
| Cardiovascular | Troponin I/T, CRP, BNP | MNPs, graphene oxide, nanocomposites | Electrochemical immunosensors; lateral flow assays | Rapid POC diagnostics; ultrasensitive detection | [36] |
| Infectious diseases | SARS-CoV-2 antigen, bacterial DNA/RNA, pathogen antibodies | AuNPs, AgNPs, magnetic beads, silica NPs | LFA, nucleic acid hybridization assays, colorimetric platforms | Fast detection, portable, minimal sample processing | [37] |
| Neurodegenerative | Amyloid-beta, tau, α-synuclein | Graphene QDs, nanowires, MNPs | Electrochemical and optical sensors using aptamers | Ultra-low detection limits; suitable for CSF and plasma samples | [38] |
| Diabetes | Glucose, insulin, HbA1c, adiponectin | ZnO NPs, AgNPs, CNTs | Non-enzymatic glucose sensors, wearable electrochemical platforms | Real-time monitoring, improved electrode stability | [39] |

## 5 Future Perspectives and Challenges

- Advancements in nanoparticle-based biosensing: Innovations include biodegradable nanomaterials, wearable sensors, and integrated lab-on-chip platforms.
- Integration of AI and machine learning in biosensing: Data-driven algorithms enhance pattern recognition, sensitivity optimization, and biomarker panel analysis.
- Regulatory challenges and commercialization: Standardization, reproducibility, and clinical validation are critical for translating biosensors to market.
- Emerging trends in multi-modal biosensing: Combining optical, magnetic, and electrochemical signals for comprehensive and robust diagnostics.

## 6 Conclusion

Nanoparticle-based biosensors offer powerful tools for early and sensitive detection of disease biomarkers. By leveraging unique nanoscale properties, they have revolutionized optical, electrochemical, and immunoassay detection platforms. With ongoing technological advancements and integration with AI, these biosensors hold promise for routine clinical diagnostics, personalized therapies, and global health monitoring.

## References

1. Bodaghi A, Fattahi N, Ramazani A (2023) Biomarkers: promising and valuable tools towards diagnosis, prognosis and treatment of Covid-19 and other diseases. Heliyon 9(2):e13323
2. Henry NL et al (2022) Biomarkers for systemic therapy in metastatic breast cancer: ASCO guideline update. J Clin Oncol 40(27):3205–3221
3. Jalalvand AR, Karami MM (2025) Roles of nanotechnology in electrochemical sensors for medical diagnostic purposes: a review. Sens Bio-Sens Res 47:100733
4. Jain KK (2008) Nanomedicine. In: Shoseyov O, Levy I (eds) NanoBioTechnology: BioInspired devices and materials of the future. Humana Press, Totowa, pp 303–327
5. Wang M et al (2024) Advancements in magnetic nanoparticle-based biosensors for point-of-care testing. Front Bioeng Biotechnol 12:2024
6. Yang Q et al (2024) Plasmonic nanomaterial-enhanced fluorescence and Raman sensors: multifunctional platforms and applications. Coord Chem Rev 507:215768
7. Materón EM et al (2021) Magnetic nanoparticles in biomedical applications: a review. Appl Surf Sci Adv 6:100163
8. Malik S et al (2023) Nanomaterials-based biosensor and their applications: a review. Heliyon 9(9):e19929
9. Kumalasari MR, Alfanaar R, Andreani AS (2024) Gold nanoparticles (AuNPs): a versatile material for biosensor application. Talanta Open 9:100327
10. Rehman N, Pandey A (2022) Engineered nanoparticles as drug delivery systems. CRC Press
11. Dykman L, Khlebtsov N (2012) Gold nanoparticles in biomedical applications: recent advances and perspectives. Chem Soc Rev 41(6):2256–2282
12. Farkhani SM et al (2024) Tailoring gold nanocluster properties for biomedical applications: from sensing to bioimaging and theranostics. Prog Mater Sci 142:101229

13. Xiao-Ming M et al (2018) Progress of visual biosensor based on gold nanoparticles. Chin J Anal Chem 46(1):1–10
14. Lee J-H et al (2018) Application of gold nanoparticle to plasmonic biosensors. Int J Mol Sci 19(7):2021
15. Pourmadadi M et al (2023) Properties and application of carbon quantum dots (CQDs) in biosensors for disease detection: a comprehensive review. J Drug Delivery Sci Technol 80:104156
16. Resch-Genger U et al (2008) Quantum dots versus organic dyes as fluorescent labels. Nat Methods 5(9):763–775
17. Eivazzadeh-Keihan R et al (2022) Applications of carbon-based conductive nanomaterials in biosensors. Chem Eng J 442:136183
18. Parmanik A, Bose A (2025) Chapter 8—Magnetic nanoparticles used in biosensors. In: Hasnain MS, Nayak AK, Aminabhavi TM (eds) Materials and components of biosensors in healthcare. Academic Press, pp 165–174
19. Sun X et al (2025) Recent advances in fluorescent nanomaterials designed for biomarker detection and imaging. Mater Today Bio 32:101763
20. Fu L et al (2023) Plasmonic nanoparticle-enhanced optical techniques for cancer biomarker sensing. Biosensors (Basel) 13(11)
21. Zhang W et al (2024) Biomarker detection based on nanoparticle-induced ultrasonic Rayleigh scattering. Microsyst Nanoeng 10(1):182
22. Medintz IL et al (2005) Quantum dot bioconjugates for imaging, labelling and sensing. Nat Mater 4(6):435–446
23. Anker JN et al (2008) Biosensing with plasmonic nanosensors. Nat Mater 7(6):442–453
24. Langer J et al (2020) Present and future of surface-enhanced Raman scattering. ACS Nano 14(1):28–117
25. Justino CIL et al (2010) Review of analytical figures of merit of sensors and biosensors in clinical applications. TrAC Trends Anal Chem 29(10):1172–1183
26. Lasia, A., A. Lasia, Electrochemical impedance spectroscopy and its applications, book, Springer, 2014.
27. Labib M, Sargent EH, Kelley SO (2016) Electrochemical methods for the analysis of clinically relevant biomolecules. Chem Rev 116(16):9001–9090
28. Tyagi S, Kramer FR (1996) Molecular beacons: probes that fluoresce upon hybridization. Nat Biotechnol 14(3):303–308
29. Das B et al (2021) Nanozymes in point-of-care diagnosis: an emerging futuristic approach for biosensing. Nano-Micro Lett 13(1):193
30. Pan M et al (2017) Nanomaterials for electrochemical immunosensing. Sensors 17. https://doi.org/10.3390/s17051041
31. Xu M et al (2025) Recent advances in nanomaterial-based optical biosensors and their biomedical and biopharmaceutical applications. J Pharm Analys:101349
32. Hemdan M et al (2025) Recent advances in nano-enhanced biosensors: innovations in design, applications in healthcare, environmental monitoring, and food safety, and emerging research challenges. Sens Bio-Sens Res 48:100783
33. Khoris IM et al (2019) Enhanced colorimetric detection of norovirus using in-situ growth of Ag shell on Au NPs. Biosens Bioelectron 126:425–432
34. Ardekani LS, Thulstrup PW (2022) Gold nanoparticle-mediated lateral flow assays for detection of host antibodies and COVID-19 proteins. Nanomaterials (Basel) 12(9)
35. Kumar V et al (2019) Advanced functional structure-based sensing and imaging strategies for cancer detection: possibilities, opportunities, challenges, and prospects. Adv Funct Mater 29(16):1807859
36. Gerdan Z, Saylan Y, Denizli A (2024) Biosensing platforms for cardiac biomarker detection. ACS Omega 9(9):9946–9960
37. Wang C et al (2021) Point-of-care diagnostics for infectious diseases: from methods to devices. Nano Today 37:101092
38. Yadav S et al (2025) Nanobiosensors in neurodegenerative disease diagnosis: a promising pathway for early detection. Digit Health 11:20552076251342457
39. Tonelli D et al (2023) Focus review on nanomaterial-based electrochemical sensing of glucose for health applications. Nano 13. https://doi.org/10.3390/nano13121883

# Challenges, Limitations and Social Considerations

# Personalized Cancer Therapy: Technical and Clinical Challenges

Amrendra Singh, Laiba Noor, Vibhuti Joshi, and Arun Upadhyay

**Abstract**

Personalized cancer therapy represents a transformative approach to oncology, tailoring treatments based on individual genetic and molecular profiles. However, its implementation faces significant clinical and technical challenges that hinder widespread adoption. Multiple technical challenges, for example, problems in biomarker validation, tumor heterogeneity, and integration of patients' data, hinder its widespread application. The complex process of multi-omics, requirements for high-throughput sequencing, and the limitations of artificial intelligence prediction models further complicate the development of successful personalized medicines, forcing continuous adaptation of new strategies. Other challenges include, but are not limited to, accessibility barriers, high cost, and multiple regulatory constraints. The integration of biomarker-driven strategies into clinics needs robust validation protocols and new trial designs that can ensure efficacy. Likewise, immune evasion complicates responses to immunotherapy, particularly in "cold" tumors. Also, the ethical concerns regarding patients' genomic data privacy may further slow down the viability of personalized therapy. In this chapter, we explore the technical and clinical obstacles in personalized oncology, highlighting emerging tools, for example, AI-driven bio-

These authors contributed equally

A. Singh · A. Upadhyay (✉)
Department of Bioscience and Biomedical Engineering, Indian Institute of Technology Bhilai, Durg, Chhattisgarh, India
e-mail: arun@iitbhilai.ac.in

L. Noor · V. Joshi (✉)
Department of Biotechnology, Bennett University, Greater Noida, Uttar Pradesh, India
e-mail: vibhuti.joshi@bennett.edu.in



M. K. Ghosh, M. Basu (eds.), *Personalized Cancer Therapy*,
https://doi.org/10.1007/978-981-95-6845-1_18

marker discovery, patients' molecular profiling, and multiple adaptive treatment strategies. Addressing these challenges may play a crucial role in improving outcomes and advancing the field of precision oncology.

**Keywords**

Cancer · Personalized therapy · Cancer medicine

## Abbreviations

| | |
|---|---|
| AI | Artificial Intelligence |
| ACT | Adoptive cell therapy |
| BRAF | B-Raf proto-oncogene, serine/threonine kinase |
| CAR | Chimeric Antigen Receptor |
| EGFR | Epidermal Growth Factor Receptor |
| FDA | Food and Drug Administration |
| GIST | Gastrointestinal stromal tumor |
| HGP | Human Genome Project |
| HIV | Human Immunodeficiency Virus |
| IL | Interleukin |
| ICIs | Immune checkpoint inhibitors |
| MAPK | Mitogen-Activated Protein Kinase |
| mAB | Monoclonal antibodies |
| NK | Natural Killer cells |
| NSCLC | Non-Small Cell Lung Cancer |
| NGS | Next-Generation Sequencing |
| PI3K | Phosphoinositide 3-kinase |
| PLCγ | Phospholipase C gamma |
| PKC | Protein Kinase C |
| PM | Personalized Medicine |
| STAT | Signal Transducer and Activator of Transcription |
| TCR | T-cell receptor |
| Th1 | T helper 1 cell |
| TGFβ | Transforming Growth Factor beta |
| 5-AZA | 5-azacytidine |

## 1 Fundamentals of Personalized Medicine

Personalized medicine (PM), a revolutionary approach in medical science, is expanding beyond the treatment of just some diseases to becoming a preferred method for addressing a wide range of ailments. It is often used interchangeably with precision medicine; however, some argue that the two are distinct or that personalized medicine represents an earlier version of precision medicine, encompassing broader aspects of treatment. In simple terms, personalized medicine considers an individual's unique traits, including lifestyle and preferences, while precision medicine focuses on biological data like genes and proteins to target treatments more effectively [20, 74]. Several key elements contribute to personalized medicine, including an individual's genetics and genomics, epigenetic modifications, tissue markers, diagnostic methods, personality traits, behavior, environmental exposure, and access to healthcare [32]. In 2003, the successful completion of the Human Genome Project marked the first major step toward the development of personalized medicine. This was further strengthened by the International HapMap Project (2010), which provided comprehensive genetic data on various diseases [60].

Personalized medicine has the potential to minimize adverse patient outcomes, reduce overall treatment costs, and streamline the medication process, eliminating inefficiencies in the traditional trial-and-error approach [52]. The foundation of personalized medicine lies in the active alliance between patients and doctors, which can identify the root cause of disease and improve the outcomes of disease management [76]. Personalized medicine does not necessarily develop entirely new medications, therapeutics, or medical devices for each patient. Instead, it majorly focuses on utilizing the existing resources in the most effective combination to optimally address a patient's specific requirements by accurately identifying smaller subgroups of patients and tailoring their treatment strategies accordingly [71]. According to the International Consortium for Personalised Medicine (ICPerMed), successful applications of personalized medicine have been documented for various diseases and conditions, including Juvenile Idiopathic Arthritis (JIA), Familial Hypercholesterolemia (FH), cardiovascular diseases (CVDs), severe combined immunodeficiency (SCID), hemophilia, human immunodeficiency virus (HIV), and multiple forms of cancer. The US-FDA continues to push for the advancement in precision medicine, so over 20% of FDA-approved drugs each year belong to this category [34, 48, 68, 78].

Personalized medicine offers tailored treatment plans, which can enhance outcomes by utilizing therapeutic interventions optimally. It has impacted the treatment methodologies in diseases such as cancer, rare genetic disorders, and multiple infectious diseases. In recent years, PM has made a significant impact on cystic fibrosis and cancer treatment. In cystic fibrosis, precision medicine helped identify different mutations in the CFTR gene, which led to the development of drugs that can specifically target the defective proteins, improving patients' health [24]. In cancer, precision medicine has also made great promises by revolutionizing treatment by delivering therapies like targeted kinase inhibitors, immune-therapies, and personalized chemotherapy. These treatments are largely based on tumor genetics, not on its anatomical location. The advances of artificial intelligence (AI), particularly its immense success in big data analysis and widespread applications in medical science, may boost the impact of personalized medicine [66]. AI-powered tools and platforms can enable medical experts/clinicians to effectively categorize patients based on their genomic profiling, first by establishing correlations from existing cases and then determining the most suitable treatment regimen. The ability to analyze huge datasets for disease prediction and/or prevention is the most important contribution that AI-driven tools have made in this direction [61, 82]. While AI can play a major role in advancing PM, it also raises several concerns about patients' data privacy and security risks. Beyond data-related challenges, many other issues, for example, high cost, limited accessibility, and difficulties in clinical implementation, remain significant hurdles [53, 75].

Despite these obstacles, personalized medicine, driven by clinical and technological advancements, is reshaping healthcare, offering improved treatment outcomes with lower costs and fewer side effects. As highlighted earlier, it is increasingly favored in oncology, leading to the advancements like precision cancer therapies. In the following sections of this chapter, we will explore current personalized cancer treatments, the clinical and technical challenges they face, and their future prospects in the evolving landscape of precision oncology.

## 2 Personalized Therapy in Cancer

Integrating personalized medicine into cancer research holds great promise for advancing cancer treatment. High-throughput technologies, especially multi-omics approaches, like genomics, proteomics, and metabolomics, have uncovered a wide range of molecular alterations that can serve as potential targets for new drug development. Several compounds, either already approved or currently in clinical trials, are enabling more tailored and effective therapies [31]. Mutations in genes, such as PIK3CA, FGFR, BRAF, and NTRK, are some of the key alterations being targeted, facilitating slowing down the cancer progression. Targeted therapy approach has led to a major advancement by enabling treatment strategies, tailored to individual patients; for example, imatinib for CML (chronic myeloid leukemia), endocrine therapy for breast cancer, and EGFR inhibitors for lung cancer. Building on these advancements, personalized medicine can be broadly categorized into distinct therapeutic strategies, including targeted drug therapy and immunotherapy. As summarized in Fig. 1, many such examples highlight the transformative potential of precision medicine [54, 84].

### 2.1 Targeted Drug Therapy

Targeted therapies are based on blocking specific proteins or modifying biological pathways that are associated with cancer progression. These molecular targets, such as receptors, growth factors, kinases, or proteins, involved in apoptosis and angiogenesis, are often overexpressed or mutated in cancer cells [55]. Since these targets can differ even among individuals with the same cancer type, tumor samples obtained through biopsy or surgery are analyzed to identify actionable targets. This enables a more tailored and effective treatment strategy, enhancing therapeutic precision [41]. PIK3CA mutations are common changes identified in many solid tumors, which can directly affect the PIK3/AKT/mTOR signaling system [33].

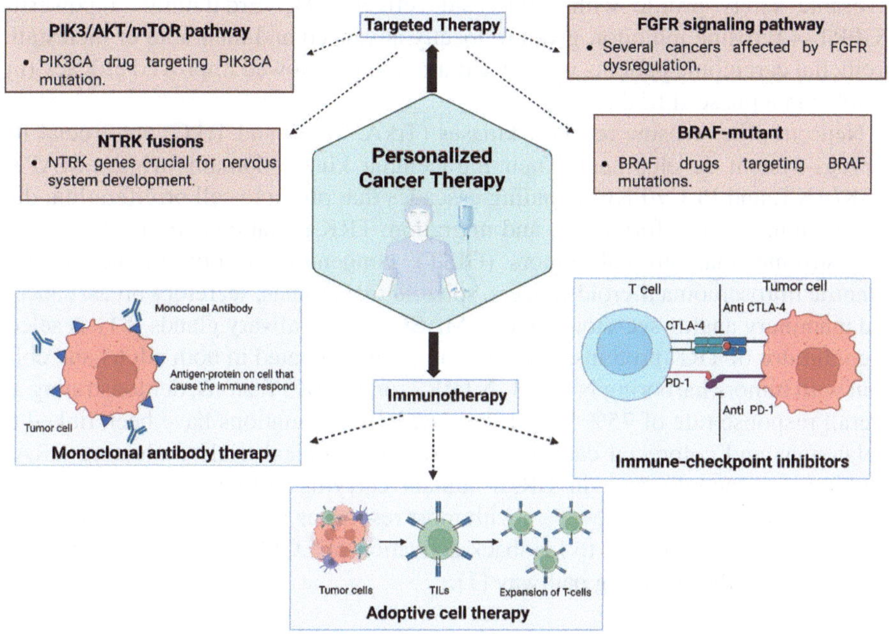


**Fig. 1** Schematic representation of various personalized cancer therapy approaches that include molecular and immune profiling to tailor treatments specific to individual tumor characteristics

Many clinical trials were conducted or are currently ongoing to evaluate the function of PIK3CA inhibitors in solid tumors. In a phase III trial involving patients with PIK3CA-mutant luminal breast cancer who had progressed on hormonal therapy, administration of alpelisib, a selective p110α PIK3CA inhibitor, in addition to fulvestrant, significantly enhanced antitumor efficacy [4]. Additionally, there are more targeted PI3K inhibitors currently in clinical trials, such as GDC-0077, MEN1611 (PA799) in breast cancer, and AMG319 in hematological malignancies, which have been evaluated for tumors harboring PIK3CA mutations [27, 38]. Similarly, AMG 319, disrupts regulatory T cell (Treg)-mediated immunosuppression and augments $CD8^+$ T-cell-mediated antitumor responses in hematological malignancies. On the other hand, CH5132799 is an oral, potent pan-PI3K inhibitor designed to broadly inhibit multiple PI3K isoforms, offering potential benefits in tumors with complex pathway dysregulation [8].

The fibroblast growth factor receptor (FGFR) signaling pathway is critical for cellular survival and proliferation. Activation of these receptors by fibroblast growth factors (FGFs) initiates downstream signaling cascades essential for various cellular functions [21]. Futibatinib (a FGFR2 inhibitor) has shown potential improvement in patients with bile duct cancer [36]. Fumibatinib, taken orally, caused a notable dose-dependent decrease in tumor growth in many FGFR-driven human tumor xenograft models [79]. Moreover, tumor reduction was linked with sustained, dose-dependent inhibition of FGFR. Compared to chemotherapy, the drug also demonstrated a more favorable safety profile with milder side effects [94]. Erdafatinib, a powerful FGFR1–4 tyrosine inhibitor, given to locally advanced and incurable or metastatic urothelial carcinoma patients with FGFR alterations, showed improved efficacy (up to 40%) in a phase II trial [46].

Neurotrophic tyrosine receptor kinases (TrkA, TrkB, and TrkC) are crucial for nervous-system development. Their intracellular kinase domains trigger MAPK, PI3K/AKT, and PLCγ/PKC signaling cascades that promote cell proliferation, differentiation, synapse formation, and migration. TRK mutations are most common in gastrointestinal stromal tumors (GIST), congenital mesoblastic nephromas, infantile fibrosarcoma thyroid cancers, spitzoid melanomas, secretory breast cancer, and mammary analog secretory cancer (MASC) of the salivary glands [51]. A selective inhibitor of TRK proteins, Larotrectinib, was evaluated in both adults and children with tumors harboring NTRK1, NTRK2, or NTRK3 fusions, demonstrating an overall response rate of 75% [23]. Likewise, BRAF mutations have been linked to melanomas and colorectal cancers [90]. Multiple clinical trials evaluating BRAF inhibitors as monotherapy in BRAF-mutant carrying colorectal cancer patients reported response rates below 5%, with most responses being short-lived. The limited efficacy was attributed to feedback activation of EGFR, which led to reactivation of the MAPK signaling pathway [1].

## 2.2 Immunotherapy

Cancer immunotherapy prepares the patient's immune system to recognize and target cancer cells specifically. It has become a transformative approach over the years, particularly in precision oncology. This promising intervention skillfully utilizes the patient's own immune cells, antibodies, and cytokines, along with vaccinations, to kill the cancer cells specifically, offering effective treatment strategies compared to traditional methods [47]. Various immunotherapy approaches developed so far include monoclonal antibodies, adoptive cell therapy (ACT), and cytokine therapies, which show great promise in cancer therapy.

Monoclonal antibodies (mAbs) are engineered to target cancer cell-specific antigens, making them a highly precise treatment option. However, while they are a key component of precision medicine, they are not necessarily personalized for each patient. Instead, monoclonal antibodies are developed to target specific biomarkers found in certain types of cancer, meaning they can be used for a broader group of patients who share similar molecular characteristics [91]. The novel "hybridoma" technology fuses lymphocytes and myeloma cell lines to develop novel cell lines that can produce antibodies [56]. The mAbs bind to specific antigens overexpressed by tumor cells, leading to tumor cell death through multiple pathways. The development of immune checkpoint inhibitors (ICIs) seems to be a promising immunotherapy method. Over the past few decades, the FDA has approved multiple drugs, including durvalumab, pembrolizumab, nivolumab, ipilimumab, avelumab, cemiplimab, and atezolizumab, which have demonstrated significant effectiveness in selected patients with advanced-stage disease. Some patients have even achieved durable complete remission, highlighting the transformative potential of ICIs in cancer treatment [22]. ICI therapy is mainly developed from the concept that T cells have evolutionarily conserved inhibitory signals, known as "checkpoints," which regulate T-cell activation [87].

Adoptive cell therapy (ACT) is a cutting-edge treatment strategy to specifically target and eliminate tumor cells. Immune cells collected from a patient's blood or tissue are expanded in vitro to be reinfused back into the patient's blood circulation [72]. These are genetically modified cells for recognizing and targeting tumor-specific antigens. ACT includes tumor-infiltrating lymphocyte (TIL) therapy, chimeric antigen receptor (CAR) T-cell therapy, engineered T-cell receptor (TCR) therapy, and natural killer (NK) cell therapy [92]. TIL-based ACT utilizes T-cells because they are known to infiltrate the patient's tumor naturally. Autologous cells (e.g., T-cells) are collected from the patient, and are expanded and activated in the lab, and then reinfused. This method has shown encouraging outcomes in the treatment of metastatic melanoma, nasopharyngeal, and cervical carcinoma [3, 14, 28, 81].

Another ACT known as CAR T-cell therapy uses genetically modified patient T-lymphocytes to identify and target antigens found on cancer cells. Adoptive T-cell therapy has also notably achieved high rates of durable complete response in hematologic malignancies, with treatment-resistant patients. The FDA has approved another CAR T-cell therapy, Kymriah™, for treating paediatric and young patients with refractory or relapsed B-cell precursor acute lymphoblastic leukemia. Similarly, Yescarta™ treatment is recommended for adults with diffuse large B-cell lymphoma [6, 11]. On the other hand, TCR therapy employs retroviruses to integrate a new TCR transgene into the T-cell genome, enabling the recognition of antigens that are highly expressed across various cancers [35]. TCR therapy has been implemented in both hematologic and solid tumors. Ongoing trials are examining therapeutic efficacy, toxicity related to treatment, and the binding affinity to tumor-specific antigens in selected patients with a high tumor burden [39, 58].

Similarly, NK cells, the cytotoxic lymphocytes, provide an appealing option for cancer treatment. Haploidentical NK cells combined with recombinant human interleukin-15 were used to treat relapsed/refractory acute myeloid leukemia; achieved complete remission in 32% of patients. Ongoing clinical trials are assessing CAR-NK cells for the treatment of hematologic (NCT03056339, NCT00995137) and solid tumors (NCT03656705, NCT03383978) [19, 26, 45, 50, 64]. Cytokine therapy is another breakthrough in immunotherapy. Cytokines are small proteins essential for cell signaling, have been engineered and used therapeutically to strengthen the immune system by precisely modulating immune responses [15]. IL-2 is one of the earliest and best-known molecules used for cytokine therapy. It is used to stimulate T cell activation and proliferation, particularly in the treatment of melanoma and metastatic renal cell carcinoma [30].

The development of IL-2 variants by advancements in cytokine engineering, such as the IL-2 "superkine," which binds more firmly to the IL-2 receptor on effector T cells and NK cells while simultaneously reducing the activation of regulatory T cells and boost their antitumor activity [42]. Moreover, NK cells have been engineered by attaching them to tumor-targeting antibodies. This fusion improves the targeting of IL-12 to tumor sites and overcomes several challenges in cytokine therapy. These immunocytokines mark a new age of cytokine therapeutics that combine the potent immune-stimulating properties of cytokines with the specificity of monoclonal antibodies to provide a more targeted treatment to cancer patients [83].

All the above-mentioned therapies depict the tremendous effort to introduce personalized cancer medicine by targeting multiple aspects of cancer, despite the fact that there is a lot that needs to be done before personalized medicine becomes the first choice of treatment for cancer patients. Further sections of this chapter elaborate on the challenges we are facing in this effort to make personalized cancer medicine popular.

### 2.3 Other Methods of Personalized Cancer Treatment Strategies

Multiple emerging personalized medicine approaches are currently being developed and used for cancer treatment by studying the tumor's unique molecular characteristics for each patient. For example, biomarker-guided chemotherapy refines traditional chemotherapeutic regimens by the identification of specific genetic markers. For instance, screening for DPYD gene mutations before prescribing 5-fluorouracil (5-FU) or capecitabine helps clinicians avoid severe toxicities, allowing dose customization or alternative therapies [17]. Similarly, in companion diagnostics, lab tests are developed alongside specific drugs to determine whether a patient benefits from a treatment, ensuring that the drugs are administered only to those with the appropriate molecular target [86]. In addition to small drug molecules, gene therapy and RNA-based treatments are other frontiers in cancer care, which apply to correct or silence aberrant genes responsible for oncogenesis [89]. These therapies hold immense promise for hematologic malignancies and rare genetic tumors.

Personalized cancer vaccines, particularly neoantigen-based vaccines, utilize sequencing methods to design immunogens based on the patient's tumor mutations. These tailored vaccines aim to prime the immune system to recognize and destroy malignant cells bearing unique neoantigens, offering a highly specific intervention with potentially minimal off-target effects [2, 43]. Liquid biopsy is another tool that enables real-time monitoring of tumor evolution through circulating tumor DNA (ctDNA) analysis from blood samples, facilitating early identification of resistance mutations, which supports timely therapeutic adjustments without performing invasive tissue biopsies [49, 65]. Altogether, these strategies explain a shift from population-averaged treatments toward individualized regimens, where each patient's genomic, transcriptomic, and immunologic signatures guide care decisions. However, the implementation of these innovations in real life faces challenges, including high costs, data interpretation complexity, and the need for advanced diagnostic infrastructure. In the following sections, we will discuss complications in adapting personalized medicine approaches and ways to address those obstacles.

## 3 Clinical Challenges in Personalized Cancer Therapy

Despite the progress made, personalized medicine faces challenges (see Table 1). The most notable issues are complex cancer genetics and tumor heterogeneity, which can easily diminish the effectiveness of targeted therapies [12]. Cancer cannot be defined as a single disease; instead, it is caused by a number of mutations and perturbance in multiple signaling pathways. Tumors show intratumoral heterogeneity, i.e., different areas of the same tumor can have different mutations, which makes it hard to effectively target all the relevant pathways and mechanisms [85]. The variability and heterogeneity obscure the treatment selection and may often lead to drug resistance. Intratumoral heterogeneity leads to drug resistance, making it difficult to achieve uniform therapeutic efficacy [16, 70]. In addition, the epigenetic and transcriptomic modifications create subpopulations of cells with varying drug sensitivities [59]. The tumor microenvironment contains varying oxygen levels, a highly diverse immune infiltration, and stromal interactions, which all may contribute to therapeutic resistance. Temporal evolution increases the complexity severalfold because tumors continuously adapt and thus render many effective therapies

**Table 1** Personalized medicine interventions: their challenges and possible solutions

| Method/strategy | Challenges | Solutions |
|---|---|---|
| Targeted therapy | Tumor heterogeneity leads to drug resistance | Leverage liquid biopsies to detect heterogeneity and guide combination therapies as first-line treatment |
| | Off-target effects and toxicity | Design highly selective agents and optimize dosing strategies to minimize systemic toxicity. |
| Immunotherapy | Immune evasion mechanisms by tumors | Combine with checkpoint inhibitors or immune adjuvants |
| | Limited efficacy in "cold" tumors lacking immune cell infiltration | Use immune modulators or tumor microenvironment remodeling agents |
| Biomarker-guided chemotherapy | Lack of validated predictive biomarkers | Establish large-scale biomarker validation cohorts<br>Integrate multi-omics data for predictive modeling |
| Companion diagnostics | Regulatory and cost barriers | Standardize protocols and incentivize integration in clinical trials |
| Gene therapy | Delivery efficiency and immune responses | Use viral and non-viral delivery systems with high specificity and low immunogenicity |
| RNA-based interventions | Instability and targeted delivery challenges | Employ chemical modifications to enhance RNA stability and specificity |
| Personalized cancer vaccines | Identification of tumor-specific neoantigens | Use AI and high-throughput sequencing to identify and validate neoantigens rapidly |
| Liquid biopsies | Sensitivity in detecting low-frequency mutations | Enhance assay sensitivity using PCR or NGS-based methods and validate across longitudinal patient samples |

less potent over time [18]. In order to overcome these challenges, many combination therapies have been designed and AI-driven precision medicine is currently explored to develop interventions based on tumor evolution over time [44].

In cancer cells, the drug resistance mainly develops by on-target or off-target mechanisms. In on-target resistance, cells develop mutations directly in the genes, which are intended targets of the drug, reducing the drug's effectiveness. For example, mutations in the EGFR gene make treatment ineffective in lung cancer [13]. Similarly, in off-target resistance, cancer cells compensate for the effects of the drug by modulating alternative pathways. The tumors can activate some survival pathways to bypass a blocked signaling pathway. For example, the MET D1228N mutation in non-small cell lung cancer (NSCLC) has been identified as a mechanism of acquired resistance to Crizotinib. The D1228N mutation alters the MET kinase domain, which reduces the drug's binding efficiency and causes a persistent activation of downstream signaling pathways [88, 95].

Immunotherapy is currently used as a treatment strategy. It basically strengthens the immune system. Interestingly, the cancer cells may develop resistance against immunotherapy as well. In fact, histological transformation could also be a factor leading to drug resistance. For example, NSCLC to SCLC transformation in EGFR-mutant lung adenocarcinoma may lead to diminishing the effects of the targeted therapy [62]. Likewise, epigenetic alterations also help in immune escape either by upregulating immunosuppressive pathways, like TGF-β, B7-H4, and IL-10, or by downregulating co-stimulatory molecules, such as CD28 and CD83, interfering with immune recognition [37, 93]. Notably, 5-azacytidine (a DNA methyltransferase inhibitor) enhances the tumor immunogenicity by upregulating antigen processing and cytokine expression [73]. It also restores MHC class I antigen presentation, increases T-cell infiltration, and stimulates interferon signaling, positioning epigenetic therapy as a promising approach to sensitize tumors to immunotherapy, particularly in immunosuppressive cancers [25, 77]. Similarly, CTLA-4 and DcR3 overexpression in prostate cancer leads to immune evasion in 90% of patients by suppressing T-cell activation and interfering with immune signaling, while PD-1 upregulation further blocks immune surveillance [9]. Although CTLA-4 and PD-1 blockade improves immune responses, the tumor's immunologically "cold" nature, characterized by low lymphocyte infiltration, makes immunotherapy less effective.

## 4 Technical Challenges in Personalized Cancer Medicine

Although personalized medicine holds great promise in cancer treatment, several technical challenges may further hinder its widespread implementation (see Fig. 2). For example, data collection presents a significant challenge in personalized medicine, primarily due to concerns regarding reliability, reproducibility, and accessibility. Accurate collection of patient data and innovation in diagnostic devices are essential for effective treatment. Moreover, variability in diagnostic machines introduces significant errors that can affect clinical data integrity and overall precision in biomarker identification, leading to inconsistencies in personalized treatment

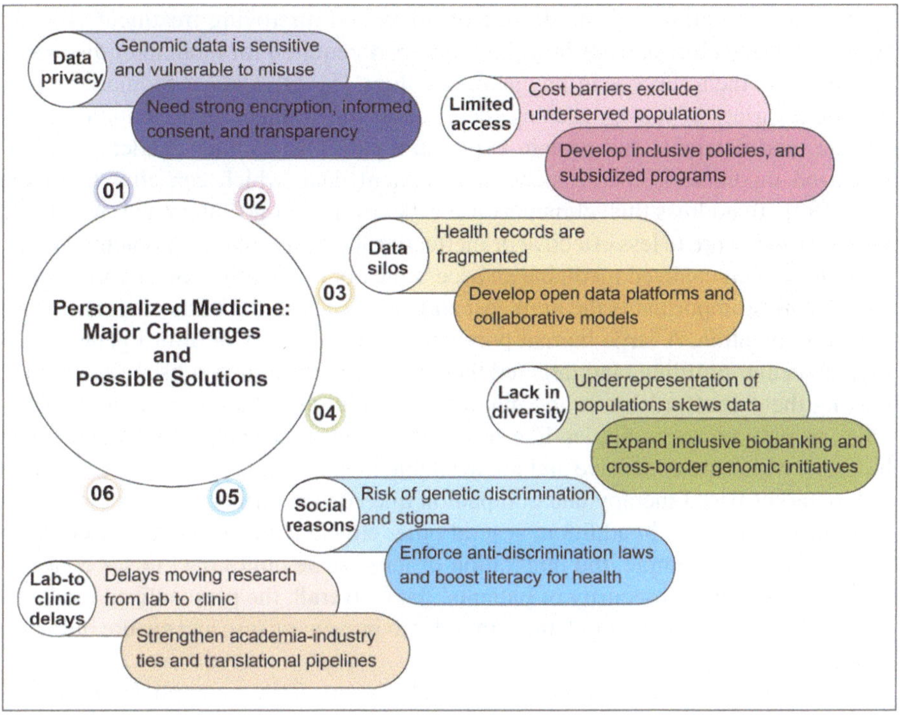


**Fig. 2** A summary of challenges and possible solutions of personalized medicine

strategies [57]. In addition, the health data varies in form, including observational records, clinical evaluations, procedural instructions, and human-generated inputs, each susceptible to errors. One of the most challenging aspects of personalized medicine is the evolution of data collection protocols, which change over time, influencing the accuracy and standardization of patient records [61]. Transitioning from paper to digital records may also introduce layers of complexity in data extraction as well as its interpretation. Sometimes, inconsistencies in questioning by a medical professional and responses recorded from patients can also lead to inaccuracies [7]. Another major concern is patients' data integrity, which remains at risk due to unstandardized frameworks. A common source code could be a possible solution to address at least a few of these issues. This practice also enhances transparency and validation [67].

Due to continuous evolution of tumor cells, their molecular tracking may be crucial for optimizing treatment strategies. A detailed molecular profiling could be accomplished by high-throughput functional genomic approaches for precise characterization of various mutations in the tumor [63]. Tumor progression is usually governed by one (or a few) driver mutations, which actively contribute to tumor growth and survival. Therefore, targeting the driver mutation containing gene (or protein) is very crucial for optimizing the treatment strategy [69]. During drug development in PM, accurate analysis and identification of biomarkers is also a crucial step in optimizing the dosage of drugs and improving treatment efficacy. However, many clinical trials lack the framework required for biomarker discovery, which limits their success in precision oncology [5, 29]. Cancer biomarkers may include a variety of factors causing the disease, for example, gene mutations, epigenetic modifications, and protein expression patterns. Another challenge in personalized medicine is inconsistencies in patient data, which can alter treatment plans [80]. To address this, clinicians use a "learn-as-you-go" approach, which can minimize exposure to less effective or ineffective therapies, reducing potential harm.

To make personalized medicine an accessible option, multiple other barriers persist. The most important is the high cost and its inaccessibility to the majority of the human population. A large human population cannot go for or afford genetic testing, molecular profiling, and targeted therapy, as they require a substantial amount, making them unaffordable to the majority. In addition to that, precision oncology relies heavily on technologies like next-generation sequencing (NGS) and liquid biopsies, which are expensive and not available in many parts of the globe. The cost of biomarker-based therapy and companion diagnostics further restricts their overall affordability [10]. In addition, several other ethical concerns pose barriers to its application. For example, the integration of large-scale omics data raises concerns about the privacy and security of patients' data. Overall, the personalized medicine approach needs stringent regulations in order to prevent misuse and ensure informed consent.

Several technological and developmental barriers further complicate the adoption, which include but are not limited to diagnostic tools, like integrated multi-omics, single-cell sequencing, and biomarker discovery. Companion diagnostics is another approach that assists in PM; it includes tests that can identify patients who are likely to benefit from the proposed targeted therapies [40]. Lack of standardized protocols for real-time molecular data integration into decision-making processes adds to the complexity and hence delays the progress in precision oncology. Many insurance providers may not cover these treatments, mainly due to their high costs and uncertain outcomes in terms of long-term benefits. One major obstacle in transitioning from conventional treatments to PM is underdeveloped health infrastructure and lack of specialized training for oncologists. All these challenges collectively make the implementation of personalized medicine therapy difficult and unfeasible.

## 5 Conclusions and Future Direction

Precision oncology is a transformative approach that aims to understand the disease with accuracy at a much finer detail than traditional methods. It not only reads illnesses from observable symptoms but also incorporates high-throughput technologies, such as multi-omics, and AI-based tools for making critical decisions. It looks into the genetic factors, molecular features, and biological components of an individual's pathology to segregate and define its uniqueness. By doing so, clinicians/doctors can formulate treatment plans in a highly precise manner, leading to improved outcomes and fewer side effects. Advances in omics technologies are the core of personalized medicine. Our ability to read and analyze the entire genome of the patient has opened the door for precision medicine that relies on understanding and pinpointing individual mutations contributing to the onset or driving the progression of diseases. PM also extends beyond omics and includes many other technologies, such as advanced imaging, bioinformatics, and AI-driven tools that can read large datasets and help make decisions in a quick time. Altogether, using multiple technologies on one platform enables the segregation of subgroups within diseases that would otherwise be treated uniformly. This way, we can develop a targeted approach to improve the effectiveness of therapies. PM matches the biology of the patient's specific condition instead of a "one-size-fits-all" model.

In summary, by weaving together advances in molecular diagnostics, targeted therapeutics, and precision immunology, personalized medicine is developing a new course in oncology, one where treatment is not merely disease-specific, but person-specific. The challenges are there, but a combined approach from different domains and the incorporation of AI and other tools may help address many of those. More research is required to cut down costs and make the technology accessible to those living in underdeveloped and low-income countries or those who cannot afford these treatments. Many steps need to be taken at the policy level, so the new advances in cancer treatment can be made affordable and accessible to millions of those who are suffering from cancer.

**Declarations**

**Author's Contributions** A.U. and V.J. conceived the idea and review structure. A.S. and L.N. wrote the manuscript. A.N., L.N., V.J., and A.U. revised and edited the manuscript. All the authors have read, agreed, and approved the final draft of the manuscript.

**Funding** A.U. thanks the Indian Institute of Technology, Bhilai, for providing the Research Initiation Grant (IITBH/R&D/2024–25/386). V.J. thanks Bennett University for providing the seed grant to conduct research.

**Data Availability** Figure 1 has been drawn using the Biorender (https://www.biorender.com) platform.

**Conflict of Interest** The authors have no conflict of interest.

## References

1. Aiman W, Ali MA, Jumean S, Asfeen U, Garcia J, Quirem M, Ahmad A, Rayad MN, Alkhlaifat O, Al Omour B, Chemarthi VS, Maroules M, Guron G, Shaaban H (2023) BRAF inhibitors in BRAF-mutated colorectal cancer: a systematic review. J Clin Med 13(1). https://doi.org/10.3390/jcm13010113
2. Aljabali AAA, Hamzat Y, Alqudah A, Alzoubi L (2025) Neoantigen vaccines: advancing personalized cancer immunotherapy. Explor Immunol 5:1003190. https://doi.org/10.37349/ei.2025.1003190

3. Andersen R, Donia M, Ellebaek E, Borch TH, Kongsted P, Iversen TZ, Hölmich LR, Hendel HW, Met Ö, Andersen MH, Thor Straten P, Svane IM (2016) Long-lasting complete responses in patients with metastatic melanoma after adoptive cell therapy with tumor-infiltrating lymphocytes and an attenuated IL2 regimen. Clin Cancer Res 22(15):3734–3745. https://doi.org/10.1158/1078-0432.Ccr-15-1879
4. André F, Ciruelos E, Rubovszky G, Campone M, Loibl S, Rugo HS, Iwata H, Conte P, Mayer IA, Kaufman B, Yamashita T, Lu YS, Inoue K, Takahashi M, Pápai Z, Longin AS, Mills D, Wilke C, Hirawat S, Juric D (2019) Alpelisib for PIK3CA-mutated, hormone receptor-positive advanced breast cancer. N Engl J Med 380(20):1929–1940. https://doi.org/10.1056/NEJMoa1813904
5. Antoniou M, Kolamunnage-Dona R, Wason J, Bathia R, Billingham C, Bliss JM, Brown LC, Gillman A, Paul J, Jorgensen AL (2019) Biomarker-guided trials: challenges in practice. Contemp Clin Trials Commun 16:100493. https://doi.org/10.1016/j.conctc.2019.100493
6. Awasthi R, Maier HJ, Zhang J, Lim S (2023) Kymriah® (tisagenlecleucel) - an overview of the clinical development journey of the first approved CAR-T therapy. Hum Vaccin Immunother 19(1):2210046. https://doi.org/10.1080/21645515.2023.2210046
7. Ayaad O, Alloubani A, ALhajaa EA, Farhan M, Abuseif S, Al Hroub A, Akhu-Zaheya L (2019) The role of electronic medical records in improving the quality of health care services: comparative study. Int J Med Inform 127:63–67. https://doi.org/10.1016/j.ijmedinf.2019.04.014
8. Blagden S, Omlin A, Josephs D, Stavraka C, Zivi A, Pinato DJ, Anthoney A, Decordova S, Swales K, Riisnaes R, Pope L, Noguchi K, Shiokawa R, Inatani M, Prince J, Jones K, Twelves C, Spicer J, Banerji U (2014) First-in-human study of CH5132799, an oral class I PI3K inhibitor, studying toxicity, pharmacokinetics, and pharmacodynamics, in patients with metastatic cancer. Clin Cancer Res 20(23):5908–5917. https://doi.org/10.1158/1078-0432.Ccr-14-1315
9. Bou-Dargham MJ, Sha L, Sang QA, Zhang J (2020) Immune landscape of human prostate cancer: immune evasion mechanisms and biomarkers for personalized immunotherapy. BMC Cancer 20(1):572. https://doi.org/10.1186/s12885-020-07058-y
10. Bourgeois A, Horrill T, Mollison A, Stringer E, Lambert LK, Stajduhar K (2024) Barriers to cancer treatment for people experiencing socioeconomic disadvantage in high-income countries: a scoping review. BMC Health Serv Res 24(1):670. https://doi.org/10.1186/s12913-024-11129-2
11. Brown CE, Alizadeh D, Starr R, Weng L, Wagner JR, Naranjo A, Ostberg JR, Blanchard MS, Kilpatrick J, Simpson J, Kurien A, Priceman SJ, Wang X, Harshbarger TL, D'Apuzzo M, Ressler JA, Jensen MC, Barish ME, Chen M et al (2016) Regression of glioblastoma after chimeric antigen receptor T-cell therapy. N Engl J Med 375(26):2561–2569. https://doi.org/10.1056/NEJMoa1610497
12. Burrell RA, McGranahan N, Bartek J, Swanton C (2013) The causes and consequences of genetic heterogeneity in cancer evolution. Nature 501(7467):338–345. https://doi.org/10.1038/nature12625
13. Chhouri H, Alexandre D, Grumolato L (2023) Mechanisms of acquired resistance and tolerance to EGFR targeted therapy in non-small cell lung cancer. Cancers (Basel) 15(2). https://doi.org/10.3390/cancers15020504
14. Comoli P, Pedrazzoli P, Maccario R, Basso S, Carminati O, Labirio M, Schiavo R, Secondino S, Frasson C, Perotti C, Moroni M, Locatelli F, Siena S (2005) Cell therapy of stage IV nasopharyngeal carcinoma with autologous Epstein-Barr virus-targeted cytotoxic T lymphocytes. J Clin Oncol 23(35):8942–8949. https://doi.org/10.1200/jco.2005.02.6195
15. Conlon KC, Miljkovic MD, Waldmann TA (2019) Cytokines in the treatment of cancer. J Interf Cytokine Res 39(1):6–21. https://doi.org/10.1089/jir.2018.0019
16. de Bruin EC, McGranahan N, Mitter R, Salm M, Wedge DC, Yates L, Jamal-Hanjani M, Shafi S, Murugaesu N, Rowan AJ, Grönroos E, Muhammad MA, Horswell S, Gerlinger M, Varela I, Jones D, Marshall J, Voet T, Van Loo P et al (2014) Spatial and temporal diversity in genomic instability processes defines lung cancer evolution. Science 346(6206):251–256. https://doi.org/10.1126/science.1253462

17. de Moraes FCA, de Almeida Barbosa AB, Sano VKT, Kelly FA, Burbano RMR (2024) Pharmacogenetics of DPYD and treatment-related mortality on fluoropyrimidine chemotherapy for cancer patients: a meta-analysis and trial sequential analysis. BMC Cancer 24(1):1210. https://doi.org/10.1186/s12885-024-12981-5
18. de Visser KE, Joyce JA (2023) The evolving tumor microenvironment: from cancer initiation to metastatic outgrowth. Cancer Cell 41(3):374–403. https://doi.org/10.1016/j.ccell.2023.02.016
19. Del Zotto G, Marcenaro E, Vacca P, Sivori S, Pende D, Della Chiesa M, Moretta F, Ingegnere T, Mingari MC, Moretta A, Moretta L (2017) Markers and function of human NK cells in normal and pathological conditions. Cytometry B Clin Cytom 92(2):100–114. https://doi.org/10.1002/cyto.b.21508
20. Delpierre C, Lefèvre T (2023) Precision and personalized medicine: what their current definition says and silences about the model of health they promote. Implication for the development of personalized health. Front Sociol 8:1112159. https://doi.org/10.3389/fsoc.2023.1112159
21. Dieci MV, Arnedos M, Andre F, Soria JC (2013) Fibroblast growth factor receptor inhibitors as a cancer treatment: from a biologic rationale to medical perspectives. Cancer Discov 3(3):264–279. https://doi.org/10.1158/2159-8290.Cd-12-0362
22. Dobosz P, Dzieciątkowski T (2019) The intriguing history of cancer immunotherapy. Front Immunol 10:2965. https://doi.org/10.3389/fimmu.2019.02965
23. Drilon A, Laetsch TW, Kummar S, DuBois SG, Lassen UN, Demetri GD, Nathenson M, Doebele RC, Farago AF, Pappo AS, Turpin B, Dowlati A, Brose MS, Mascarenhas L, Federman N, Berlin J, El-Deiry WS, Baik C, Deeken J et al (2018) Efficacy of Larotrectinib in TRK fusion-positive cancers in adults and children. N Engl J Med 378(8):731–739. https://doi.org/10.1056/NEJMoa1714448
24. Dugger SA, Platt A, Goldstein DB (2018) Drug development in the era of precision medicine. Nat Rev Drug Discov 17(3):183–196. https://doi.org/10.1038/nrd.2017.226
25. Ebelt ND, Zuniga E, Johnson BL, Diamond DJ, Manuel ER (2020) 5-Azacytidine potentiates anti-tumor immunity in a model of pancreatic ductal adenocarcinoma. Front Immunol 11:538. https://doi.org/10.3389/fimmu.2020.00538
26. Ehrend E, Manns P, Herkt SC, Lee S, Romanski A, Burger M, Steinbach J, Wels WS, Bönig H, Tonn T (2024) Real world product quality data of car-engineered Nk cells for intracranial injection in patients with recurrent Her2-positive glioblastoma from the ongoing Car2brain clinical trial. Cytotherapy *26*(6, Supplement):S9–S10. https://doi.org/10.1016/j.jcyt.2024.03.017
27. Fiascarelli A, Merlino G, Capano S, Talucci S, Bisignano D, Bressan A, Bellarosa D, Carrisi C, Paoli A, Bigioni M, Tunici P, Irrissuto C, Salerno M, Arribas J, de Stanchina E, Scaltriti M, Binaschi M (2023) Antitumor activity of the PI3K δ-sparing inhibitor MEN1611 in PIK3CA mutated, trastuzumab-resistant HER2 + breast cancer. Breast Cancer Res Treat 199(1):13–23. https://doi.org/10.1007/s10549-023-06895-2
28. Forget MA, Haymaker C, Hess KR, Meng YJ, Creasy C, Karpinets T, Fulbright OJ, Roszik J, Woodman SE, Kim YU, Sakellariou-Thompson D, Bhatta A, Wahl A, Flores E, Thorsen ST, Tavera RJ, Ramachandran R, Gonzalez AM, Toth CL et al (2018) Prospective analysis of adoptive TIL therapy in patients with metastatic melanoma: response, impact of anti-CTLA4, and biomarkers to predict clinical outcome. Clin Cancer Res 24(18):4416–4428. https://doi.org/10.1158/1078-0432.Ccr-17-3649
29. Fountzilas E, Tsimberidou AM (2018) Overview of precision oncology trials: challenges and opportunities. Expert Rev Clin Pharmacol 11(8):797–804. https://doi.org/10.1080/17512433.2018.1504677
30. Fyfe G, Fisher RI, Rosenberg SA, Sznol M, Parkinson DR, Louie AC (1995) Results of treatment of 255 patients with metastatic renal cell carcinoma who received high-dose recombinant interleukin-2 therapy. J Clin Oncol 13(3):688–696. https://doi.org/10.1200/jco.1995.13.3.688
31. Gambardella V, Tarazona N, Cejalvo JM, Lombardi P, Huerta M, Roselló S, Fleitas T, Roda D, Cervantes A (2020) Personalized medicine: recent progress in cancer therapy. Cancers (Basel) 12(4). https://doi.org/10.3390/cancers12041009
32. Goetz LH, Schork NJ (2018) Personalized medicine: motivation, challenges, and progress. Fertil Steril 109(6):952–963. https://doi.org/10.1016/j.fertnstert.2018.05.006

33. Goncalves MD, Hopkins BD, Cantley LC (2018) Phosphatidylinositol 3-kinase, growth disorders, and cancer. N Engl J Med 379(21):2052–2062. https://doi.org/10.1056/NEJMra1704560
34. Gooderham MJ, Pink AE, Simpson EL, Silverberg JI, Güler E, Watkins M (2023) Abrocitinib 100 mg once daily for moderate-to-severe atopic dermatitis: a review of efficacy and safety, and expert opinion on use in clinical practice. Dermatol Ther (Heidelb) 13(9):1893–1907. https://doi.org/10.1007/s13555-023-00948-6
35. Govers C, Sebestyén Z, Coccoris M, Willemsen RA, Debets R (2010) T cell receptor gene therapy: strategies for optimizing transgenic TCR pairing. Trends Mol Med 16(2):77–87. https://doi.org/10.1016/j.molmed.2009.12.004
36. Goyal L, Meric-Bernstam F, Hollebecque A, Valle JW, Morizane C, Karasic TB, Abrams TA, Furuse J, Kelley RK, Cassier PA, Klümpen HJ, Chang HM, Chen LT, Tabernero J, Oh DY, Mahipal A, Moehler M, Mitchell EP, Komatsu Y et al (2023) Futibatinib for FGFR2-rearranged intrahepatic cholangiocarcinoma. N Engl J Med 388(3):228–239. https://doi.org/10.1056/NEJMoa2206834
37. Grosche L, Knippertz I, König C, Royzman D, Wild AB, Zinser E, Sticht H, Muller YA, Steinkasserer A, Lechmann M (2020) The CD83 molecule - an important immune checkpoint. Front Immunol 11:721. https://doi.org/10.3389/fimmu.2020.00721
38. Hanan EJ, Braun MG, Heald RA, MacLeod C, Chan C, Clausen S, Edgar KA, Eigenbrot C, Elliott R, Endres N, Friedman LS, Gogol E, Gu XH, Thibodeau RH, Jackson PS, Kiefer JR, Knight JD, Nannini M, Narukulla R et al (2022) Discovery of GDC-0077 (Inavolisib), a highly selective inhibitor and degrader of mutant PI3Kα. J Med Chem 65(24):16589–16621. https://doi.org/10.1021/acs.jmedchem.2c01422
39. Johnson LA, Morgan RA, Dudley ME, Cassard L, Yang JC, Hughes MS, Kammula US, Royal RE, Sherry RM, Wunderlich JR, Lee CC, Restifo NP, Schwarz SL, Cogdill AP, Bishop RJ, Kim H, Brewer CC, Rudy SF, VanWaes C et al (2009) Gene therapy with human and mouse T-cell receptors mediates cancer regression and targets normal tissues expressing cognate antigen. Blood 114(3):535–546. https://doi.org/10.1182/blood-2009-03-211714
40. Kang SL, Kwon JY, Kim SM (2024) Insights into post-marketing clinical validation of companion diagnostics with reference to the FDA, EMA, PMDA, and MFDS. Mol Ther Methods Clin Dev 32(4):101346. https://doi.org/10.1016/j.omtm.2024.101346
41. Krzyszczyk P, Acevedo A, Davidoff EJ, Timmins LM, Marrero-Berrios I, Patel M, White C, Lowe C, Sherba JJ, Hartmanshenn C, O'Neill KM, Balter ML, Fritz ZR, Androulakis IP, Schloss RS, Yarmush ML (2018) The growing role of precision and personalized medicine for cancer treatment. Technology (Singap World Sci) 6(3–4):79–100. https://doi.org/10.1142/s2339547818300020
42. Levin AM, Bates DL, Ring AM, Krieg C, Lin JT, Su L, Moraga I, Raeber ME, Bowman GR, Novick P, Pande VS, Fathman CG, Boyman O, Garcia KC (2012) Exploiting a natural conformational switch to engineer an interleukin-2 'superkine'. Nature 484(7395):529–533. https://doi.org/10.1038/nature10975
43. Li X, You J, Hong L, Liu W, Guo P, Hao X (2023) Neoantigen cancer vaccines: a new star on the horizon. Cancer Biol Med 21(4):274–311. https://doi.org/10.20892/j.issn.2095-3941.2023.0395
44. Liao J, Li X, Gan Y, Han S, Rong P, Wang W, Li W, Zhou L (2022) Artificial intelligence assists precision medicine in cancer treatment. Front Oncol 12:998222. https://doi.org/10.3389/fonc.2022.998222
45. Liu E, Marin D, Banerjee P, Macapinlac HA, Thompson P, Basar R, Nassif Kerbauy L, Overman B, Thall P, Kaplan M, Nandivada V, Kaur I, Nunez Cortes A, Cao K, Daher M, Hosing C, Cohen EN, Kebriaei P, Mehta R et al (2020) Use of CAR-transduced natural killer cells in CD19-positive lymphoid tumors. N Engl J Med 382(6):545–553. https://doi.org/10.1056/NEJMoa1910607
46. Loriot Y, Necchi A, Park SH, Garcia-Donas J, Huddart R, Burgess E, Fleming M, Rezazadeh A, Mellado B, Varlamov S, Joshi M, Duran I, Tagawa ST, Zakharia Y, Zhong B, Stuyckens K, Santiago-Walker A, De Porre P, O'Hagan A et al (2019) Erdafitinib in locally advanced

or metastatic urothelial carcinoma. N Engl J Med 381(4):338–348. https://doi.org/10.1056/NEJMoa1817323

47. Luo N (2023) Advances in targeted immunotherapy in cancers. Int J Mol Sci 24(24). https://doi.org/10.3390/ijms242417475
48. Ma C, Chang J, Yu B (2024a) Sunlenca® (Lenacapavir): A first-in-class, long-acting HIV-1 capsid inhibitor for treating highly multidrug-resistant HIV-1 infection. Acta Pharm Sin B 14(12):5512–5514. https://doi.org/10.1016/j.apsb.2024.08.009
49. Ma L, Guo H, Zhao Y, Liu Z, Wang C, Bu J, Sun T, Wei J (2024b) Liquid biopsy in cancer current: status, challenges and future prospects. Signal Transduct Target Ther 9(1):336. https://doi.org/10.1038/s41392-024-02021-w
50. Marin D, Li Y, Basar R, Rafei H, Daher M, Dou J, Mohanty V, Dede M, Nieto Y, Uprety N, Acharya S, Liu E, Wilson J, Banerjee P, Macapinlac HA, Ganesh C, Thall PF, Bassett R, Ammari M et al (2024) Safety, efficacy and determinants of response of allogeneic CD19-specific CAR-NK cells in CD19(+) B cell tumors: a phase 1/2 trial. Nat Med 30(3):772–784. https://doi.org/10.1038/s41591-023-02785-8
51. Märkl B, Hirschbühl K, Dhillon C (2019) NTRK-fusions - a new kid on the block. Pathol Res Pract 215(10):152572. https://doi.org/10.1016/j.prp.2019.152572
52. Mathur S, Sutton J (2017) Personalized medicine could transform healthcare. Biomed Rep 7(1):3–5. https://doi.org/10.3892/br.2017.922
53. Mawkili WA (2025) The future of personalized medicine in Saudi Arabia: opportunities and challenges. Saudi Med J 46(1):19–25. https://doi.org/10.15537/smj.2025.46.1.20240700
54. Millner LM, Strotman LN (2016) The future of precision medicine in oncology. Clin Lab Med 36(3):557–573. https://doi.org/10.1016/j.cll.2016.05.003
55. Min HY, Lee HY (2022) Molecular targeted therapy for anticancer treatment. Exp Mol Med 54(10):1670–1694. https://doi.org/10.1038/s12276-022-00864-3
56. Mitra S, Tomar PC (2021) Hybridoma technology; advancements, clinical significance, and future aspects. J Genet Eng Biotechnol 19(1):159. https://doi.org/10.1186/s43141-021-00264-6
57. Molla G, Bitew M (2024) Revolutionizing personalized medicine: synergy with multi-omics data generation, main hurdles, and future perspectives. Biomedicine 12(12). https://doi.org/10.3390/biomedicines12122750
58. Morgan RA, Dudley ME, Wunderlich JR, Hughes MS, Yang JC, Sherry RM, Royal RE, Topalian SL, Kammula US, Restifo NP, Zheng Z, Nahvi A, de Vries CR, Rogers-Freezer LJ, Mavroukakis SA, Rosenberg SA (2006) Cancer regression in patients after transfer of genetically engineered lymphocytes. Science 314(5796):126–129. https://doi.org/10.1126/science.1129003
59. Mroz EA, Rocco JW (2017) The challenges of tumor genetic diversity. Cancer 123(6):917–927. https://doi.org/10.1002/cncr.30430
60. O'Shaughnessy KM (2006) HapMap, pharmacogenomics, and the goal of personalized prescribing. Br J Clin Pharmacol 61(6):783–786. https://doi.org/10.1111/j.1365-2125.2006.02683.x
61. Onciul R, Tataru CI, Dumitru AV, Crivoi C, Serban M, Covache-Busuioc RA, Radoi MP, Toader C (2025) Artificial intelligence and neuroscience: transformative synergies in brain research and clinical applications. J Clin Med 14(2). https://doi.org/10.3390/jcm14020550
62. Oser MG, Niederst MJ, Sequist LV, Engelman JA (2015) Transformation from non-small-cell lung cancer to small-cell lung cancer: molecular drivers and cells of origin. Lancet Oncol 16(4):e165–e172. https://doi.org/10.1016/s1470-2045(14)71180-5
63. Özdoğan M, Papadopoulou E, Tsoulos N, Tsantikidi A, Mariatou VM, Tsaousis G, Kapeni E, Bourkoula E, Fotiou D, Kapetsis G, Boukovinas I, Touroutoglou N, Fassas A, Adamidis A, Kosmidis P, Trafalis D, Galani E, Lypas G, Orhan B et al (2021) Comprehensive tumor molecular profile analysis in clinical practice. BMC Med Genet 14(1):105. https://doi.org/10.1186/s12920-021-00952-9
64. Pan S, Wang F, Jiang J, Lin Z, Chen Z, Cao T, Yang L (2023) Chimeric antigen receptor-natural killer cells: a new breakthrough in the treatment of solid tumours. Clin Oncol (R Coll Radiol) 35(3):153–162. https://doi.org/10.1016/j.clon.2022.10.019

65. Pandey S, Yadav P (2025) Liquid biopsy in cancer management: integrating diagnostics and clinical applications. Pract Lab Med 43:e00446. https://doi.org/10.1016/j.plabm.2024.e00446
66. Parekh AE, Shaikh OA, Simran, Manan S, Hasibuzzaman MA (2023) Artificial intelligence (AI) in personalized medicine: AI-generated personalized therapy regimens based on genetic and medical history: short communication. Ann Med Surg (Lond) 85(11):5831–5833. https://doi.org/10.1097/ms9.0000000000001320
67. Pasquier T, Lau MK, Trisovic A, Boose ER, Couturier B, Crosas M, Ellison AM, Gibson V, Jones CR, Seltzer M (2017) If these data could talk. Sci Data 4(1):170114. https://doi.org/10.1038/sdata.2017.114
68. Perche PO, Cook MK, Feldman SR (2023) Abrocitinib: a new FDA-approved drug for moderate-to-severe atopic dermatitis. Ann Pharmacother 57(1):86–98. https://doi.org/10.1177/10600280221096713
69. Peters L, Venkatachalam A, Ben-Neriah Y (2024) Tissue-predisposition to cancer driver mutations. Cells 13(2). https://doi.org/10.3390/cells13020106
70. Ramirez M, Rajaram S, Steininger RJ, Osipchuk D, Roth MA, Morinishi LS, Evans L, Ji W, Hsu CH, Thurley K, Wei S, Zhou A, Koduru PR, Posner BA, Wu LF, Altschuler SJ (2016) Diverse drug-resistance mechanisms can emerge from drug-tolerant cancer persister cells. Nat Commun 7:10690. https://doi.org/10.1038/ncomms10690
71. Redekop WK, Mladsi D (2013) The faces of personalized medicine: a framework for understanding its meaning and scope. Value Health 16(6 Suppl):S4–S9. https://doi.org/10.1016/j.jval.2013.06.005
72. Rosenberg SA, Restifo NP (2015) Adoptive cell transfer as personalized immunotherapy for human cancer. Science 348(6230):62–68. https://doi.org/10.1126/science.aaa4967
73. Roulois D, Loo Yau H, Singhania R, Wang Y, Danesh A, Shen SY, Han H, Liang G, Jones PA, Pugh TJ, O'Brien C, De Carvalho DD (2015) DNA-demethylating agents target colorectal cancer cells by inducing viral mimicry by endogenous transcripts. Cell 162(5):961–973. https://doi.org/10.1016/j.cell.2015.07.056
74. Shlyakhto EV (2022) Scientific basics of personalized medicine: realities and opportunities. Her Russ Acad Sci 92(6):671–682. https://doi.org/10.1134/s1019331622060041
75. Sigman M (2018) Introduction: personalized medicine: what is it and what are the challenges? Fertil Steril 109(6):944–945. https://doi.org/10.1016/j.fertnstert.2018.04.027
76. Simmons LA, Wolever RQ, Bechard EM, Snyderman R (2014) Patient engagement as a risk factor in personalized health care: a systematic review of the literature on chronic disease. Genome Med 6(2):16. https://doi.org/10.1186/gm533
77. Símová J, Polláková V, Indrová M, Mikyšková R, Bieblová J, Stěpánek I, Bubeník J, Reiniš M (2011) Immunotherapy augments the effect of 5-azacytidine on HPV16-associated tumours with different MHC class I-expression status. Br J Cancer 105(10):1533–1541. https://doi.org/10.1038/bjc.2011.428
78. Smith RA, Raugi DN, Nixon RS, Seydi M, Margot NA, Callebaut C, Gottlieb GS (2024) Antiviral activity of Lenacapavir against human immunodeficiency virus type 2 (HIV-2) isolates and drug-resistant HIV-2 mutants. J Infect Dis 229(5):1290–1294. https://doi.org/10.1093/infdis/jiad562
79. Sootome H, Fujita H, Ito K, Ochiiwa H, Fujioka Y, Ito K, Miura A, Sagara T, Ito S, Ohsawa H, Otsuki S, Funabashi K, Yashiro M, Matsuo K, Yonekura K, Hirai H (2020) Futibatinib is a novel irreversible FGFR 1-4 inhibitor that shows selective antitumor activity against FGFR-deregulated tumors. Cancer Res 80(22):4986–4997. https://doi.org/10.1158/0008-5472.Can-19-2568
80. Stefanicka-Wojtas D, Kurpas D (2023) Personalised medicine-implementation to the health-care system in Europe (focus group discussions). J Pers Med 13(3). https://doi.org/10.3390/jpm13030380
81. Stevanović S, Draper LM, Langhan MM, Campbell TE, Kwong ML, Wunderlich JR, Dudley ME, Yang JC, Sherry RM, Kammula US, Restifo NP, Rosenberg SA, Hinrichs CS (2015) Complete regression of metastatic cervical cancer after treatment with human papillomavirus-

targeted tumor-infiltrating T cells. J Clin Oncol 33(14):1543–1550. https://doi.org/10.1200/jco.2014.58.9093
82. Taherdoost H, Ghofrani A (2024) AI's role in revolutionizing personalized medicine by reshaping pharmacogenomics and drug therapy. Intell Pharm 2(5):643–650. https://doi.org/10.1016/j.ipha.2024.08.005
83. Tatsumi T, Takehara T, Yamaguchi S, Sasakawa A, Miyagi T, Jinushi M, Sakamori R, Kohga K, Uemura A, Ohkawa K, Storkus WJ, Hayashi N (2007) Injection of IL-12 gene-transduced dendritic cells into mouse liver tumor lesions activates both innate and acquired immunity. Gene Ther 14(11):863–871. https://doi.org/10.1038/sj.gt.3302941
84. Tsimberidou AM, Fountzilas E, Nikanjam M, Kurzrock R (2020) Review of precision cancer medicine: evolution of the treatment paradigm. Cancer Treat Rev 86:102019. https://doi.org/10.1016/j.ctrv.2020.102019
85. Upadhyay A (2021) Cancer: an unknown territory; rethinking before going ahead. Genes Dis 8(5):655–661. https://doi.org/10.1016/j.gendis.2020.09.002
86. Valla V, Alzabin S, Koukoura A, Lewis A, Nielsen AA, Vassiliadis E (2021) Companion diagnostics: state of the art and new regulations. Biomark Insights 16:11772719211047763. https://doi.org/10.1177/11772719211047763
87. Waldman AD, Fritz JM, Lenardo MJ (2020) A guide to cancer immunotherapy: from T cell basic science to clinical practice. Nat Rev Immunol 20(11):651–668. https://doi.org/10.1038/s41577-020-0306-5
88. Wang Y, Chen Z, Han X, Li J, Guo H, Shi J (2021) Acquired MET D1228N mutations mediate Crizotinib resistance in lung adenocarcinoma with ROS1 fusion: a case report. Oncologist 26(3):178–181. https://doi.org/10.1002/onco.13545
89. Yadav S, Shekhawat M, Jahagirdar D, Kumar Sharma N (2017) Natural and artificial small RNAs: a promising avenue of nucleic acid therapeutics for cancer. Cancer Biol Med 14(3):242–253. https://doi.org/10.20892/j.issn.2095-3941.2017.0038
90. Yaeger R, Chatila WK, Lipsyc MD, Hechtman JF, Cercek A, Sanchez-Vega F, Jayakumaran G, Middha S, Zehir A, Donoghue MTA, You D, Viale A, Kemeny N, Segal NH, Stadler ZK, Varghese AM, Kundra R, Gao J, Syed A et al (2018) Clinical sequencing defines the genomic landscape of metastatic colorectal cancer. Cancer Cell 33(1):125–136.e123. https://doi.org/10.1016/j.ccell.2017.12.004
91. Zahavi D, Weiner L (2020) Monoclonal antibodies in cancer therapy. Antibodies (Basel) 9(3). https://doi.org/10.3390/antib9030034
92. Zaretsky JM, Garcia-Diaz A, Shin DS, Escuin-Ordinas H, Hugo W, Hu-Lieskovan S, Torrejon DY, Abril-Rodriguez G, Sandoval S, Barthly L, Saco J, Homet Moreno B, Mezzadra R, Chmielowski B, Ruchalski K, Shintaku IP, Sanchez PJ, Puig-Saus C, Cherry G et al (2016) Mutations associated with acquired resistance to PD-1 blockade in melanoma. N Engl J Med 375(9):819–829. https://doi.org/10.1056/NEJMoa1604958
93. Zeng Z, Wei F, Ren X (2020) Exhausted T cells and epigenetic status. Cancer Biol Med 17(4):923–936. https://doi.org/10.20892/j.issn.2095-3941.2020.0338
94. Zhou Y, Peng S, Wang H, Cai X, Wang Q (2024) Review of personalized medicine and pharmacogenomics of anti-cancer compounds and natural products. Genes (Basel) 15(4). https://doi.org/10.3390/genes15040468
95. Zhu J, Chen J, Liu W, Zhang J, Gu Y (2024) Mutation of MET D1228N as an acquired potential mechanism of Crizotinib resistance in NSCLC with MET Y1003H mutation. Lung Cancer (Auckl) 15:123–128. https://doi.org/10.2147/lctt.S467584

# Successfully Financing Personalized Cancer Drug Therapy *for All*: A Social Policy Perspective

Christian Aspalter

**Abstract**

The paper first delivers a reality check and lessons about current choices, perceived ones and not yet envisioned but very efficient and practical ones, in deciding for the financing method and financing origins with regard to personalized cancer drug therapies. People's health is invaluable. Personalized cancer drug therapies are getting relatively super cheap only if they are financed and owned by the state entirely. Market approaches will exclude most of those people who need personalized cancer drug therapies and then most of them will die (unnecessarily prematurely). There is a choice between (1) deaths of millions versus (2) market profits and misplaced/mis-perceived market beliefs. This article by a global social policy scholar and economist clearly and only chooses to help (and make sure to) prevent the deaths of millions, as much as possible, with the policy choices, financing method choices, at hand—following fully the principles of feasibility, efficiency and most important of all effectiveness.

**Keywords**

Social policy · Cancer drugs · Financing · Financial & behavioural incentives

C. Aspalter (✉)
Department of Social Sciences, Beijing Normal-Hong Kong Baptist University, Zhuhai, China
e-mail: christian@bnbu.edu.cn

 
M. K. Ghosh, M. Basu (eds.), *Personalized Cancer Therapy*,
https://doi.org/10.1007/978-981-95-6845-1_19

This chapter puts forward a generally new approach in normative (evaluative-cum-prescriptive) social policy theory, that of axiomatic derivation of particular social policy solutions. The purpose here is not double-checking, but rather it is pedagogical and explanatory in axiomatic terms. It shows the underlying, supportive logic behind some new, some formerly not thought about policy solutions and their strategies, whose outcomes fully work, but whose political and intellectual support has *generally not yet* been established. This is a new approach to convey, or also possibly devise, new and sometimes very non-traditional and thus very path-breaking solutions, paradigms and their strategies.

The same method of axiomatic derivation (or 'the method of working with axioms') can be running (as usual in mathematics and quantum physics) forward (or 'upward'), from axioms to theories, or from established theories backward ('downward') to their underlying axioms, or again first backward and then forward (down and then up) to perhaps not only check on the fundamental axioms, but also to potentially find new theorems or new theories in the process of applying this down-up method. This method can be applied especially in causal (explanatory) theory building (e.g. [4, 5]), but also as here in normative (prescriptive) theory building.

In normative social policy, there is no such thing as an insurmountable problem. If a problem is insurmountable for any expert, this depends on the lack of solutions known to or accepted by the expert/scientist, or the lack of innovative capacity to form and devise new policy solutions, or, as in our case, new financing, ownership and management solutions. The real problem only comes with a lack of political will [6] to implement the solutions prescribed and the implementation itself (e.g. delivery systems, corruption, etc.).

At times, time paired with patience is needed to think all options through, and explore all options perceivable (cf. Aristotle cited in Bauböck [10]: 41), this applies, e.g. to both (a) technological unemployment across the globe due to new mass high-end digital solutions, new AI applications especially when paired with broadly accessible new quantum computation capabilities (cf. [7]), and (b) the seemingly impossible mission to find trillions of dollars for personal cancer drug therapy for all people that are being diagnosed to have cancer, or could/would if they were checked for it, in large countries and across the world, as is our mission and purpose here.

In science, mission-based development policy is the causal reason for scientists not only to be able to devise ways of going to the Moon and the Mars, for staying alive during the journey, and to make it back to Earth safely, but also to make this happen in formerly seemingly impossible time limit, and with formerly impossible seeming budget/financing constraints. Having set up and accepted a 'grand mission' first will open up ways to finance, ways to manage formerly thought impossible human feats.

Mission-oriented normative economic policy theory has been made now known worldwide by world-leading economist Mariana Mazzucato [29]. This is the way forward also for social policy, where the government has the greatest innovation and implementation powers in terms of efficiency and effectiveness by way of economies of scale (esp. when talking about large, and super-large countries) (cf. [26]). This in the past has included the invention of the Internet, space technologies, and fundamental research, like CERN in Geneva, or new scaled-up solutions or solutions that only work when government supports and/or mandates, or finances, them—e.g. provident fund systems in the case of Singapore [25, 48] or universal basic income [2].

In normative social policy theory, the many findings and lessons provided by Mariana Mazzucato have not been debated and hence not been understood yet (cf., e.g. [50]).

**Lesson 1: 'Private Versus Public' Debate/Logic Is Superfluous and Hence Entirely Unwarranted.** Mazzucato [26] was very clear, the lines between public and private are blurry, they are intertwined, and to increasing degrees indistinguishable, when looked at closer. Even though, if one would and could determine this distinction with ease, it does not and would not matter in health care, education, infrastructure, medicine, public safety, the achievement of long and healthy lives. Who cares the building is owned by whom, the public or a private owner, when one is learning in an educational institution or when one is having a surgery? It is not part of education/learning or health care: the ownership, who owns what. The artificial distinction between public and private, be it both financing and provision, or just one of the two, is distracting and obstructing policy making, the discourse in and the research focuses chosen, the conclusions not reached and reached in those research papers/studies, and hence, of course, it also obstructs social policy suggestions, proposed and implemented solutions, and for sure also the outcomes. In health care policy, it is always best if the government can control the management and provision of health care institutions in order to cut health care administration costs, and hence not inflate, but reduce and prevent overall administrative and other costs in health care provision (e.g. esp. the cases of health care system effectiveness and efficiency in Singapore and Taiwan, and the case of already super-aged Japan, when compared to the United States with 17% of GDP spent on health care, cf. [9, 8, 38, 1]).

**Lesson 2: Socializing Risks and Privatizing Rewards Is Neoliberal Madness and a Crime Against the Poor and People in Need.** It cannot be stated in clearer words, as Mazzucato did back in 2013. The government provides research, by financing education and research, especially fundamental research of all kind, including especially also of course medical and pharmaceutical research. Land given to hospitals, hospitals built, public loans and finance provided, subsidies, research grants, salaries of scientists in public institutions, all these are given by the public, i.e. paid by the taxpayer, which means the working and middle classes, as the rich hide their wealth and incomes, in trust funds, overseas, in companies, stock and property ownership. The outcomes are known, when privatization takes place in health care, social insurance, criminal justice, etc. and elsewhere: a robbery is taking place from the poor and the working classes who pay taxes at their grocery store, at the gas station, on their salaries and investments, to highest possible degree, in many overlaying multiple layers, in devastating cumulative fashion over time. Companies profit immensely, after the general public has made their tax contributions. The profits of the super-rich directly stem from these very tax contributions, from management fees, profit taking, highest salaries and bonuses paid to executives, and lowest salaries paid to workers (cf., e.g. [11–13, 30]). All this happens, and still subsides (and bailouts) from the public coffers are handed out to constantly sweeten the deal, the massive robbery of public taxes paid, in form of (a) *privatized services/licenses provided* (the research for which was enabled and financed by governmental subsidies and direct government finance of education and research institutions and their staff salaries, including indirect government financing of underlying fundamental research) and/or (b) *publicly mandated and publicly protected services and goods provided*, and (c) *the publicly ensured monopolies (or oligopolies) thereof.*

**Lesson 3: How to Value Health, Non-Illness, Happiness and Long Healthy Lives.** Mazzucato [27] took a magnifying glass perspective and inspected the confusions that are created in the capital system, selling destruction of property value creation as profit or success. The same also holds true for the incarceration of millions, for profit of many super-rich and their companies and corporations [31, 40], and with it there is the destruction of lives, their spirit and competencies/capabilities, and most important the killing of opportunities and eternal condemnation to exposure to drugs and crime, in and then out of (and back and forth from to) prisons. Health is invaluable, health of populations is invaluable. Quality of life paired with long lives, long safe and happy lives, is invaluable (cf. [39]). The constant mantra of GDP, i.e. the sum of (usually just estimated) economic transactions in a certain geographical area in a year, is killing people around the world, literally by not being able to afford checkups, drugs and therapies. It is also killing countless more millions, every year around the globe, by way of denying them 'value' in public discourse, the political and scientific/professional discourse, along which financial decisions, policy decisions are being made—thus perpetuating millions of premature *otherwise-preventable* deaths and miseries.

**Lesson 4: The Mission to Save Lives Is the Mission.** Millions of children die every year of hunger around the world. Billions of people live in poverty. Tens of millions of people die because they cannot get medicine or medical treatment. *It does not have to be like this*! Mazzucato's [28] book holds up the embodiment of that very conclusion. If global social and political will is there, if we come around and realize, that all this does not have to be, that we chase ghosts in form of imagined myths of hidden efficiency in market places, in companies, in corporations that determine and coin their efficiency in terms how much money they can rip off from the government, from the taxes paid, to be paid, fees collected and mandated, from the general population, the families and their children, from the poor and the working poor, and all of the middle classes in any perceivable way (cf., [19]).

As our job is not to please capitalism, or GDP, we need to learn what GDP is all about. Government spending is also GDP, transactions by governments, transactions mandated by governments are all part of GDP. The welfare state is GDP. But GDP is not welfare, or the welfare state, of course. Also, we need to understand that we do not need to make the super-super-rich extra super-super-rich. We need to find finance for personalized cancer drug therapy, lots of it (trillions), and the problem is not finding the money, but seeing it, grasping the possibility that it is right there in front of the government's eyes, but '*custom*' works against this, as history and old habits of thinking are travelling in our veines and in our brains and culture ([32, 33]). There is lack of preparedness to look into the other, another, direction, when one needs to look for money. This is called the dominant capitalist common sense [17], these are the dominant power relations [14], and dominant human entanglements.

Benefits are not the answer, 300 or 3,000 Euro here or there is not the answer. We need tens of thousands of Euros for each cancer patient. Piketty [34–36] has told what is going on, Holcombe [19, 20] has why. The super-super-rich and the super-powerful play the game of extraction, resources from the masses, this system only benefits the super-super-rich, it is very complex, and hides in plain sight (cf. also [44–46 and 24]). Why the middle class cannot afford personalized cancer drugs, why the government forces itself to accept exorbitantly high market prices for cancer drugs, why the government sponsors fundamental research and let all the large pharmaceutical and medical corporations in the medical-industrial complex reap all the sky-high profits? The super-super-rich want it, and they learned to control the ways of government officers, politicians, anyone who graduates from an economics program at university—their thinking, their discourse, their evaluation procedures, and thus their subsequent decisions and actions.

Nobel Prize winner Paul A. Samuelson wrote, 'I don't care who writes a nation's laws—or crafts its advanced treatises—if I can write its economics textbooks' (cited in [42]). This is referring especially to his 1948 textbook that changed how the post–World War II Western world thought and thinks about economic exchange, the economics system and the imagined role of markets [43], and the allowed/sanctioned role of governments [18].

Based on the above said, we may proceed with our logic (logic is always subjective, semantics are always subjective, even when based on or mingled with mathematics, which in itself is a mix of the subjective and the objective at best, cf. [15]).

*Axiom 1*: When provided by 'privately controlled' markets, personalized cancer drug therapies provided by corporations that are owned and controlled by the super-super-rich will be super expensive, in the range of tens of thousands and hundreds of thousands of Euros [49].

*Axiom 2*: If personalized cancer drugs therapies are provided on super massive scales, universally for free for all, then the variable costs are being lowered to very affordable (for society, for the government) levels, as fix costs (including especially, e.g. research and development costs) are being divided up by additional tens of millions of personalized cancer drug therapy cases (per year), and fix costs make up the vast proportion of overall costs.

*Axiom 3*: In 'privately controlled' markets, people who need cancer drugs, for the very most part, cannot afford them.

*Axiom 4*: When, by way of forcing people to pay the *highest* social insurance contributions/taxes or mandated private insurance contributions/taxes, money is taken on a grand scale to finance cancer drug therapies owned by corporations of the super-super-rich in 'privately controlled' markets, economies are slowing down or reversing their growth trends (into negative territory). As labour costs get too expensive, people lose their jobs *en mass*, companies close *en mass*, and yet many more people are losing their jobs. Social insurance contributions have the same effects as taxes, they are taxes. Then, when this happens, people immediately have again less money to save for health care expenditures, to live a healthy life, and a spiralling downward trend in society, economy and in terms of health outcomes is the result. The commanding case of Argentina in the 1990s shall be a lesson what too high social insurance costs do to economic development: economic collapse, plus social and political unrest.

*Axiom 5*: People with cancer who do not get cancer drugs, will for the very most part die.

*Axiom 6*: If governments subscribe blindly to capitalism, millions of people in each part of the world die every year very much preventable deaths.

*Axiom 7*: If scientists and government officers, and politicians included, are not looking at other options, they will not consider, or understand them, and for sure not choose them.

*Axiom 8*: Governments need to think out of the mental box (that is currently full of neoliberal ideas) to solve the conundrum of financing large-scale enterprise of developing (researching), producing and delivering personalized cancer drug therapies.

**Solution 1** Governments do it themselves, that is, governments take ownership of patents and/or the companies that hold them, and governments grant rights to all its citizens for free-of-charge cancer drug therapies which are then fully financed by taxing the super-super-rich—(1) with a *marginal net* wealth tax over, e.g. 50 million Euros, taxed on past wealth, a one-time tax, of, e.g. up to 90 or 100%; and/or personal *marginal* income tax on the super-rich over five million Euros per year of up to around 80%; and/or (2) with a tax on all capital transactions—i.e. with a marginal tax of each transaction over, e.g. one million Euro, of, e.g. 1% (including currency exchange and any bank or private transfers), and/or (3) nationalizing the medical-industrial complex, then these (newly) nationally owned entities are run 100-percent like private corporations/companies in terms of efficiency and cost-saving, and maximizing production output and quality. Here, executives are kept in place, with relative moderate salaries however, and with or without moderate bonuses, while keeping the competitive spirit of performance amongst them alive. That is, this changes the ownership but not management. The costs for personalized cancer drug therapies are hence minimized, especially as now all cancer patients are eligible for drugs that are free of charge, because of (a) the economies of scale, and because of (b) saving costs for profits, highest salaries and bonuses, marketing, and fancy administrative costs, and this cost saving is making drug therapies per unit (per person) much cheaper. The amplitude of these cost saving effects mostly depends on the size of the country, its cancer rates, and the amount of fix costs (for developing the drugs mainly).

**Solution 2** Using the powers of the government to lower prices, by central negotiation by one government agency (for all of the country, for all health care providers) immediate and strict (low) price caps, as well as quick introduction of generic drugs (cf., e.g. [16, 23]).

**Solution 3** Thinking about other solutions (communism, neoliberalism, mercantilism, utopianism, dystopianism, etc.), and then, subsequently, circling back to *Solution 1* (as '*all lives*' must be saved/rescued, especially also in all poor countries or countries with large poor or very poor populations) or *Solution 2* (*if* health insurance systems or health care provident fund systems are well developed and *if* the country is relatively rich, and *if* the government is choosing *not* to save scores/millions of people from very much preventable deaths by providing free-of-charge personalized cancer drug therapies for all people with cancer).

## 1 Conclusions

There is no Left or Right (cf. [19, 20]). There is solving the problem, or not solving the problem. There is no black and white (no 'good and evil'). It is 'not' about choosing communism versus rule by the super-super-rich on the conditions set by the super-super-rich and the super-powerful (that are controlled by the super-super-rich, the billionaires control politicians and no the other way around). There are other options, winning and successful option, such as e.g. lowering the taxes for the working classes and lowering the taxes for the small and medium enterprises. This very well can be financed (paid for) by taxing the super-super-rich and the super-rich. It turns 'poverty- and misery-blind capitalism' into 'poverty and misery-preventing capitalism', where all the work and investment incentives are fully in place.

The right low percentages for taxation and even, as suggested by the author elsewhere, no taxation at all (cf. [2, 3]) are 'enabled' by e.g. fully (in the fullest sense of the word) taxing the net wealth and the net incomes of the super-super-rich and the super-rich, and/or taxing capital transactions, and so forth.

Middle of the road, gives half of cancer rates and cancer deaths that are preventable—not the full prevention of all cancer that can be prevented. Thus, middle of the road solutions are not recommendable, as they are not ethical. Neoclassical solutions are also unethical and hence not acceptable in any way, when it comes to cancer drug therapies in general, as with only a few cancer drug providers in the market, each and every one of them is de facto a 'monopolist', as the doctor only can decide which drug to use first, and only can decide the order of switching from the first to the last of the couple of general cancer drugs that are available today (cf., e.g. [37, 41]).

We have to think out of the box, as Huang, Sherraden and Sherraden ([21, 22]) did, financing for health care needs to become a *public good*, while they had provident fund systems in mind, we here in this article have the financing of personalized cancer drug therapies in mind. By changing ownership, we have changed financing, and turned cancer drugs into a public good, for the good of society. With this different ownership structure in place, the economy nevertheless is a vibrant and still fully capitalist economy—i.e. transactions (GDP) will be in fact higher, with more people living longer, more salaries are being paid, and more (government) investments being made, and the production of personalized cancer drug therapies being scaled up massively.

To be sure, more people are spending more (people living longer spend more), there are more salaries paid and more investments made Plus there is a greater scale of cancer drug production. These are all positive and directly causal factors for increasing levels of GDP, which signifies nothing less and nothing more than the sum of 'all economic transactions' in a given geographic area in a given time.

Personalized and other cancer drug therapies now can be fully provided to all in need, based on public financing by taxing the net wealth and the net income super-rich and the super-super-rich, and capital transfers of the rich, and by public ownership of the very same companies, with the very same managers (mostly) and the very same scientists working on the very same cancer drug therapies. The ownership of the companies are now in the hands of governments, profits are not needed for the super-super-rich, but company results are used just as before, as incentives for the management to do their jobs of controlling costs (this works in Singapore; see, e.g. Temasek, which is a state-owned entity, but run like a private corporation; cf. [47]).

## References


1. Aspalter C (2007) The Asian cure for health care. Far Eastern Econ Rev 170(9):56–59
2. Aspalter C (2023) Ten worlds of welfare capitalism: a global data analysis. Springer, Cham
3. Aspalter C (2025) Quantum economics: the new economics of communication, power and power relations. Edward Elgar, Cheltenham
4. Aspalter C (2026a) Evolutionary social policy: its 8 axioms and 6 theorems. In: Aspalter C (ed) Elgar research encyclopedia of social policy. Edward Elgar, Cheltenham
5. Aspalter C (2026b) Standard axioms of the forces of human coordination: a quantum economics point of view. In: Aspalter C (ed) Elgar research encyclopedia of social policy. Edward Elgar, Cheltenham
6. Aspalter C (2026c) Political will and social policy. In: Aspalter C (ed) Elgar research encyclopedia of social policy. Edward Elgar, Cheltenham
7. Aspalter C (2026d) AI and social policy. In: Aspalter C (ed) Elgar research encyclopedia of social policy. Edward Elgar, Cheltenham
8. Aspalter C, Teguh-Pripadi K, Gauld R (2017) Health care systems in developing countries in Asia. Routledge, London
9. Aspalter C, Uchida Y, Gauld R (2012) Health care systems in Europe and Asia. Routledge, London
10. Bauböck R (2008) Normative political theory and empirical research. In: Porta D, Keating M (eds) Approaches and methodologies in the social sciences. Cambridge University Press, Cambridge
11. Borzutsky S (1991) The Chicago boys, social security and welfare in Chile. In: Glennerster H, Midgley J (eds) The radical right and the welfare state. Barnes and Noble, Savage
12. Borzutsky S (1997) Privatizing social security: relevance of the Chilean experience. In: Midgley J, Sherraden M (eds) Alternatives to social security. Auburn House, London
13. Borzutsky S (2002) Vital connections: politics, social security, and inequality in Chile. University of Notre Dame Press, Notre Dame
14. Foucault M (1975) Surveiller et punir. Gallimard, Paris
15. Frenkel E (2023) Infinity, string theory, death, the self, theories of everything, podcast by Curt Jaimungal. https://www.youtube.com/watch?v=n_oPMcvHbAc
16. Goldstein DA et al (2016) Global differences in cancer drug prices: a comparative analysis. J Clin Oncol 34(18). https://doi.org/10.1200/JCO.2016.34.18_suppl.LBA6500
17. Gramsci A (2011) Prison notebooks, Vol. 1-3. Columbia University Press, New York
18. Hayek FAv (1960) The constitution of liberty. University of Chicago Press, Chicago
19. Holcombe RG (2018) Political capitalism: how political influence is made and maintained. Cambridge University Press, Cambridge
20. Holcombe RG (2022) Capitalism, cronyism, and inequality. In: Manish GP, Miller SC (eds) Capitalism and inequality: the role of state and market. Routledge, Oxon
21. Huang J, Sherraden M, Sherraden MS (2021a) We have clean water and clean air: why not clean finance, too? A vision for inclusion and equity. In: Boshara R, Rademacher A (eds) The future of building wealth: brief essays on the best ideas to build wealth—for everyone. Federal Reserve Bank and Aspen Institute. https://futureofwealth.org/wp-content/uploads/2021/09/Sec-2.pdf
22. Huang J, Sherraden MS, Sherraden M (2021b) Toward finance as a public good, CSD Working Paper, No. 21-03, Center for Social Development, Washington University. https://doi.org/10.7936/p8dd-p256
23. Janssen Daalen JM et al (2021) Determinants of drug prices: a systematic review of comparison studies. BMJ Open 11(7):e046917. https://doi.org/10.1136/bmjopen-2020-046917
24. Jayasuriya S (2017) The end of the Kuznets Curve: explaining Piketty's argument. https://impact.monash.edu/economy
25. Low L, Aw T-C (1997) Housing a healthy, educated, and wealthy nation through the CPF. The Institute of Policy Studies, Singapore

26. Mazzucato M (2013) The entrepreneurial state: debunking public vs. private sector myths. Anthem, London
27. Mazzucato M (2018) The value of everything: making and taking in the global economy. Public Affairs, New York
28. Mazzucato M (2021) Mission economy: a moonshot guide to changing capitalism. Penguin, London
29. Mazzucato M (2023) Governing the economics of the common good: from correcting market failures to shaping collective goals. J Econ Policy Reform. https://doi.org/10.1080/17487870.2023.2280969
30. Mazzucato M, Collington R (2023) The big con: how the consulting industry weakens our businesses, infantilizes our governments and warps our economies. Penguin, London
31. Nathan S (2025) Prison privatization in the United Kingdom. https://www.iprt.ie/site/assets/files/5894/cpch_14__prison_privatisation_in_the_uk.pdf
32. Nietzsche FW (1887) Zur Genealogie der Moral: Eine Streitschrift. C.G. Naumann, Leipzig
33. Nietzsche FW (2008) [1878] Human, all too human: a book for free spirits. Wordsworth Editions, Hertfordshire
34. Piketty T (2006) The Kuznets curve yesterday and tomorrow. In: Banerjee AV, Benabou R, Mookherjee D (eds) Understanding poverty. Oxford University Press, Oxford
35. Piketty T (2014) [2013] Capital in the twenty-first century. Harvard University Press, Cambridge
36. Piketty T (2020) Capital and ideology. Harvard University Press, Cambridge
37. Rajkumar SV (2020) The high cost of prescription drugs: causes and solutions. Blood Cancer J 10(6):71. https://doi.org/10.1038/s41408-020-0338-x
38. Reid TR (2010) The healing of America: a global quest for better, cheaper, and fairer health care. Penguin, New York
39. Rojo-Pérez F, Fernández-Mayoralas M (2021) Handbook of active aging and quality of life. Springer, Cham
40. Scott D, Coyle M (eds) (2020) The Routledge international handbook of penal abolition. Routledge, London
41. Siddiqui M, Rajkumar SV (2012) The high cost of cancer drugs and what we can do about it. Mayo Clin Proc 87(10):935–943. https://doi.org/10.1016/j.mayocp.2012.07.007
42. Siegel R (2009) Economist Paul Samuelson remembered. https://www.npr.org/2009/12/14/121433434/economist-paul-samuelson-remembered
43. Stiglitz JE (2002) There is no invisible hand. The Guardian, December 20
44. Stiglitz JE (2010) Freefall: America, free markets, and the sinking of the world economy. W.W. Norton, New York
45. Stiglitz JE (2019) People, power, and profits. W.W. Norton, New York
46. Stiglitz JE (2024) The road to freedom. W.W. Norton, New York
47. Temasek (2025) website, https://www.temasek.com.sg/en/about-us/corporate-governance
48. Teo Y (2023) The Singaporean welfare state system: a special focus on public housing and the Central Provident Fund. In: Aspalter C (ed) The Routledge international handbook to welfare state systems: towards global social policy science. Routledge, Oxon
49. Vokinger KN (2020) Determinants of cancer drug pricing and how to overcome the cancer premium. Cell 186:1528–1531
50. Wendt C, Bambra C (2020) From ideal types to health care systems typologies. In: Aspalter C (ed) Ideal types in comparative social policy. Routledge, Oxon

# Role of Health Workers in Personalized Medicine

Mausam Prajapati, Palmi Modi, and Bhumi M. Shah

**Abstract**

Personalized medicine, also known as precision medicine, is reshaping modern healthcare by moving away from the traditional standardized treatment model toward customized care tailored to individual patients. This approach integrates cutting-edge technologies like genomics, proteomics, and bioinformatics to design prevention, diagnosis, and treatment strategies that align with a person's unique genetic profile, lifestyle, and environmental exposures. The goal is to deliver more accurate, efficient, and effective healthcare interventions that meet the specific needs of each patient. Health professionals, including nurses, pharmacists, physicians, genetic counselors, and community health workers, play vital roles in its implementation. Working collaboratively, they develop and deliver personalized care plans, enhance disease detection, optimize therapeutic responses, and provide continuous monitoring and support. Interdisciplinary collaboration is crucial for effective communication, coordination, and care delivery. Despite its promise, personalized medicine faces challenges such as limited access to advanced technologies, ethical concerns about genetic data privacy, and disparities in healthcare access, particularly for underserved populations. Additionally, many health workers lack adequate training in precision medicine, highlighting the need for improved education programs. Future directions focus on expanding training, developing supportive policies, and engaging communities to ensure broader access. As genomic innovations continue, personalized medicine holds the potential to significantly enhance healthcare outcomes and reduce health disparities globally, provided current barriers are effectively addressed.

M. Prajapati · P. Modi · B. M. Shah (✉)
Department of Pharmaceutical Chemistry, L.J. Institute of Pharmacy, L.J. University, Ahmedabad, Gujarat, India



M. K. Ghosh, M. Basu (eds.), *Personalized Cancer Therapy*,
https://doi.org/10.1007/978-981-95-6845-1_20

**Keywords**

Personalized medicine · Health worker · Genomic Medicine · Interdisciplinary collaboration · Precision healthcare · Patient-centered care

## 1 Introduction

> *Personalized medicine (PM)* is a medical approach that supports healthcare professionals in designing treatment strategies tailored to each patient's unique characteristics, including their genetic makeup, lifestyle, and environmental factors. It aims to enhance therapeutic effectiveness while minimizing side effects by accounting for individual variability in genes, proteins, and other biological markers.

In recent years, PM has received considerable attention in both academic research and public health discourse. Its primary objective is to enhance the efficiency of healthcare by delivering tailored treatments based on an individual's unique genetic and biological profile. Although the importance of PM has been increasingly recognized and accompanied by substantial investments from public research institutions and pharmaceutical companies, its global adoption remains limited. Fundamentally, it involves customizing treatment based on a patient's unique genetic, molecular, and clinical profile [1]. The primary aim of personalized medicine is to integrate genomic data for disease prediction, diagnosis, and prevention, thereby guiding informed therapeutic decisions. This approach is further supported by advancements in molecular diagnostics and genomic technologies, enabling a deeper understanding of the human genome and exome. These developments have significantly contributed to the emergence of "Systems Medicine" and "Systems Biology," enhancing the accuracy and precision of diagnostic and therapeutic strategies [2]. PM involves the adaptation of dosage forms through various innovative approaches, such as 3D-printed tablets, film-based formulations, mini pills, and more. Notably, many of these technologies are already in clinical use, particularly in the field of immunotherapy. Such innovations hold significant potential for improving treatment outcomes and may even lead to complete disease remission [3].

### 1.1 Importance of Health Workers in Precision Medicine

The rapid advancements in precision medicine are transforming current and future healthcare systems. Central to the successful implementation of these scientific breakthroughs are healthcare professionals, whose contributions are indispensable in translating research into clinical practice. Health workers occupy a pivotal role across various domains of precision medicine. Among their many responsibilities, data collection is fundamental. Clinicians and nurses systematically gather patients' medical histories and biological samples, which serve as critical inputs for genomic research.

Genetic counsellors play a key role in interpreting genomic test results, helping patients comprehend the implications for their health and guiding disease management decisions. In clinical settings, doctors and pharmacists utilize this genomic information to tailor treatment plans, especially within pharmacogenomics, to optimize drug selection and dosage for individual patients [4]. Additionally, community health workers act as vital bridges between healthcare systems and the public. They promote health awareness, disseminate information, and facilitate participation in community-based research. Patient navigators complement these efforts within clinical environments by supporting patients in managing their healthcare journey. They enhance communication, address health literacy barriers, and provide reassurance, ensuring patients can more effectively access and benefit from precision medicine [5].

## 2 The Evolution of Personalized Medicine

### 2.1 Transitioning from Generalized to Personalized Healthcare

PM represents more than just a contemporary trend, and it marks a significant evolution in how we diagnose and treat disease, moving away from generalized approaches toward more accurate and individualized care. Historically, medical science has progressed from metaphysical concepts to a deep molecular and systems-level understanding of disease mechanisms [6]. The limitations of the traditional "one-size-fits-all" treatment models are increasingly evident, underscoring the need for more personalized approaches [7]. By tailoring drug selection and dosage to individual factor, such as disease mechanisms, pharmacokinetics, and pharmacogenomics, clinicians can improve treatment efficacy while minimizing adverse effects. This shift is being accelerated by advancements in data analytics, physiologically based pharmacokinetic (PBPK) modeling, and adaptive clinical trial designs. Looking ahead, future strategies are expected to incorporate omics-based technologies, gene therapies, and advanced disease modeling tools to further enhance the precision and personalization of care [8].

### 2.2 Technologies and Advances in Personalized Medicine

PM remains a central objective in modern healthcare; however, its full potential is often limited by the complexity of diseases and an incomplete understanding of the underlying biological factors [9]. Recent technological advancements, particularly in liquid biopsies and next-generation DNA sequencing, are transforming this field. These innovations facilitate precise biomarker discovery, real-time disease monitoring, and the development of personalized therapies based on individual genetic profiles. By integrating these tools, precision medicine is increasingly positioned to provide optimized, tailored care specific to each patient's unique biological makeup.

A range of emerging technologies is shaping the future of personalized medicine. One such innovation is *liquid biopsy*, a minimally invasive technique used to analyze circulating tumor cells (CTCs). Unlike traditional tissue biopsies, liquid biopsies capture CTCs from both primary and metastatic sites, offering a more comprehensive genetic overview of tumor progression. For instance, in colorectal cancer, metastatic CTCs expressing CEA but not CD45 have been associated with a higher risk of relapse, in contrast to primary CTCs. Another transformative approach is *immunotherapy*, which leverages the body's immune system to selectively target cancer cells. A pivotal breakthrough occurred in 1963 with the discovery of tumor-specific serum antibodies, revealing that cancers express unique antigens. This finding paved the way for therapies using engineered immune cells or monoclonal antibodies to precisely target tumor markers, reducing side effects and improving outcomes. *MicroRNA (miRNA) therapies* also show promise by correcting abnormal gene regulation. Two key strategies include restoring downregulated tumor-suppressor miRNAs using synthetic mimics and silencing oncogenic miRNAs through anti-miRNA oligonucleotides (AMOs), which inhibit their interaction with target mRNAs. The *CRISPR-Cas9 gene-editing system* further exemplifies precision medicine's potential, allowing accurate and long-lasting genetic modifications. Using guide RNAs, researchers can correct disease-causing mutations—for example, editing the $\beta$-globin gene in embryos to treat $\beta$-thalassemia highlights CRISPR's potential in managing inherited disorders [9].

These advances, combined with tools such as artificial intelligence (AI), wearable devices, and omics-based data analysis, enable real-time disease detection, biomarker discovery, and therapy optimization. However, to fully realize the promise of personalized medicine, integrated biomarker strategies and restructured clinical trial models are essential for adaptive and effective drug evaluation [10].

## 2.3 Integration into Clinical Practice

The transition to PM necessitates the integration of molecular insights into clinical care, prioritization of patient input, infrastructure upgrades, and equitable access to advanced technologies. Key strategies include strengthening health data systems, aligning clinical practices with genomic science, and creating scalable frameworks that support the widespread adoption of personalized care models [11]. The advancement of PM is further driven by promoting education, empowerment, and value. However, a significant barrier to its implementation is the limited awareness among both patients and healthcare professionals. Overcoming this challenge requires the development of reliable, accessible, and regularly updated educational resources that clearly explain the core principles and technologies of PM. It is equally important to provide patients with transparent information about treatment options and the privacy of their genetic data.

Emerging genetic counseling policies are helping individuals make informed decisions about the use and sharing of their molecular information, thereby reducing risks such as genetic discrimination or insurance denial. Widespread adoption of

PM also depends on demonstrating its clinical effectiveness and economic value. Stakeholders—particularly payers and providers—require robust, data-driven evidence before supporting new models of care. Collaborative efforts among healthcare professionals, payers, and developers can strengthen these evaluations. Tailoring evidence to meet the needs of different stakeholders and incorporating real-world clinical data are crucial for aligning reimbursement policies and fostering the adoption of personalized care. Nonetheless, several challenges persist, including low genomic literacy among both patients and providers, the absence of comprehensive clinical guidelines, concerns about data privacy, and equity issues related to cost. Addressing these barriers is essential for the ethical, responsible, and accessible integration of genomics into routine healthcare [12].

## 3 Key Roles of Health Professionals in Personalized Medicine

Health professionals, including nurses, pharmacists, physicians, community health workers, and genetic counsellors, play pivotal roles in the successful implementation of PM. Their combined expertise ensures that PM strategies are integrated into clinical practice effectively and equitably.

### 3.1 Nurses

As the largest sector of the healthcare workforce, nurses play a vital role in advancing and delivering precision health. Their holistic perspective and close patient interactions uniquely position them as leaders in achieving the goals of the Personalized Medicine Initiative (PMI), particularly through care coordination and patient-centered approaches [13].

With the expansion of electronic health records, nurses are increasingly involved in interpreting clinical, genetic, and patient-reported data to guide care decisions. Their ability to synthesize this information and communicate it clearly to both patients and healthcare teams is essential. By considering biological, social, economic, and environmental factors, nurses offer a truly personalized approach to health. Incorporating omics data, such as genomics and epigenetics alongside lifestyle and environmental exposures, nurses help identify complex interactions influencing disease risk and treatment response. They play a critical role in pharmacogenetics and pharmacogenomics, assessing how genetic variations affect drug metabolism and individual responses to medications [14]. In oncology, for example, this has enabled more accurate tumor profiling and the application of targeted therapies.

Nurses also assist patients in understanding genetic testing, anticipating potential outcomes, and navigating treatment options. They often collaborate with genetic counselors, helping to bridge service gaps and ensure comprehensive patient support. Additionally, nurses are key educators informing patients and families about omics-based testing, collecting feedback on emerging technologies, and tailoring care to individual needs. To fully realize the promise of PM, nurses must work collaboratively within multidisciplinary teams, supporting early detection, patient education, and risk assessment. Given the increasing demand for genetic counseling, nurses are well-positioned to assume critical roles that promote equitable, informed, and personalized healthcare for all individuals [15].

## 3.2 Pharmacists

Pharmacists play a pivotal role in multidisciplinary healthcare teams, contributing significantly to the management of chronic diseases, medication reconciliation, and patient education. Their responsibilities extend to administering immunizations, thereby advancing public health initiatives and improving vaccination rates [16]. This evolving role underscores a broader shift toward patient-centred care, with increased focus on optimizing clinical outcomes, personalizing treatment regimens, and promoting preventive healthcare.

Pharmacy has long embraced principles of PM, traditionally tailoring treatments based on factors such as age, weight, renal function, and potential drug interactions. Today, pharmacists continue to support PM through active participation in pharmacy and therapeutics committees and the development of clinical decision-support tools. Their specialized knowledge enables them to recommend gene-drug pairings for institutional formularies and to advise prescribers using robust, evidence-based data. Additionally, Collaborative Practice Agreements (CPAs) have expanded pharmacists' scope, allowing them to initiate or adjust therapies according to established protocols. This collaborative model improves both the efficiency and quality of patient care [17].

Telepharmacy has emerged as a crucial resource for rural and underserved populations, enabling pharmacists to deliver remote consultations, conduct medication reviews, and support treatment adherence. These services have proven effective in managing chronic diseases and reducing hospital readmissions [16]. For instance, Arnall J.R. et al. reported that a resident clinical pharmacist participating in a Precision Medicine Program collaborated with oncologists to optimize drug strategies, facilitate medication access, and counsel 14 cancer patients both in person and remotely, ultimately improving treatment safety and efficacy [18].

## 3.3 Physicians

With the rise of PM, physicians face increasing legal risks. Past patterns show that technological advances in medicine can lead to heightened liability. Factors contributing to this include inconsistent adoption of PM, rapid growth of new data, and the thousands of deaths linked to drug side effects with possible genetic causes. Changes in malpractice law, such as the erosion of the locality rule and growing emphasis on reasonableness, further elevate the risk [19]. For example, if a physician fails to recommend BRCA1/2 testing to a woman with potential but unclear risk factors, they could be held liable—even if guidelines do not mandate testing, especially in cases involving limited family history, like adoption or small families [20].

### 3.4 Genetic Counsellors

Genetic counsellors play a critical role in bridging the gap between complex genomic information and clinical care. With advancements in genomics, technology, and bioinformatics, the field of genetic counseling has been evolving rapidly [21]. Due to the complexity of genomic data, there is a growing need for genetic counselors to specialize alongside medical subspecialties such as cardiology or gastroenterology. These specialties help guide testing decisions, such as choosing between gene panels, whole genome sequencing (WGS), or single-gene tests, and also support the interpretation of uncertain or negative results, evaluation of cost-effectiveness, and consideration of psychosocial impacts. Their involvement can improve care, reduce unnecessary testing, and enhance clinical decision-making, especially in cases involving rare or complex disorders [22].

Kohler J.N. et al. analyzed the roles of 19 genetic counselors in a multisite Undiagnosed Diseases Network (UDN) study, revealing diverse responsibilities across institutions. Despite variations in titles and duties, core themes of collaboration, clinical care, and data curation emerged. Notably, 50% of the counselors split their time between the UDN and other professional roles [23].

### 3.5 Community Health Workers (CHWs)

The Community Health Worker (CHW) model addresses healthcare disparities by employing culturally competent lay health workers to connect patients with providers, education, and services. CHWs enhance patient engagement by tailoring care to individual needs [24]. They have proven effective in managing both communicable and noncommunicable diseases, including diabetes, hypertension, smoking cessation, and more. While the World Health Organization (WHO) has issued guidelines to optimize CHW programs, evidence of their specific capacity for cancer screening in low- and middle-income countries (LMICs) remains limited and underreported in existing research. Hand T. et al., in a scoping review, analyzed the roles of CHWs in breast cancer screening across diverse cultural and geographic contexts, comparing their impact in high-income and LMIC settings. The review highlighted CHWs' contributions in three key areas: education, screening facilitation, and patient navigation. As global cancer rates rise and survival data remain limited, integrating CHWs into healthcare systems may improve outcomes in resource-constrained environments [25]. Supporting this, Keegan C.N. et al. found that telehealth-based training for CHWs improved glycemic control and care navigation, although further longitudinal research is needed to confirm long-term outcomes [26]. Given their pivotal role in public health, CHWs are well-positioned to enhance health literacy, promote engagement in precision medicine and research, and bridge gaps between communities and healthcare providers, making structured CHW programs a vital national priority for improving health outcomes.

## 4 Core Responsibilities of Health Workers

### 4.1 Developing Individualized Treatment Plans

Patient-centred care emphasizes the vital role of healthcare professionals in conducting comprehensive assessments and risk profiling. This approach identifies individual risk factors, comorbidities, and lifestyle characteristics that influence overall health, forming the foundation for personalized care plans. By incorporating lifestyle modifications such as dietary changes, increased physical activity, and stress reduction providers promote holistic care while empowering patients to take an active role in their well-being [27]. Furthermore, patient education and improved access to healthcare resources enhance self-management by increasing treatment adherence and awareness of early warning signs. Healthcare professionals also play a key role in leveraging telemedicine and remote monitoring technologies, which facilitate continuous engagement, enable real-time interventions, and help minimize complications or unnecessary hospital visits [28]. A valuable yet underutilized tool within this framework is the Shared Care Plan (SCP). Although often seen as resource-intensive and relatively simple in structure, SCPs offer significant benefits, including the promotion of personalized care, respect for patient autonomy, and support for provider decision-making. To fully realize their potential, healthcare professionals must lead the integration of SCPs into routine care, establishing them as a standard component of collaborative, patient-focused healthcare [29].

### 4.2 Implementing Preventive Strategies

Healthcare professionals play a crucial role in clinical prevention and the overall health of the population. Their responsibilities extend beyond traditional clinical settings, making it essential for all clinicians to understand and collaborate with public health systems [30]. Preventive strategies can be categorized into three levels. Primary prevention involves proactive measures such as vaccination, lifestyle modification, and substance use regulation to prevent the onset of disease. Secondary prevention focuses on early detection and prompt intervention—such as cancer screenings—to slow disease progression. In this context, healthcare professionals can educate patients on genetic risks, including polygenic risk scores (PRS), which enhance the precision of preventive strategies. Tertiary prevention targets individuals already affected by disease, aiming to improve quality of life by reducing complications and restoring function through rehabilitation, which may include emerging personalized and technological therapies [31]. To strengthen this framework, it is essential to integrate preventive medicine and population health into healthcare education. Interprofessional training fosters a collaborative approach to care, while appropriate reimbursement for preventive counseling and chronic disease management ensures long-term sustainability. Empowering healthcare professionals in this way is key to advancing personalized, preventive care and improving health outcomes at the community level [32].

### 4.3 Enhancing Disease Detection and Diagnosis

Patients play an important role by sharing their insights and experiences, which inform clinical decision-making and help identify diagnostic errors. Healthcare providers must foster environments that encourage patient engagement through tools such as health IT systems, which enhance transparency and provide access to health records. A multidisciplinary approach—encompassing physicians, nurses, pharmacists, radiologists, pathologists, social workers, and patient navigators—is essential for ensuring diagnostic accuracy. Emphasizing teamwork and the recognition of roles across disciplines can significantly enhance diagnostic quality [33]. AI technologies and predictive modeling support early disease detection by identifying risks and guiding preventive care. While machine learning tools such as decision trees, support vector machines (SVMs), and random forests enable data-driven diagnostics, they should complement, not replace, clinical judgment [34]. Additionally, advanced molecular diagnostic tools like Surface-Enhanced Raman Scattering (SERS) offer unmatched sensitivity. Recent innovations using antibody-conjugated nanostructures have enabled SERS to detect trace cancer biomarkers with 1000-fold greater sensitivity than ELISA, marking a major advancement in noninvasive diagnostics [35].

### 4.4 Optimizing Treatment and Drug Response

Healthcare professionals play a central role in ensuring safe, effective, and individualized prescribing practices. Prescribing optimization remains a key focus in healthcare research, emphasizing the need for appropriate medication use tailored to individual patients. Appropriateness in prescribing includes more than pharmacological accuracy; it incorporates clinician and patient preferences, shared decision-making, and consideration of nondrug interventions. With the support of advanced analytics, clinicians can assess treatment responses and patient outcomes to enhance therapeutic efficacy while optimizing resource allocation [28, 36]. A notable example is the dose optimization of erdafitinib, approved for advanced urothelial carcinoma with Fibroblast Growth Factor Receptor (FGFR) alterations. In a multiphase clinical trial (BLC2001), multiple dosing regimens were evaluated based on serum phosphate levels and patient tolerance. The final approved regimen is 8 mg daily with potential increase to 9 mg, balanced effectiveness with safety, demonstrating a 32% objective response rate. Healthcare professionals were instrumental in adjusting doses based on patient-specific biomarkers and tolerability [37]. In parallel,

Srilakshmi V. et al. introduced an automated neural architecture search (NAS) framework for drug response prediction. Using Random Search, Q-Learning, and Bayesian Optimization, NAS outperformed traditional models, supporting healthcare professionals in advancing personalized medicine [38].

### 4.5 Monitoring, Follow-Up, and Patient Support

Remote monitoring technologies and telemedicine have revolutionized patient care by enabling continuous engagement and real-time health tracking. These tools empower healthcare professionals to monitor patient progress remotely, adjust treatment plans proactively, and intervene early to prevent complications and hospital readmissions. Medication management and adherence support programs led by care teams further reduce noncompliance and adverse drug events, enhancing long-term disease control [28]. Effective chronic disease management also relies on coordinated, interdisciplinary care. Collaboration among physicians, specialists, nurses, and support services ensures a patient-centered approach that addresses physical, emotional, and social health needs. Fairbrother P. et al. highlighted that patients and providers value relationship-based continuity in telemonitoring. Familiarity with patients helps professionals contextualize biometric data and apply clinical judgment more effectively. Local integration of services was favored over centralized systems due to the benefits of personalized care [39]. In cancer follow-up, Jacobs L.A. et al. emphasized the expanding role of primary care providers. Shared care models between oncologists and primary providers support recurrence monitoring, management of treatment effects, and psychosocial care, reinforcing the importance of healthcare professionals in survivorship planning [40].

## 5 Interdisciplinary Collaboration in Personalized Medicine

### 5.1 Team-Based Approach

Many healthcare professionals from various fields collaborate interprofessionally to provide coordinated, high-quality care. Rooted in mutual respect, trust, and effective communication, this approach addresses the growing complexity of modern healthcare, especially with chronic conditions like chronic obstructive pulmonary disease (COPD). Multidisciplinary teams, including doctors, nurses, pharmacists, therapists, and social workers, offer a holistic view of patient care by integrating medical, psychological, and social support. This coordinated model reduces fragmented care, medical errors, and treatment duplication while enhancing outcomes and satisfaction. For example, care coordinators play a key role in managing COPD by ensuring proper follow-up and patient education, which reduces hospital readmissions. Emphasizing patient-centred care, this model involves patients in decision-making, shifting away from traditional top-down practices. Studies show that patient engagement in collaborative care leads to greater satisfaction and better adherence to treatment. Open communication among providers and with patients fosters transparency, shared goals, and trust, which are key to delivering effective, personalized healthcare [41].

## 5.2 Communication and Coordination

Promoting interdisciplinary collaboration in healthcare relies on clear, structured communication. Standardized tools like Situation, Background, Assessment, Recommendation (SBAR), checklists, and templates help ensure consistent information sharing and reduce errors. Digital technologies such as Electronic Health Records (EHR) and Health Information Exchanges (HIE) allow real-time access to patient data, enhancing decision-making and care continuity. Telehealth platforms further support remote collaboration. Regular interprofessional meetings or case conferences encourage open dialogue and shared problem-solving. Clearly defined roles and responsibilities foster accountability and mutual respect, improving coordination. Building trust through team-building, training, and consistent interaction enhances communication and collaboration. Constructive feedback mechanisms and regular performance reviews help refine communication practices. Cultural competence training enables healthcare providers to navigate diverse backgrounds, promoting inclusivity and understanding. Finally, involving patients in care decisions supports shared goals and tailored interventions. Patient engagement not only boosts satisfaction but also leads to better adherence and outcomes, reinforcing the purpose of collaborative care [41].

## 5.3 Case Studies of Collaborative Success

In the genomic era, inherited conditions like Long QT syndrome (LQTS) highlight the necessity of interdisciplinary care, integrating genetic, psychological, ethical, and medical expertise. The Montefiore–Einstein Centre for Cardio-Genetics (MECC) exemplifies this approach by providing comprehensive, personalized care for families with hereditary cardiac risks.

*Case 1*: A teenage girl experienced a fainting episode and was assessed by MECC's team. Though genetic tests for LQTS were negative, a psychologist discreetly suspected pregnancy, which was later confirmed. The team's early rapport allowed for emotional support when the news emerged. Due to geographic constraints, the patient sought care from a local obstetrician unfamiliar with LQTS. MECC's cardiologist collaborated with her obstetrician, ensuring safe pregnancy management. Postpartum, follow-up care continued at MECC.

*Case 2*: A 12-year-old boy with confirmed LQTS faced maternal resistance to lifestyle changes needed for treatment. MECC's psychologist discovered the boy exaggerated symptoms to avoid baseball and felt alienated by medication use. With this insight, a tailored care plan was formed [42]

In another study by de Rijke et al. on Alzheimer's care emphasized personalized medicine and patient-orchestrated care. While professionals supported shared decision-making, they noted challenges in implementation and varied views on patient involvement, underscoring the need for flexible, patient-centred approaches [43].

## 6 Challenges in Implementing Personalized Medicine

### 6.1 Technological and Infrastructure Barriers

The clinical adoption of personalized and precision medicine is increasingly constrained by technological and infrastructure-related challenges. Legacy regulatory systems, built around one-size-fits-all models and linear clinical trial designs, struggle to accommodate platform-based diagnostics, AI/machine learning (ML)-driven analytics, and individualized treatment regimens. Current infrastructure lacks the flexibility to integrate real-time omics data, dynamic electronic medical records (EMRs), and adaptive trial frameworks. AI-guided tools like CURATE.AI demonstrate potential for precise, patient-specific dosing, but demand interoperable data ecosystems and high-quality, standardized inputs—still underdeveloped in most healthcare systems. Regulatory frameworks must shift from static endpoints to adaptive, Bayesian-driven methodologies that reflect patient variability. Moreover, the clinical trial landscape requires transformation to support decentralized designs, biomarker-guided enrollment, and real-world data integration. Global initiatives such as AI Singapore and the EU's Innovative Medicines Initiative (IMI) aim to address these barriers by modernizing digital infrastructure, improving preclinical data quality, and enabling AI-enhanced drug development. Streamlined strategies must now focus on scalable infrastructure and regulatory agility to support precision care at the population scale [9].

### 6.2 Ethical and Privacy Concerns

The implementation of personalized treatment planning demands robust ethical and data privacy safeguards to uphold patient autonomy and trust. Informed consent is central, ensuring patients understand the implications of personalized care. Transparency in data collection, usage, and sharing is vital, alongside clear organizational policies and patient access to information. Bias mitigation is crucial, as AI-driven tools may unintentionally reinforce discrimination based on sensitive attributes. Healthcare providers must embrace data minimization, collecting only essential information, and anonymizing it to reduce reidentification risks. Regulatory compliance with laws such as HIPAA and GDPR is nonnegotiable. Ongoing ethical oversight by review boards is essential to evaluate emerging concerns. Effective incident response strategies must be in place to address data breaches swiftly, while comprehensive employee training helps prevent breaches stemming from human error. Finally, maintaining detailed audit trails ensures accountability and supports forensic analysis in case of security incidents. Together, these measures create a secure, ethical foundation for personalized healthcare [28].

### 6.3 Equitable Access and Health Disparities

Equitable access to personalized medicine is essential to reduce health disparities and ensure that all populations benefit from advancements in individualized care [6]. However, challenges remain in integrating diverse personalized medical strategies and fragmented health records, particularly in underserved communities. Effective management of personal health data, linking lab results with digital resources, can support the foundation for lifelong electronic health records. Precision medicine, grounded in genomics and technology, has made significant strides, yet it often neglects social determinants and cultural contexts. The exclusion of minority groups and older adults in clinical research further exacerbates disparities. To address this, the National Institutes of Health (NIH) mandates age diversity in trials, though functional status often matters more than chronological age [24]. Bridging these gaps requires strategies that reflect real-world diversity and promote health equity. The future of precision medicine must prioritize global inclusion by expanding personalized care to marginalized and resource-constrained settings, transforming innovation into truly equitable, patient-centred healthcare [28].

### 6.4 Training and Educational Gaps

A major barrier to implementing precision medicine lies in the significant education and training gaps among healthcare professionals, particularly in pharmacogenomics. Most clinicians lack practical exposure during their education and feel unprepared to interpret or communicate genomic data to patients. Pharmacists, though uniquely positioned to lead pharmacogenomics integration, face limited postgraduate opportunities, with very few accredited residencies producing qualified experts. Online courses offer continuing education, but often fall short in closing knowledge gaps. Institutional support for pharmacogenomics education is often minimal, making large-scale implementation difficult. Additionally, the complexity of genomic data and existing clinical workloads makes integration challenging without coordinated workflows and interprofessional collaboration. There is also no consensus on which professionals should act on test results. Standardized pharmacogenomics training across healthcare disciplines is essential to ensure team-based, patient-centered care. Empowering clinicians and patients through education can enhance shared decision-making, improve medication adherence, and ultimately optimize outcomes in personalized treatment strategies [44].

# 7 Future Directions and Recommendations

## 7.1 Enhancing Health Worker Education and Training

Future-oriented educational frameworks are increasingly emphasizing competencies essential for implementing personalized precision medicine. These include genomics, digital health, bioinformatics, data science, AI, and ethical-legal considerations, alongside cross-cutting skills like communication, leadership, and collaboration. Programs such as EATRIS's summer school and ERASMUS+ integrate clinical research with digital and ethical training, highlighting a multidisciplinary, data-driven approach. National strategies, like Navarra's in Spain, prioritize training in omics, Information and Communication Technologies (ICTs), and evidence-based methodologies. As healthcare shifts from disease-focused to health-oriented models, equipping professionals with advanced technological and ethical competencies is vital to meet future challenges and personalize prevention, diagnosis, and treatment effectively [45].

## 7.2 Policy and Regulatory Considerations

Regulatory and policy frameworks play a critical role in addressing ethical, legal, and social challenges in personalized medicine as access to sensitive genetic data grows. Initiatives like the ELSI program, launched alongside the Human Genome Project, underscore the need for structured governance. Key policy measures include risk-based informed consent, deployment of trusted intermediaries to regulate data access, and comprehensive legal protections—as exemplified by Switzerland's legislation safeguarding dignity, privacy, and data use. Guidelines must also address the needs of genetic minorities to prevent exclusion and discrimination. Policymakers must define clear boundaries between rational and irrational genetic-based decisions to ensure equity and trust [6].

## 7.3 Expanding Access Through Community Engagement

Expanding access to PM requires rebuilding trust through meaningful community engagement. Models like CHWs and community champions—who reflect the culture, language, and lived experiences of the populations they serve—help bridge gaps in care. Early inclusion of communities in research design and implementation fosters transparency and relevance. Initiatives like NIH's "All of Us" and NHGRI's partnerships with Alaskan Native tribes exemplify this approach. Sustained collaboration, lay-friendly communication, and equitable benefit-sharing are essential. Multisectoral partnerships across academia, industry, healthcare, and government further enable inclusive, culturally responsive PM that serves diverse populations effectively [24].

### 7.4 Embracing Innovations in Genomic Medicine

Genomic innovations are transforming healthcare by uncovering disease mechanisms and enabling targeted therapies. Next Generation Sequencing (NGS) enhances drug discovery, refines clinical trials, and supports early diagnosis and treatment monitoring. As NGS becomes more accessible, it will form the backbone of proactive disease prevention and personalized care. Its utility spans across systems and conditions, offering precise insights into patient health [46]. A clear application is in optimizing anticoagulant therapy, where genetic variations impact drug response. By identifying genes affecting metabolism and sensitivity, genetic technologies help minimize side effects and improve outcomes. These tools are key to advancing patient-centered, precision medicine [47].

## 8 Conclusion

In conclusion, the integration of personalized and precision medicine into modern healthcare marks a transformative shift toward more predictive, preventive, and patient-specific treatment strategies. By leveraging genomic insights, biomarker identification, and pharmacogenomic profiling, clinicians can tailor therapies to optimize outcomes and minimize adverse effects. These advances hold particular promise in complex fields like neurobiology and neurodegenerative diseases, including Alzheimer's, where individualized approaches are critical. For researchers and healthcare professionals alike, embracing this paradigm is essential for advancing data-driven, targeted interventions and shaping the future of therapeutic development and clinical practice.

**Declarations**

**Author's Contributions** Conceived the idea and review structure: BMS; Wrote the manuscript: MP; Revised and edited the manuscript: PM. The author has read, agreed, and approved the final draft of the manuscript.

**Funding** No funding available.

**Conflicts of Interest** There is no competing and conflict of interest.

## References

1. Schleidgen S, Klingler C, Bertram T, Rogowski WH, Marckmann G (2013) What is personalized medicine: sharpening a vague term based on a systematic literature review. BMC Med Ethics 14:1–2. https://doi.org/10.1186/1472-6939-14-55
2. Di Sanzo M, Cipolloni L, Borro M, La Russa R, Santurro A, Scopetti M, Simmaco M, Frati P (2017) Clinical applications of personalized medicine: a new paradigm and challenge. Curr Pharm Biotechnol 18(3):194–203. https://doi.org/10.2174/1389201018666170224105600

3. Muthusamy V, Venkatesan A, Radhakrishnan A, Kuppuswamy G (2021) Personalized medicine: a review on scope and its pitfall. Int J Pharm Res 13(1). https://doi.org/10.31838/ijpr/2021.13.01.141
4. Alsharhoof SM, Hassan HA, Almajhed AI, Algannam SA, Alanazi RO, Alanazi AO, Alminhaly EM, Al Ajmi FA, Al Abbas SH, ALYami AM (2024) Comprehensive review of health workers' impact on precision medicine. JoE 3(8):2358–2368. https://doi.org/10.62754/joe.v3i8.4983
5. Ramos IN, Ramos KN, Ramos KS (2019) Driving the precision medicine highway: community health workers and patient navigators. J Transl Med 17(1):85. https://doi.org/10.1186/s12967-019-1826-2
6. Fierz W (2004) Challenge of personalized health care: to what extent is medicine already individualized and what are the future trends. Med Sci Monit 10(5):111–123
7. Abrahams E, Silver M (2010) The history of personalized medicine. Integr Neurosci Pers Med:3–16
8. Lesko LJ, Schmidt S (2012) Individualization of drug therapy: history, present state, and opportunities for the future. Clin Pharmacol Ther 92(4):458–466. https://doi.org/10.1038/clpt.2012.113
9. Maggi E, Patterson NE, Montagna C (2016) Technological advances in precision medicine and drug development. Expert Rev Precis Med Drug Dev 1(3):331–343. https://doi.org/10.1080/23808993.2016.1176527
10. Ho D, Quake SR, McCabe ER, Chng WJ, Chow EK, Ding X, Gelb BD, Ginsburg GS, Hassenstab J, Ho CM, Mobley WC (2020) Enabling technologies for personalized and precision medicine. Trends Biotechnol 38(5):497–518. https://doi.org/10.1016/j.tibtech.2019.12.021
11. Pritchard DE, Moeckel F, Villa MS, Housman LT, McCarty CA, McLeod HL (2017) Strategies for integrating personalized medicine into healthcare practice. Pers Med 14(2):141–152. https://doi.org/10.2217/pme-2016-0064
12. Najafzadeh M, Davis JC, Joshi P, Marra C (2013) Barriers for integrating personalized medicine into clinical practice: a qualitative analysis. Am J Med Genet A 161(4):758–763. https://doi.org/10.1002/ajmg.a.35811
13. Fu MR, Kurnat-Thoma E, Starkweather A, Henderson WA, Cashion AK, Williams JK, Katapodi MC, Reuter-Rice K, Hickey KT, de Mendoza VB, Calzone K (2020) Precision health: a nursing perspective. Int J Nurs Sci 7(1):5–12. https://doi.org/10.1016/j.ijnss.2019.12.008
14. Ielapi N, Andreucci M, Licastro N, Faga T, Grande R, Buffone G, Mellace S, Sapienza P, Serra R (2020) Precision medicine and precision nursing: the era of biomarkers and precision health. Int J Gen Med:1705–1711. https://doi.org/10.2147/IJGM.S285262
15. Vorderstrasse AA, Hammer MJ, Dungan JR (2014) Nursing implications of personalized and precision medicine. Semin Oncol Nurs 30(2):130–136. https://doi.org/10.1016/j.soncn.2014.03.007
16. Duhaim TZ, Al-Maliki MA, Al-Shahri TI, Zahrani HY, Aljalis MI, Nawab NA, Alghamdi TA, Al-Surayye'a RA, Al-Battah HM, Hakimi AM, Alshedoukhi AF. (2024) The evolving role of pharmacists in patient care: trends and innovations. J Int Crisis Risk Commun Research 7(S9):1408. https://doi.org/10.63278/jicrcr.vi.742
17. Mandumpala JJ (2024) Precision medicine and clinical pharmacists. Indian J Pharma Drug Stud 3(3)
18. Arnall JR, Petro R, Patel JN, Kennedy L (2019) A clinical pharmacy pilot within a Precision Medicine Program for cancer patients and review of related pharmacist clinical practice. J Oncol Pharm Pract 25(1):179–186. https://doi.org/10.1177/1078155217738324
19. Puryear L, Downs N, Nevedal A, Lewis ET, Ormond KE, Bregendahl M, Suarez CJ, David SP, Charlap S, Chu I, Asch SM (2018) Patient and provider perspectives on the development of personalized medicine: a mixed-methods approach. J Community Genet 9:283–291. https://doi.org/10.1007/s12687-017-0349-x
20. Marchant GE, Campos-Outcalt DE, Lindor RA (2011) Physician liability: the next big thing for personalized medicine? Pers Med 8(4):457–467. https://doi.org/10.2217/pme.11.33
21. Biesecker B (2020) Genetic counseling and the central tenets of practice. Cold Spring Harb Perspect Med 10(3):a038968. https://doi.org/10.1101/cshperspect.a038968

22. Shelton CA, Whitcomb DC (2015) Evolving roles for physicians and genetic counselors in managing complex genetic disorders. Clin Transl Gastroenterol 6(11):e124. https://doi.org/10.1038/ctg.2015.46
23. Kohler JN, Glanton E, Boyd BM, Sillari CH, Marwaha S, Network UD, Wheeler MT, Acosta MT, Adam M, Adams DR, Agrawal PB (2022) Genetic counselor roles in the undiagnosed diseases network research study: clinical care, collaboration, and curation. J Genet Couns 31(2):326–337. https://doi.org/10.1002/jgc4.1493
24. Ory MG, Adepoju OE, Ramos KS, Silva PS, Vollmer DD (2023) Health equity innovation in precision medicine: current challenges and future directions. Front Public Health 11:1119736. https://doi.org/10.3389/fpubh.2023.1119736
25. Hand T, Rosseau NA, Stiles CE, Sheih T, Ghandakly E, Oluwasanu M, Olopade OI (2021) The global role, impact, and limitations of Community Health Workers (CHWs) in breast cancer screening: a scoping review and recommendations to promote health equity for all. Glob Health Action 14(1):1883336. https://doi.org/10.1080/16549716.2021.1883336
26. Keegan CN, Johnston CA, Cardenas VJ Jr, Vaughan EM (2020) Evaluating the impact of telehealth-based, diabetes medication training for community health workers on glycemic control. J Pers Med 10(3):121. https://doi.org/10.3390/jpm10030121
27. Kane EP, Collinsworth AW, Schmidt KL, Brown RM, Snead CA, Barnes SA, Fleming NS, Walton JW (2016) Improving diabetes care and outcomes with community health workers. Fam Pract 33(5):523–528. https://doi.org/10.1093/fampra/cmw055
28. Malik P, Gang U, Singh T, Vyas P, Singh S, Ramani Y (2023) Optimizing treatment planning and patient outcomes: the role of advanced analytics and personalized approaches in healthcare. Int J Mod Dev Eng Sci 2(9):20–24
29. Bevilacqua G, Bolcato M, Rodriguez D, Aprile A (2020) Shared care plan: an extraordinary tool for the personalization of medicine and respect for self-determination. Acta Biomed Ateneo Parmense 92(1):e2021001. https://doi.org/10.23750/abm.v92i1.9597
30. Zenzano T, Allan JD, Bigley MB, Bushardt RL, Garr DR, Johnson K, Lang W, Maeshiro R, Meyer SM, Shannon SC, Spolsky VW (2011) The roles of healthcare professionals in implementing clinical prevention and population health. Am J Prev Med 40(2):261–267. https://doi.org/10.1016/j.amepre.2010.10.023
31. Pastorino R, Loreti C, Giovannini S, Ricciardi W, Padua L, Boccia S (2021) Challenges of prevention for a sustainable personalized medicine. J Pers Med 11(4):311. https://doi.org/10.3390/jpm11040311
32. Gjødsbøl IM, Winkel BG, Bundgaard H (2021) Personalized medicine and preventive health care: juxtaposing health policy and clinical practice. Crit Public Health 31(3):327–337. https://doi.org/10.1080/09581596.2019.1685077
33. Ball JR, Miller BT, Balogh EP (eds) (2015) Improving diagnosis in health care. The National Academies Press, Washington, DC
34. Kolawole TO, Mustapha AY, Mbata AO, Olamide B, Tomoh AY, Kelvin-Agwu MC (2025) A systematic review of predictive analytics applications in early disease detection and diagnosis. J Eng Technol 10(3):4265–4283
35. Chisanga M, Muhamadali H, Ellis DI, Goodacre R (2019) Enhancing disease diagnosis: biomedical applications of surface-enhanced Raman scattering. Appl Sci 9(6):1163. https://doi.org/10.3390/app9061163
36. Topinková E, Baeyens JP, Michel JP, Lang PO (2012) Evidence-based strategies for the optimization of pharmacotherapy in older people. Drug Aging 29:477–494. https://doi.org/10.2165/11632400-000000000-00000
37. Fourie Zirkelbach J, Shah M, Vallejo J, Cheng J, Ayyoub A, Liu J, Hudson R, Sridhara R, Ison G, Amiri-Kordestani L, Tang S (2022) Improving dose-optimization processes used in oncology drug development to minimize toxicity and maximize benefit to patients. J Clin Oncol 40(30):3489–3500. https://doi.org/10.1200/JCO.22.00371
38. Srilakshmi V, Padmini G, Sreenidhi G, Venkata Ramana B, Preetham Reddy GJ (2025) Neural architecture search-driven optimization of deep learning models for drug response prediction. Procedia Comput Sci 252:172–181. https://doi.org/10.1016/j.procs.2024.12.019

39. Fairbrother P, Ure J, Hanley J, McCloughan L, Denvir M, Sheikh A, McKinstry B, Telescot Programme Team (2014) Telemonitoring for chronic heart failure: the views of patients and healthcare professionals–a qualitative study. J Clin Nurs 23(1–2):132–144. https://doi.org/10.1111/jocn.12137
40. Jacobs LA, Shulman LN (2017) Follow-up care of cancer survivors: challenges and solutions. Lancet Oncol 18(1):e19–e29. https://doi.org/10.1016/S1470-2045(16)30386-2
41. Fatimah S, Alanazi S, Alenezi M, Abdulmajeed H, Aldhafeeri B, Sawadi I, Alnaim F, Alasmari AS, Asiri MY, Almoafa YA (2024) Comprehensive collaboration between different health care providers as family medicine, pharmacy, emergency care, laboratory technicians, dentistry and rehabilitation services. J Int Crisis Risk Commun Res 7(S6):2047
42. Erskine KE, Griffith E, DeGroat N, Stolerman M, Silverstein LB, Hidayatallah N, Wasserman D, Paljevic E, Cohen L, Walsh CA, McDonald T (2013) An interdisciplinary approach to personalized medicine: case studies from a cardiogenetics clinic. Pers Med 10(1):73–80. https://doi.org/10.2217/pme.12.108
43. De Rijke TJ, Vasseur D, van der Flier WM, Minkman MM, Rhodius-Meester HF, Verwey NA, Smets EM, Visser LN (2025) Exploring interdisciplinary perspectives on the implementation of personalized medicine and patient-orchestrated care in Alzheimer's disease: a qualitative study within the ABOARD research project. J Alzheimers Dis 105(1):120–133. https://doi.org/10.1177/13872877251326166
44. Ta R, Cayabyab MA, Coloso R (2019) Precision medicine: a call for increased pharmacogenomic education. Pers Med 16(3):233–245. https://doi.org/10.2217/pme-2018-0107
45. Martin-Sanchez F, Lázaro M, López-Otín C, Andreu AL, Cigudosa JC, Garcia-Barbero M (2023) Personalized precision medicine for health care professionals: development of a competency framework. JMIR Med Educ 9:e43656. https://doi.org/10.2196/43656
46. Goswami J (2022) Genomic innovation democratizes precision medicine. J Prec Med 8:3
47. Navaei O (2025) Personalized medicine: tailoring treatment plans based on genetic profile. Eurasian J Chem Med Pet Res 4(1):200–222

# Novel Innovations and Future Perspectives

# Emerging Technologies and Approaches

Sayani Ghosh, Sunny Kumar, and Mrinal K. Ghosh

**Abstract**

Oncology is transforming from a one-size-fits-all paradigm to a highly customized one wherein thorough molecular and functional characterization of individual tumors are required for treatment choices. Precision oncology is changing because of developments in single-cell and spatial multi-omics, liquid biopsy technologies, patient-derived functional models, customized immunotherapies, radiotheranostics, and artificial intelligence (AI). Recent developments in these fields are summarized here. In the upcoming years, advancements will occur in the realms of closed-loop precision oncology, which combines functional assessment, AI-guided therapy optimization, and continuous molecular monitoring.

**Keywords**

Emerging technologies · Omics · Oncology · Vaccine and Theranostics

## 1 Introduction

Personalized medicine emphasizes more on tailored therapeutic approaches rather than traditional treatments. It is quickly changing cancer research and clinical practice. As medications with novel mechanisms of action allow for therapeutic intervention on a large scale of biomarker-selected patient populations, the field of precision oncology is growing. The long-standing paradigm of matching a single gene to a single treatment is rapidly changing due to our growing understanding of

S. Ghosh · S. Kumar (✉) · M. K. Ghosh (✉)
Cancer Biology and Inflammatory Disorder Division, Council of Scientific and Industrial Research-Indian Institute of Chemical Biology (CSIR-IICB), TRUE Campus, Kolkata, India
e-mail: mrinalghosh@iicb.res.in

 
M. K. Ghosh, M. Basu (eds.), *Personalized Cancer Therapy*,
https://doi.org/10.1007/978-981-95-6845-1_21

the clinical implications of specific mutations, the epistatic relationship between co-occurring mutations, the role of the immune environment in therapy selection, and more. Additional medication approvals have resulted from the continuous use of composite biomarkers, such as tumor mutational burden (TMB) and microsatellite instability (MSI) status, to customize immunotherapeutic approaches to specific tumors. For instance, individuals with MSI-high advanced or recurrent endometrial malignancies were approved to receive chemotherapy and the checkpoint inhibitor dostarlimab. Future multi-omics biomarker development is probably going to be accelerated by the realization that genomic instability events play a part in oncogenesis.

Precision molecular characterization of tumors has been made possible by developments in genomics, transcriptomics, proteomics, and metabolomics. This has allowed for customized therapies based on distinct genetic and epigenetic profiles [1]. This paradigm change is being driven by emerging technologies that enable real-time monitoring of disease progression and therapeutic response, including single-cell analysis, AI, next-generation sequencing (NGS), and multi-omics integration. Additionally, cutting-edge techniques like liquid biopsy, multiplex imaging, organoid models, and CRISPR-based functional screens are improving drug discovery, early identification, and the creation of predictive biomarkers. These developments are opening the door to more potent, long-lasting, and less toxic cancer treatments by bridging the gap between lab research and clinical use. Personalized medicine presents a promising way to maximize therapy results and increase patient survival, as cancer heterogeneity continues to be a major obstacle.

## 2 High-Resolution Tumor Profiling

Biomarker identification and logical combination therapies are possible by the resolution of intra-tumoral heterogeneity, rare resistant clones, and therapy-perturbed cell states by single-cell RNA-seq and multimodal assays (ATAC-seq, CITE-seq, and lineage tracing). New "multi-ome" kits improve the connection between regulatory programs and phenotypes by capturing transcriptomes and chromatin accessibility from the same cell. Clinical effects include resistance patterns to immunological and targeted treatments, immune–tumor interaction, and minimal residual disease biology. Single-cell analysis has emerged as a growing field of study in precision medicine and life sciences due to cell heterogeneity. This subject has been revitalized by recent developments in analytical platforms and instruments. It is a result of the rise of extremely sensitive, multiplexed, high-throughput single-cell omics approaches. These omics techniques act as a link between visible cell behavior and molecular changes at the cellular level. By bridging the gap between observable cell behavior and molecular changes at the cellular level, these omics technologies help track and understanding disease progression. By examining gene expression patterns within individual cells, these approaches address a variety of complex issues and reveal the underlying mechanisms, behavior, and interactions between individual cells and the human body (Fig. 1).

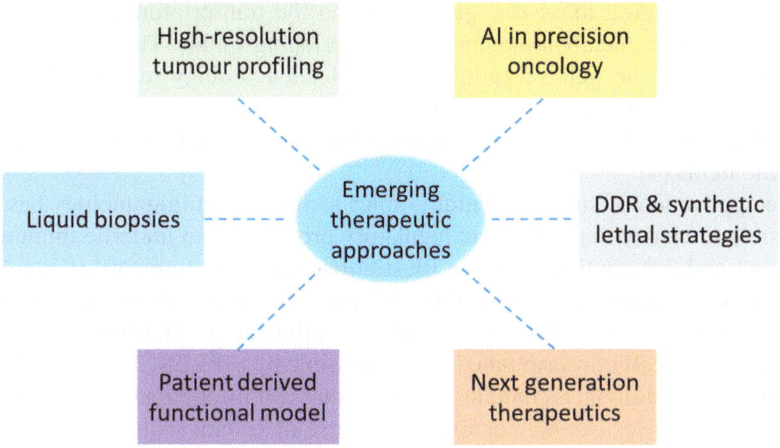


**Fig. 1** Various emerging therapeutic approaches

### 2.1 Single-Cell Multi-omics

Single-cell multi-omics comprehends the distinct molecular properties of various cell populations and reveals the heterogeneity within tumors. In contrast to typical bulk omics techniques, it examines cell populations collectively. Hence, we can find uncommon cell subpopulations that play a key function in tumor growth, metastasis, and therapeutic resistance. Additionally, single-cell multi-omics analysis offers crucial information about the immune response brought on by several therapeutic approaches, i.e., cell therapy, chemotherapy, and immune checkpoint blockade. Therefore, it aids in the comprehension of the complex tumor microenvironment and how it affects patient treatment response [2].

Single-cell multi-omics technologies are essential for assessing unique molecular characteristics originating from a single biological unit. This new method makes it possible to quantify several molecular species in a single cell, providing deep biological insights. Hence, technological developments are continuously increasing the effectiveness of these approaches at reasonable costs. One representative example is single-cell epigenomic sequencing, a novel method used to thoroughly analyze the epigenetic profile of individual cells [3]. This approach includes the examination of several epigenetic phenomena, such as histone modifications and DNA methylation patterns. It emphasizes the identification of epigenetic changes within the genome. Scientists can gain important insights into the intrinsic variety and diversity that exist within biological populations at the epigenetic level by using single-cell epigenomic sequencing. The course of cellular differentiation can differ significantly due to the intrinsic heterogeneity in epigenetic profiles among cells; this feature is made more noticeable as the human body matures and grows.

Various strategies that enable the detailed characterization of cell state and activity are included in single-cell multi-omics approaches. These techniques offer unmatched insights by concurrently integrating many single-modal omics techniques, including those that gather data on the transcriptome, genome, epigenome, proteome, metabolome, and other developing omics. These developments are transforming the field of cellular and molecular biology research. Numerous unique single-cell multi-omics techniques have been invented and developed due to research and technological advancements, which have proven useful in a variety of scientific fields [4].

The application of single-cell multi-omics techniques in immunology has made it possible to clarify the epigenetic regulatory processes that underlie immune cell activation and differentiation. As an illustration, Lan et al. transformed single-cell transcriptome sequencing using FIPRESCI, a method that combines combinatorial marker technology with microfluidic platform optimization [5]. Multi-modal analysis of the single-cell transcriptome is made possible by FIPRESCI, which solves the drawbacks of traditional techniques by increasing cell throughput by a factor of ten and cutting expenses. Another example is the use of several sequencing techniques by Xue et al. to investigate liver cancer heterogeneity at the single-cell level. For the first time, they used single-cell precision to identify five immune microenvironment subtypes of liver cancer (TIMELASER) [6]. They investigated the geographical distribution of cells, the chemokine receptor–ligand network, cellular composition, and certain genomic traits. The processes by which CCL4+ and PD-L1+ TAN promote tumor growth were validated by extensive testing. The possibility of targeting tumor-associated neutrophils as a novel immunotherapy technique was further proven by the scientists, who also built a mouse liver cancer model [7].

## 2.2 Spatial Transcriptomics and Multiplex Imaging

Immunohistochemistry (IHC) and flow cytometry are two examples of conventional techniques that offer little insight into the multi-marker, spatially complicated architecture of tumors. Although transcriptional variety has been revealed by single-cell RNA sequencing, geographical context is absent. This gap is filled by multiplex imaging, which makes large-scale spatially resolved single-cell phenotyping possible. Multiplex imaging has evolved over the years from specialized laboratory instruments to platforms used in clinical trials and translational research. By connecting tissue context with patient-specific results, these tools not only revolutionize histopathology but also open up possibilities for precision oncology [8].

An excellent method for determining the geographical distributions and heterogeneity of cancer cells inside tissues is the use of spatial transcriptomics sequencing technologies. By combining geographical coordinates within tissues with cellular transcriptome data, this state-of-the-art technology offers insights into the relationships between various cell types and the overall architecture of tissues. Spatial transcriptomics has provided a more thorough understanding of the mechanisms behind tumor initiation and progression, opening new avenues for cancer diagnosis and precision treatment. It combines bioinformatics, sequencing, biomarker analysis, and imaging to pinpoint gene expression in tissue slices. It creates gene expression maps of various tissue locations, explores the relationships between cell populations, and shows the spatial distribution of various cell types in tissues. This cutting-edge technique has enormous potential for revealing the complex mechanics behind several illnesses like cancer. The three main categories of spatial transcriptomics techniques include spatial indexing-based approaches, in situ imaging-based approaches, and laser capture microdissection-based approaches.

## 3 Liquid Biopsies

Rapid advancements in liquid "biopsy" technology, which allows for the molecular interrogation of liquid samples (often blood), have made it possible for routine clinical use in cancer patients, as well as for the rapid expansion of research capabilities that are revealing the causes of malignant growth. Liquid biopsies offer a technique for extracting tumor-derived information from bodily fluids. It is minimally invasive. Although a variety of bodily fluids can be biopsied, blood is the most often used fluid. It is possible to separate circulating tumor DNA (ctDNA) and circulating extracellular nucleic acids from blood. While ctDNA is of tumor origin, cell-free DNA (cfDNA) is DNA that is freely present in the blood and may or may not be tumor-derived. It is also possible to separate circulating tumor cells (CTCs) from blood. These are cells that tumors release into the bloodstream [9]. They typically only remain in the bloodstream for one to two and a half hours before being destroyed by the immune system, though a tiny percentage of them may survive and spread to distant metastatic locations. NGS can be used to analyze CTCs and ctDNA/cfDNA. It is also possible to evaluate the functional CTC properties, which can direct the usage of possible therapeutic drugs, as well as the protein and RNA CTC content.

The following methods have been used to assess ctDNA/cfDNA: WGBS-Seq, WES, tagged-amplicon deep sequencing (TAm-Seq), cancer personalized profiling by deep sequencing (CAPP-Seq), droplet digital polymerase chain reaction (ddPCR), beads, emulsion, amplification, and magnetics (BEAMing), and WGS. CtDNA was utilized by Ricciuti et al. to predict how non-small cell lung cancer patients would react to immunotherapy-based treatment [10]. At baseline, ctDNA was assessed in patients receiving pembrolizumab ± platinum/pemetrexed treatment. There was a strong correlation between the percentage change in tumor target lesions and the percentage change in ctDNA during the initial follow-up. Compared to those with an increase, those with a decrease in ctDNA had substantially greater response rates, longer median progression-free survival, and median overall survival. Jensen et al. assessed the response to immunotherapy using the genome instability number (GIN), which was computed using low-coverage, genome-wide sequencing of cfDNA [11]. The amount of cfDNA in the plasma and the size of the copy number changes both affect GIN, which is the total deviation of all copy number changes throughout the genome. The study discovered that there is a quick and consistent rise in blood-derived GIN in the initial weeks of treatment. This technique also identified patients with early hyper-progression.

# 4 Patient-Derived Functional Models

Cancer biology and the cancer landscape have been studied using animal models, including patient-derived organoids and patient-derived xenografts (PDX). The significance of some genes in the progression of cancer can be studied using genetically modified mice (GEM) models and allografts developed from GEM. These models, however, are unable to capture the wide range of driver mutations and profound genomic changes seen in human cancer. Mouse models created by chemical carcinogens are well-established models used to investigate the biology and etiology of cancer. Even without considering the effects of varying dosage timings and animal strains, these models create unexpected cancer landscapes that are challenging to replicate. Furthermore, the underlying oncogenesis mechanisms in the animal-derived cancer models result from their interspecies incompatibility with human disease.

## 4.1 Organoids

Organoids are in vitro structures that are cultivated in the appropriate extracellular matrix-based media [12]. They have comparatively high success rates. They can be derived from human stem cells, organ-specific progenitor cells, or even dissociated tumor tissues. The architecture and function of tumoroids are like those of primary tissues. They preserve the genetic profile, histological characteristics, mutational landscape, and even therapeutic responses. Therefore, organoids have great potential for translational research and are useful instruments for studying carcinogenesis and cancer progression in vitro. Organoids from a variety of human tumor types have been effectively created. Tumoroids stably retain the essential features of actual tumors even after prolonged passaging, and organoid formation takes less time and tissue than PDX models [13]. Organoids recreate a multitude of cellular interactions and metabolic processes present in certain organs, allowing complete testing of drug targeting and efficiency.

## 4.2 Patient-Derived Xenografts (PDX)

There has been a history of "rediscovery" for PDX models. When Rygaard and Povlsen excised a patient's colon cancer and transplanted the tumor pieces into naked mice in 1969, it was the first documented patient-derived xenograft that met the criteria for PDX models. Subsequently, researchers demonstrated that PDX models responded similarly to their counterparts in patients to chemotherapy. Primary or metastatic tumors are dissected and implanted with preserved tissue architecture to create PDX models [14]. Implanting the tumor fragments under the renal capsule, in the anterior compartment of the eye, or in the intracapsular fat pad can be done subcutaneously, orthotopically, or heterotopically. It can take a few days to several months for different tumor types to form PDX. Usually, the tumor can be removed, segmented, and reimplanted for passage when it reaches 1–2 $cm^3$ (first generation). Additionally, with passes, the PDX establishment time steadily stabilizes at 40–50 days. Conventional PDX models are usually made with immunocompromised mice, such as athymic nude mice, severe combined immunodeficiency (SCID) mice, nonobese diabetic severe combined immunodeficiency (NOD-SCID) mice, BALB/cRag2 null/IL2λreceptor null (BRG) mice, Rag-2 null/Jak3 null (BRJ) mice, and NODSCID-/IL2λ-receptor null (NSG) mice, in order to prevent tumor engraftment rejection in mouse models. Engraftment rates vary among mouse strains due to varying degrees of immunosuppression; these rates are higher in mice with weakened immune systems (BRG/BRJ>NSG>NOD/SCID>SCID>nude).

Apart from the inherent elements of the tumor microenvironment (TME), tumor-immune interaction is also linked to tumor behaviors and cancer patient survival, both of which are not reversible in general PDX. Therefore, PDX with human immune systems in working order may be an effective tool for studying tumor immunology and immunotherapy in vivo. Resources for engraftment include foetal hematopoietic tissues, hematopoietic stem and progenitor cells (HSPCs), or human peripheral blood mononuclear cells (PBMCs). Human hematopoietic stem cell engraftment may result in the production of T cells, B cells, myeloid dendritic cells, and plasmacytoid dendritic cells in irradiated NSG or BRG mice. PDX models are effective tools for examining the various states across various locations in the primary lesion and metastases because they preserve genetic profiles and intratumor heterogeneity. Braekeveldt et al., for example, removed a neuroblastoma tumor from a patient and implanted ten pieces of the tumor into ten mice independently. Curiously, the xenografts had differences in transcription and growth rate, and patient subtypes determined by these PDX models' gene clusters displayed different prognoses.

## 4.3 Tissue Engineering and 3D Bioprinting

Design optimization and customization are made possible by additive manufacturing, also referred to as 3D printing, which builds components one layer at a time from a 3D computer model [15]. These intriguing qualities have led to the rapid development of medical applications, such as 3D printed implants, prostheses, and anatomical models. Inert and acellular materials, including plastics, are typically utilized for 3D printing in the medical field. Some of those materials can be utilized for implantation because they are biocompatible, while others are biodegradable and serve as guides for soft tissue rebuilding, such as breast reconstruction following cancer surgery. Three-dimensional bioprinting is a relatively recent area of 3D printing study. To produce a live tissue, 3Dbioprinting employs 3D printing technology to print cells, as well as a supporting matrix known as bioink [16]. The term "a formulation of cells suitable for processing by an automated bio fabrication technology that may also contain biologically active components and biomaterials" is used by Groll et al. to describe bioinks, which might be repurposed as materials featuring cells. Although it appears to be a suitable tool for complicated tissue in vitro modelling, three-dimensional bioprinting presents several issues and difficulties, particularly the creation of biocompatible bioinks and their incorporation into the human body.

### 4.4 Cancer-On-a-Chip (CoC)

For various malignancies, several CoC platforms have been developed. Although the architecture of CoC chips varies according to the kind of cancer, most CoCs have numerous characteristics in common. Straight, parallel microchannels with an input and an outlet made up the earliest CoC models [17]. These microchannels are used to seed cancer mono- or cocultures, and the inlet is used to inject culture material to promote flow dynamics. These chips are frequently useful for researching the formation of early-stage tumors in response to various stimuli that are delivered into the channels, including flow, fibroblasts, endothelial cells, chemotherapeutic drugs, and others [18]. The 3D character of tumors is not adequately represented by this architecture, which limits tissue culture contact to lateral interaction between the microchannels.

Single-organ CoCs frequently seek to simulate a distinct cancer cascade for a more thorough examination or several pathways for a more all-encompassing view of a particular organ impacted by cancer. The type of cancer and the damaged organ have a significant impact on the chip design. For example, lung CoC typically consists of two microchambers divided by a permeable membrane, with endothelial cells lining one chamber to simulate the alveolus and lung cancer cells cocultured with airway epithelial cells in the other chamber. Comparably, the designs of glioblastoma CoC frequently include compartmentalized designs to replicate the core tumor microenvironment and an exterior chamber to recreate the endothelial cells, epithelial cells, and extracellular matrix [19]. Various CoCs have been created to simulate particular cancer cascades, including invasion, migration, neovascularization, and cancer development [20]. Scaffolds or hydrogels implanted with stromal and endothelial components are frequently used in vascularized systems [21]. CoCs that investigate tumor migration and extravasation in a similar manner are frequently made so that the cancer cells and endothelial cells interact in a different chamber or channel.

## 5 Next-Generation Therapeutics

Cancer vaccine is a clear way to direct human immunity's force toward the destruction of tumor cells [22]. Although there are positive indications that this goal can be accomplished, success is still elusive [23]. Patient-specific neoantigens found by sequencing and epitope prediction algorithms are the focus of customized cancer vaccinations [24]. When compared to pembrolizumab alone, the synergistic use of mRNA-4157 (V940) vaccine with pembrolizumab enhanced recurrence-free survival in high-risk resected melanoma. Radiotheranostics uses the same ligand to combine targeted radionuclide therapy and diagnostic imaging. An example is ~177Lu-PSMA-617, which showed a survival advantage in the VISION trial for metastatic castration resistant prostate cancer.

## 5.1 Personalized Cancer Vaccines

Over millennia, immunity has developed to efficiently respond to and thwart future pathogen attacks, offering a proven model for infectious disease vaccines. However, tumors are not sudden external events; rather, they develop within the host and only become clinically noticeable after escaping the immunological pressure that governs early neoplasms [25]. Furthermore, in order to facilitate escape, surviving tumor cells co-opt other cells, including immune cells, and build their inherent defences. Both innate and adaptive immune cells are impacted by the ensuing complex ecology of the immunosuppressive microenvironment, which can even extend beyond the tumor border and even change recently infiltrated cells [26]. Therefore, preventative vaccinations for infectious pathogens present different issues than therapeutic cancer vaccines, which are intended to treat the existing disease.

### 5.1.1 Neoantigen-Based Vaccine

It has been demonstrated that neoantigen-based personalized therapeutic cancer vaccines are safe, effective, and immunogenic in glioblastoma and melanoma patients. Clinical trials involving patients with different tumor types are now testing different vaccination forms and administration methods. To better understand their therapeutic action and improve vaccination tactics, a more thorough assessment of the phenotypic, functional, and long-lasting memory potential of vaccine-induced neoantigen-specific CD4+ and CD8+ T cells is necessary. To examine immunogenic neoepitopes that CD8+ T cells can recognize, neoantigen target finding is constantly being improved. Algorithms for the more difficult task of predicting CD4+ T cell neoepitopes are also emerging [27].

### 5.1.2 mRNA Vaccine

In order to trigger an immune response against the cancer, mRNA-based cancer vaccines encode tumor-specific antigens, which the patient's cells—especially antigen-presenting cells—then convert into proteins [28]. By specifically targeting cancer cells and using the body's immune system to combat the illness, this approach may produce better results with fewer side effects. The viability of mRNA-based cancer treatments has been greatly increased by recent developments in mRNA stability, delivery methods, and immunogenicity. By shielding mRNA from enzymatic breakdown, promoting its uptake by cells, and raising its expression levels, methods such as lipid nanoparticle encapsulation have increased mRNA stability and transport efficiency [29]. These developments demonstrate how mRNA technology can trigger robust and targeted immune responses, bolstering the usefulness and effectiveness seen in clinical settings.

## 5.2 Radiotheranostics

Theranostics is a fast-developing field of medicine that is a prime example of precision medicine. It involves diagnosing diseases and treating people who have them using instruments for therapy that are the same or like those used for diagnosis. Recently, theranostics was proposed as a potential priority for cancer research in the United States [30]. In keeping with the idea of precision medicine, it describes the methodical fusion of specific tests and treatments. The use of diagnostic techniques to identify patients who are candidates for a certain treatment is known as theranostics. In a larger sense, this might include using isotope-based techniques to visualize and quantify a target's expression, followed by therapy with chemicals, such as nonradioactive molecules that bind to the target. Imaging, particularly positron emission tomography (PET), enables noninvasive quantitative study of all lesions in the body, not only the biopsied ones, in contrast to the immunohistochemistry analysis of biopsies [31]. Patients with neuroendocrine tumors and prostate cancer who exhibit the proteins prostate-specific membrane antigen (PSMA) and somatostatin receptor 2 (SSTR2) in their malignancy are increasingly undergoing radiotheranostics, which use radiopharmaceuticals. The discovery that a percentage of patients with significant metastatic load exhibit both complete and partial responses is causing a noticeable buzz in the profession and is spurring the hunt for other radiotheranostics techniques. While not all patients will benefit from radiotheranostics, there is a very high chance of a response for those who pass the target-detected line. Thyroid cancer can be treated with 131I, osseous metastases can be treated with 223Ra, neuroendocrine tumors can be treated with 177Lu-DOTATATE, and prostate cancer can be treated with 177Lu-PSMA (prostate-specific membrane antigen) are some established ones [32]. The theranostic toolbox is growing with new targets, including integrins and fibroblast activation protein (FAP). Accurate patient selection and therapy monitoring are guaranteed by companion PET imaging.

# 6 DDR and Synthetic-Lethal Strategies

Loss of DNA damage repair (DDR) pathways has been identified in recent decades as an early and common occurrence in carcinogenesis, occurring in 40–50% of many cancer types. DDR-deficient tumors that rely on backup DNA repair mechanisms are the foundation of synthetic lethality in cancer treatment. The idea of synthetic lethality in cancer has been expanded to include gene pairs, wherein pharmacological suppression of one gene and deletion or mutation of the other results in the death of cancer cells while normal cells are immune to the drug's effects [33]. In BRCA1/2-defective cells, the paradigm investigation involves blocking PARP to cause cell death. ATM, ATR, CHK1/2, and WEE1 inhibitors are among the increasing number of new DDR inhibitors undergoing preclinical and clinical development since the effective use of PARP inhibitors. Synthetic lethality efficiency may be greatly increased by combining PARP inhibitors with other DDR inhibitors or by combining many elements of the same pathway. The silencing of

DDR genes by aberrant epigenetic modifications opens up new possibilities for synthetic lethal therapy in cancer as epigenetics joins Knudson's two-hit theory. Novel therapeutic treatments for cancer have been developed because of an understanding of the epigenetic modifications that cause loss-of-function. DDR and associated genes, such as BRCA1/2, MGMT, WRN, MLH1, CHFR, P16, and APC, have been reported to be often methylated in human malignancies. It is possible for both genetic and epigenetic changes to function as artificially fatal therapeutic indicators.

Notably, by focusing on their synthetic lethality partners, the synthetic lethality technique enables the therapeutic exploitation of directly difficult-to-target (hard-druggable) oncogenes and non-druggable mutant tumor suppressor genes [34]. By exploiting the intrinsic distinctions between cancer cells and healthy cells, synthetic lethality techniques can produce the effect of "target damage," something that conventional chemotherapy is unable to accomplish [35]. Hartwell was the first to use conventional genetic techniques, genomic data, and model organisms to find and validate new targets for medications that would specifically kill tumor cells with a specific molecular target. This was due to the differences in genomic features between cancer cells and healthy cells. To identify artificially lethal gene pairs in human cells, siRNA and CRISPR screenings have now been developed [36]. By examining vast amounts of cancer genomic data, Jerby-Arnon et al. recently presented a computational and bioinformatics method called the data mining synthetic lethality identification pipeline (DAISY) to find genome-wide synthetic lethal interactions. They created a synthetic lethal network with 2077 genes and 2816 synthetic lethal interactions, as well as a synthetic dosage lethal (SDL) network with 3158 genes and 3635 SDL interactions, by using DAISY to find every gene pair that is likely to be synthetically lethal in cancer [37].

## 7 AI and Digital Twins in Precision Oncology

There are a few benefits of using AI and machine learning in cancer research, chief among them being the ability to analyze information more quickly and improve the precision of clinical judgment. Known as "Intelligent Systems" (IS), the primary enabling tools used in precision, digital, and translational medicine today make use of previously unheard-of data quantities and seek to predict the factors and underlying diverse influences that are linked to patient outcomes. Given that IS performance and functionality are linked to intricate judgements about diagnosis and treatment, a wide range of patterns and characteristics found in high-dimensional data may be essential for drawing conclusions [38].

### 7.1 Deep Learning

Clinical and translational oncology have employed a variety of DL techniques that integrate genomic, transcriptomic, or histopathological data to enhance patient diagnosis, prognosis, and therapy selection. Nevertheless, despite the new DL techniques, human involvement is still essential in oncology [39]. Thus, the goal of DL is not to outperform or replace humans, but to develop decision support tools that aid medical professionals in the clinical management of cancer patients and cancer researchers in their investigation of the disease [40]. A histology image viewed at high magnification (typically 20x or 40x) can reveal millions of tiny cellular characteristics, and deep CNN models are excellent at extracting features from high-resolution image materials [41]. Deep CNNs based on histology have been effectively used to automate the grading of cancer; studies show that deep CNNs can grade prostate [42], breast [43], colon cancer [44], and lymphoma [45] with performance comparable to pathologists. Explainability techniques can help histology-based classification models by allowing pathologists to validate DL-generated predictions. Hägele & Co. utilized whole-slide lung cancer images to apply the Layerwise Relevance Propagation (LRP) technique to DL models that differentiate between malignant and healthy areas [39].

## 7.2 Digital Twins

Digital twins in oncology integrate real-time data, including tumor features, genomic profiles, and physiological markers, to model treatment results. Clinicians can use these simulations to predict how a patient's illness will develop and how they might react to different treatment approaches. To optimize irradiation regimens for high-grade gliomas, Chaudhuri et al. used a predictive digital twin [46]. This allowed for the fine-tuning of radiation dosages to maximize tumor control while minimizing collateral harm to healthy tissue. Hernandez-Boussard et al. demonstrated how digital twins can mimic the course of cancer and the consequences of treatment, enabling more complex and successful clinical judgments [47]. The following are the main growing areas of cancer applications for digital twins:

- Treatment optimization involves predicting how patients will react to immunotherapy, radiation, and chemotherapy while adjusting dosages to minimize adverse effects.
- Clinical Trial Simulations: Predicting patient reactions, optimizing trial design, and simulating clinical trials with digital twins.
- Oncologists can benefit greatly from therapeutic decision support, which provides real-time data on tumor progression and potential responses to ongoing treatments [48].

These methods seek to lower the hazards associated with conventional medicines while increasing therapeutic efficacy.

## 8 Conclusion

A revolutionary age in cancer care has begun with the introduction of new technologies and innovative personalized medicine strategies. Clinicians can now create patient-specific therapies that optimize therapeutic efficacy while minimizing toxicity by combining genetic, epigenomic, proteomic, and metabolomic findings. Innovations like single-cell profiling and NGS are transforming therapy optimization, real-time illness monitoring, and early detection. Liquid biopsies offer real-time disease surveillance. Organoid models, CRISPR-based screening, multiplex imaging, and AI-driven analytics are also helping to improve predictive biomarkers, find new treatment targets, and get a deeper mechanistic knowledge of tumor biology. AI-driven digital twins provide adaptive, patient-specific treatment; patient-derived models allow functional testing; individualized vaccinations and theranostics expand therapeutic possibilities; and single-cell and spatial omics resolve tumor complexity.

The trajectory of research and early clinical integration indicates that these approaches will be widely adopted over the future years, despite issues with cost, standardization, and equity. Widespread use is restricted by cancer heterogeneity, the high expense of sophisticated technologies, and the requirement for strong clinical confirmation. Careful consideration must also be given to ethical issues pertaining to data protection and fair access. However, personalized oncology appears to be headed in a positive direction. The combination of multi-omics data and computational modelling will probably speed up the creation of tailored, adaptive treatment plans as precision techniques continue to advance.

Finally, by tailoring treatments to each patient's own genetic and clinical profile, personalized medicine has the potential to completely transform cancer care, increasing survival and quality of life. This paradigm change reflects a more compassionate, patient-based approach to oncology in addition to scientific advancement.

**Declarations**

**Author's Contributions** Conceived the idea and review structure: SG, SK, MKG; Wrote the manuscript: SG; Revised and edited the manuscript: SG, SK, MKG. The author has read, agreed, and approved the final draft of the manuscript.

**Funding** This work is supported by the Focused Basic Research (FBR), OLP121, HCT, CSIR, Govt. of India.

**Conflicts of Interest** There is no competing and conflict of interest.

## References


1. Murciano-Goroff YR, Suehnholz SP, Drilon A, Chakravarty D (2023) Precision oncology: 2023 in review. Cancer Discov 13(12):2525–2531
2. Peng A, Mao X, Zhong J, Fan S, Hu Y (2020) Single-cell multi-omics and its prospective application in cancer biology. Proteomics 20(13):1900271. https://doi.org/10.1002/pmic.201900271
3. Lim J, Park C, Kim M, Kim H, Kim J, Lee DS (2024) Advances in single-cell omics and multiomics for high-resolution molecular profiling. Exp Mol Med 56(3):515–526
4. Liang A, Kong Y, Chen Z et al (2024) Advancements and applications of single-cell multiomics techniques in cancer research: unveiling heterogeneity and paving the way for precision therapeutics. Biochem Biophys Rep 37:101589
5. Lan W, Chen J, Liu M, et al. Deep imputation bi-stochastic graph regularized matrix factorization for clustering single-cell RNA-sequencing data. IEEE Trans Comput Biol Bioinform. Published online 2024. Accessed August 22, 2025. https://ieeexplore.ieee.org/abstract/document/10496964/
6. Xue R, Li R, Guo H et al (2016) Variable intra-tumor genomic heterogeneity of multiple lesions in patients with hepatocellular carcinoma. Gastroenterology 150(4):998–1008
7. Wu X, Yang X, Dai Y et al (2024) Single-cell sequencing to multi-omics: technologies and applications. Biomark Res 12(1):110. https://doi.org/10.1186/s40364-024-00643-4
8. Jin Y, Zuo Y, Li G et al (2024) Advances in spatial transcriptomics and its applications in cancer research. Mol Cancer 23(1):129. https://doi.org/10.1186/s12943-024-02040-9
9. Nikanjam M, Kato S, Kurzrock R (2022) Liquid biopsy: current technology and clinical applications. J Hematol Oncol 15(1):131. https://doi.org/10.1186/s13045-022-01351-y
10. Ricciuti B, Jones G, Severgnini M et al (2021) Early plasma circulating tumor DNA (ctDNA) changes predict response to first-line pembrolizumab-based therapy in non-small cell lung cancer (NSCLC). J Immunother Cancer 9(3):e001504
11. Jensen TJ, Goodman AM, Kato S et al (2019) Genome-wide sequencing of cell-free DNA identifies copy-number alterations that can be used for monitoring response to immunotherapy in cancer patients. Mol Cancer Ther 18(2):448–458
12. Qu S, Xu R, Yi G et al (2024) Patient-derived organoids in human cancer: a platform for fundamental research and precision medicine. Mol Biomed. 5(1):6. https://doi.org/10.1186/s43556-023-00165-9
13. Xu H, Jiao D, Liu A, Wu K (2022) Tumor organoids: applications in cancer modeling and potentials in precision medicine. J Hematol Oncol 15(1):58. https://doi.org/10.1186/s13045-022-01278-4
14. Liu Y, Wu W, Cai C, Zhang H, Shen H, Han Y (2023) Patient-derived xenograft models in cancer therapy: technologies and applications. Signal Transduct Target Ther 8(1):160
15. Germain N, Dhayer M, Dekiouk S, Marchetti P (2022) Current advances in 3D bioprinting for cancer modeling and personalized medicine. Int J Mol Sci 23(7):3432
16. Chung JH, Naficy S, Yue Z et al (2013) Bio-ink properties and printability for extrusion printing living cells. Biomater Sci 1(7):763–773
17. Gil JF, Moura CS, Silverio V, Gonçalves G, Santos HA (2023) Cancer models on chip: paving the way to large-scale trial applications. Adv Mater 35(35):2300692. https://doi.org/10.1002/adma.202300692
18. Abed H, Radha R, Anjum S et al (2024) Targeted Cancer Therapy-on-A-Chip. Adv Healthc Mater 13(29):2400833. https://doi.org/10.1002/adhm.202400833
19. Yi HG, Jeong YH, Kim Y et al (2019) A bioprinted human-glioblastoma-on-a-chip for the identification of patient-specific responses to chemoradiotherapy. Nat Biomed Eng 3(7):509–519
20. Sontheimer-Phelps A, Hassell BA, Ingber DE (2019) Modelling cancer in microfluidic human organs-on-chips. Nat Rev Cancer 19(2):65–81

21. Jeon JS, Bersini S, Gilardi M et al (2015) Human 3D vascularized organotypic microfluidic assays to study breast cancer cell extravasation. Proc Natl Acad Sci 112(1):214–219. https://doi.org/10.1073/pnas.1417115112
22. Lei W, Zhou K, Lei Y, Li Q, Zhu H (2025) Cancer vaccines: platforms and current progress. Mol Biomed 6(1):3. https://doi.org/10.1186/s43556-024-00241-8
23. Shemesh CS, Hsu JC, Hosseini I et al (2021) Personalized cancer vaccines: clinical landscape, challenges, and opportunities. Mol Ther 29(2):555–570
24. Sellars MC, Wu CJ, Fritsch EF (2022) Cancer vaccines: building a bridge over troubled waters. Cell 185(15):2770–2788
25. Dunn GP, Bruce AT, Ikeda H, Old LJ, Schreiber RD (2002) Cancer immunoediting: from immunosurveillance to tumor escape. Nat Immunol 3(11):991–998
26. Li H, Peng K, Yang K et al (2022) Circular RNA cancer vaccines drive immunity in hard-to-treat malignancies. Theranostics 12(14):6422
27. Blass E, Ott PA (2021) Advances in the development of personalized neoantigen-based therapeutic cancer vaccines. Nat Rev Clin Oncol 18(4):215–229
28. Yaremenko AV, Khan MM, Zhen X, Tang Y, Tao W (2025) Clinical advances of mRNA vaccines for cancer immunotherapy. Med 6(1). Accessed August 21, 2025. https://www.cell.com/med/abstract/S2666-6340(24)00456-2
29. Yao R, Xie C, Xia X (2024) Recent progress in mRNA cancer vaccines. Hum Vaccin Immunother 20(1):2307187
30. Jaffee EM, Van Dang C, Agus DB et al (2017) Future cancer research priorities in the USA: a lancet oncology commission. Lancet Oncol 18(11):e653–e706
31. Aboagye EO, Barwick TD, Haberkorn U (2023) Radiotheranostics in oncology: making precision medicine possible. CA Cancer J Clin 73(3):255–274. https://doi.org/10.3322/caac.21768
32. Pomykala KL, Hadaschik BA, Sartor O et al (2023) Next generation radiotheranostics promoting precision medicine. Ann Oncol 34(6):507–519
33. Pilié PG, Tang C, Mills GB, Yap TA (2019) State-of-the-art strategies for targeting the DNA damage response in cancer. Nat Rev Clin Oncol 16(2):81–104
34. Gao A, Guo M (2020) Epigenetic based synthetic lethal strategies in human cancers. Biomark Res 8(1):44. https://doi.org/10.1186/s40364-020-00224-1
35. McLornan DP, List A, Mufti GJ (2014) Applying synthetic lethality for the selective targeting of cancer. N Engl J Med 371(18):1725–1735. https://doi.org/10.1056/NEJMra1407390
36. Huang A, Garraway LA, Ashworth A, Weber B (2020) Synthetic lethality as an engine for cancer drug target discovery. Nat Rev Drug Discov 19(1):23–38
37. Jerby-Arnon L, Pfetzer N, Waldman YY et al (2014) Predicting cancer-specific vulnerability via data-driven detection of synthetic lethality. Cell 158(5):1199–1209
38. Capobianco E (2022) High-dimensional role of AI and machine learning in cancer research. Br J Cancer 126(4):523–532
39. Zhu W, Xie L, Han J, Guo X (2020) The application of deep learning in cancer prognosis prediction. Cancer 12(3):603
40. Walsh S, De Jong EEC, Van Timmeren JE et al (2019) Decision support systems in oncology. JCO Clin Cancer Inform 3:1–9. https://doi.org/10.1200/CCI.18.00001
41. Maggiori E, Tarabalka Y, Charpiat G, Alliez P (2017) High-resolution image classification with convolutional networks. In: 2017 IEEE international geoscience and remote sensing symposium (IGARSS). IEEE, pp 5157–5160. Accessed August 22, 2025. https://ieeexplore.ieee.org/abstract/document/8128163/
42. Ryu HS, Jin MS, Park JH et al (2019) Automated Gleason scoring and tumor quantification in prostate core needle biopsy images using deep neural networks and its comparison with pathologist-based assessment. Cancer 11(12):1860
43. Ehteshami Bejnordi B, Mullooly M, Pfeiffer RM et al (2018) Using deep convolutional neural networks to identify and classify tumor-associated stroma in diagnostic breast biopsies. Mod Pathol 31(10):1502–1512
44. Vuong TLT, Lee D, Kwak JT, Kim K (2020) Multi-task deep learning for colon cancer grading. In: 2020 International conference on electronics, information, and communica-

tion (ICEIC). IEEE, pp 1–2. Accessed August 22, 2025. https://ieeexplore.ieee.org/abstract/document/9051305/
45. El Achi H, Khoury JD (2020) Artificial intelligence and digital microscopy applications in diagnostic hematopathology. Cancer 12(4):797
46. Chaudhuri A, Pash G, Hormuth DA et al (2023) Predictive digital twin for optimizing patient-specific radiotherapy regimens under uncertainty in high-grade gliomas. Front Artif Intell 6:1222612
47. Hernandez-Boussard T, Macklin P, Greenspan EJ et al (2021) Digital twins for predictive oncology will be a paradigm shift for precision cancer care. Nat Med 27(12):2065–2066
48. Giansanti D, Morelli S (2025) Exploring the potential of digital twins in cancer treatment: a narrative review of reviews. J Clin Med 14(10):3574

# Applications of Bioinformatics and Computational Biology

Shweta Singh, Mohini Milmile, Pandey Ajay Sanjay, and Ruchi Chawla

**Abstract**

Personalized cancer therapy is transforming oncology by tailoring treatment paradigms to patient's unique molecular tumor profiles. This chapter focuses on using next-generation sequencing (NGS), single-cell RNA sequencing, and liquid biopsies, combined with computational biology, to tackle tumor heterogeneity and personalize precision medicine. Clinicians can define actionable mutations and develop personalized treatments by deciphering genomic, transcriptomic, proteomic, and epigenomic changes. Bioinformatics is essential to analyze complex data. The processes include variant calling, pathway analysis, neoantigen prediction, and clinical records integration, which are now integral to personalized oncology pipelines. The role of machine learning (ML) and artificial intelligence (AI) is to improve this process by determining oncogenic drivers, drug response predictions, and patient stratification. PERCEPTION and PanDrugs exemplify how therapeutic optimization surpasses guidelines using AI-derived insights.

Multi-omics integration enhances resolution by correlating several molecular levels, where emergent biomarkers and resistance mechanisms are uncovered. Computational models like MOFA and netDx enable patient stratification and predictive modeling of outcomes. NCI-MATCH, WINTHER, and I-PREDICT trial case studies illustrate the clinical utility of this approach. Challenges are ongoing; however, with data standardization, interpretability of models, and access parity, there has been an enhancement in the prediction efficiency. Ethical considerations about privacy and informed consent are paramount given the growing clinical use of genomic data. Directions for future development are

S. Singh · M. Milmile · P. A. Sanjay · R. Chawla (✉)
Department of Pharmaceutical Engineering and Technology, Indian Institute of Technology (BHU), Varanasi, India
e-mail: rchawla.phe@iitbhu.ac.in



M. K. Ghosh, M. Basu (eds.), *Personalized Cancer Therapy*,
https://doi.org/10.1007/978-981-95-6845-1_22

broadening diverse omics databases, enhancing transparency in AI, and creating predictive models like digital twins for adaptive therapy. Computer-aided personalized cancer treatments provide a paradigm shift toward effective, accurate, and patient-specific oncologic treatment.

**Keywords**

Precision oncology · Multi-omics integration · Bioinformatics · Next-Generation Sequencing (NGS) · Computational oncology · Personalized cancer therapy

## 1 Introduction

Cancer encompasses a broad spectrum of disorders that can develop virtually in any tissue, each shaped by unique molecular alterations and evolutionary dynamics. One of the significant challenges in cancer management is tumor heterogeneity—there is a high degree of variability not only between patients but also within a single tumor. This complexity makes cancer difficult to diagnose and treat effectively, as uniform therapies often fail to address the malignant subtypes, enabling resistant cells to survive and spread [1, 2].

The rise of personalized cancer therapy represents a paradigm shift from traditional, uniform treatments toward approaches tailored to each patient's specific genetic, molecular, and clinical features. Targeted treatments like trastuzumab and imatinib have revolutionized the management of HER2-positive breast cancer and BCR-ABL-positive leukemia, illustrating the power of aligning drugs with specific molecular drivers [3, 4]. However, in order to tackle the complexity of cancer, there is a need for multilayered molecular data—genomics, transcriptomics, proteomics—and sophisticated computational methods to interpret them [5].

Computational biology and bioinformatics have become indispensable in managing and interpreting the massive volumes of data generated by cancer research. Data from flagship projects—including The Cancer Genome Atlas and the International Cancer Genome Consortium—demonstrate that the vast bulk of tumors possess several key "driver" mutations amid numerous benign "passenger" variants. Distinguishing these drivers is a central bioinformatics challenge, with integrative analyses showing that approximately 91% of tumors contain at least one known driver alteration. At the same time, a minority remain unexplained, highlighting both the power and the ongoing challenge of cancer genomics.

Precision oncology leverages molecular markers to guide treatment. For example, KRAS mutations in colorectal cancer predict lack of response to EGFR-targeted therapies, enabling more effective, individualized care. High-throughput genomic assays are increasingly used in clinics to match patients with suitable drugs or clinical trials, with studies indicating that over 75% of cancers harbor changes linked to available therapies [6]. However, robust computational pipelines are essential: careful comparison of tumor and normal DNA is needed to avoid misclassifying benign germline variants as actionable mutations, thus preventing inappropriate treatments.

In summary, computational biology and bioinformatics are critical for translating biological data into clinical decision-making in cancer. By uncovering new biomarkers and actionable targets, these disciplines form the backbone of precision medicine. This chapter thus helps explore the applications and implications of computational approaches in personalized cancer therapy.

## 2 Personalized Therapeutics Approaches in Oncology

Personalized oncology applies molecular profiling of tumors to customize treatment plans, with biomarker-guided strategies enhancing therapeutic precision. This approach shifts from the traditional "one-size-fits-all" chemotherapy model to therapies based on specific genetic or phenotypic features [7]. Targeted therapies and immunotherapies have improved outcomes in defined patient subsets.

### 2.1 Historical Milestones in Targeted Therapy

In the 2000s, targeted cancer therapy—drugs that target particular genetic drivers of tumor growth—revolutionized oncology. The monoclonal antibody trastuzumab (Herceptin) targeting the breast cancer HER2 receptor was an early breakthrough. Trastuzumab was FDA-approved in 1998 after a significant study demonstrated that combining it with chemotherapy improved survival in HER2-positive metastatic breast cancer [3]. This showed that choosing patients based on a tumor biomarker (HER2 overexpression) might improve therapeutic outcomes. Another breakthrough was imatinib (Gleevec), a small-molecule BCR-ABL tyrosine kinase inhibitor. Targeting the disease-causing fusion oncoprotein, imatinib showed extraordinary response rates in chronic myeloid leukaemia [4]. Targeting an oncogenic driver may mitigate a deadly malignancy. Imatinib's success inspired a rush of kinase inhibitor research throughout oncology. Researchers found activating mutations in the EGFR gene in a subgroup of lung adenocarcinomas in the early 2000s, explaining their significant responses to gefitinib and erlotinib [8]. EGFR mutation testing became standard in non-small cell lung cancer (NSCLC) in 2004–2005, and EGFR inhibitors outperformed chemotherapy in EGFR-mutant patients, a molecularly stratified therapeutic breakthrough. Subsequent milestones shortly followed: ALK gene fusions were found in lung malignancies and effectively targeted by the ALK inhibitor crizotinib, with an early study in 2010 indicating significant response rates in ALK-positive patients [9]. Likewise, in melanoma, identifying the BRAF V600E mutation led to the BRAF inhibitor vemurafenib, which in 2011 was demonstrated to enhance survival in BRAF-mutant metastatic melanoma, altering therapy for that illness [10]. Each of these successes—HER2 in breast cancer, BCR-ABL in CML, EGFR and ALK in lung cancer, BRAF in melanoma—gave a blueprint for matching targeted medications to the individuals most likely to benefit, therefore inaugurating the age of precision oncology. By the 2010s, dozens of targeted medicines (against VEGF, HER2, RAF, MEK, PARP, etc.) were licensed, generally accompanying companion diagnostic tests to detect cancers with the required target or mutation [11]. Cumulatively, these accomplishments revealed that leveraging tumor genomes might generate strong medicines that are more successful and frequently less toxic than generic chemotherapy, so long as the correct patients are picked [12]. Targeted therapy has therefore become a cornerstone of customized cancer treatment, and is consistently increasing, as novel driver mutations and pathways are being discovered.

## 2.2 Biomarkers in Personalized Cancer Therapy

Biomarkers are measurable indicators of a biological state or condition, and, in oncology, they play a pivotal role in tailoring treatments to patients. In personalized cancer therapy, biomarkers can stratify patients likely to benefit from a particular treatment (predictive biomarkers) or indicate the overall prognosis regardless of therapy (prognostic biomarkers). The advent of high-throughput omics and bioinformatics has greatly expanded the universe of candidate biomarkers, moving beyond classical single-gene markers to complex molecular signatures derived from large datasets [13]. Table 1 shows key molecular alterations, such as protein overexpression, gene mutations, or mRNA expression levels, that are used as biomarkers in various cancer types.

## 2.3 Biomarker-Driven Clinical Trials

Biomarkers like gene mutations, protein expression, and tumor burden are used as a guide for personalized cancer therapy. Their diversity challenges traditional trials, necessitating adaptive designs such as basket, umbrella, and N-of-1 studies. These innovative frameworks are essential for translating molecular insights into clinically effective precision oncology strategies. Oncology has embraced personalized medicine through biomarker-driven clinical trial designs, notably basket and umbrella trials [7]. Basket trials enrol patients with a common molecular alteration across cancer types, as shown by vemurafenib's efficacy in nonmelanoma cancers with BRAFV600 mutations [26]. Umbrella trials, such as Lung-MAP, assign patients to treatment arms based on distinct biomarkers within a single cancer type, allowing parallel evaluation of targeted therapies.

These designs have enabled tumor-agnostic drug approvals, where treatments are based on biomarkers rather than cancer histology. The FDA approved pembrolizumab for MSI-H tumors and Larotrectinib for NTRK fusions, following impressive basket trial results [11, 12]. Trials like NCI-MATCH integrate genomic screening to assign patients to appropriate therapies, supported by adaptive statistical designs and validated companion diagnostics [27].

**Table 1** Clinically relevant molecular biomarkers in cancer and their associated targeted therapies

| Biomarker (Molecular Alteration) | Cancer | Associated targeted therapy | Ref |
|---|---|---|---|
| PD-L1 overexpression (protein) | Multiple cancers (e.g., NSCLC, melanoma) | PD-1/PD-L1 pathway inhibitors like the monoclonal antibody pembrolizumab | [13, 14] |
| EGFR activating mutation (DNA) | NSCLC | Inhibitors of EGFR tyrosine kinase activity, such as Osimertinib | [8] |
| BRCA1/2 loss-of-function mutation (DNA) | Breast and ovarian cancers | PARP inhibitors (e.g., Olaparib) | [15] |
| High expression of recurrence-associated genes (mRNA) | Breast cancer | Recurrence prediction, chemotherapy decision (e.g., Oncotype DX, MammaPrint) | [16] |
| High Tumor Mutational Burden (TMB) | Multiple cancers | Checkpoint inhibitors (e.g., PD-1 blockade) | [14] |
| DNA damage repair gene alterations (BRCA2, ATM) | Prostate cancer | PARP inhibitors and platinum chemotherapy | [17] |
| Multi-omics immune signatures (e.g., RNA + TMB) | Immunotherapy-eligible tumors | Improved immunotherapy prediction | [18] |
| Network dysregulation biomarkers (PPI networks) | Various cancers | Network-based targeted therapies | [19] |
| HER2 overexpression (protein) | Breast cancer | HER2-targeted therapies (e.g., trastuzumab) | [20] |
| KRAS mutation (DNA) | Colorectal cancer | Predicts resistance to EGFR inhibitors | [21] |
| Anaplastic lymphoma kinase (ALK) rearrangement (DNA) | NSCLC | Inhibitors of ALK (e.g., crizotinib) | [22] |
| BRAF V600E mutation (DNA) | Melanoma, colorectal cancer | BRAF inhibitors (e.g., vemurafenib) | [10] |
| TP53 mutation (DNA) | Various cancers | Prognostic marker; potential for targeted therapies | [23] |
| Circulating Tumor Cells (CTCs) | Breast, prostate, colorectal cancers | Prognostic marker; monitoring treatment response | [24] |
| Microsatellite Instability-High (MSI-H) | Colorectal, endometrial, gastric cancers | Predictive of response to immunotherapy (e.g., pembrolizumab) | [11] |
| Oncotype DX gene expression profile | Breast cancer | Predicts benefit from chemotherapy | [25] |

### 2.4 N-of-1 and Adaptive Therapy Paradigms

N-of-1 trials and adaptive therapy are emerging paradigms reshaping personalized cancer treatment. This trial is conducted on an individual participant, enabling intra-individual comparison of therapeutic outcomes, allowing clinicians to test tailored therapies sequentially and refine treatment based on real-time responses [28]. These trials are particularly valuable for rare cancers or patients with unique genomic profiles. The I-PREDICT study exemplified this approach, using individualized therapy regimens based on genomic profiling for advanced cancer patients, with dynamic adjustments to optimize outcomes [29].

In contrast to curative approaches, adaptive therapy seeks to stabilize tumor burden by sustaining a balance between sensitive and resistant cell populations. Based on evolutionary dynamics, it avoids the selective pressure of continuous high-dose treatment that favors resistant clones [30]. Instead, intermittent or dose-adjusted strategies preserve sensitive cells, which suppress resistant populations. Preclinical studies and early trials, such as one involving metastatic prostate cancer, have shown prolonged disease control using adaptive approaches [31, 32]. This strategy requires real-time monitoring and mathematical models to guide dosing decisions [30], offering a promising alternative to traditional aggressive therapy, especially in resistance-prone cancers. However, Basket trials can be limited by small subpopulations for rare biomarkers, and N-of-1 trials, though being of an appealing nature, face statistical and logistical challenges in drawing broadly applicable conclusions.

## 3 Next-Generation Sequencing in Cancer Genomics

The advent of NGS has enabled comprehensive characterization of tumor genomes, catalyzing the evolution of individualized cancer treatment strategies. By rapidly sequencing DNA and RNA, NGS enables comprehensive profiling of mutations, gene expression, copy number alterations, and other genomic features that drive cancer. In clinical practice, NGS is deployed at various scales—from examining a handful of genes to entire genomes—depending on the informational needs and practical considerations. The main sequencing approaches used in cancer genomics are the analytic steps of variant calling and interpretation, and emerging applications like liquid biopsies. The table outlines key features, benefits, limitations, and applications of major NGS approaches, including Whole-Genome Sequencing (WGS), Whole-Exome Sequencing (WES), Targeted Panel Sequencing, and Variant Calling and Annotation. These techniques differ in genomic coverage, cost, sensitivity, and clinical utility, and are selected based on diagnostic or research needs (Table 2).

**Table 2** Comparison of Next Generation Sequencing (NGS) techniques used in cancer genomics

| Techniques | Key features | Advantages | Limitations | Use case | Ref |
|---|---|---|---|---|---|
| Whole-Genome Sequencing (WGS) | Sequences the entire genome; detects all mutation types, including structural and noncoding. | Comprehensive; reveals novel alterations; suitable for structural variant detection. | High cost, complex data, lower coverage, harder to interpret. | Research: complex/ undiagnosed cancers; rare mutation discovery. | [33, 34] |
| Whole-Exome Sequencing (WES) | Targets the exonic regions of genes, constituting approximately 1–2% of the human genome. | Lower cost, deep coverage, and well-supported pipelines. | Misses noncoding/ structural variants; variable coverage. | Broad tumor profiling in research and clinic; rare cancers. | [33, 34] |
| Targeted Panel Sequencing | Limited gene sets; deep sequencing of known mutations. | Cost-effective, fast, straightforward interpretation, and deep coverage. | Limited scope; misses novel/ rare alterations. | Routine diagnostics, liquid biopsies, and therapy guidance. | [34] |
| Variant Calling and Annotation | Processes raw NGS data to detect mutations; adds functional context. | Detects SNVs, indels, CNVs, structural variants; uses knowledge bases for interpretation. | Interpretation can be ambiguous; it depends on the quality of the reference data. | Core to all sequencing-based diagnostics, guides therapy. | [13, 27] |
| Liquid Biopsy/ ctDNA | Detects tumor DNA fragments in blood. | Noninvasive; real-time monitoring; systemic view of tumor genomics. | Low ctDNA fraction; high sensitivity needed; risk of false positives. | Genotyping, resistance tracking, minimal residual disease, adaptive therapy. | [35, 36] |

## 4 Role of Computational Analysis in Cancer Genomics

With sequencing strategies generating massive data, computational analysis is needed to extract meaningful insights. The deluge of data from tumor genomic profiling requires advanced computational analysis to extract actionable insights. Beyond processing individual cases, computational methods enable discovery by synthesizing data across large cohorts and applying algorithms to identify patterns not apparent to humans. In cancer genomics, computational analysis plays multiple roles: discovering new driver mutations, deciphering mutational signatures of tumor etiology and evolution, and, increasingly, deploying AI to interpret genomic data and predict outcomes.

### 4.1 Data-Driven Discovery of Driver Mutations

Computational genomics has revolutionized the identification of cancer driver mutations—genetic alterations that promote tumorigenesis—by distinguishing them from abundant passenger mutations. Algorithms like MutSig assess mutation recurrence against background rates to highlight potential drivers. Integrative efforts, such as the Pan-Cancer analysis of ~9000 tumor exomes using 26 algorithms, identified 299 high-confidence driver genes and over 3400 candidate missense mutations [33]. While many were known (e.g., TP53, KRAS, EGFR), novel context-specific drivers emerged. This data-driven approach underpins precision oncology by informing drug targeting and clinical trial eligibility based on computationally predicted driver profiles, even in genes not traditionally associated with cancer.

Computational tools refine driver mutation identification within known genes by assessing the functional impact of specific variants. Algorithms like CHASM and HotSpot3D integrate ML, conservation, and structural data to distinguish oncogenic mutations (e.g., pathogenic TP53 variants) from benign ones. Network-based approaches detect driver pathways by analyzing mutual exclusivity or co-occurrence, revealing interchangeable oncogenic events (e.g., RAF or MEK mutations in MAPK signalling). These insights are increasingly used clinically to guide diagnostics and targeted therapy.

### 4.2 Mutational Signatures and Clonal Evolution

Mutational signatures are distinct patterns in tumor DNA that reflect prior DNA damage and repair processes. Computational methods like non-negative matrix factorization have identified recurrent signatures linked to UV exposure, tobacco smoke, and mismatch repair deficiency [37]. These signatures serve as molecular records, guiding personalized therapy. For example, tumors with mismatch repair deficiency respond well to PD-1 inhibitors [11], while homologous recombination deficiency (HRD) signatures predict PARP inhibitor sensitivity, even in non-BRCA-mutant tumors [38]. Tools like SigProfiler and deconstructSigs match tumor mutational spectra to known signatures, providing diagnostic and therapeutic insights. APOBEC-related signatures may reveal ongoing mutagenesis, and BRCA-like signatures can expand PARP inhibitor eligibility. These signatures also help diagnose unknown primary cancers and monitor therapy-induced changes (e.g., new signatures after chemotherapy).

Beyond mutational processes, computational models reconstruct tumor clonal evolution using variant allele frequencies and copy number data. Tools like PyClone and PhyloWGS identify subclones, revealing trunk (foundational) and branch (later) mutations. Clinically, subclonal driver mutations (e.g., EGFR in 30% of cells) may predict partial therapy responses or resistance. The TRACERx study found that high intratumoral heterogeneity correlates with worse outcomes [39]. Clonal tracking supports adaptive therapy strategies—modulating treatment to maintain sensitive clones and suppress resistant ones. Real-time monitoring of ctDNA can guide early therapy changes. Computational models simulate clonal dynamics, informing sequencing and dosing decisions [30]. Real-time clonal tracking and AI-driven analysis enable dynamic, personalized cancer therapy, thereby anticipating tumor evolution and resistance.

## 4.3 AI and ML for Genomic Interpretation in Precision Oncology

The vast and multifaceted nature of cancer genomics has necessitated the integration of AI and ML into personalized oncology strategies. These technologies can identify patterns and infer insights from high-dimensional datasets, far beyond manual capabilities, making them indispensable for interpreting genomic, transcriptomic, and clinical data. ML models are widely used to classify variants as pathogenic or benign by learning from curated datasets of driver versus passenger mutations. Tools like CHASM apply supervised learning on features such as evolutionary conservation, amino acid properties, and structural context [40]. Deep learning models further enhance this by integrating protein sequence and structural data to predict oncogenicity of novel mutations, even those rarely observed. These models outperform traditional rule-based scores (e.g., SIFT, PolyPhen) by capturing higher order biological interactions. As AI tools become more robust and validated, they are increasingly integrated into clinical variant annotation pipelines [41].

AI is vital in analyzing single-cell RNA sequencing (scRNA-seq) data due to its size and complexity. Unsupervised learning methods (e.g., clustering, UMAP) reveal intra-tumoral heterogeneity, identifying malignant subclones, immune infiltration states, and stromal populations; these insights guide therapy, e.g., identifying immunosuppressive microenvironments that may benefit from checkpoint blockade. AI also performs pseudotime analysis, reconstructing cellular differentiation trajectories, revealing points of therapeutic vulnerability. For example, expression of drug efflux pumps in a resistant subclone can be flagged through scRNA-seq AI analysis. The PERCEPTION pipeline [42] uses ML and transfer learning on single-cell transcriptomes to predict drug responses by integrating patient-specific and large-scale screening data, outperforming previous bulk-level models.

Natural Language Processing (NLP), a subfield of AI, helps manage information overload by extracting actionable insights from scientific literature and clinical trial databases. NLP can detect associations between rare mutations (e.g., FGFR alterations) and therapies, aiding clinical decision-making. Tools like OncoKB and CIViC use AI for automated evidence curation [43]. While not AI itself, high-performance computing (HPC) underpins scalable AI applications in genomics. Analysis of whole-genome data, 3D protein modeling, and digital pathology demand enormous computational power. Platforms like DRAGEN (FPGA-accelerated) significantly reduce variant calling time, while GPU-based cloud platforms enable rapid execution of deep learning models, making real-time AI-assisted tumor profiling feasible [41].

AI augments oncologist decision-making by prioritizing actionable mutations, predicting prognosis, and identifying therapy-resistant clones. For instance, AI might flag a TP53 mutation, a gene expression profile, and a clinical history that, together, predict poor prognosis—informing intensified monitoring or treatment. Importantly, AI recommendations still require clinical validation, highlighting the synergy between computational tools and expert judgment. AI is trained on unbiased datasets, validating predictions across diverse populations, and gaining regulatory approval, and, providentially, precedents exist—e.g., FDA-approved AI tools in pathology and radiology. In genomics, AI-driven decision support tools are progressing toward software-as-a-medical-device (SaMD) status.

## 5 Convergence of Computational Biology and Multi-omics Approaches in Precision Cancer Therapy

Combining computational biology with multi-omics data has revolutionized cancer research and precision medicine by enabling systems-level insights into tumor biology. Advances in high-throughput profiling now allow simultaneous measurement of genomic, transcriptomic, proteomic, metabolomic, and epigenomic layers from cancer samples. Computational integration of these diverse datasets reveals complex interactions and molecular mechanisms that underlie tumorigenesis, drug response, and resistance; however, insights are often missed when analyzing single data types [44]. Figure 1 illustrates the integration of genomics, transcriptomics, proteomics, metabolomics, and epigenomics to decode complex disease mechanisms. Through computational analysis, it reveals insights such as predictive markers and tumor immune dynamics. The outcome supports customized treatment strategies within a personalized healthcare framework.

Key frameworks for multi-omics integration include unsupervised learning techniques and network-based approaches. Multi-Omics Factor Analysis (MOFA) employs factor analysis to uncover latent patterns of variation shared across diverse omics datasets, while netDX builds patient similarity networks from each omic modality and integrates them to predict clinical outcomes [45]. These methods can correlate DNA mutations with gene expression and phosphoproteomic data to identify driver pathways. For instance, combining omics layers can uncover co-altered gene and protein networks associated with survival—systems-level biomarkers not apparent in single-modality studies.

Public repositories like TCGA and ICGC offer standardized, Findable, Accessible, Interoperable, Reusable (FAIR) multi-omics datasets, enabling reproducible analyses at scale [46]. These initiatives support integration by enforcing common data formats, ontologies, and metadata standards, allowing computational pipelines to align samples across omics types. Clinical outcome integration with omics data enhances the discovery of prognostic and predictive signatures. For example, linking treatment response to integrated omics profiles allows identification of biomarkers that guide patient stratification and therapy selection [13]. Computational workflows are now capable of modeling how genomic alterations

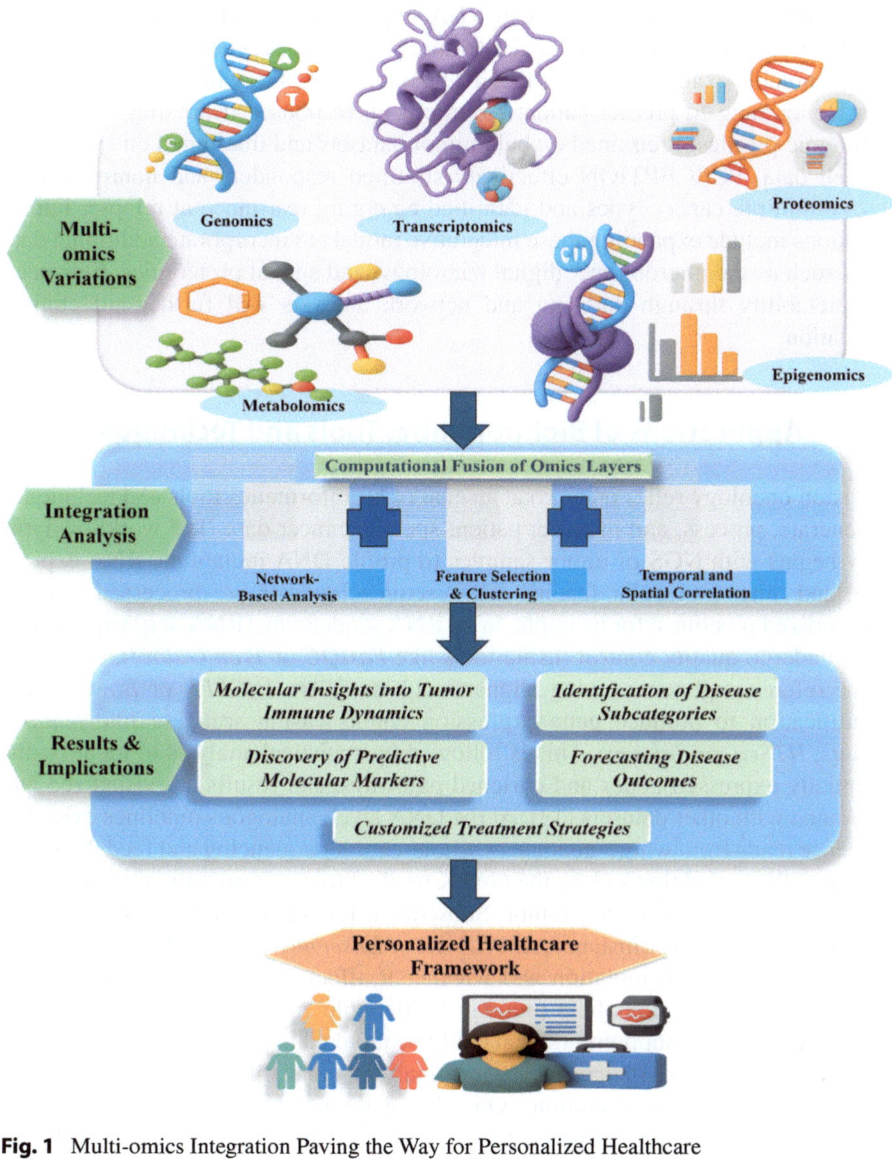


**Fig. 1** Multi-omics Integration Paving the Way for Personalized Healthcare

translate into transcriptomic shifts, protein abundance changes, and ultimately clinical phenotypes.

A cutting-edge development in multi-omics is the inclusion of single-cell and spatially resolved omics. Unlike bulk sequencing, which averages signals across heterogeneous tumor cell populations, single-cell technologies reveal cell-type-specific variations that drive therapy response or resistance. Computational approaches now integrate single-cell RNA-seq data with bulk omics and clinical outcomes to map cell-level molecular features to patient prognosis.

One notable example is the PERCEPTION pipeline [42], which applies ML and transfer learning to predict patient-specific drug responses from single-cell transcriptomic profiles. Pretrained on bulk tumor datasets and fine-tuned on sparse single-cell data, PERCEPTION effectively stratified responders and nonresponders across multiple cancer types and identified emerging resistance at relapse. Future directions include expanding these integrative models to incorporate additional data types such as the microbiome, digital pathology, and spatial proteomics. Enhanced interpretability through pathway and network analysis will further aid clinical translation.

## 6 Applications of Bioinformatics Tools and Techniques

Precision oncology relies on a broad arsenal of bioinformatics tools and techniques to generate, process, and interpret patient-specific cancer data. The workflow typically begins with NGS of tumor samples to profile DNA mutations, RNA expression, and other molecular features. Raw sequencing data are processed through standardized pipelines: for example, in an RNA sequencing (RNA-seq) experiment, reads undergo quality control (using tools like *FastQC* or *Trim Galore*), alignment to the reference genome (using aligners such as *STAR*, *HISAT2*, or *Bowtie*), and quantification to produce gene expression counts (using software like *featureCounts*, *HTSeq*, or *Salmon*). This is followed by statistical analysis to identify differentially expressed genes and enriched pathways, and results are visualized and integrated with other datasets [13]. At the DNA level, analogous pipelines exist that sequence reads from whole-genome or whole-exome sequencing and feed them into variant-calling workflows (e.g., the GATK toolkit) to detect somatic mutations and copy number alterations in a tumor. Subsequent tools perform variant annotation and interpretation—for instance, *ANNOVAR* or *Ensembl VEP* will predict the functional impact of each mutation and whether it affects known cancer genes [47]. These variant annotations help pinpoint "actionable" mutations, such as those in oncogenes or tumor suppressors, that could be targeted by a specific drug. In parallel, *copy-number* and *structural variant* analysis tools (like FACETS or Manta) identify larger genomic alterations. Once key genomic changes are identified, they are often mapped onto pathways: for example, if a mutation activates the PI3K/AKT pathway, that pathway can become a focus for therapy. Integrative tools like Gene Set Enrichment Analysis and Ingenuity Pathway Analysis support biological inference from complex omics data through network-based approaches. These tools identify which cellular pathways are perturbed in a patient's tumor, suggesting possible mechanism-based treatment strategies. Overall, a foundational toolkit in computational oncology spans from data processing programs (for sequence alignment, quantification, etc.) to analytical libraries in R/Python for statistics and ML, all orchestrated to turn raw omics data into clinically meaningful information [13].

Beyond these general pipelines, numerous specialized tools have been developed for particular applications in personalized cancer therapy. One critical class of tools is for cancer genome interpretation: platforms like cBioPortal and the Genomic Data Commons (GDC) portal allow researchers and clinicians to explore a patient's genomic alterations in the context of large cancer datasets [13]. These platforms integrate data from thousands of patients and provide interactive visualizations of mutations, copy number changes, and clinical annotations. By querying a patient's mutation profile against databases of known cancer drivers (e.g., COSMIC) and druggable targets, such tools can highlight which mutations are of potential clinical significance. Other databases like ClinVar and OncoKB provide curated knowledge on the pathogenicity of variants and their level of evidence as therapeutic biomarkers.

Data integration tools also play an important role in combining different data types for analysis; for instance, the R/Bioconductor package TCGAbiolinks can gather DNA, RNA, and clinical data from TCGA for integrated analysis [48]. Additionally, network visualization tools such as Cytoscape are commonly used to depict molecular interaction networks or gene co-expression networks inferred from multi-omics data [49]. These networks can help identify central "hub" genes or proteins that might be effective drug targets. As the scale of cancer datasets has grown, researchers have also embraced ML techniques to find patterns that human analysis might miss. In precision oncology, ML algorithms can be trained on genomic and clinical data to predict outcomes like drug sensitivity. Languages such as Python and R offer rich ecosystems for this: Python's *scikit-learn* and deep learning frameworks (TensorFlow, PyTorch) and R's ML packages (caret, randomForest, etc.) are frequently used to build predictive models. For example, ML classifiers have been applied to gene expression data to predict which patients will respond to immunotherapy versus those who will not, sometimes using complex feature sets spanning tumor mutation burden, neoantigen load, and immune gene expression signatures. An emerging area is AI-driven pathology, where computational tools analyze histopathology images to infer molecular features (e.g., predict mutation status from H&E slides)—effectively offer a suitable modality to integrate [41]. All these tools and techniques, from basic processing to advanced AI, form a layered ecosystem that enables the translation of raw biological data into actionable clinical insights.

A particularly important suite of bioinformatics tools in personalized cancer therapy focuses on neoantigen prediction and immunotherapy. Neoantigens are tumor-specific peptides arising from somatic mutations and represent critical targets for the development of personalized cancer vaccines and adoptive T cell therapies. The process of identifying candidate neoantigens is computationally intensive and involves multiple steps [50]. Initially, tumor sequencing reveals somatic mutations, notably nonsynonymous single-nucleotide variants (SNVs) or short indels. Subsequent human leukocyte antigen (HLA) typing—via sequencing or in silico approaches—is essential, as the HLA profile determines antigen presentation to cytotoxic T cells. The mutated protein sequences are then processed in silico into all possible peptide fragments (typically 8–11 amino acids for class I HLA) that encompass the mutation. These peptides are evaluated by MHC binding prediction algorithms—tools like NetMHCpan, MHCflurry, or PickPocket use ML trained on large datasets of peptide-HLA binding assays to predict how strongly each mutant peptide will bind to the patient's HLA molecules. Peptides predicted to bind with high affinity are potential neoantigens. Additional filters assess whether the peptide is likely to be produced and presented by the cell's antigen-processing machinery (e.g., using predictors for proteasomal cleavage and TAP transport). Finally, the candidate neoantigens are prioritized by combining factors such as binding affinity, peptide abundance (sometimes integrating RNA expression levels of the mutant gene), and similarity to self-peptides. To streamline the neoantigen workflow, integrated tool suites have been developed. One such platform is pVACtools, which includes modules for each step of neoantigen identification and a web-based visualization interface. The recent tool pVACview provides an interactive dashboard for reviewing predicted neoantigens, allowing users to visually prioritize candidates by various criteria (binding affinity, clonal abundance, etc.). This addresses a practical challenge: after running prediction algorithms, scientists must sift through hundreds of peptide hits—pVACview helps pinpoint the most promising targets for experimental validation [51]. Immunogenomics tools exemplify how computational techniques enable a new modality of personalized therapy: tailoring immunotherapies to the unique mutant antigens of an individual's cancer.

Another critical application of bioinformatics in personalized therapy is drug prioritization and treatment matching. Oncologists are often confronted with a list of genomic alterations in a patient's tumor, only a subset of which may be clinically actionable (i.e., targetable by an approved or experimental drug). Decision-support tools have been developed to interpret these genomic profiles and suggest potential therapies. One example is PanDrugs, a computational platform that takes as input the set of genomic alterations (mutations, copy-number changes, fusions) identified in a tumor and outputs a ranked list of anticancer drugs tailored to those alterations. PanDrugs hosts one of the most expansive databases of drug–gene interactions, spanning regulatory-approved treatments and experimental agents in ongoing trials. The tool assigns scores based on the "gene cancer relevance" (how crucial a mutated gene is to cancer, using databases of cancer drivers) and "drug feasibility" (availability and efficacy evidence for drugs against that target) [52]. In effect, PanDrugs will interpret a mutation in, say, the PIK3CA gene and identify PI3K inhibitors that could be beneficial, or if it finds an ALK gene fusion, it will prioritize ALK inhibitors. Importantly, PanDrugs goes beyond one-to-one gene–drug matches by considering pathway context and multi-gene markers—for instance, co-occurring mutations in a pathway might make a case for combination therapy or indicate a bypass route of resistance. In a proof-of-concept, the creators of PanDrugs applied it to The Cancer Genome Atlas cohort and showed it could reproduce known standard-of-care recommendations and also suggest options for cases with no obvious treatment [52]. There are other similar tools: IBM's Watson for Oncology attempted to integrate genomic data with literature to recommend therapies, and academic efforts like Molecular Tumor Board (MTB) platforms have been developed to systematize how complex genomic data is discussed by clinicians. Moreover, clinical knowledge bases such as Clinical Interpretation of Variants in Cancer (CIViC) and

OncoKB are often embedded in these tools to provide evidence levels for each biomarker–drug association [43]. As the field evolves, AI-based systems are beginning to assist in therapy matching by learning from the growing corpus of clinical outcomes: for example, ML models have been trained on past patients' genomic profiles and treatment responses to predict which new patients might respond to a given drug [53]. While such approaches are still experimental, they represent the next generation of bioinformatics-driven decision support. Regardless of the specific software, the end goal is the same: to help the treating physician navigate the complexity of a patient's tumor genomics and select the therapy (or combination) most likely to yield benefit. By leveraging these bioinformatics tools and techniques, personalized cancer therapy can move beyond intuition and single-marker decisions to a more systematic, data-driven paradigm where each treatment choice is underpinned by integrative analysis of the patient's unique cancer "blueprint." (Fig. 2).

## 7 Case Studies and Real-World Applications

Table 3 summarizes major initiatives and programs advancing precision oncology through genomic profiling, AI-driven therapy selection, and adaptive trial designs. These efforts span umbrella and basket trials, institutional sequencing programs, and computational platforms that integrate multi-omics or simulate tumor evolution. Collectively, they demonstrate the clinical feasibility and impact of personalized cancer treatment, highlighting the role of AI and bioinformatics in decision-making, and emphasize the importance of large-scale data sharing for

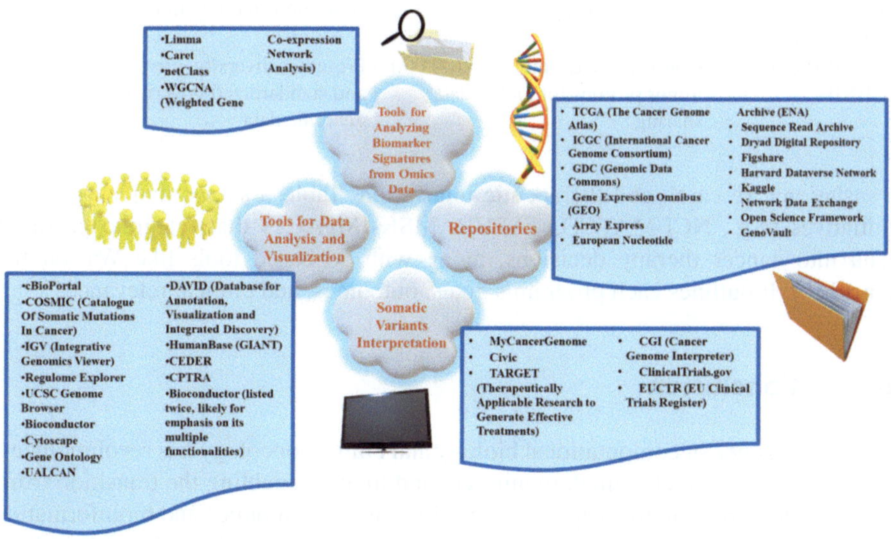


**Fig. 2** Bioinformatics tools and repositories supporting biomarker discovery, variant interpretation, data visualization, and clinical integration in precision oncology workflows

**Table 3** Precision oncology trials and platforms linking genomic profiling to clinical outcomes

| Initiative/Program | Description | Key outcomes/Insights | Ref |
|---|---|---|---|
| NCI-MATCH | Umbrella trial assigning patients with refractory cancers to biomarker-matched therapies based on genomic sequencing | 1473 patients enrolled in 38 substudies; confirmed feasibility of multi-arm, biomarker-based design | [54] |
| WINTHER | Trial integrating DNA and RNA profiling to match patients to therapies | RNA-based profiling increased the benefit rate to 26.2%; the primary endpoint not met | [55] |
| TAPUR | Basket trial by ASCO offering off-label targeted drugs based on genomic alterations | Positive disease control in 15 cohorts; notable outcomes with Olaparib (BRCA1/2) and palbociclib (CDKN2A) | [56] |
| MSK-IMPACT | Prospective sequencing program at Memorial Sloan Kettering using a targeted panel | 36.7% patients had actionable alterations; 11% enrolled in matched trials | [57, 58] |
| Dana-Farber Profile | Large-scale tumor profiling initiative using the OncoPanel assay (~400 genes) | Over 50,000 samples profiled; data used for treatment and research | [59] |
| Watson for Oncology | AI-based tool trained on guidelines and literature to recommend cancer therapies | ~80% concordance with expert boards; limited real-world performance | [60] |
| PERCEPTION | AI model using single-cell RNA-seq to predict therapy response/resistance | Outperformed bulk-based models in predicting therapy outcomes | [61] |
| PanDrugs | Bioinformatics platform for drug repurposing using tumor-specific profiles | Ranks therapies beyond standard care using gene–drug interaction data | [52] |
| Adaptive Therapy Simulations | Uses tumor evolution models to personalize drug dosing schedules | Shown to prolong control vs. continuous therapy in models | [62] |
| AACR Project GENIE | Consortium assembling >100,000 tumor profiles for global data sharing | Promotes diversity, privacy, and standardized governance of genomic data | [63] |

translational research and therapeutic innovation. Table 3 shows major trials and initiatives (e.g., NCI-MATCH, TAPUR, MSK-IMPACT) that integrate genomic data into cancer therapy decisions, along with AI-based tools like Watson for Oncology. It outlines each program's goals, outcomes, and clinical relevance.

## 8 Conclusion

The convergence of computational biology and clinical oncology has revolutionized the way cancers are classified, monitored, and treated, enabling the transition from generalized to personalized approaches. High-throughput omics and bioinformatics have facilitated deep tumor profiling, allowing therapies to be tailored based on genomic, transcriptomic, and proteomic markers. Clinical trials and institutional sequencing programs have confirmed the value of biomarker-guided therapy in improving patient outcomes. The dynamic feedback between clinical practice and research continues to refine treatment strategies, making real-world data a cornerstone of discovery pipelines.

AI, ML, and systems biology are gaining prominence as transformative tools in next-generation cancer research and therapeutic development. AI/ML models are increasingly capable of integrating complex, heterogeneous datasets—including genomic sequences, imaging, and health records—to generate actionable predictions. When combined with systems biology frameworks, these tools can simulate cellular behavior and tumor evolution, supporting real-time adjustments to treatment plans. This fusion allows for proactive therapy design, resistance forecasting, and optimization of combination strategies.

However, full clinical translation demands progress in healthcare policy and education. Regulatory bodies must establish standards for genomic diagnostics and AI-based tools while ensuring equitable access and data privacy. Medical education must evolve to train clinicians in interpreting genomic data, using computational decision aids, and engaging patients in personalized care. Multidisciplinary collaboration, including bioinformaticians and systems modelers, is now central to oncology practice.

Looking ahead, the concept of "digital twins"—virtual replicas of a patient's cancer integrating molecular and clinical data—promises a paradigm shift. These models could allow simulation of treatment responses before clinical application, enabling truly adaptive and individualized therapy. Beyond digital twins, AI-generated therapeutics, real-time monitoring implants, and predictive molecular diagnostics may transform prevention and care. In essence, personalized cancer therapy is entering an era defined by computational precision, systems integration, and patient-specific adaptability, marking a profound evolution in modern oncology.

## References

1. Dagogo-Jack I, Shaw AT (2018) Tumour heterogeneity and resistance to cancer therapies. Nat Rev Clin Oncol [Internet] 15(2):81–94. Available from: https://www.nature.com/articles/nrclinonc.2017.166
2. Gerlinger M, Rowan AJ, Horswell S, Larkin J, Endesfelder D, Gronroos E (2012) Intratumor heterogeneity and branched evolution revealed by multiregion sequencing. N Engl J Med 366(10):883–892
3. Slamon DJ, Clark GM, Wong SG et al (1987) HER2/neu amplification in breast cancer. Science 235(4785):177–182
4. Druker BJ, Talpaz M, Resta DJ et al (2001) Efficacy and safety of a specific inhibitor of the BCR-ABL tyrosine kinase in chronic myeloid leukaemia. N Engl J Med 344(14):1031–1037
5. Garcia-del Rio L, Diaz-Rodriguez P, Landin M (2021) Design of novel orotransmucosal vaccine-delivery platforms using artificial intelligence. Eur J Pharm Biopharm [Internet] 159:36–43. Available from: https://linkinghub.elsevier.com/retrieve/pii/S0939641120303751
6. Senft D, Leiserson MDM, Ruppin E, Ronai ZA (2017) Precision oncology: the road ahead. Trends Mol Med [Internet] 23(10):874–898. Available from: https://linkinghub.elsevier.com/retrieve/pii/S1471491417301430

7. Aaltonen LA, Abascal F, Abeshouse A, Aburatani H, Adams DJ, Agrawal N et al (2020) Pan-cancer analysis of whole genomes. Nature [Internet] 578(7793):82–93. Available from: https://www.nature.com/articles/s41586-020-1969-6
8. Lynch TJ, Bell DW, Sordella R, Gurubhagavatula S, Okimoto J, Brannigan X et al (2004) Activating mutations in the epidermal growth factor receptor underlying responsiveness of non-small-cell lung cancer to gefitinib. N Engl J Med 350(21):2129–2139
9. Kwak EL, Bang YJ, Camidge DR, Shaw AT, Solomon B, Maki RG et al (2010) Anaplastic lymphoma kinase inhibition in non–small-cell lung cancer. New Engl J Med [Internet] 363(18):1693–1703. Available from: http://www.nejm.org/doi/abs/10.1056/NEJMoa1006448
10. Chapman PB, Hauschild A, Robert C, Haanen JB, Ascierto PA, Larkin J et al (2011) Improved survival with vemurafenib in melanoma with BRAF V600E mutation. N Engl J Med 364(26):2507–2516
11. Le DT, Uram JN, Wang HS et al (2015) PD-1 blockade in tumors with mismatch-repair deficiency. N Engl J Med 372(26):2509–2520
12. Drilon A, Laetsch TW, Kummar S, DuBois SG, Lassen U, Demetri GD et al (2018) Efficacy of Larotrectinib in TRK fusion-positive cancers in adults and children. N Engl J Med 378(8):731–739
13. Clark AJ, Lillard JW (2024) A comprehensive review of bioinformatics tools for genomic biomarker discovery driving precision oncology. Genes [Internet] 15(8):1036. Available from: https://www.mdpi.com/2073-4425/15/8/1036
14. Chan TA, Yarchoan TJ, Anagnostou V et al (2019) Development of tumor mutation burden as an immunotherapy biomarker: utility for the oncology clinic. Ann Oncol 30(1):44–56
15. Bryant HE, Schultz N, Thomas H, Parker KA, Flower D, Lopez E et al (2005) Specific killing of BRCA2-deficient tumours with inhibitors of poly(ADP-ribose) polymerase. Nature 434(7035):913–917
16. Sotiriou C, Pusztai L (2009) Gene-expression signatures in breast cancer. N Engl J Med [Internet] 360(8):790–800. Available from: http://www.nejm.org/doi/abs/10.1056/NEJMra0801289
17. Mateo J, Zafeiriou Z, de Bono HS et al (2015) DNA-repair defects and Olaparib in metastatic prostate cancer. N Engl J Med 373(18):1697–1708
18. Cristescu R, Mogg BM, Ayers X et al (2018) Pan-tumor genomic biomarkers for PD-1 checkpoint blockade–based immunotherapy. Science 362(6411):eaar3593
19. Park S, Jeong KJ, Kim Y et al (2017) Network-based identification of genetic influences on disease phenotypes. Nat Commun 8(1):1439
20. Baselga J (2010) Treatment of HER2-overexpressing breast cancer. Ann Oncol [Internet] 21:vii36–vii40. Available from: https://linkinghub.elsevier.com/retrieve/pii/S0923753419397248
21. Amado RG, Wolf M, Peeters M, Van Cutsem E, Siena S, Freeman DJ et al (2008) Wild-type KRAS is required for Panitumumab efficacy in patients with metastatic colorectal cancer. J Clin Oncol [Internet] 26(10):1626–1634. Available from: https://ascopubs.org/doi/10.1200/JCO.2007.14.7116
22. Wang L, Li G, Ezeana CF, Ogunti R, Puppala M, He T et al (2022) An AI-driven clinical care pathway to reduce 30-day readmission for chronic obstructive pulmonary disease (COPD) patients. Sci Rep 12(1):20633
23. Olivier M, Hollstein M, Hainaut P (2010) TP53 mutations in human cancers: origins, consequences, and clinical use. Cold Spring Harb Perspect Biol [Internet] 2(1):a001008. Available from: http://cshperspectives.cshlp.org/lookup/doi/10.1101/cshperspect.a001008
24. Cristofanilli M, Budd GT, Ellis MJ, Stopeck A, Matera J, Miller MC et al (2004) Circulating tumor cells, disease progression, and survival in metastatic breast cancer. N Engl J Med [Internet] 351(8):781–791. Available from: http://www.nejm.org/doi/abs/10.1056/NEJMoa040766
25. Paik S, Shak S, Tang G, Kim C, Baker J, Cronin M et al (2004) A multigene assay to predict recurrence of tamoxifen-treated, node-negative breast cancer. N Engl J Med [Internet] 351(27):2817–2826. Available from: http://www.nejm.org/doi/abs/10.1056/NEJMoa041588

26. Hyman DM, Taylor BS, Baselga J (2017) Implementing genome-driven oncology. Cell 168(4):584–599
27. Chakravarty D, Gao J, Phillips SM et al (2017) OncoKB: a precision oncology knowledge base. JCO Precis Oncol 2017:1–16
28. Schork NJ (2015) Personalized medicine: time for one-person trials. Nature [Internet] 520(7549):609–611. Available from: https://www.nature.com/articles/520609a
29. Kato S, Kimura T, Kiyo M, Abrams J, Devine L, Armenian S et al (2020) Real-world data from a molecular tumor board demonstrates improved outcomes with a precision N-of-one strategy. Nat Commun 11:4965
30. Gatenby RA, Silva AS, Gillies RJ, Frieden BR (2009) Adaptive therapy. Cancer Res 69(11):4894–4903
31. Zhang J, Liu X, Liao L, Choi SH, Yoon YK, Wong KF et al (2017) Integrating evolutionary dynamics into treatment of metastatic castrate-resistant prostate cancer. Nat Commun 8:181
32. Enriquez-Navas PM, Kam Y, Das T, Hassan S, Silva A, Foroutan P et al (2016) Exploiting evolutionary principles to prolong tumor control in preclinical models of breast cancer. Sci Transl Med [Internet] 8(327). Available from: https://www.science.org/doi/10.1126/scitranslmed.aad7842
33. Bailey MH, Tokheim C, Porta-Pardo E et al (2018) Comprehensive characterization of cancer driver genes and mutations. Cell 173(2):371–385.e18
34. Arlt M, Hutter B, Hüllein J, Richter D, Schneider MMK, Hlevnjak M et al (2025) Translational and clinical comparison of whole genome and transcriptome to panel sequencing in precision oncology. NPJ Precis Oncol 9(1):9
35. Wan JCM, Massie C, Garcia-Corbacho J, Mouliere F, Brenton JD, Caldas C et al (2017) Liquid biopsies come of age: towards implementation of circulating tumour DNA. Nat Rev Cancer [Internet] 17(4):223–238. Available from: https://www.nature.com/articles/nrc.2017.7
36. Satoh T, Yasui H, Satake H, Oki E, Kato T, Ohta T et al (2025) Targeted therapy guided by circulating tumor DNA analysis in advanced gastrointestinal tumors (GOZILA study). Nat Med 31(2):165–175
37. Alexandrov LB, Nik-Zainal S, Wedge DC (2013) Signatures of mutational processes in human cancer. Nature 500(7463):415–421
38. Davies H, Glodzik D, Morganella S, Yates LR, Staaf J, Zou X et al (2017) HRDetect is a predictor of BRCA1 and BRCA2 deficiency based on mutational signatures. Nat Med [Internet] 23(4):517–525. Available from: https://www.nature.com/articles/nm.4292
39. Jamal-Hanjani M, Wilson GA, McGranahan N, Birkbak NJ, Hiley CT, Court EC et al (2017) Tracking the evolution of non-small-cell lung cancer. N Engl J Med 376(22):2109–2121
40. Carter H, Chen S, Isik L, Tyekucheva S, Velculescu VE, Kinzler KW et al (2009) Cancer-specific high-throughput annotation of somatic mutations: computational prediction of driver missense mutationss. Cancer Research [Internet] 69(16):6660–6667. Available from: https://aacrjournals.org/cancerres/article/69/16/6660/549888/Cancer-Specific-High-Throughput-Annotation-of
41. Lin PC, Tsai YS, Yeh YM, Shen MR (2022) Cutting-edge AI technologies meet precision medicine to improve cancer care. Biomolecules 12(8):1133
42. Sinha S, Vegesna R, Mukherjee S, Kammula AV, Dhruba SR, Wu W et al (2024) {PERCEPTION} predicts patient response and resistance to treatment using single-cell transcriptomics of their tumors. Nat Cancer 5(6):938–952
43. Griffith M, Spies NC, Krysiak K, McMichael JF, Coffman AC, Danos AM et al (2017) CIViC is a community knowledgebase for expert crowdsourcing the clinical interpretation of variants in cancer. Nat Genet [Internet] 49(2):170–174. Available from: https://www.nature.com/articles/ng.3774
44. Vlachavas EI, Bohn J, Ückert F, Nürnberg S (2021) A detailed catalogue of multi-omics methodologies for identification of putative biomarkers and causal molecular networks in translational cancer research. Int J Mol Sci 22(6):2822
45. Argelaguet R, Velten B, Arnol D, Dietrich S, Zenz T, Marioni JC et al (2018) Multi-omics factor analysis—a framework for unsupervised integration of multi-omics data sets. Mol Syst Biol [Internet] 14(6). Available from: https://www.embopress.org/doi/10.15252/msb.20178124

46. Wilkinson MD, Dumontier M, Aalbersberg IJ, Appleton G, Axton M, Baak A et al (2016) The {FAIR} Guiding Principles for scientific data management and stewardship. Sci Data 3:160018
47. Wang K, Li M, Hakonarson H (2010) ANNOVAR: functional annotation of genetic variants. Nucleic Acids Res 38(16):e164
48. Colaprico A, Silva TC, Olsen C, Garofano L, Cava C, Garolini D et al (2016) TCGAbiolinks: an R/Bioconductor package for integrative analysis of TCGA data. Nucleic Acids Res 44(8):e71
49. Shannon P, Markiel A, Ozier O, Baliga NS, Wang JT, Ramage D et al (2003) Cytoscape: a software environment for integrated models of biomolecular interaction networks. Genome Res 13(11):2498–2504
50. Richters MM, Xia H, Campbell KM, Gillanders WE, Griffith OL, Griffith M (2019) Best practices for bioinformatic characterization of neoantigens for clinical utility. Genome Med 11(1):56
51. Xia Y, Rao R, Xiong M, He B, Zheng B, Jia Y (2024) AI-driven CRISPR analysis for high-throughput HPV detection. ACS Anal Chem [Internet]. Available from: https://pubs.acs.org/doi/full/10.1021/acs.analchem.3c04363
52. Piñeiro-Yáñez E, Reboiro-Jato M, Gómez-López G, Perales-Patón J, Troulé K, Rodriguez JM et al (2018) PanDrugs: a novel method to prioritize anticancer drug treatments according to individual genomic data. Genome Med 10(1):41
53. Yuan Y et al (2023) Artificial intelligence in precision oncology: applications and challenges. Cancer Commun 43:433–457
54. O'Dwyer PJ, Gray RJ, Flaherty KT, Chen AP, Li S, Wang V et al (2023) The NCI-MATCH trial: lessons for precision oncology. Nat Med [Internet] 29(6):1349–1357. Available from: https://www.nature.com/articles/s41591-023-02379-4
55. Rodon J, Soria-Comes BS, Tabernero J et al (2019) Genomic and transcriptomic profiling expands precision cancer medicine: the WINTHER trial. Nat Med 25:751–758
56. Mangat PK, Halabi S, Bruinooge SS et al (2018) TAPUR: real-world targeted therapy trial. J Clin Oncol 36(15_suppl):TPS2600
57. Zehir A, Benayed R, Shah RH et al (2017) Mutational landscape of metastatic cancer revealed from prospective clinical sequencing of 10,000 patients. Nat Med 23(6):703–713
58. mskcc.org/msk-impact [Internet]. Available from: https://www.mskcc.org/msk-impact
59. dana-farber.org. Available from: https://www.dana-farber.org/research/featured/precision-cancer-medicine
60. Jie Z, Zhiying Z, Li L (2021) A meta-analysis of Watson for Oncology in clinical application. Sci Rep [Internet] 11(1):5792. Available from: https://www.nature.com/articles/s41598-021-84973-5
61. Sinha S, Vegesna R, Mukherjee S et al (2024) PERCEPTION predicts patient response and resistance to treatment using single-cell transcriptomics of their tumors. Nature Cancer 5(6):938–952
62. Strobl MAR, Gallaher J, West J, Robertson-Tessi M, Maini PK, Anderson ARA (2022) Spatial structure impacts adaptive therapy by shaping intra-tumoral competition. Commun Med [Internet] 2(1):46. Available from: https://www.nature.com/articles/s43856-022-00110-x
63. Pugh TJ, Bell JL, Bruce JP, Doherty GJ, Galvin M, Green MF et al (2022) AACR project GENIE: 100,000 cases and beyond. Cancer Discov [Internet] 12(9):2044–2057. Available from: https://aacrjournals.org/cancerdiscovery/article/12/9/2044/708766/AACR-Project-GENIE-100-000-Cases-and-BeyondGENIE

# Artificial Intelligence (AI) and Machine Learning (ML) in Personalized Medicine

Indranil Paul


## Abstract

Personalized medicine aims to tailor disease prevention, diagnosis, and treatment to individual patients by leveraging genetic, clinical, and lifestyle data. Artificial intelligence (AI) and machine learning (ML) have emerged as powerful tools for rapidly analyzing complex biomedical datasets, thereby advancing precision healthcare. This chapter explores the applications, methodologies, challenges, and future directions of AI/ML in personalized medicine. Key AI techniques, including supervised and unsupervised learning, deep learning, and reinforcement learning, are being applied to diverse data sources such as genomics, electronic health records (EHRs), medical imaging, and wearable devices. These models enhance disease diagnosis, risk prediction, and personalized treatment strategies, particularly in fields like oncology, cardiology, and neurology. Additionally, AI is transforming drug discovery, clinical trial optimization, and precision public health initiatives. However, challenges such as data privacy, algorithmic bias, model interpretability, and regulatory constraints must be addressed. Ethical concerns are also discussed, alongside potential solutions like explainable AI (XAI) and federated learning (FL). Looking ahead, future trends include AI-powered digital twins, multi-omics data integration, and real-time patient monitoring systems. As AI continues to evolve, it promises to drive personalized medicine toward more precise, efficient, and scalable healthcare solutions.



I. Paul (✉)
Program in Quantitative Oncology, Division of Oncological Sciences, Knight Cancer Research Building, Oregon Health & Science University, Portland, OR, USA





M. K. Ghosh, M. Basu (eds.), *Personalized Cancer Therapy*,
https://doi.org/10.1007/978-981-95-6845-1_23

**Keywords**

Personalized Medicine · Artificial Intelligence (AI) · Machine Learning (ML) · Genomics · Deep Learning · Drug Discovery · Clinical Trials · Federated Learning · Biomarkers

# 1 Introduction

## 1.1 The Evolution of Personalized Medicine

The field of medicine has long relied on a "one-size-fits-all" approach, where treatments and interventions are designed based on population averages rather than individual variability. While this approach has yielded significant advancements, it often fails to account for the unique genetic, environmental, and lifestyle factors that influence disease susceptibility, progression, and response to therapy. Personalized medicine, also known as precision medicine, represents a paradigm shift in healthcare, aiming to tailor prevention, diagnosis, and treatment strategies to the individual characteristics of each patient [1]. This shift is driven by the recognition that biological heterogeneity underpins the variability in disease presentation and therapeutic outcomes.

The advent of genomics has been a cornerstone in the evolution of personalized medicine. The completion of the Human Genome Project in 2003 marked a turning point, enabling researchers to decode the genetic basis of diseases and identify biomarkers that predict disease risk and treatment response [2]. Biomarkers, such as genetic mutations, protein expressions, and metabolic signatures, provide critical insights into the molecular mechanisms of diseases, allowing clinicians to stratify patients into subgroups based on their likelihood of benefiting from specific therapies [3]. Beyond genomics, the integration of multi-omics data—including transcriptomics, proteomics, metabolomics, and epigenomics—has further enriched our understanding of disease biology. These layers of biological information, when combined with clinical and lifestyle data, create a comprehensive profile of an individual's health status, paving the way for truly personalized care.

However, the realization of personalized medicine faces significant challenges. The sheer volume, complexity, and heterogeneity of biomedical data require advanced analytical tools to extract meaningful insights. Traditional statistical methods are often inadequate for handling the high-dimensional and nonlinear relationships inherent in multi-omics datasets [4]. Moreover, the integration of diverse data types, such as genomic data, electronic health records (EHRs), medical imaging, and real-time data from wearable devices, demands innovative approaches to data harmonization and analysis. This is where artificial intelligence (AI) and machine learning (ML) come into play [1].

## 1.2 The Role of AI and ML in Personalized Medicine

AI and ML have emerged as transformative technologies in healthcare, offering unprecedented capabilities for analyzing complex datasets and generating actionable insights [5]. At their core, AI/ML algorithms are designed to identify patterns, correlations, and predictive signals within data, enabling the development of models that can inform clinical decision-making. These technologies are particularly well-suited to the challenges of personalized medicine, where the goal is to translate vast amounts of heterogeneous data into precise, individualized treatment strategies (Fig. 1).

One of the key strengths of AI/ML lies in their ability to handle big data. The healthcare industry generates enormous volumes of data daily, from genomic sequences and medical images to EHRs and wearable device outputs. AI/ML algorithms can process and analyze these datasets at scale, uncovering hidden relationships that would be impossible to detect using traditional methods [6]. For example, supervised learning algorithms can predict patient outcomes based on historical data, while unsupervised learning techniques can identify novel patient subgroups with distinct molecular profiles. Deep learning, a subset of ML, has shown remarkable success in analyzing unstructured data, such as medical images and free-text clinical notes, further expanding the scope of personalized medicine [7]. The synergy between big data analytics and personalized treatment decisions is a defining feature of AI/ML-driven precision healthcare. By integrating data from multiple sources, AI/ML models can generate holistic patient profiles that capture the interplay between genetic, environmental, and lifestyle factors. These profiles enable clinicians to make more informed decisions, such as selecting the most effective

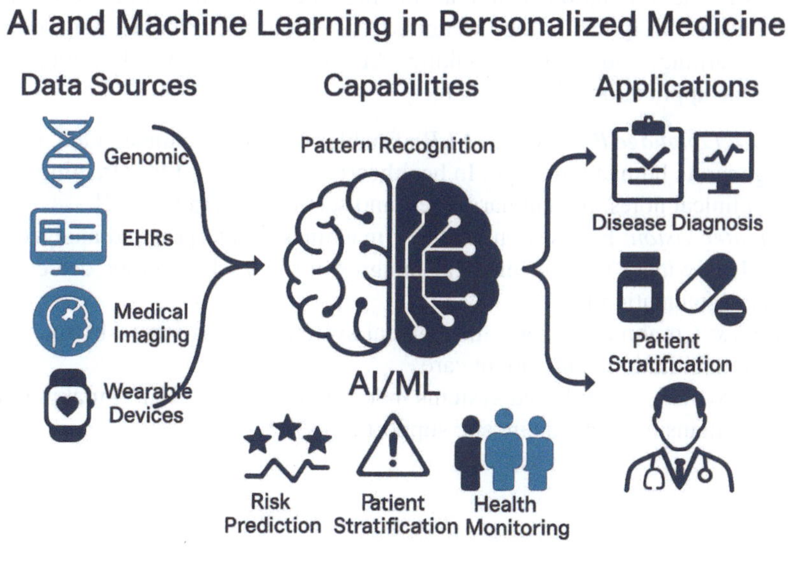


**Fig. 1** AI/ML workflow in personalized medicine

drug for a patient based on their genetic makeup or predicting the risk of adverse drug reactions [8]. In oncology, for instance, AI/ML models are being used to match patients with targeted therapies based on tumor genomics, significantly improving treatment outcomes [9].

Moreover, AI/ML is accelerating the discovery of novel biomarkers and therapeutic targets. By analyzing large-scale omics datasets, these technologies can identify molecular signatures associated with disease progression or treatment response, facilitating the development of new diagnostic tools and therapies [10]. In drug discovery, AI/ML is streamlining the identification of drug candidates and optimizing clinical trial designs, reducing the time and cost of bringing new treatments to market [11].

Despite their immense potential, the adoption of AI/ML in personalized medicine is not without challenges. Issues such as data privacy, algorithmic bias, model interpretability, and regulatory compliance must be addressed to ensure the ethical and equitable use of these technologies. Nevertheless, the integration of AI/ML into personalized medicine represents a transformative opportunity to revolutionize healthcare, moving us closer to a future where treatments are not only precise but also predictive, preventive, and participatory [2, 3].

## 2 Foundations of AI and ML in Healthcare

### 2.1 Understanding Artificial Intelligence and Machine Learning

#### 2.1.1 Defining AI: Overview of AI Subfields

AI refers to the simulation of human intelligence in machines, enabling them to perform tasks that typically require human cognition, such as reasoning, problem-solving, learning, and decision-making. AI encompasses several subfields, each with unique applications in healthcare:

- *Natural Language Processing (NLP)*: Enables machines to understand, interpret, and generate human language. In healthcare, NLP is used for extracting insights from clinical notes, patient narratives, and scientific literature [12].
- *Computer Vision*: Focuses on enabling machines to interpret and analyze visual data. It is widely used in medical imaging for tasks like tumor detection and organ segmentation [13].
- *Robotics*: Combines AI with mechanical systems to perform tasks such as surgery, rehabilitation, and patient care.
- *Expert Systems*: Rule-based systems that mimic human decision-making in specific domains, such as diagnostic support tools.

#### 2.1.2 ML Paradigms: Supervised, Unsupervised, and Reinforcement Learning

ML, a subset of AI, involves training algorithms to learn patterns from data and make predictions or decisions. ML paradigms include:

- *Supervised Learning*: Involves training models on labeled datasets, where input–output pairs are known. It is used for classification (e.g., disease diagnosis) and regression (e.g., predicting patient outcomes). Common algorithms include logistic regression, support vector machines (SVMs), and random forests [14].
- *Unsupervised Learning*: Deals with unlabeled data, aiming to identify hidden structures or patterns. Techniques like clustering (e.g., k-means and hierarchical clustering) and dimensionality reduction (e.g., PCA) are used for patient stratification and biomarker discovery [15].
- *Reinforcement Learning*: Involves an agent learning to make decisions by interacting with an environment and receiving feedback through rewards or penalties. It is applied in optimizing treatment protocols and robotic surgery [16].

#### 2.1.3 Deep Learning: Introduction to CNNs, RNNs, Transformers

Deep learning, a specialized branch of ML, uses neural networks with multiple layers to model complex patterns in data. Key architectures include:

- *Convolutional Neural Networks (CNNs)*: Designed for image recognition and processing, CNNs automatically learn spatial hierarchies of features from input images. They are widely used in medical imaging for tasks like tumor detection and classification [17].
- *Recurrent Neural Networks (RNNs)*: Ideal for sequential data, such as time-series or text, RNNs maintain a form of memory through connections between nodes in a temporal sequence. Variants like Long Short-Term Memory (LSTM) networks are used for predicting disease progression and analyzing genomic sequences [18].
- *Transformers*: Initially developed for NLP tasks, transformers leverage self-attention mechanisms to handle dependencies in data sequences. They are highly effective for tasks requiring context understanding, such as analyzing clinical notes and multi-omics data [19].

### 2.2 Data Types in Personalized Medicine

#### 2.2.1 Genomic Data: DNA Sequencing, SNPs, Gene Expression

Genomic data, derived from DNA sequencing, provides insights into an individual's genetic makeup. Key components include:

- *Single Nucleotide Polymorphisms (SNPs)*: Variations at a single position in the DNA sequence, used to identify genetic predispositions to diseases.
- *Gene Expression*: Measures the activity of genes, providing insights into cellular functions and disease mechanisms [20].

- *Whole Genome Sequencing (WGS)*: Captures the complete DNA sequence, enabling comprehensive analysis of genetic variations.

### 2.2.2 Electronic Health Records (EHRs): Structured and Unstructured Clinical Data

EHRs contain a wealth of patient information, including:

- *Structured Data*: Demographics, lab results, medications, and diagnoses [21].
- *Unstructured Data*: Clinical notes, imaging reports, and discharge summaries. NLP tools are used to extract meaningful insights from unstructured text.

### 2.2.3 Medical Imaging: MRI, CT, Histopathology Slides

Medical imaging data provides visual representations of anatomical and pathological conditions. Key modalities include:

- *Magnetic Resonance Imaging (MRI)*: Used for soft tissue visualization, such as brain and musculoskeletal imaging.
- *Computed Tomography (CT)*: Provides detailed cross-sectional images of internal organs and structures [22].
- *Histopathology Slides*: Microscopic images of tissue samples, used for cancer diagnosis and grading.

### 2.2.4 Wearables and IoT Devices: Continuous Monitoring of Vital Signs

Wearable devices and IoT technologies enable real-time monitoring of vital signs, such as:

- Heart rate, blood pressure, and glucose levels [23].
- Physical activity and sleep patterns. This data is invaluable for managing chronic diseases and detecting early warning signs.

### 2.2.5 Lifestyle and Environmental Data: Nutrition, Activity, Social Determinants of Health

Lifestyle and environmental factors play a critical role in health outcomes. Key data types include:

- *Nutrition*: Dietary habits and nutrient intake.
- *Physical Activity*: Exercise patterns and sedentary behavior.
- *Social Determinants of Health*: Socioeconomic status, education, and environmental exposures [24].

## 2.3 Key ML Techniques in Personalized Medicine

### 2.3.1 Supervised Learning: Logistic Regression, SVMs, Random Forests, Neural Networks

Supervised learning techniques are widely used for predictive modeling in personalized medicine:

- *Logistic Regression*: Used for binary classification tasks, such as predicting disease presence or absence.
- *Support Vector Machines (SVMs)*: Effective for high-dimensional data, used in cancer classification and drug response prediction [25].
- *Random Forests*: Ensemble methods that combine multiple decision trees to improve accuracy and reduce overfitting.
- *Neural Networks*: Capable of modeling complex relationships, used for tasks like genomic data interpretation and patient outcome prediction.

### 2.3.2 Unsupervised Learning: Clustering (k-means, Hierarchical), PCA

Unsupervised learning techniques help identify patterns and groupings within data:

- *Clustering*: Techniques like k-means and hierarchical clustering group similar data points, aiding in patient stratification and biomarker discovery [15].
- *Principal Component Analysis (PCA)*: Reduces data dimensionality, simplifying analysis while retaining essential information.

### 2.3.3 Deep Learning: CNNs for Imaging, RNNs for Sequential Patient Data

Deep learning techniques are particularly impactful in personalized medicine:

- *CNNs for Imaging*: Used for analyzing medical images, such as detecting tumors in radiology and pathology [17].
- *RNNs for Sequential Data*: Suitable for modeling time-series data, such as predicting disease progression and treatment outcomes [18].

### 2.3.4 Tools and Technologies

Several tools and frameworks facilitate the application of AI/ML in personalized medicine:

- *Python Libraries*: Scikit-learn, TensorFlow, PyTorch, and Keras for model development [26].
- *Genomic Analysis Tools*: GATK, PLINK, and Bioconductor for processing and analyzing genomic data.
- *Medical Imaging Platforms*: MONAI, 3D Slicer, and ITK for image analysis.
- *Cloud Computing*: AWS, Google Cloud, and Microsoft Azure for scalable data storage and processing [27].

# 3 AI/ML Applications in Personalized Medicine

Recent advancements in AI and ML have ushered in a new era of personalized medicine, a discipline focused on tailoring medical decisions, treatments, and preventive strategies to individual patient characteristics. By leveraging complex datasets—including genomic, transcriptomic, proteomic, and clinical data—AI algorithms can provide more accurate disease diagnosis, predict disease risk and progression, optimize treatments, accelerate drug discovery, and inform public health strategies. This chapter explores these AI-driven applications, highlighting both emerging trends and tangible case studies that illustrate the transformative potential of ML in personalized medicine.

## 3.1 Disease Diagnosis and Prediction

### 3.1.1 Early Disease Detection: AI-Driven Biomarker Discovery

Early detection of disease is critical for improving patient outcomes and reducing healthcare costs. AI-driven biomarker discovery utilizes advanced algorithms—such as deep learning and natural language processing—to sift through large-scale omics data (genomics, proteomics, metabolomics) and identify subtle biological indicators of disease [10]. For instance, ML models can analyze transcriptomic data to isolate gene expression signatures that differentiate diseased from healthy tissues. These models often apply feature-selection techniques to rank biomarkers in terms of their predictive power.

Moreover, AI can integrate imaging data, electronic health records (EHRs), and patient-generated data from wearable devices to develop composite biomarkers [28]. By combining multiple data modalities, these algorithms enhance sensitivity and specificity for early detection. An example is the use of convolutional neural networks (CNNs) to interpret medical images—like MRIs and CT scans—to detect microcalcifications in breast tissue or subtle lesions in the lung that may be early indicators of malignancy [29]. This integrated, data-driven approach can transform the diagnosis process, significantly reducing the time and cost of identifying individuals at high risk.

### 3.1.2 Predicting Disease Risk and Progression: AI in Alzheimer's, Diabetes, and Cancer

Beyond early detection, AI models excel at predicting not only disease onset but also disease trajectory. In Alzheimer's disease, for example, ML algorithms can analyze neuroimaging data and longitudinal clinical assessments to detect early cognitive decline patterns [30]. By forecasting the progression rate of dementia, these systems help clinicians make more informed decisions about interventions and care strategies.

In diabetes management, AI tools analyze continuous glucose monitoring data, lifestyle factors, and genetic predispositions to identify patients at high risk for complications like nephropathy, retinopathy, or cardiovascular events [31]. By predicting these outcomes with reasonable accuracy, clinicians can intervene early, adjusting medications or lifestyle recommendations.

In oncology, AI-driven risk stratification models leverage genomic profiling, tumor marker expression, and clinical metadata to predict cancer progression and recurrence [9]. These models guide personalized monitoring schedules, enabling more frequent surveillance in high-risk individuals and reducing unnecessary testing in lower risk groups. Ultimately, AI-based risk and progression modeling empowers healthcare providers to deliver more precise care, potentially improving survival rates and patient quality of life.

#### 3.1.3 Case Study: AI-Powered Cancer Detection Models

A notable illustration of AI in disease diagnosis is the deployment of cancer detection models trained on large patient datasets. For instance, Google Health's deep learning model for breast cancer screening was trained on tens of thousands of mammograms. The algorithm demonstrated improved sensitivity and specificity compared to human radiologists in several studies [32]. Similarly, AI tools have shown promise in detecting malignant lung nodules in CT scans and classifying skin lesions (e.g., melanoma vs. benign nevi) with dermatology-level accuracy [33].

In clinical practice, these AI models can act as second readers, offering real-time feedback to radiologists and pathologists, reducing diagnostic errors. As these tools evolve, they incorporate additional patient information—such as family history, genetic mutations, and lifestyle factors—to yield even more nuanced risk assessments and diagnostic suggestions. Thus, AI-powered cancer detection models highlight the synergy of large-scale data, ML algorithms, and clinical expertise in personalizing cancer care.

### 3.2 Personalized Treatment and Drug Optimization

#### 3.2.1 Drug Selection Based on Genetics: Pharmacogenomics and Precision Dosing

Pharmacogenomics combines pharmacology and genomics to understand how genetic variability influences individual drug responses. ML has become an integral part of pharmacogenomics research, enabling the discovery of genetic variants linked to drug metabolism, efficacy, and toxicity [34]. By integrating patient genotypes with clinical and demographic data, ML models can predict the optimal drug dosage or select an alternative medication to minimize adverse effects.

These predictive algorithms often incorporate single nucleotide polymorphism (SNP) data along with liver enzyme activity and kidney function indicators. For example, warfarin dosing algorithms can account for genetic variants in the CYP2C9 and VKORC1 genes, leading to more accurate dosing recommendations and fewer bleeding complications [35]. In oncology, targeted therapies are selected by identifying mutations in tumors (e.g., EGFR or ALK mutations in lung cancer) that correspond to specific targeted drugs. AI models can streamline this process by sifting through molecular profiles to quickly pinpoint the most effective therapy for an individual patient.

### 3.2.2 Predicting Drug Response and Adverse Reactions: AI Models Analyzing Drug–Gene Interactions

A key challenge in personalized medicine is anticipating a patient's response to a particular drug regimen. Traditional approaches rely on population-level trial data that may not accurately reflect individual variations. ML models that integrate EHR data, genomic information, and drug–gene interaction databases can generate personalized risk profiles [36]. These models highlight potential side effects—such as cardiotoxicity or hepatotoxicity—as well as suboptimal efficacy based on a patient's metabolic pathways or concurrent medications.

By continuously learning from real-world evidence, including adverse event reports, these AI systems adapt and refine their predictions over time. This iterative process can greatly reduce the trial-and-error approach in medication selection and dosage adjustment, significantly improving patient outcomes and lowering healthcare costs.

### 3.2.3 AI in Oncology: Tumor Profiling and Personalized Cancer Therapy

Cancer therapy is among the most promising areas for AI-driven personalized medicine. Tumor profiling integrates genomic sequencing, transcriptomics, and proteomics to generate a comprehensive molecular signature of a patient's cancer [9]. ML models process these high-dimensional data to identify oncogenic drivers and potential therapeutic targets. This approach underpins the concept of precision oncology, wherein treatments such as immunotherapies or targeted kinase inhibitors are matched to specific tumor characteristics.

Additionally, AI-based tools can predict therapy resistance by examining mutational patterns that emerge over time, guiding clinicians to adapt treatments before resistance becomes clinically apparent [37]. For instance, in non-small cell lung cancer, AI algorithms can detect emerging mutations (e.g., T790M in EGFR) that may prompt a switch to a next-generation inhibitor. The result is more sustained therapeutic responses and a better quality of life.

### 3.2.4 Case Study: IBM Watson for Oncology

IBM Watson for Oncology emerged as a pioneer in AI-assisted treatment planning. By ingesting vast amounts of clinical guidelines, medical literature, and patient data, Watson aimed to provide oncologists with evidence-based treatment recommendations. When a physician inputs a patient's clinical profile—such as tumor type, stage, and relevant biomarkers—Watson's AI engine proposes a prioritized list of therapeutic regimens aligned with the latest research. Although early evaluations highlighted challenges in integrating such a complex tool into clinical workflows, Watson demonstrated the feasibility and potential utility of AI-driven precision oncology [38]. This case underscores the importance of robust data curation, continuous updates from new research, and user-friendly interfaces to translate AI insights into actionable clinical decisions.

## 3.3 AI in Drug Discovery and Development

### 3.3.1 AI for Target Identification: Predicting Drug Mechanisms Using ML

Identifying viable molecular targets is a critical early step in drug development. AI-driven target identification methods employ supervised and unsupervised learning to analyze large biological datasets—genomic data, protein interaction networks, and disease-specific omics—to reveal novel targets and drug mechanisms of action [10]. For example, unsupervised clustering algorithms can uncover subgroups of patients who share common molecular signatures and are responsive to certain classes of compounds. This segmentation refines drug discovery efforts, ensuring that new therapeutics address the underlying biology.

Natural language processing (NLP) can also play a role by mining the scientific literature to map out gene-disease and protein-disease associations [12]. By synthesizing evidence from thousands of papers, AI systems can highlight previously overlooked connections, accelerating the pace of hypothesis generation. These computational predictions are then validated through in vitro and in vivo experiments, forming a more targeted and cost-effective approach to early drug discovery.

### 3.3.2 Virtual Drug Screening: Deep Learning-Driven Molecular Docking

Virtual screening harnesses the power of computational modeling to evaluate how small molecules might bind to a biological target. Traditional in silico approaches often rely on physics-based docking simulations, but integrating deep learning into these workflows enhances their accuracy and speed [11]. Deep learning models can learn molecular representations (e.g., SMILES strings and 3D conformations) and predict binding affinities, drastically reducing the number of candidate molecules that need to be tested in the lab [39].

With the ability to screen millions of compounds in a fraction of the time of traditional high-throughput wet-lab methods, AI-driven virtual drug screening expedites lead identification and optimization. Furthermore, generative adversarial networks (GANs) or graph neural networks (GNNs) can design novel molecular structures with desired properties, potentially producing drug candidates that would be difficult or impossible to uncover using conventional methods.

### 3.3.3 AI in Clinical Trials: Optimizing Patient Recruitment, Trial Design, and Predictive Analytics

Clinical trial success depends heavily on selecting the right patient cohort, optimizing trial design, and monitoring patient responses throughout the study. AI can streamline patient recruitment by matching trial eligibility criteria with patient profiles extracted from EHRs, reducing the time and cost of recruiting the appropriate participants [40]. This targeted approach fosters higher success rates, as patients are more likely to benefit from and adhere to the study's treatment protocols.

Additionally, AI tools can assist in adaptive trial designs, where ML models analyze interim results to modify dosing strategies or stratify patient groups [41]. Predictive analytics further help forecast patient dropouts, adverse events, and overall trial success, allowing sponsors to intervene proactively. By leveraging real-time data, clinical trials become more efficient, ethical, and cost-effective.

### 3.3.4 Case Study: AI-Powered COVID-19 Drug Repurposing

During the COVID-19 pandemic, the urgency to find effective treatments accelerated the adoption of AI for drug repurposing. Several research consortia used ML algorithms to sift through databases of approved drugs, analyzing structural similarities, molecular docking results, and real-world patient outcomes to identify potential antiviral candidates [42]. Models highlighted compounds like remdesivir, hydroxychloroquine, and others for rapid clinical evaluation. Although results were mixed, these efforts demonstrated the power of AI to rapidly generate hypotheses and support global collaborative research in times of crisis. They also underscored the importance of real-world evidence, data sharing, and robust model validation.

## 3.4 Personalized Prevention and Wellness

### 3.4.1 Risk Factor Analysis: AI in Preventive Healthcare Recommendations

Personalized prevention involves identifying risk factors for chronic diseases—such as obesity, cardiovascular disease, or mental health disorders—before they manifest clinically. By analyzing EHRs, genomic information, and lifestyle data (diet, physical activity, and sleep patterns), AI models can highlight an individual's primary risk contributors [24]. For instance, a ML algorithm might identify that a patient with specific genetic variants, sedentary lifestyle, and borderline metabolic markers is at high risk for type 2 diabetes.

Armed with these insights, clinicians and health coaches can develop targeted interventions that address the specific risk factors of each patient. Whether suggesting a tailored exercise regimen, nutrition plan, or stress management techniques, the goal is to slow or prevent disease onset. Over time, continuous patient monitoring through wearables or mobile apps can feed data back into the AI model, refining risk predictions and offering updated, data-driven recommendations [23].

### 3.4.2 AI-Powered Health Coaching: Wearables and Personalized Lifestyle Interventions

The convergence of wearable technology and AI is redefining the concept of health coaching. Today's wearable devices measure heart rate, sleep quality, activity levels, and even certain biochemical markers. These real-time data streams are processed by AI algorithms that can personalize feedback and interventions. For instance, an AI-driven coaching app may recommend a low-impact workout routine if it detects suboptimal recovery metrics or signs of stress in the user's physiological data [43].

Beyond exercise, AI can also guide dietary choices by analyzing patterns in blood glucose readings, meal composition, and metabolic responses. Some systems use reinforcement learning to adapt recommendations dynamically, reinforcing positive behavioral changes and offering real-time incentives (like virtual rewards or gamification elements) to sustain user engagement. As a result, AI-powered health coaching facilitates continuous, granular support for individuals seeking to optimize their long-term health and well-being.

### 3.4.3 Case Study: AI for Cardiovascular Disease Prevention

ML-driven risk calculators for cardiovascular disease have been integrated into clinical practice in several health systems. These tools pull in patient data—age, sex, blood pressure, lipid profiles, smoking status, and in some cases genetic markers—to predict the likelihood of adverse cardiac events. The Framingham Risk Score laid the groundwork for such calculators; today's AI-enhanced models can incorporate far more variables and complex relationships, further refining individual risk estimates [44].

Some programs also integrate wearable data—like heart rate variability and activity levels—to monitor for early signs of atrial fibrillation or other cardiac anomalies. In high-risk patients, ML algorithms can prompt timely interventions, such as recommending a cardiology consult or adjusting antihypertensive medication. Overall, the integration of AI into cardiovascular disease prevention exemplifies how personalized prevention strategies can be scaled to large populations.

## 3.5 AI in Precision Public Health

### 3.5.1 Population Health Analytics: AI-Driven Disease Surveillance and Intervention Strategies

Precision public health extends the principles of personalized medicine to entire populations, focusing on targeted interventions that address community-specific health challenges. By merging large epidemiological datasets with patient-level data, AI algorithms can identify geographic hotspots for certain diseases or detect emerging outbreaks [45]. Early warning systems use ML techniques to analyze social media trends, environmental data, and even mobility patterns, offering predictive insights into disease spread and potential health risks.

Public health agencies can then deploy resources more effectively—allocating vaccines, staffing healthcare workers, or implementing localized prevention campaigns—based on AI-derived predictions [46]. By refining these analytics in near-real time, policymakers can intervene proactively, mitigating the severity of outbreaks or addressing social determinants of health that contribute to disparities.

### 3.5.2 Personalized Vaccination and Pandemic Management: AI-Powered Epidemiological Forecasting

Vaccination programs traditionally follow a one-size-fits-all schedule. However, AI can refine this approach, identifying population subgroups that may require different immunization strategies based on immune profiles, comorbidities, or risk of exposure. For instance, during a pandemic, ML models might suggest prioritizing not only the elderly but also certain communities with higher transmission rates due to occupational or social factors [47].

AI-driven epidemiological forecasting models simulate how infectious diseases spread under various intervention scenarios (e.g., masking, physical distancing, and school closures). These simulations inform policymakers about the potential outcomes of policy decisions, helping to strike a balance between public health measures and societal impact. As demonstrated during COVID-19, real-time data collection (testing results and contact tracing data) paired with AI modeling can substantially influence and refine policymaking throughout a health crisis [48].

### 3.5.3 Case Study: AI in COVID-19 Response and Outbreak Prediction

During the COVID-19 pandemic, multiple AI-based platforms emerged to track and predict outbreak trends. Projects like BlueDot used AI-driven text mining to detect early signals of the novel coronavirus spread, while other teams built ML dashboards integrating confirmed cases, mobility data, and hospital capacity [48]. These tools helped health authorities anticipate infection surges, allocate resources, and design targeted lockdown or vaccination strategies.

In some regions, AI-enabled contact tracing apps used Bluetooth signals to identify potential exposures, aiming to contain transmission without blanket lockdown measures. Though these technologies faced challenges related to privacy and data accuracy, they underscored the potential for AI to guide both individual-level and population-level interventions. Moving forward, the lessons learned from COVID-19 demonstrate how AI can bolster global preparedness for future pandemics and epidemics.

### 3.5.4 Conclusion

AI and ML have emerged as catalysts in the evolution of personalized medicine, reshaping how diseases are diagnosed, treated, and prevented. From the early detection of complex disorders through biomarker discovery to the refinement of therapeutic regimens via pharmacogenomic insights, AI is enhancing the precision and efficiency of clinical care. In drug discovery and development, ML reduces the time and cost associated with target identification, virtual screening, and clinical trials. Beyond acute care, AI-powered health coaching and risk factor analysis promote personalized prevention strategies for chronic disease management. Lastly, the integration of AI into population health initiatives holds promise for more targeted public health interventions, as evidenced by pandemic response strategies.

Despite the transformative potential of AI, challenges remain. Data privacy, bias in training datasets, and regulatory oversight must be managed carefully to ensure equitable and safe deployment. Ongoing collaboration between clinicians, researchers, data scientists, policymakers, and patients will be vital to harness the full potential of AI in personalized medicine. As these technologies continue to mature, the vision of truly individualized healthcare—where each patient receives the right intervention at the right time—edges closer to reality, promising improved outcomes and a more sustainable healthcare system for all (Table 1).

# 4 Challenges and Ethical Considerations

## 4.1 Data Challenges

### 4.1.1 Data Quality and Interoperability: Issues with Fragmented and Inconsistent Healthcare Data

The effectiveness of AI/ML models in personalized medicine hinges on the availability of high-quality, standardized data. However, healthcare data is often fragmented across institutions, stored in incompatible formats (e.g., EHRs from different vendors), and riddled with inconsistencies such as missing entries or coding errors [49]. For example, lab results may use different units or terminologies across hospitals, complicating large-scale analysis. Efforts like Fast Healthcare Interoperability Resources (FHIR) aim to standardize data exchange, but adoption remains uneven. Poor interoperability limits the ability to aggregate datasets, undermining the development of robust AI models that require diverse, representative data.

### 4.1.2 Bias in Training Data: Risk of AI Models Reinforcing Healthcare Disparities

AI models trained on biased or nonrepresentative datasets risk perpetuating healthcare inequities. For instance, genomic databases like the UK Biobank predominantly include data from individuals of European ancestry, leading to models that perform poorly for underrepresented racial or ethnic groups [50]. Similarly, diagnostic algorithms trained on imaging data from lighter skinned populations may misdiagnose conditions like skin cancer in darker skinned patients. These biases can exacerbate existing disparities in healthcare access and outcomes. Mitigating bias requires intentional curation of diverse datasets and algorithmic fairness techniques, such as reweighting training samples or adversarial debiasing.

## 4.2 Privacy and Security Concerns

### 4.2.1 Regulatory Frameworks: GDPR, HIPAA, and Patient Data Protection

Patient privacy is paramount in healthcare, but AI's reliance on large datasets conflicts with stringent regulations like the EU's General Data Protection Regulation (GDPR) and the US Health Insurance Portability and Accountability Act (HIPAA) [51]. These frameworks mandate strict controls over data anonymization, storage,

**Table 1** Comparison of some common ML algorithms for disease prediction

| ML algorithm | Category | Use cases in disease prediction | Performance considerations |
|---|---|---|---|
| Logistic REGRESSION | Supervised | Binary disease classification (e.g., presence/absence of disease), risk factor analysis. | Simple, interpretable, good baseline. May struggle with complex, nonlinear relationships. |
| Support Vector Machines (SVMs) | Supervised | Cancer classification, drug response prediction, and high-dimensional genomic data analysis. | Effective in high-dimensional spaces, robust to outliers. Can be computationally intensive for large datasets. |
| Random Forests | Supervised | Disease diagnosis, risk prediction, and identifying important features (e.g., biomarkers). | High accuracy handles complex data, resistant to overfitting, and provides feature importance. Less interpretable than simpler models. |
| K-Nearest Neighbors (KNN) | Supervised | Disease classification (e.g., patient subgrouping), pattern recognition. | Simple to understand, no explicit training phase. Computationally expensive for large datasets, sensitive to irrelevant features. |
| Naive Bayes | Supervised | Diagnostic support, text classification (e.g., clinical notes analysis). | Fast, efficient, and works well with small datasets. Assumes independence of features, which is often not true in real-world data. |
| Deep Learning (CNNs) | Supervised | Medical image analysis (tumor detection, lesion classification), histopathology. | Excellent at learning complex patterns from raw image data, with high accuracy. Requires large datasets and significant computational resources, less interpretable. |
| Deep Learning (RNNs/LSTMs) | Supervised | Analyzing sequential patient data (EHRs, disease progression), genomic sequences. | Good for time-series and sequential data, captures temporal dependencies. Can be complex to train, prone to vanishing/exploding gradients. |
| Transformers | Supervised | Natural Language Processing (analyzing clinical notes), multi-omics data integration. | Highly effective for understanding context and relationships in sequences, powerful for large textual and multimodal data. Computationally very intensive, large data requirements. |
| Clustering (e.g., K-Means, Hierarchical) | Unsupervised | Patient stratification (identifying novel patient subgroups), biomarker discovery. | Useful for finding hidden patterns and natural groupings in unlabeled data. Requires domain knowledge for interpreting clusters, sensitivity to initial conditions. |
| Principal Component Analysis (PCA) | Unsupervised | Dimensionality reduction for high-dimensional data (e.g., genomics) and data visualization. | Reduces noise and complexity, helps in visualizing data. Information loss can occur, and interpretation of principal components can be challenging. |

and sharing. However, anonymization is not foolproof—reidentification risks persist, especially when combining datasets. Balancing data utility with privacy remains a key challenge, particularly in cross-border collaborations where regulatory standards vary (Fig. 2).

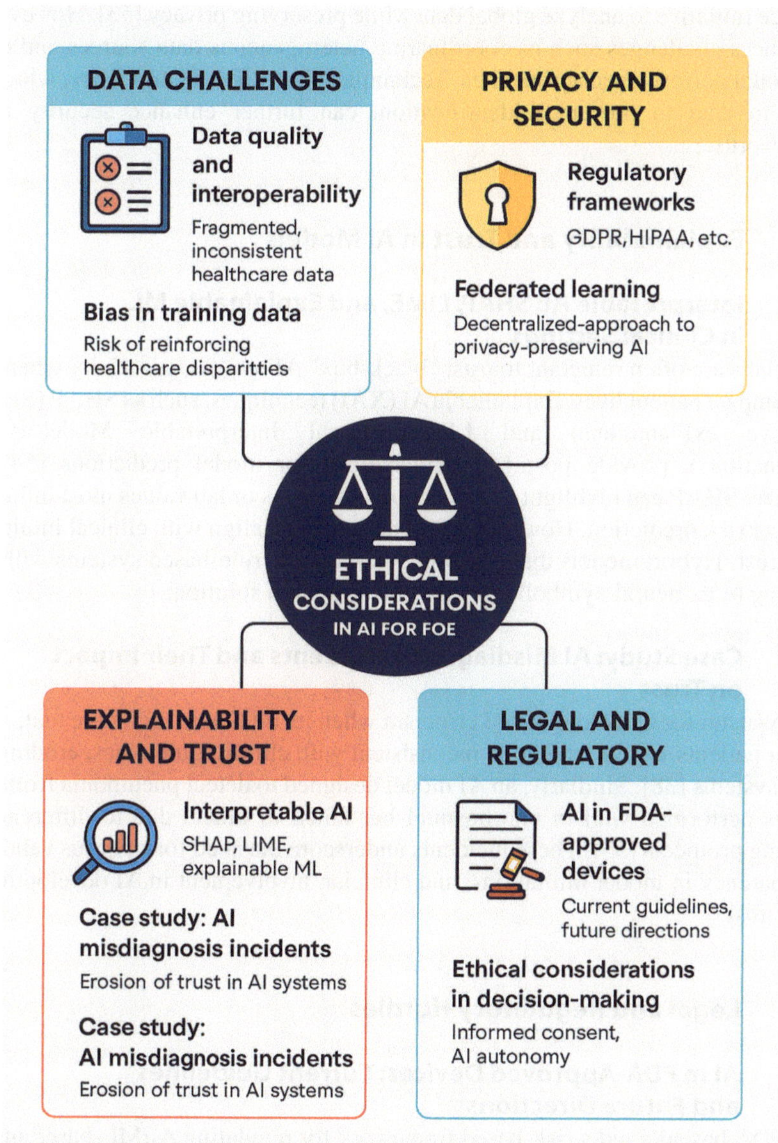


**Fig. 2** Ethical considerations in AI for personalized medicine. FOE Fields of Endeavour

### 4.2.2 Federated Learning: A Decentralized Approach to Privacy-Preserving AI

Federated learning (FL) addresses privacy concerns by training AI models across decentralized datasets without transferring raw data [52]. For example, hospitals can collaboratively train a model on their local EHRs, sharing only model updates rather than patient records. FL has been used in projects like the COVID-19 Open Science Initiative to analyze global data while preserving privacy [53]. However, FL introduces challenges such as coordinating heterogeneous data sources and ensuring model performance across sites. Techniques like differential privacy, which add noise to data to prevent reidentification, can further enhance security in FL frameworks.

## 4.3 Explainability and Trust in AI Models

### 4.3.1 Interpretable AI: SHAP, LIME, and Explainable ML in Clinical Settings

Clinicians are often reluctant to trust "black-box" AI models, especially when decisions impact patient lives. Explainable AI (XAI) techniques, such as SHAP (SHapley Additive exPlanations) and LIME (Local Interpretable Model-agnostic Explanations), provide post hoc explanations for model predictions [54]. For instance, SHAP can highlight which genomic variants or lab values most influenced a cancer risk prediction. However, explanations must align with clinical intuition to gain trust. Hybrid models that combine interpretable rule-based systems with deep learning (e.g., neural-symbolic AI) are emerging as a solution.

### 4.3.2 Case Study: AI Misdiagnosis Incidents and Their Impact on Trust

IBM Watson for Oncology faced criticism when its treatment recommendations for cancer patients were found to be inconsistent with clinical guidelines, eroding trust in AI systems [38]. Similarly, an AI model designed to detect pneumonia from chest X-rays performed well in one hospital but failed in others due to differences in imaging protocols [55]. These incidents underscore the need for rigorous validation, transparency in model limitations, and clinician involvement in AI development to build trust.

## 4.4 Legal and Regulatory Hurdles

### 4.4.1 AI in FDA-Approved Devices: Current Guidelines and Future Directions

The FDA has adopted a risk-based framework for regulating AI/ML-based medical devices, categorizing them into Software as a Medical Device (SaMD). Examples include IDx-DR, an AI system for diabetic retinopathy screening [56]. However, most AI tools are locked post-deployment, unable to adapt to new data. The FDA's

2021 action plan proposes a "predetermined change control plan" to allow iterative updates, but balancing innovation with safety remains contentious.

### 4.4.2 Ethical Considerations in AI Decision-Making: Informed Consent and AI Autonomy

When AI influences treatment decisions, patients have a right to know how and why recommendations are made. However, explaining complex models to nonexperts is challenging. Should patients consent to AI-driven care? Additionally, as AI gains autonomy (e.g., robotic surgery systems), questions arise about liability for errors [57]. Ethicists argue for human oversight as a safeguard, ensuring AI remains a decision-support tool rather than a replacement for clinicians.

## 4.5 Accessibility and Cost Considerations

### 4.5.1 Making AI-Driven Personalized Medicine Scalable and Affordable

The high cost of AI development and deployment risks widening the gap between resource-rich and resource-poor healthcare systems. For example, whole-genome sequencing and AI-based tumor profiling remain prohibitively expensive in low-income countries. Open-source tools like TensorFlow and initiatives such as the Global Alliance for Genomics and Health aim to democratize access [58]. Public–private partnerships, like the NIH's All of Us Research Program, also prioritize inclusivity by collecting diverse data. However, sustainable funding models and infrastructure investments are critical to ensure equitable adoption.

### 4.5.2 Case Study: Rwanda's Drone-Delivered Blood and AI-Powered Diagnostics

Rwanda partnered with Zipline to use drones for blood delivery in remote areas, supported by AI algorithms that predict demand based on historical data [59]. Similarly, AI tools like Ada Health provide low-cost symptom-checking apps in regions with limited healthcare access. These examples highlight the potential for AI to bridge gaps in care, but scalability requires addressing systemic issues like Internet connectivity and workforce training.

### 4.5.3 Conclusion

The integration of AI/ML into personalized medicine offers transformative potential but demands careful navigation of technical, ethical, and societal challenges. Addressing data biases, ensuring privacy, fostering trust through explainability, and promoting equitable access are prerequisites for ethical AI adoption. Collaborative efforts among technologists, clinicians, policymakers, and patients will be essential to harness AI's power responsibly, ensuring it serves as a force for equity rather than exclusion in the future of healthcare.

# 5 Future Trends and Opportunities

## 5.1 Next-Generation AI/ML Technologies

### 5.1.1 Federated Learning: Privacy-Preserving AI for Distributed Healthcare Data

FL is poised to revolutionize healthcare by enabling collaborative AI model training across institutions without sharing raw patient data [52]. Instead, models are trained locally on decentralized datasets, and only aggregated updates are exchanged. This approach addresses privacy concerns while leveraging diverse data sources, such as genomic databases from multiple hospitals. For example, the COVID-19 Open Science Consortium used FL to analyze global clinical data, accelerating insights into patient outcomes [53]. FL also supports rare disease research by pooling fragmented datasets while complying with GDPR and HIPAA. However, challenges like data heterogeneity and computational overhead remain, driving innovations in adaptive aggregation algorithms and hybrid FL frameworks.

### 5.1.2 Graph Neural Networks in Drug Discovery: AI for Modeling Biological Interactions

Graph neural networks (GNNs) are emerging as a transformative tool for modeling complex biological systems, where interactions (e.g., protein–protein networks and drug–target binding) are represented as graphs [60]. Unlike traditional methods, GNNs capture relational patterns in molecular structures and biological pathways, enabling more accurate predictions of drug efficacy and toxicity. For instance, GNNs have been used to identify novel antibiotics by analyzing chemical compound libraries and predicting interactions with bacterial targets. Projects like AlphaFold (DeepMind) and Drug Repurposing Knowledge Graph (Stanford) highlight GNNs' potential to accelerate drug discovery and repurposing, particularly for understudied diseases.

## 5.2 Multi-Omics Integration for Personalized Healthcare

AI is breaking down silos between genomics, proteomics, metabolomics, and epigenomics, enabling holistic insights into disease mechanisms. Advanced ML models, such as multimodal deep learning, integrate multi-omics data to uncover biomarkers and therapeutic targets that single-omics approaches miss. For example, researchers at the Broad Institute combined genomic and proteomic data to identify dysregulated pathways in autoimmune diseases, paving the way for personalized immunotherapies. Tools like Multi-Omics Factor Analysis (MOFA+) use AI to disentangle latent factors driving disease heterogeneity, enabling precise patient stratification. The integration of real-time wearable data with multi-omics profiles will further refine dynamic health monitoring, ushering in an era of "live" personalized medicine.

### 5.3 AI-Enabled Digital Twins in Medicine

Digital twins—virtual replicas of patients—are leveraging AI to simulate disease progression and treatment responses. By combining genomic data, EHRs, and wearable inputs, AI models create dynamic patient avatars that predict outcomes under different interventions. For example, Cardiac Digital Twins simulate heart function to optimize treatment plans for arrhythmia patients, while oncology researchers use tumor digital twins to test chemotherapy regimens in silico. The EU's EDITH Initiative aims to build a federated ecosystem of digital twins for chronic disease management. Challenges include scaling computational models and ensuring real-time data synchronization, but advancements in edge computing and quantum ML could soon make patient-specific simulations routine in clinical practice.

### 5.4 AI in Population Health and Global Health Equity

AI is increasingly being harnessed to address healthcare disparities by tailoring interventions to underserved populations [58]. For example, Google Health's AI-powered screening tools for diabetic retinopathy have been deployed in rural India, where specialist access is limited. Similarly, Zebra Medical Vision uses AI to analyze low-cost chest X-rays for tuberculosis detection in sub-Saharan Africa. Initiatives like the WHO's AI for Health prioritize ethical AI deployment in low-resource settings, focusing on culturally adapted algorithms and open-source platforms. However, achieving global equity requires addressing systemic biases in training data and infrastructure gaps. Collaborative efforts, such as NITI Aayog's AI healthcare mission in India, demonstrate how public–private partnerships can democratize AI-driven care.

The future of AI in personalized medicine is marked by groundbreaking technologies—from federated learning and GNNs to digital twins—that promise to bridge the gap between individual care and population health. Multi-omics integration will unlock deeper biological insights, while AI's role in global health equity will depend on inclusive innovation and cross-sector collaboration. As these trends converge, they will redefine healthcare as predictive, preventive, and universally accessible, transforming patient outcomes worldwide (Fig. 3).

## 6 Conclusion

### 6.1 Recap of AI/ML's Transformative Role

AI and ML have fundamentally reshaped personalized medicine, transitioning healthcare from a reactive, population-based model to a proactive, patient-centric paradigm [5]. By analyzing vast, heterogeneous datasets—including genomics, electronic health records (EHRs), medical imaging, and wearable device

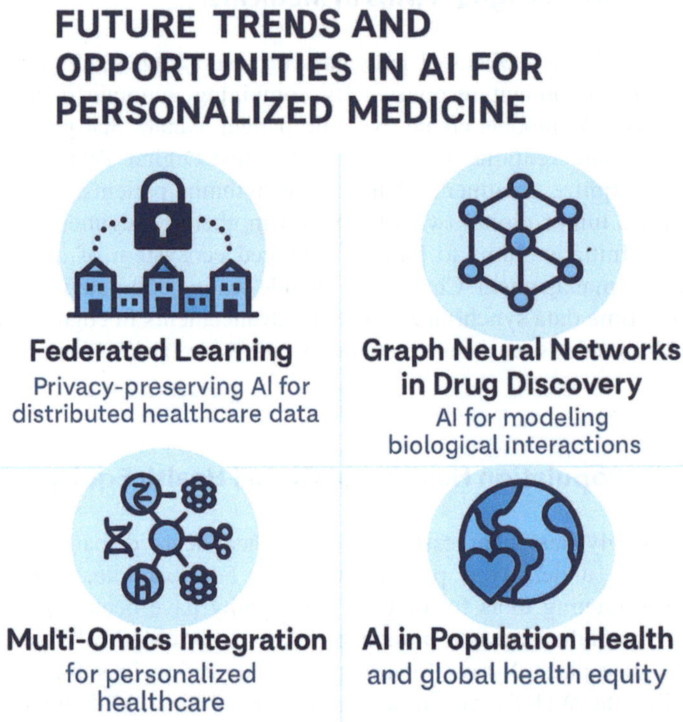


**Fig. 3** Future trends of AI/ML in personalized medicine

outputs—AI/ML has enabled unprecedented precision in disease diagnosis, risk prediction, and treatment optimization. For instance, AI-driven tools like Google's DeepMind have enhanced early cancer detection through advanced imaging analysis [32], while IBM Watson for Oncology has demonstrated the potential of AI in tailoring cancer therapies to individual genetic profiles [38]. In drug discovery, platforms such as Atomwise leverage deep learning to accelerate the identification of novel therapeutics and federated learning frameworks have facilitated global collaborations without compromising patient privacy [52]. These advancements underscore AI's capacity to decode biological complexity, offering clinicians actionable insights that improve outcomes and reduce trial-and-error medicine.

## 6.2 Addressing the Challenges

The integration of AI into healthcare is not without hurdles. Data bias remains a critical concern, as models trained on nonrepresentative datasets risk exacerbating health disparities, particularly for underrepresented racial or ethnic groups [50]. Mitigating this requires intentional curation of diverse data and fairness-aware algorithms. Privacy and security challenges, amplified by regulations like GDPR and HIPAA, demand innovative solutions such as federated learning and differential privacy to balance data utility with confidentiality [51]. Equally pressing are regulatory and ethical gaps, where existing frameworks struggle to keep pace with AI's rapid evolution. Transparent validation processes, explainable AI (XAI) tools like SHAP and LIME, and updated guidelines for adaptive AI systems are essential to build trust among clinicians and patients [54]. Addressing these challenges is not optional—it is a prerequisite for ethical, equitable, and effective AI deployment.

## 6.3 The Future of AI in Healthcare

Looking ahead, AI promises to propel personalized medicine into a new era of predictive and preventive care. Emerging technologies like AI-enabled digital twins will simulate patient-specific responses to treatments, while multi-omics integration will unravel the interplay of genomics, proteomics, and metabolomics for holistic health insights. On a global scale, AI-driven initiatives, such as low-cost diagnostic tools for tuberculosis in resource-limited settings, highlight its potential to democratize healthcare access. Realizing this vision demands interdisciplinary collaboration—bridging data scientists, clinicians, ethicists, and policymakers—to ensure AI innovations are scalable, equitable, and patient-centered. As AI continues to evolve, its greatest impact will lie not merely in advancing technology but in fostering a future where precision medicine is synonymous with universal, compassionate, and proactive healthcare. The journey ahead is complex, but the rewards—a world where every patient receives care tailored to their unique biology—are within reach.

**Declarations**

**Author's Contributions** Conceived the idea and review structure: IP; Wrote the manuscript: IP; Revised and edited the manuscript: IP. The author has read, agreed, and approved the final draft of the manuscript.

**Data Availability** Figures were generated using GPT-4o based on detailed, domain-specific prompts. All AI-generated content was thoroughly reviewed, edited, and validated by the authors to ensure scientific accuracy.

**Conflicts of Interests** There is no conflict of interest.

## References

1. Liao W-L, Tsai F-J (2013) Personalized medicine: a paradigm shift in healthcare. Biomedicine 3:66–72. https://doi.org/10.1016/j.biomed.2012.12.005
2. Aworunse OS, Adeniji O, Oyesola OL et al (2018) Genomic interventions in medicine. Bioinforma Biol Insights 12:1177932218816100. https://doi.org/10.1177/1177932218816100

3. Ahmad A, Imran M, Ahsan H (2023) Biomarkers as biomedical bioindicators: approaches and techniques for the detection, analysis, and validation of novel biomarkers of diseases. Pharmaceutics 15:1630. https://doi.org/10.3390/pharmaceutics15061630
4. Jiang Z, Zhang H, Gao Y, Sun Y (2025) Multi-omics strategies for biomarker discovery and application in personalized oncology. Mol Biomed 6:115. https://doi.org/10.1186/s43556-025-00340-0
5. Fahim YA, Hasani IW, Kabba S, Ragab WM (2025) Artificial intelligence in healthcare and medicine: clinical applications, therapeutic advances, and future perspectives. Eur J Med Res 30:848. https://doi.org/10.1186/s40001-025-03196-w
6. Ahmed MK, Rozario E, Mohonta SC et al (2025) Leveraging Big Data analytics for personalized cancer treatment: an overview of current approaches and future directions. J Eng 2025:9928467. https://doi.org/10.1155/je/9928467
7. Tripathi S, Tabari A, Mansur A et al (2024) From machine learning to patient outcomes: a comprehensive review of AI in pancreatic cancer. Diagnostics 14:174. https://doi.org/10.3390/diagnostics14020174
8. Abbaoui W, Retal S, El Bhiri B et al (2024) Towards revolutionizing precision healthcare: a systematic literature review of artificial intelligence methods in precision medicine. Inform Med Unlocked 46:101475. https://doi.org/10.1016/j.imu.2024.101475
9. Huhulea EN, Huang L, Eng S et al (2025) Artificial intelligence advancements in oncology: a review of current trends and future directions. Biomedicine 13:951. https://doi.org/10.3390/biomedicines13040951
10. Ligero M, El Nahhas OSM, Aldea M, Kather JN (2025) Artificial intelligence-based biomarkers for treatment decisions in oncology. Trends Cancer 11:232–244. https://doi.org/10.1016/j.trecan.2024.12.001
11. Serrano DR, Luciano FC, Anaya BJ et al (2024) Artificial intelligence (AI) applications in drug discovery and drug delivery: revolutionizing personalized medicine. Pharmaceutics 16:1328. https://doi.org/10.3390/pharmaceutics16101328
12. Van Bulck L, Reading Turchioe M, Topaz M, Song J (2024) Exploring the full potential of the electronic health record: the application of natural language processing for clinical practice. Eur J Cardiovasc Nurs 24:332–337. https://doi.org/10.1093/eurjcn/zvae091
13. Olveres J, González G, Torres F et al (2021) What is new in computer vision and artificial intelligence in medical image analysis applications. Quant Imaging Med Surg 11:3830–3853. https://doi.org/10.21037/qims-20-1151
14. Rashidi HH, Albahra S, Robertson S et al (2023) Common statistical concepts in the supervised Machine Learning arena. Front Oncol 13:1130229. https://doi.org/10.3389/fonc.2023.1130229
15. Trezza A, Visibelli A, Roncaglia B et al (2024) Unsupervised learning in precision medicine: unlocking personalized healthcare through AI. Appl Sci 14:9305. https://doi.org/10.3390/app14209305
16. Jayaraman P, Desman J, Sabounchi M et al (2024) A primer on reinforcement learning in medicine for clinicians. NPJ Digit Med 7:337. https://doi.org/10.1038/s41746-024-01316-0
17. Jia H, Zhang J, Ma K et al (2024) Application of convolutional neural networks in medical images: a bibliometric analysis. Quant Imaging Med Surg 14:3501–3518. https://doi.org/10.21037/qims-23-1600
18. Shobayo O, Saatchi R (2025) Developments in deep learning artificial neural network techniques for medical image analysis and interpretation. Diagnostics 15:1072. https://doi.org/10.3390/diagnostics15091072
19. Nerella S, Bandyopadhyay S, Zhang J et al (2024) Transformers and large language models in healthcare: a review. Artif Intell Med 154:102900. https://doi.org/10.1016/j.artmed.2024.102900
20. Lonsdale J, Thomas J, Salvatore M et al (2013) The genotype-tissue expression (GTEx) project. Nat Genet 45:580–585. https://doi.org/10.1038/ng.2653
21. Seinen TM, Kors JA, van Mulligen EM, Rijnbeek PR (2025) Using structured codes and free-text notes to measure information complementarity in electronic health records: feasibility and validation study. J Med Internet Res 27:e66910. https://doi.org/10.2196/66910

22. Hussain S, Mubeen I, Ullah N et al (2022) Modern diagnostic imaging technique applications and risk factors in the medical field: a review. Biomed Res Int 2022:5164970. https://doi.org/10.1155/2022/5164970
23. Mamdiwar SD, RA, Shakruwala Z et al (2021) Recent advances on IoT-assisted wearable sensor systems for healthcare monitoring. Biosensors 11:372. https://doi.org/10.3390/bios11100372
24. Graham H, White PCL (2016) Social determinants and lifestyles: integrating environmental and public health perspectives. Public Health 141:270–278. https://doi.org/10.1016/j.puhe.2016.09.019
25. Ozer ME, Sarica PO, Arga KY (2020) New machine learning applications to accelerate personalized medicine in breast cancer: rise of the support vector machines. OMICS J Integr Biol 24:241–246. https://doi.org/10.1089/omi.2020.0001
26. Sharma D, Anabala M, Jain VV et al (2025) Computational landscape in drug discovery: from AI/ML models to translational application. Scientifica 2025:1688637. https://doi.org/10.1155/sci5/1688637
27. Thallam NST (2023) Comparative analysis of public cloud providers for Big Data analytics: AWS, azure, and Google Cloud. Int J AI BigData Comput Manag Stud 4:18–29. https://doi.org/10.63282/3050-9416.IJAIBDCMS-V4I3P103
28. Ye J, Woods D, Jordan N, Starren J (2024) The role of artificial intelligence for the application of integrating electronic health records and patient-generated data in clinical decision support. AMIA Summits Transl Sci Proc 2024:459–467
29. Nomani A, Ansari Y, Nasirpour MH et al (2022) PSOWNNs-CNN: a computational radiology for breast cancer diagnosis improvement based on image processing using machine learning methods. Comput Intell Neurosci 2022:5667264. https://doi.org/10.1155/2022/5667264
30. Borchert RJ, Azevedo T, Badhwar A et al (2023) Artificial intelligence for diagnostic and prognostic neuroimaging in dementia: a systematic review. Alzheimers Dement 19:5885–5904. https://doi.org/10.1002/alz.13412
31. Iftikhar M, Saqib M, Qayyum SN et al (2024) Artificial intelligence-driven transformations in diabetes care: a comprehensive literature review. Ann Med Surg 86:5334–5342. https://doi.org/10.1097/MS9.0000000000002369
32. McKinney SM, Sieniek M, Godbole V et al (2020) International evaluation of an AI system for breast cancer screening. Nature 577:89–94. https://doi.org/10.1038/s41586-019-1799-6
33. Geppert J, Asgharzadeh A, Brown A et al (2024) Software using artificial intelligence for nodule and cancer detection in CT lung cancer screening: systematic review of test accuracy studies. Thorax 79:1040–1049. https://doi.org/10.1136/thorax-2024-221662
34. Haga SB (2024) Artificial intelligence, medications, pharmacogenomics, and ethics. Pharmacogenomics 25:611–622. https://doi.org/10.1080/14622416.2024.2428587
35. Xue L, He S, Singla RK et al (2024) Machine learning guided prediction of warfarin blood levels for personalized medicine based on clinical longitudinal data from cardiac surgery patients: a prospective observational study. Int J Surg Lond Engl 110:6528–6540. https://doi.org/10.1097/JS9.0000000000001734
36. Ranjan S, Singh A, Yadav R (2025) Artificial intelligence in precision medicine and patient-specific drug design. Biomed Pharmacol J 18:283–294
37. Chang T-G, Park S, Schäffer AA et al (2025) Hallmarks of artificial intelligence contributions to precision oncology. Nat Cancer 6:417–431. https://doi.org/10.1038/s43018-025-00917-2
38. Zhou N, Zhang C, Lv H et al (2019) Concordance study between IBM Watson for oncology and clinical practice for patients with cancer in China. Oncologist 24:812–819. https://doi.org/10.1634/theoncologist.2018-0255
39. Abeer ANMN, Urban NM, Weil MR et al (2024) Multi-objective latent space optimization of generative molecular design models. Patterns 5. https://doi.org/10.1016/j.patter.2024.101042
40. Lu X, Yang C, Liang L et al (2024) Artificial intelligence for optimizing recruitment and retention in clinical trials: a scoping review. J Am Med Inform Assoc JAMIA 31:2749–2759. https://doi.org/10.1093/jamia/ocae243

41. Badani A, de Moraes FY, Vollmuth P et al (2025) AI and innovation in clinical trials. Npj Digit Med 8:683. https://doi.org/10.1038/s41746-025-02048-5
42. Ozdemir ES, Ranganathan SV, Nussinov R (2022) How has artificial intelligence impacted COVID-19 drug repurposing and what lessons have we learned? Expert Opin Drug Discov 17:1061–1065. https://doi.org/10.1080/17460441.2022.2128333
43. Tahri Sqalli M, Al-Thani D (2020) Evolution of wearable devices in health coaching: challenges and opportunities. Front Digit Health 2:545646. https://doi.org/10.3389/fdgth.2020.545646
44. Wang Y, Aivalioti E, Stamatelopoulos K et al (2025) Machine learning in cardiovascular risk assessment: towards a precision medicine approach. Eur J Clin Investig 55:e70017. https://doi.org/10.1111/eci.70017
45. Feng M, Mishra S, Ma J et al (2026) Artificial intelligence for public health can harness data for healthier populations. Nat Health 1–4. https://doi.org/10.1038/s44360-025-00005-w
46. Panteli D, Adib K, Buttigieg S et al (2025) Artificial intelligence in public health: promises, challenges, and an agenda for policy makers and public health institutions. Lancet Public Health 10:e428–e432. https://doi.org/10.1016/S2468-2667(25)00036-2
47. El Arab RA, Alkhunaizi M, Alhashem YN et al (2025) Artificial intelligence in vaccine research and development: an umbrella review. Front Immunol 16:1567116. https://doi.org/10.3389/fimmu.2025.1567116
48. Feng S, Feng Z, Ling C et al (2021) Prediction of the COVID-19 epidemic trends based on SEIR and AI models. PLoS One 16:e0245101. https://doi.org/10.1371/journal.pone.0245101
49. Kristiansen TB, Kristensen K, Uffelmann J, Brandslund I (2022) Erroneous data: the Achilles' heel of AI and personalized medicine. Front Digit Health 4. https://doi.org/10.3389/fdgth.2022.862095
50. Norori N, Hu Q, Aellen FM et al (2021) Addressing bias in big data and AI for health care: a call for open science. Patterns 2:100347. https://doi.org/10.1016/j.patter.2021.100347
51. Yadav N, Pandey S, Gupta A et al (2023) Data privacy in healthcare: in the era of artificial intelligence. Indian Dermatol Online J 14:788–792. https://doi.org/10.4103/idoj.idoj_543_23
52. Abbas SR, Abbas Z, Zahir A, Lee SW (2024) Federated learning in smart healthcare: a comprehensive review on privacy, security, and predictive analytics with IoT integration. Healthcare 12:2587. https://doi.org/10.3390/healthcare12242587
53. Rieke N, Hancox J, Li W et al (2020) The future of digital health with federated learning. Npj Digit Med 3:119. https://doi.org/10.1038/s41746-020-00323-1
54. Rudin C (2019) Stop explaining black box machine learning models for high stakes decisions and use interpretable models instead. Nat Mach Intell 1:206–215. https://doi.org/10.1038/s42256-019-0048-x
55. Zech JR, Badgeley MA, Liu M et al (2018) Variable generalization performance of a deep learning model to detect pneumonia in chest radiographs: a cross-sectional study. PLoS Med 15:e1002683. https://doi.org/10.1371/journal.pmed.1002683
56. Shah S, Thakor P, Shah A, Singh OV (2025) Software as Medical Devices: requirements and regulatory landscape in the United States. Expert Rev Med Devices 22:1201–1214. https://doi.org/10.1080/17434440.2025.2561918
57. Ploug T, Jørgensen RF, Motzfeldt HM et al (2025) The need for patient rights in AI-driven healthcare—risk-based regulation is not enough. J R Soc Med 01410768251344707. https://doi.org/10.1177/01410768251344707
58. Victor A (2025) Artificial intelligence in global health: an unfair future for health in Sub-Saharan Africa? Health Aff Sch 3:qxaf023. https://doi.org/10.1093/haschl/qxaf023
59. Amukele T (2022) Using drones to deliver blood products in Rwanda. Lancet Glob Health 10:e463–e464. https://doi.org/10.1016/S2214-109X(22)00095-X
60. Zhang X-M, Liang L, Liu L, Tang M-J (2021) Graph neural networks and their current applications in bioinformatics. Front Genet 12:690049. https://doi.org/10.3389/fgene.2021.690049

# Potential Breakthroughs and Future Research Directions

Nilanjana Das, Sunny Kumar, and Malini Basu

**Abstract**

Precision medicine is revolutionizing the landscape of cancer diagnosis, prognosis, and therapy by tailoring treatment strategies to the unique genetic, molecular, and environmental characteristics of individual patients. This chapter explores the most promising breakthroughs that are redefining cancer care, including advances in multi-omics integration, liquid biopsy technologies, artificial intelligence (AI)–driven diagnostics, and next-generation targeted therapies. We delve into the clinical impact of CRISPR-based gene editing, tumor microenvironment modulation, and personalized immunotherapies such as neoantigen vaccines and CAR-T cell platforms. The chapter also highlights current limitations—such as tumor heterogeneity, resistance mechanisms, and disparities in biomarker access—and outlines future research priorities aimed at addressing these challenges. These include the development of dynamic biomarkers, improved patient stratification models, and global data-sharing frameworks. By examining both current innovations and future trajectories, this chapter provides a comprehensive outlook on how precision medicine is poised to transform oncology from reactive care to proactive and predictive intervention.

N. Das · S. Kumar
Cancer Biology and Inflammatory Disorder Division, Council of Scientific and Industrial Research-Indian Institute of Chemical Biology (CSIR-IICB), TRUE Campus, Kolkata, West Bengal, India

M. Basu (✉)
Department of Microbiology, Dhruba Chand Halder (DCH) College, University of Calcutta, Kolkata, West Bengal, India



M. K. Ghosh, M. Basu (eds.), *Personalized Cancer Therapy*,
https://doi.org/10.1007/978-981-95-6845-1_24

**Keywords**

High-throughput sequencing (HTS) · Sequencing · AI in personalized medicine · Single-cell in situ transcriptomics · Microbiome profiling

## 1 Introduction

Personalized or precision medicine is an emerging approach in oncology, tailoring cancer therapy and prevention based on patients' genetic backgrounds, the molecular characteristics of tumors, immune environments, and patient lifestyles. This approach aims to target specific oncogenic molecules and improve therapeutic outcomes by considering a person's unique disease profile. If we trace back through the history of cancer treatment, in 2600 BC, Egyptian physicians used poultices and incisions to treat tumors [1]. Other ancient civilizations used localized and whole-body heat therapies. With advancements of social and scientific understanding in the second century AD, surgery took over the main treatment modality, an early step toward modern oncology. But the foundation of modern cancer treatment began with the discovery of X-rays, the use of transplantable tumor models, and Halsted's radical mastectomy [2–4]. Paul Ehrlich was the first to use animal models to screen drugs for therapeutic potential and coined the term "chemotherapy." It has brought the treatment modalities one step ahead. By the 1960s, surgery and radiotherapy were standard regimens, but the sign of micro-metastasis led to a jump forward; the combination chemotherapy became a key strategy. However, these treatment modalities were a generalized approach that lacked personalization. The advancement of cancer treatment has seen milestones in recent times, including the approval of BCG for bladder cancer in 1990 [5], Rituximab as the first targeted therapy for B-cell lymphomas in 1997 [6], Trastuzumab for HER2-positive breast cancer, and kinase inhibitors, like Imatinib and Gefitinib, modified treatment modalities for chronic myeloid leukemia (CML) and non-small cell lung cancer (NSCLC) [7, 8]. In the 2000s, when cancer metastasis was the main focus, the introduction of Bevacizumab as an anti-angiogenic changed the situation [9]. The 2010s were marked by a focus on immunotherapy, where CAR-T cell therapy gained approval [10–12]. The modification from an organ-centric concept toward in-depth molecular analysis is driving a personalized approach. Several advanced tools, including next-generation sequencing (NGS) and RNA sequencing, have greatly enhanced the detection of cancer-specific molecular markers and their altered states. Detection of gene amplifications, mutations, and fusions has shifted the perspective toward personalized treatment.

## 2 Emerging Frontiers in Personalized Oncology

Over the past few decades, efforts have been made to unravel the mechanisms of carcinogenesis, focusing on single-omics approaches. In contrast, the recent time multi-OMICS approach covering the multiple steps of the central dogma is prioritized. These approaches involve the investigation of the cancer cell/tissue at multiple biological levels to uncover the underlying molecular mechanisms that cause different phenotypic manifestations, metastasis, and angiogenesis. Genomic sequences became the most available OMICS data for examining different types of cancer. High-resolution understanding of the nucleotide arrangement within an organism's genome is essential for recognizing its genetic structure, functionality, and evolutionary history. This high-throughput or NGS offers an efficient method for thoroughly identifying and analyzing entire genomes with exceptional throughput, precision, and scalability in a short period [13, 14]. The data generated holds tremendous promise for improving healthcare systems, creating new personalized treatments.

## 2.1 High-Throughput Sequencing (HTS)

High-throughput sequencing (HTS) encompasses technologies that enable rapid, precise, and cost-effective sequencing of large amounts of DNA or RNA. HTS is categorized into three generations based on technological advancements: second, third, and fourth. Early HTS methods, such as sequencing by hybridization (SBH) [15] and Massively Parallel Signature Sequencing (MPSS) [16], utilized oligonucleotides and microbead technology to achieve ultra-high-throughput mRNA sequencing, producing sequence signatures from numerous small DNA segments. Polony sequencing emerged as another method using emulsion PCR to generate cloned DNA templates with high accuracy but faced challenges related to process complexity and throughput. While these early techniques have been largely phased out, they established a foundation for modern advancements. Subsequent methods like sequencing by synthesis, pyrosequencing, and sequencing by ligation introduced improvements in throughput, read length, and accuracy, marking significant progress in the field [15].

## 2.2 Second-Generation Sequencing

Rapid whole-genome sequencing and genetic breakthroughs are made possible by second-generation sequencing, a major breakthrough in DNA sequencing technologies that optimizes speed, throughput, and accuracy. Key examples include Roche 454 pyrosequencing, Illumina Solexa technology, and ABI/SOLiD sequencing by ligation [2]. Illumina sequencing is a leading method within this technology, integrating sample amplification with sequencing by synthesis (SBS). The process consists of four key steps: sample preparation, cluster generation, sequencing, and data analysis. In sample preparation, purified DNA is fragmented and linked with adapters—PCR adapters for amplification and index adapters for multiplexing—enabling the simultaneous sequencing of multiple samples. After constructing the libraries, they are denatured and introduced to flow cells that assist in base pairing. The flow cell amplifies DNA through PCR and creates single strands that hybridize with oligonucleotides, forming DNA clusters. Sequencing begins with the hybridization of a primer to these clusters, followed by adding fluorescently labeled nucleotides. The emitted light is recorded for accurate base-calling. Unique indices facilitate demultiplexing and organization of sequencing files. By aligning contiguous sequences with reference genomes, researchers can efficiently identify genetic variants [17].

The principle of Roche 454 sequencing involves the amplification of DNA fragments captured on microbeads using emulsion, followed by pyrosequencing that involves detecting released pyrophosphate and light emission upon nucleotide incorporation. Roche 454 sequencing offers advantages in accurately sequencing longer reads simultaneously compared to Illumina. However, it has notable drawbacks, including lower output, issues with homopolymers, and high reagent costs. While it can sequence longer DNA reads with minimal GC bias, its disadvantages include lower throughput, higher error rates, and operational costs, which have ultimately led to its discontinuation. Both platforms require DNA fragmentation and adapter ligation, but Roche 454's process necessitates skilled personnel due to its emulsion PCR and bead enrichment steps. SOLiD DNA sequencing involves fragmenting DNA and attaching adapters to both ends. Fragmentation can be done via nebulization, sonication, or digestion. Adapters at the 5′ and 3′ ends are essential for attaching DNA to the glass slide and for sequencing initiation. This technique amplifies each DNA fragment, producing many identical templates. The sequencing begins with the addition of a fluorescently labelled 8-mer primer, where the first two nucleotides bind to the template DNA. DNA ligase then ligates the primer, and silver ions cleave the phosphorothioate linkage, releasing the fluorescent dye for measurement. This process is repeated until full hybridization occurs, and the fluorescent signals are converted into digital data for sequencing interpretation. SOLiD sequencing offers high accuracy, particularly for longer fragments, by using dual base encoding. However, it faces challenges like high costs, the need for high-quality DNA samples, and complex library preparation [15, 18, 19] (Table 1).

## 2.3 Third-Generation Sequencing

It offers a distinct approach by allowing the sequencing of nucleic acids from longer samples at a single-molecule level, resulting in long read outputs in a shorter time. Major platforms like PacBio, Helicos, and Nanopore sequence purified nucleic acids directly, while RNA polymerase-based sequencing maps the movement of RNA polymerase on DNA using molecular tweezers. PacBio Sequencing developed by Pacific Biosciences, PacBio sequencing, or single-molecule real-time (SMRT) sequencing observes real-time nucleotide incorporation by DNA polymerase as it synthesizes complementary DNA strands using zero-mode waveguides (ZMWs). These ZMWs host single DNA polymerase molecules and DNA template strands, to which adapters are added for circularization. Nucleotides labeled with fluorescent markers are introduced, emitting signals as DNA polymerase adds them, allowing the generation of DNA sequences. While PacBio provides advantages like

**Table 1** Impact of various sequencing generations on cancer precision medicine

| Technology | Description | Impact on oncology |
|---|---|---|
| High-throughput sequencing | This method allows parallel sequencing of millions of DNA fragments quickly and affordably | Enables comprehensive tumor profiling, helping identify mutations, fusions, and structural variations critical for therapy selection |
| Second-generation sequencing | Includes platforms like Illumina, characterized by massively parallel short-read sequencing, offering high accuracy at low cost | Forms the backbone of most clinical NGS panels for detecting actionable mutations across multiple cancers |
| Third-generation sequencing | PacBio and Oxford Nanopore sequencing technologies that read longer DNA fragments with greater structural clarity | Improves detection of complex structural variants and epigenetic changes, beneficial for cancers with large rearrangements |
| Fourth-generation sequencing | Often refers to ultra-fast, real-time, and portable sequencing technologies, including integration with AI for real-time analysis | Promises bedside genomic diagnostics and faster decision-making for personalized treatment in acute care settings |
| Massive parallel spatially resolved sequencing | Integrates transcriptomic or genomic data with spatial localization within tissue context, preserving microenvironmental information | Reveals spatial tumor heterogeneity, enhancing understanding of tumor-immune interactions and guiding immunotherapy |
| Single-cell in situ transcriptomics | Profiles gene expression in individual cells within intact tissue samples, preserving spatial and cellular resolution | Uncovers cellular diversity within tumors, tracks evolution and treatment resistance mechanisms at the single-cell level |

real-time sequencing at single-nucleotide resolution and long read throughput, it has limitations such as higher error rates compared to short-read techniques like Illumina. The complexity of sample preparation, needing high-purity, high molecular weight DNA, along with a complicated workflow and data analysis, can make it less favorable, especially given its higher cost per sample. Nonetheless, PacBio has proven effective in identifying mutations linked to antibiotic resistance in emerging superbugs. Helicos sequencing was developed by Helicos Biosciences, which enables direct sequencing of cellular nucleic acids at single-nucleotide resolution. Key steps include sample preparation, immobilization, sequencing-by-synthesis, fluorescence imaging, and data analysis. The target DNA is fragmented and has a poly-A tail added for hybridization to a poly-T sequence on the flow cell. DNA polymerase and fluorescently labeled nucleotides are added sequentially, and the flow cell is monitored for fluorescent signals after each extension round, continuing until sequencing is complete [20, 21].

To improve treatment options and outcomes for patients with advanced cancer, the Belgian Society of Medical Oncologists (BSMO) PRECISION working group launched a nationwide initiative [22]. That programme performed NGS-based comprehensive genomic profiling (CGP) technology to test 324 cancer genes in solid tumors. The programme highlighted pathogenic short variants (SVs), select rearrangements, copy number alterations (CNAs), along with multifaceted biomarkers microsatellite instability (MSI), tumor mutational burden (TMB), and genomic loss of heterozygosity (gLOH) [22]. The program's conclusion demonstrated that patients who were matched to targeted treatment across tumor types had significantly longer PFS and OS. In another prospective study involving 1000 biopsy-confirmed cancer patients, CGP proved superior to traditional single-biomarker analysis. CGP approach identified a significantly higher percentage of druggable genes—47%—compared to smaller panels that captured only 14%. Results detected a mean of 1.7 mutations per sample and 1747 genomic alterations. The patient identified tumor-agnostic markers and detected alterations in homologous recombination, treated with immunotherapy and platinum or PARP inhibitors, respectively. Besides these, significant mutations were recognized in different key genes, including EGFR, KRAS/BRAF, and PIK3CA. Among 618 patients, only 44% participated in the multidisciplinary molecular tumor board (MTB). The overall change in therapy based on CGP in the clinical cohort was 43%. Follow-up showed 71% of patients remained alive, highlighting the critical role CGP and MTB play in advancing personalized treatment strategies in cancer care [23].

## 2.4 Fourth-Generation Sequencing

These technologies represent an exciting advancement in genomics, allowing for the integration of spatial and sequence information, significantly impacting research and diagnostics across various biomedical fields. Notable methods include massive parallel spatially resolved sequencing and single-cell in situ transcriptomics, both enabling direct sequencing of genetic material within preserved cells and tissues [22].

## 2.5 Massive Parallel Spatially Resolved Sequencing

This technique, based on spatial transcriptomics, utilizes microarray chips with numerous spots (100 μm in diameter) that contain sequencing adapters and barcoded oligo(dT) capture probes. Messenger RNA from tissue samples hybridizes with these probes, which then act as primers for cDNA synthesis. Libraries are analyzed using NGS methods. Notably, 10X Genomics has enhanced this approach through its Visium technology, increasing throughput and sensitivity. A significant advantage is its ability to uncover novel spatially regulated genes, but challenges remain in sample preparation and data analysis [24, 25].

## 2.6 Single-Cell In Situ Transcriptomics

This advanced method enables gene expression profiling at the single-cell level while preserving tissue architecture. Key steps involve tissue fixation and cDNA preparation from individual cells, using barcoded or fluorescent probes for sequencing or imaging to connect spatially expressed genes to molecular identities. Despite challenges in sample preparation and limitations to specific gene sets, single-cell transcriptomics (scRNA-seq) allows detailed examination of genotype–phenotype connections. Minor changes in a single cell, like mutations in cancer cells leading to treatment resistance, can significantly alter tumor composition. ScRNA-seq has gained traction in biomedical research, showcasing considerable future potential [26, 27].

A recent study analyzed the transcriptomic landscape of hepatocellular carcinoma (HCC) from samples of six HCC patients and a control with non-carcinoma liver. It highlights significant intratumoral heterogeneity driven by diverse HCC cells, which impact various cellular networks. HCC cells influence tumor-associated macrophages (TAMs), leading to pro-tumorigenic characteristics and impairing M1-type TAMs' immune response due to metabolic disturbances. The research reveals a shared origin for cirrhotic and HCC lesions, indicating parallel clonal evolution. Endothelial cells maintained conserved phenotypes while exhibiting varied functions based on context. Additionally, niche genes associated with rapid recurrence were identified, emphasizing HCC cells' critical role in shaping tumor ecosystems [22, 23, 28] (Table 2).

# 3 Microbiome Profiling and Its Implications in Treatment

The microbiome plays a crucial role in cancer treatment, causing dysbiosis and severe complications. Studies show a link between gut microbiome imbalance and treatment-related toxicity. Microbiota modulation, such as probiotic supplementation and fecal microbiota transplantation, aims to increase bacterial diversity and improve treatment response. Understanding the complex relationship between microbiome and host can help integrate a microbiome-based approach into clinical practice, leading to diagnostic and prognostic microbial biomarkers. Microorganisms in tumors play a crucial role in the tumor microenvironment, influencing tumor development and progression. The tumor microbiome, malignant cells, and nonmalignant components are integral parts of this environment. ROS produced by various immune cells reduce immune responses, while certain bacteria in tumors can trigger ROS production by these cells. Intrinsic and extrinsic factors, including genetics, location, gender, age, environmental exposure, and hormones, influence cancer risk. A healthy lifestyle and diet can help prevent cancer, while unhealthy habits and obesity are linked to its development. Dietary patterns also affect gut microbiome composition. Frying food can produce carcinogenic acrylamide, disrupting gut microbiome balance. Interestingly, not all high-fat diets that correlate with obesity have pro-tumorigenic effects, as fish oil high-fat diets did not accelerate mammary

**Table 2** Comparison of various sequencing generations in terms of definition, cost, timing, and use in cancer precision medicine

| Technology | Description | Cost and time | Effectiveness and use case |
|---|---|---|---|
| High-throughput sequencing | Parallel sequencing of millions of DNA fragments; foundational to modern genomics | ~$15–100/sample; turnaround: 4–10 days | Highly cost-effective for large-scale tumor profiling and surveillance |
| Second-generation sequencing | Short-read sequencing (e.g., Illumina); high accuracy, low cost | ~$25–100/sample; 1–2 weeks depending on depth | Excellent for mutation detection; cost-effective for clinical diagnostics |
| Third-generation sequencing | Long-read, single-molecule sequencing (e.g., PacBio, ONT) | ~$100–500/sample; 1–3 days for library prep + sequencing | Better for structural variants and epigenetics; cost-effective for >60 samples |
| Fourth-generation sequencing | In situ sequencing within cells/tissues; real time, spatially aware | Still emerging; cost varies widely; time: 1–2 weeks, depending on protocol | Promising for clinical diagnostics and biomarker validation; not yet widely adopted |
| Massive parallel spatially resolved sequencing | Combines sequencing with spatial tissue context | High cost (~$200–1000/ sample); time: 1–2 weeks | Powerful for tumor microenvironment studies; less cost-effective for routine use |
| Single-cell in situ transcriptomics | Gene expression profiling at single-cell resolution within tissues | ~$25–173/sample depending on method; time: 1–2 weeks | Extremely informative for tumor heterogeneity; cost-effective with multiplexing (e.g., BRB-seq) |

tumor growth in a murine model. A meta-analysis by Papadimitriou et al. found that vegetarians and vegans have a lower cancer incidence. The study found that dairy products and calcium were inversely linked to colorectal cancer, while alcohol consumption was positively associated with several types of cancer, including breast and liver cancers. Recent studies have shown that certain pathogenic microbes play a role in the development and progression of cancer. They do this by affecting the DNA in host somatic cells, disrupting the cell cycle, increasing cell proliferation, and damaging the processes responsible for apoptosis (programmed cell death). It is estimated that nearly 20% of all cancers may be linked to microbial infections [29–31].

Certain bacteria produce toxins that lead to chronic inflammation. However, infections are prone to cancer, as both microbial genetics and host genetics play a role in cancer risk. While unfavorable microbiota metabolites may be pro-carcinogenic and cause dsDNA breaks, beneficial metabolites can have anticancer effects. Short-chain fatty acids (SCFAs) like acetate, propionate, and butyrate, primarily produced by *Faecalibacterium prausnitzii,* affect gut health and can impact other organs via blood circulation. Chemotherapy-induced mucositis is lessened, and intestinal integrity is preserved with butyrate. Through the fermentation of dietary fibers and starches, Aecalibacterium prausnitzii is the primary colonic microbiota producer of butyrate. According to an intriguing 2020 study, there was no discernible correlation between acetate and survival, but lower serum levels of butyrate and propionate were associated with longer progression-free survival in melanoma patients. Reuterine, a metabolite from the probiotic *Lactobacillus reuteri,* reduces proliferation and increases apoptosis in CRC cells by supporting protein oxidation, inhibiting ribosomal biogenesis, and protein translation. In contrast, secondary bile acids from gut microbes like Clostridium species have carcinogenic properties, leading to colon carcinogenesis through oxidative stress, DNA damage, and resistance to cell death. While research has focused on gastrointestinal cancers and their association with the gut microbiome, studies are increasingly exploring the effects of bacteria, LPS, and microbiota-derived metabolites on cancer development across various types [18, 31–33].

The microbiome significantly influences cancer treatment, with imbalances linked to treatment-related toxicity. Modulating the microbiome through methods like probiotics and fecal microbiota transplantation can enhance bacterial diversity and treatment responses. The relationship between the microbiome and host is complex, and understanding it may lead to new microbial biomarkers for diagnosis and prognosis. Tumor-associated microbes impact tumor growth and immune responses, affected by factors such as genetics, lifestyle, and diet. For instance, unhealthy habits and certain cooking methods can disrupt gut microbiome balance, while a healthy diet may reduce cancer risk. Specific bacteria and toxins can promote inflammation and alter cellular processes, contributing to cancer development. Yet, not all microbial interactions result in cancer; both microbial and host genetics matter. Beneficial metabolites from gut bacteria, like short-chain fatty acids, can have protective effects against cancer [30, 34].

## 4 Integration of AI in Personalized Medicine Development

Machine learning tools can optimize treatment regimens by learning from patient responses and adjusting dosing in real-time, notably in chemotherapy for cancer. AI integrates data to create personalized treatment plans that evolve with new information. This capability helps identify patients likely to benefit from specific therapies, reducing poor outcomes and treatment costs, especially relevant for checkpoint inhibitor immunotherapies with low overall response rates (~20%). Liu et al. developed a logistic regression classifier to predict resistance against PD-1 inhibitors in melanoma patients, while Johannet et al. used convolutional neural networks on histopathology data to forecast immunotherapy responses. Rapid assessment of treatment efficacy is crucial, but traditional manual evaluations can be challenging, particularly with atypical disease progression in checkpoint inhibitors. Dercle et al. showcased machine learning models predicting response sensitivity to various cancer therapies from CT scans. Additionally, machine learning approaches like

CURATE.AI can dynamically adjust drug dosages using real-time data. Although cell lines may have limitations, they provide valuable data for AI training, as demonstrated by Iorio et al., who used 1001 cancer cell lines to build models predicting drug efficacy based on genomic features [35–37].

One challenge in AI models is the lack of interpretability. Kuenzi et al. addressed this with DrugCell, a deep learning model that unveils biological mechanisms behind predictions and can identify synergistic drug combinations, validated through patient-derived xenograft models. Personalized medicine tailors treatments to individual patient characteristics, notably through pharmacogenomics, which uses AI algorithms to predict responses to medications based on genetic makeup. This approach assists in selecting appropriate drugs and dosages, especially for conditions like depression, where traditional trial-and-error methods can be challenging. Additionally, AI factors in lifestyle elements, patient preferences, and social determinants of health to provide a holistic view of patient well-being. Real-time data from wearable devices and health apps further enhance treatment personalization, while machine learning helps identify patterns in patient preferences for medications [35, 36].

## 5 AI in Formulation and Delivery of Drugs

Drug formulation and delivery issues have long beset the pharmaceutical sector. Trial-and-error processes are sometimes costly and time-consuming in traditional ways to optimize formulations and delivery modalities. Drug formulations are optimized using AI-generated predictive models, which ensure that active components are delivered to the target spot in the body as effectively as feasible. AI algorithms, for example, can forecast a drug's release profile from controlled-release drug formulations that offer a consistent therapeutic effect over time. AI can also be used to design drug delivery systems. It is possible to develop more precise and efficient drug delivery technologies by forecasting how these systems will interact with the body [38, 39].

For pharmaceutical formulations to be stable, bioavailable, and effective, excipients are essential. Choosing the best combinations has historically involved a lot of testing, but AI-driven models—in particular, machine learning algorithms—can evaluate big datasets and forecast the best excipient concentrations and combinations for the required disintegration and dissolution periods. In 3D-printed dosage forms, AI enhances manufacturing by personalizing medicine, optimizing drug release profiles, and managing challenges related to complexity and dose control. Additionally, AI can predict drug–excipient interactions crucial for stability and efficacy, using tools like DE-INTERACT, which has demonstrated high accuracy in validating incompatibilities through conventional analytical techniques. Overall, machine learning also aids in predicting critical quality attributes of solid dosage forms based on manufacturing techniques [40–42].

The solubility and bioavailability of drugs are essential for their therapeutic effects. Poor water-solubility can hinder bioavailability by preventing adequate dissolution in the gastrointestinal tract, affecting oral absorption and therapeutic effectiveness—a challenge seen in nearly 40% of newly developed drugs. Predicting drug solubility in aqueous media is vital in early development, guiding solubilization strategies. Machine learning models analyze large datasets of molecular properties and solubility data to identify patterns not evident through traditional methods. When a drug candidate shows poor solubility, AI can help design strategies like solid dispersions, complexation, or the use of surfactants to enhance solubility. Solubility, dissolution rate, permeability, and first-pass metabolism are some of the parameters that ultimately affect bioavailability. A drug's pharmacokinetic profile and rate of absorption can be predicted using machine learning by combining data from multiple trials [43, 44].

Liposomal, nanoparticle, dendrimer, polyplex, transferosome, and nano self-emulsifying systems are some of the nanocarriers that are utilized to enhance the delivery of nanomedicines. Drugs with a poor physicochemical profile can be triggered through a variety of physiological barriers, including the blood–brain barrier, stratum corneum, or intestinal epithelium, by using nanocarriers to deliver the drug to a specific area of the body at a higher concentration, maximizing its effect and minimizing its side effects on other areas. This is particularly useful when targeting drugs to cancer cells or infectious diseases. The size, shape, surface charge, and material composition of the nanoparticle all influence its circulation duration, cellular absorption, and biodistribution, making the design of an effective nanomedicine-based drug delivery system difficult. For example, the circulation time increases with decreasing size. AI models simplify the development, distribution, and optimization of nanomedicines, making them a game-changing technology. AI algorithms can find nanoparticle designs that optimize tumor targeting and reduce off-target effects by training the model on experimental data. AI models can also predict the most effective ligand combinations, improving accuracy and efficacy, by looking at data on receptor expression patterns and focusing on ligands with the highest binding affinities [45, 46].

Microfluidic devices are miniaturized fluidic circuits that manipulate liquids at the nanoliter scale, allowing for precise control and optimization in nanomedicine quality and encapsulation. The integration of AI enhances these devices by optimizing their design and performance, reducing development time, and lowering production costs. Machine learning models can predict outcomes based on various parameters, while AI accelerates Computational Fluid Dynamics simulations by forecasting flow patterns. By learning from experimental data, AI can generate innovative microfluidic architectures tailored for specific nanomedicine processes. Additionally, incorporating sensors enables real-time monitoring, anomaly detection, and dynamic adjustments, ensuring consistent quality in nanomedicine fabrication [47, 48].

## 6 Global Collaboration

Personalized medicine relies on collecting and utilizing genomic and clinical data from both patients and healthy individuals. To enhance the value of this data, it is essential to foster a culture that encourages proper sharing within the scientific, medical, and patient communities. An important regulatory question is how to facilitate data sharing among various stakeholders—patients, physicians, hospitals, pharmaceutical researchers, and health insurers. This challenge extends beyond individual countries, affecting data exchange in high-, low-, and middle-income economies. Data sharing is a critical foundation for the advancement of precision medicine, enabling the integration of diverse datasets—such as genomic, clinical, and lifestyle data—to develop personalized healthcare solutions. Despite extensive efforts and numerous guidelines issued over the past two decades, data sharing remains limited due to persistent barriers such as concerns over autonomy, privacy, and data quality. These issues, though central to ethical data use, are often overemphasized in policy, creating obstacles rather than enabling sharing. The article highlights further challenges, including policy fragmentation, lack of coordination among organizations, and insufficient emphasis on key enablers like interoperability, attribution, and public engagement. To address these gaps, the authors propose extending data-sharing guidelines with targeted strategies. These include implementing Dynamic Consent to give participants ongoing control; deploying Privacy-Preserving Technologies like blockchain; encouraging Reciprocity-Based Access Models to incentivize data contributions; adopting Interoperability Standards for seamless integration; establishing Attribution Mechanisms to recognize data curators; promoting Participatory Governance to involve communities in oversight; and developing Cross-Data-Type Regulations that create unified frameworks for all health data. Collectively, these measures aim to build a more effective, secure, and trustworthy ecosystem for data sharing in precision medicine [45, 49, 50].

Developing effective AI machine learning workflows in pharmaceuticals begins with data collection and cleaning, as the model's quality depends on data integrity. Addressing noise and biases helps avoid underfitting and overfitting, which can impede the model's learning. Techniques like cross-validation are essential for this purpose. Choosing the right model involves evaluating metrics such as Area Under the Receiver Operator Curve (AUROC) and Area Under the Precision-Recall Curve (AUPRC), especially in cases of class imbalance. Validating models on independent datasets ensures generalizability, and model development must adapt to concept drift over time. In drug discovery, a lengthy process typically taking about 15 years, AI can streamline identifying and developing new pharmaceutical compounds. Traditional screening is labor-intensive, while AI-driven virtual screening employs machine learning to analyze large datasets quickly, predicting biological activities and prioritizing compounds based on their therapeutic potential. AI also aids in designing novel drug molecules through deep learning and generative adversarial networks (GANs), enhancing both the safety and efficacy profiles. Predictive techniques like quantitative structure–activity relationship (QSAR) modeling significantly reduce time and costs in validating new compounds. Platforms like Atom-wise and Benevolent AI focus on identifying and prioritizing drug targets to expedite the drug discovery process. However, challenges including AI model interpretability, data infrastructure, and ethical issues like data privacy must be addressed for the successful integration of AI into this field. Another application for sequencing databases is predicting unwanted toxicities, especially concerning healthy tissue during Phase I trials. Analyzing RNA and protein expression data from normal tissues can help, with the Genotype-Tissue Expression (GTEx) project providing RNA sequencing data from over 11,000 samples across 53 human tissues. However, GTEx lacks many tissue types, which could lead to significant toxicity from drug targets in those tissues. Despite being valuable, existing databases have limitations, including the need for complex bioinformatics algorithms and the existence of data silos. Clinical annotations may also be imperfect due to privacy and security challenges, making data integration difficult. For example, it can be hard to link TCGA sequencing results to clinical trial data from the same patients. Big data in this field raises challenges, famously summed up as "garbage in—garbage out." To tackle these issues, core clinical data elements must be standardized and interconnected. A recent multistakeholder working group identified a required minimum dataset that could facilitate faster translation of precision medicine to the clinic. For the existing public databases, efforts are underway to create algorithms for interconnecting isolated datasets and developing more advanced analytics tools. Integrating common data elements from different databases may help optimize future data collection. In conclusion, to maximize sequencing data's potential in precision medicine, we need efficient analytical tools to predict outcomes and prioritize treatments. Achieving the goal of providing the right drug to the right patient at the right time hinges on timely and accurate data access for clinicians [49].

## 7 Vision for the Next Decade: Roadmap for Personalized Cancer Therapy

Personalized medicine enhances treatment effectiveness while reducing side effects. The global personalized medicine market is USD 330.8 Billion in 2023 and is projected to reach USD 746.01 Billion by 2032, growing at a CAGR of 10.7% from 2024 to 2032. This market is driven by advancements in genomics, biotechnology, AI, and big data analytics, promising more accurate and tailored diagnoses and treatments. Genetic Testing and Diagnostic Tools are crucial for identifying genetic predispositions and mutations, aiding physicians in creating effective treatment plans for patients, especially in oncology and rare diseases. Therapeutic drugs, including biologics, gene therapies, and targeted therapies, are increasingly used for complex conditions like cancer and autoimmune diseases, tailored to the patient's genetic and molecular profile. Biomarkers help detect and monitor disease progression, while companion diagnostics identify patients likely to benefit from specific therapies. Digital health tools, such as mobile apps and wearables, play a significant role in personalized medicine by allowing real-time health monitoring to optimize treatments and dosages. Personalized treatments are designed for individuals based on their genetic profile, targeting the root causes of diseases at a molecular level. This patient-centered approach emphasizes unique characteristics for a more holistic healthcare experience. Personalized medicine is mainly used in oncology for targeted cancer treatments focusing on specific mutations, in cardiology for tailored therapies addressing genetically linked cardiovascular diseases, in neurology for managing neurodegenerative conditions like Alzheimer's and Parkinson's, and in infectious diseases to tackle antibiotic resistance and viral infections through personalized approaches [51].

Technological advancements in the last decade have significantly boosted personalized medicine. Key innovations include NGS, which enables rapid and affordable genome mapping for identifying disease-related genetic mutations; AI and machine learning, which enhance treatment predictions by analyzing extensive genetic data; CRISPR gene editing, allowing precise genome modifications for treating genetic disorders; and wearable health devices that monitor vital signs to personalize chronic disease management. Artificial intelligence is transforming drug discovery, with companies like IBM Watson and BenevolentAI analyzing large datasets to identify new drug candidates for personalized therapies. Additionally, gene therapy is emerging as a significant focus, especially with CRISPR technology, which allows the treatment of genetic disorders at their source. Genetic testing and diagnostics play a crucial role in personalized medicine, encompassing various methods such as Whole Genome Sequencing (WGS), which maps an individual's genetic material, and Polymerase Chain Reaction (PCR) testing for detecting mutations and infections, particularly in oncology. Biomarker identification is also vital for monitoring diseases and predicting treatment responses. The therapeutic landscape in personalized medicine includes targeted therapies that focus on specific disease-related molecules and immunotherapies that utilize the immune system to combat cancer. Gene therapy aims to correct defective genes contributing to diseases. Digital health tools are increasingly important for managing patient health, particularly in chronic conditions and mental health. Devices such as fitness trackers and glucose monitors facilitate proactive healthcare. The market can be segmented by product, including personalized medicine diagnostics (genetic testing and DIC diagnostics), therapeutics (pharmaceuticals, genomic medicine, and medical devices), personalized medical care (telemedicine and health IT), and personalized nutrition [20, 21, 24, 25].

## 8 Conclusion

The future of cancer care is being radically reshaped by precision medicine, offering a paradigm shift from generalized treatments to highly individualized therapeutic strategies. By leveraging innovations in high-throughput and NGS, artificial intelligence, microbiome profiling, and advanced drug delivery systems, clinicians and researchers are gaining unprecedented insights into tumor biology, treatment response, and resistance mechanisms. The integration of multi-omics data with AI-driven analytics enables more accurate diagnosis, real-time treatment monitoring, and rational drug design, ensuring that therapies are tailored to the molecular and immunological makeup of each patient. Despite significant progress, challenges remain—including tumor heterogeneity, disparities in access to precision tools, limited clinical implementation of advanced sequencing platforms, and ethical concerns around data sharing. However, these barriers are increasingly being addressed through global collaboration, standardization efforts, and regulatory innovation.

In the coming decade, the convergence of genomics, machine learning, systems biology, and patient-centric digital tools is expected to democratize access to personalized oncology. As sequencing technologies become faster and more affordable, and as AI continues to mature, personalized cancer care will move closer to real-time, predictive, and preventive intervention. This evolution holds the promise of not only improving survival outcomes but also enhancing the quality of life for patients worldwide—ushering in an era where treatment is not just about curing the disease, but about treating the individual.

**Declarations**

**Author's Contributions** Conceived the idea and review structure: ND, SK, MB; Wrote the manuscript: ND; Revised and edited the manuscript: SK, MB. The author has read, agreed, and approved the final draft of the manuscript.

**Funding** This work is supported by the Department of Health Research (DHR: R.12013/37/2024-HR), ICMR, Govt. of India.

**Conflicts of Interest** There is no competing and conflict of interest.

## References

1. Metwaly AM, Ghoneim MM, Eissa IbrahimH et al (2021) Traditional ancient Egyptian medicine: a review. Saudi J Biol Sci 28(10):5823–5832. https://doi.org/10.1016/j.sjbs.2021.06.044
2. Shinde S, Bigogno CM, Simmons A et al (2025) Precision oncology through next generation sequencing in hepatocellular carcinoma. Heliyon 11(3):e42054. https://doi.org/10.1016/j.heliyon.2025.e42054
3. Kumar S, Chatterjee M, Ghosh P, Ganguly KK, Basu M, Ghosh MK (2023) Targeting PD-1/PD-L1 in cancer immunotherapy: an effective strategy for treatment of triple-negative breast cancer (TNBC) patients. Genes Dis 10(4):1318–1350
4. Kumar S, Basu M, Ghosh MK (2024) E3 ubiquitin ligases and deubiquitinases in colorectal cancer: emerging molecular insights and therapeutic opportunities. Biochim Biophys Acta BBA-Mol Cell Res 1871:119827
5. Jian N, Yu L, Ma L, Zheng B, Huang W. BCG therapy in bladder cancer and its tumor microenvironment interactions. Staley C, ed. Clin Microbiol Rev. Published online March 20, 2025:e00212–24. https://doi.org/10.1128/cmr.00212-24
6. Pierpont TM, Limper CB, Richards KL (2018) Past, present, and future of rituximab—the world's first oncology monoclonal antibody therapy. Front Oncol 8:163. https://doi.org/10.3389/fonc.2018.00163
7. Bradley R, Braybrooke J, Gray R et al (2021) Trastuzumab for early-stage, HER2-positive breast cancer: a meta-analysis of 13 864 women in seven randomised trials. Lancet Oncol 22(8):1139–1150. https://doi.org/10.1016/S1470-2045(21)00288-6

8. Sunder SS. Adverse effects of tyrosine kinase inhibitors in cancer therapy: pathophysiology, mechanisms and clinical management. Published online 2023
9. Lambert AW, Pattabiraman DR, Weinberg RA (2017) Emerging biological principles of metastasis. Cell 168(4):670–691. https://doi.org/10.1016/j.cell.2016.11.037
10. Jogalekar MP, Rajendran RL, Khan F, Dmello C, Gangadaran P, Ahn BC (2022) CAR T-cell-based gene therapy for cancers: new perspectives, challenges, and clinical developments. Front Immunol 13:925985. https://doi.org/10.3389/fimmu.2022.925985
11. Sterner RC, Sterner RM (2021) CAR-T cell therapy: current limitations and potential strategies. Blood Cancer J 11(4):69. https://doi.org/10.1038/s41408-021-00459-7
12. Uscanga-Palomeque AC, Chávez-Escamilla AK, Alvizo-Báez CA et al (2023) CAR-T cell therapy: from the shop to cancer therapy. Int J Mol Sci 24(21):15688. https://doi.org/10.3390/ijms242115688
13. Badve S, Kumar GL (2018) Predictive biomarkers in oncology: applications in precision medicine. Springer. Accessed 23 Apr 2025. https://books.google.com/books?hl=en&lr=&id=Gvx9DwAAQBAJ&oi=fnd&pg=PR7&dq=Predictive+biomarkers+in+precision+medicine+in+cancer&ots=2nZRahhkJL&sig=BeQBUmoNQAwwpmh9_sJIz-agspg
14. Ashley EA (2015) The precision medicine initiative: a new national effort. JAMA 313(21):2119–2120
15. Satam H, Joshi K, Mangrolia U et al (2023) Next-generation sequencing technology: current trends and advancements. Biology 12(7):997. https://doi.org/10.3390/biology12070997
16. Reinartz J, Bruyns E, Lin JZ et al (2002) Massively parallel signature sequencing (MPSS) as a tool for in-depth quantitative gene expression pro®ling in all organisms. Brief Funct Genom 1(1):95–104
17. Galeş LN, Păun MA, Butnariu I et al (2025) Next-generation sequencing in oncology—a guiding compass for targeted therapy and emerging applications. Int J Mol Sci 26(7):3123. https://doi.org/10.3390/ijms26073123
18. Ghosh MK, Kumar S, Ghosh S, Basu M (2024) The Dichotomous Role of CHIP in Cancer. In: Handbook of proteases in cancer. CRC Press, pp 299–316. Accessed 13 May 2025. https://www.taylorfrancis.com/chapters/edit/10.1201/9781003394716-25/dichotomous-role-chip-cancer-mrinal-ghosh-sunny-kumar-sayani-ghosh-malini-basu
19. Kumar S, Basu M, Ghosh MK (2022) Chaperone-assisted E3 ligase CHIP: a double agent in cancer. Genes Dis 9(6):1521–1555
20. Nakagawa H, Fujita M (2018) Whole genome sequencing analysis for cancer genomics and precision medicine. Cancer Sci 109(3):513–522. https://doi.org/10.1111/cas.13505
21. Dong L, Wang W, Li A et al (2015) Clinical next generation sequencing for precision medicine in cancer. Curr Genomics 16(4):253–263
22. Gibbs SN, Peneva D, Cuyun Carter G et al (2023) Comprehensive review on the clinical impact of next-generation sequencing tests for the management of advanced cancer. JCO Precis Oncol 7:e2200715. https://doi.org/10.1200/PO.22.00715
23. Ghosh M, Lingaraju SM, C.R K et al (2025) Comprehensive genomic profiling reveals a unique genomic landscape in solid tumors in an Indian cancer cohort of 1000 patients: a single institutional experience. Sci Rep 15(1):12455. https://doi.org/10.1038/s41598-025-94762-z
24. Benotmane JK, Kueckelhaus J, Will P et al (2023) High-sensitive spatially resolved T cell receptor sequencing with SPTCR-seq. Nat Commun 14(1):7432
25. Chiang DY, Getz G, Jaffe DB et al (2009) High-resolution mapping of copy-number alterations with massively parallel sequencing. Nat Methods 6(1):99–103
26. Nath A, Bild AH (2021) Leveraging single-cell approaches in cancer precision medicine. Trends Cancer 7(4):359–372
27. Kim J, Park J (2021) Single-cell transcriptomics: a novel precision medicine technique in nephrology. Korean J Intern Med 36(3):479
28. Ye J, Lin Y, Liao Z et al (2024) Single cell-spatial transcriptomics and bulk multi-omics analysis of heterogeneity and ecosystems in hepatocellular carcinoma. NPJ Precis Oncol 8(1):262
29. Cullin N, Antunes CA, Straussman R, Stein-Thoeringer CK, Elinav E (2021) Microbiome and cancer. Cancer Cell 39(10):1317–1341

30. Helmink BA, Khan MW, Hermann A, Gopalakrishnan V, Wargo JA (2019) The microbiome, cancer, and cancer therapy. Nat Med 25(3):377–388
31. Lee KA, Shaw HM, Bataille V, Nathan P, Spector TD (2020) Role of the gut microbiome for cancer patients receiving immunotherapy: dietary and treatment implications. Eur J Cancer 138:149–155
32. Ghosh MK, Kumar S, Begam S, Ghosh S, Basu M (2024) GBM immunotherapy: exploring molecular and clinical frontiers. Life Sci 356:123018
33. Sarkar S, Kumar S, Saha G, Basu M, Ghosh MK (2024) Glioma nanotherapy: unleashing the synergy of dual-loaded DIM and TMZ. Int J Pharm 665:124697
34. Elinav E, Garrett WS, Trinchieri G, Wargo J (2019) The cancer microbiome. Nat Rev Cancer 19(7):371–376
35. Rehan H (2024) Advancing cancer treatment with ai-driven personalized medicine and cloud-based data integration. J Mach Learn Pharm Res 4(2):1–40
36. Sherani AMK, Khan M, Qayyum MU, Hussain HK (2024) Synergizing AI and healthcare: pioneering advances in cancer medicine for personalized treatment. Int J Multidiscip Sci Arts 3(2):270–277
37. Ahn JS, Shin S, Yang SA et al (2023) Artificial intelligence in breast cancer diagnosis and personalized medicine. J Breast Cancer 26(5):405
38. Serrano DR, Luciano FC, Anaya BJ et al (2024) Artificial intelligence (AI) applications in drug discovery and drug delivery: revolutionizing personalized medicine. Pharmaceutics. 16(10):1328
39. Panchpuri M, Painuli R, Kumar C. Artificial intelligence in smart drug delivery systems: a step toward personalized medicine. RSC Pharm. Published online 2025. Accessed 5 Aug 2025. https://pubs.rsc.org/en/content/articlehtml/2025/pm/d5pm00089k
40. Jiang J, Ma X, Ouyang D, Williams RO III (2022) Emerging artificial intelligence (AI) technologies used in the development of solid dosage forms. Pharmaceutics 14(11):2257
41. Poudel I, Mita N, Babu RJ (2024) 3D printed dosage forms, where are we headed? Expert Opin Drug Deliv 21(11):1595–1614. https://doi.org/10.1080/17425247.2024.2379943
42. Alexis F, Rhee JW, Richie JP, Radovic-Moreno AF, Langer R, Farokhzad OC (2008) New frontiers in nanotechnology for cancer treatment. In: Urologic oncology: seminars and original investigations, vol 26. Elsevier, pp 74–85. Accessed 28 July 2025. https://www.sciencedirect.com/science/article/pii/S1078143907000920
43. Wu K, Kwon SH, Zhou X et al (2024) Overcoming challenges in small-molecule drug bioavailability: a review of key factors and approaches. Int J Mol Sci 25(23):13121
44. Arav Y (2024) Advances in modeling approaches for oral drug delivery: artificial intelligence, physiologically-based pharmacokinetics, and first-principles models. Pharmaceutics 16(8):978
45. Adir O, Poley M, Chen G et al (2020) Integrating artificial intelligence and nanotechnology for precision cancer medicine. Adv Mater 32(13):1901989. https://doi.org/10.1002/adma.201901989
46. Sacha GM, Varona P (2013) Artificial intelligence in nanotechnology. Nanotechnology 24(45):452002
47. Noor J, Chaudhry A, Batool S (2023) Microfluidic technology, artificial intelligence, and biosensors as advanced technologies in cancer screening: a review article. Cureus 15(5). Accessed 5 Aug 2025. https://www.cureus.com/articles/157516-microfluidic-technology-artificial-intelligence-and-biosensors-as-advanced-technologies-in-cancer-screening-a-review-article.pdf
48. Fetah KL, DiPardo BJ, Kongadzem E et al (2019) Cancer modeling-on-a-chip with future artificial intelligence integration. Small 15(50):1901985. https://doi.org/10.1002/smll.201901985
49. Rubin EH, Allen JD, Nowak JA, Bates SE (2014) Developing precision medicine in a global world. Clin Cancer Res 20(6):1419–1427
50. Drake TM, Knight SR, Harrison EM, Søreide K (2018) Global inequities in precision medicine and molecular cancer research. Front Oncol 8:346
51. Edsjö A, Holmquist L, Geoerger B et al (2023) Precision cancer medicine: concepts, current practice, and future developments. J Intern Med 294(4):455–481. https://doi.org/10.1111/joim.13709

MIX
Papier aus verantwortungsvollen Quellen
Paper from responsible sources
FSC® C105338

If you have any concerns about our products,
you can contact us on
**ProductSafety@springernature.com**

In case Publisher is established outside the EU,
the EU authorized representative is:
**Springer Nature Customer Service Center GmbH**
**Europaplatz 3, 69115 Heidelberg, Germany**

Printed by Libri Plureos GmbH
in Hamburg, Germany